Auditory Sound Transmission
An Autobiographical Perspective

Auditory Sound Transmission
An Autobiographical Perspective

Jozef J. Zwislocki
Institute for Sensory Research
Syracuse University

Psychology Press
Taylor & Francis Group
HOVE AND NEW YORK

Copyright © 2002 by Lawrence Erlbaum Associates, Inc.
All rights reserved. No part of this book may be reproduced in any form, by photostat, microform, retrieval system, or any other means, without prior written permission of the publisher.

First published by
Lawrence Erlbaum Associates, Inc., Publishers
10 Industrial Avenue
Mahwah, NJ 07430

First issued in paperback 2013

This edition published 2013 by Psychology Press
27 Church Road, Hove East Sussex BN3 2FA
711 Third Avenue New York, NY 10017

Psychology Press is an imprint of the Taylor & Francis Group, an informa business

Cover design by Kathryn Houghtaling Lacey

Library of Congress Cataloging-in-Publication Data

Zwislocki, Jozef J., 1922–
Auditory sound transmission : an autobiographical perspective / Jozef J. Zwislocki.
    p.   cm.
Includes bibliographical references and index.
ISBN 0-8058-0679-2 (cloth : alk. paper)
ISBN 978-0-415-64601-7 (Paperback)

1. Zwislocki, Jozef J., 1922–   2. Audiologists—United States
    —Biography.   I. Title.
RF38.Z95 A3   2002
612.8'5'092—dc21
[B]   2001040388
                                                      CIP

To know and understand

Dedicated to the memory of my grandfather,
Ignacy Moscicki

Contents

Preface ix

1 The Path of Sound 1

2 The Outer Ear 8
Measurement of Sound Pressure Transformation 9
Modeling the Outer Ear Acoustics 15

3 The Middle Ear 27
Measurement of the Acoustic Impedance at the Tympanic Membrane 29
Analysis of the Middle Ear Function 54
Middle Ear Transfer Function Versus the Threshold of Audibility 83

4 The Cochlea Simplified by Death 89
Functional Anatomy and Physical Constants 92
Mathematical Analysis 110

5 Live Cochlea: Physical Constants and Fundamental Characteristics — 175

Physical Parameter Values of the Gerbil Tectorial Membrane   178
Physical Parameter Values Associated With OHC Stereocilia   210
Physical Parameter Values of the Basilar Membrane and Cochlear Wave Velocity   212
Some Fundamental Characteristics of Cochlear Dynamics   231

6 Live Cochlea: Analysis — 280

A Simple Mechanical Model   286
Dynamics of the Stereocilia-Tectorial Membrane Complex   291
Dynamics of the Cochlear Partition   310
Passive Cochlear Dynamics   322
Active Feedback Effect   363
Cochlear Compression   375

7 Pitch and Loudness Codes — 390

Place Code for Pitch   392
Cochlear Loudness Code   400

Author Index — 411

Subject Index — 417

Preface

This book is intended as a culmination of my life's research on sound transmission in the human ear. I have labored on the subject, on and off, since 1945 and touched upon many of its facets. Information gained as a result lies scattered in many articles belonging to many journals and does not lend itself to a convenient synthesis. Much of this information is bound to become lost unless it is made part of a global picture. I now feel ready to paint such a picture.

It is not my intention to present a passive review of my past work, however. Original insights have been modified according to the current state of our knowledge, and some results that remained unpublished because of time constraints and others so new that they did not yet find their way to referred journals have been included. The latter are parts of a revolution in our concepts of cochlear mechanics, which started in the late 1960s and is not yet entirely finished. The picture that is emerging is much more complicated than the classical picture whose simplicity we have to reluctantly abandon. I attempted to present some of its parts at several meetings and in related articles, but always felt that I was unable to do them justice without being able to present the picture as a whole. Consequently, I deferred full disclosure of several findings until this book.

The field of auditory science is predominantly experimental, and mathematical theory is often ignored or at least distrusted. In part, this is justified because the history of auditory research is full of examples of unrealistic mathematical and conceptual models that ignore existing experimental evidence and contradict fundamental physical laws. Nevertheless, advanced science cannot exist without mathematics. Mathematics is the universal language of quantitative science and serves as a universal glue for binding pieces of experimental evidence together. This is how I have used it in this book.

I have chosen this moment in time for writing the book for several reasons. The least scientific is my age that suggests to me that I should begin wrapping up my affairs and take a last look at the fascinating world around me. Perhaps the most scientific is the state of the science concerning auditory sound transmission. A coherent picture of the process seems finally to have emerged, and I anticipate that its main features will change little in the coming years. Of course, much cleanup work remains to be done. Past research has shown that knowledge of the sound transmission process, including the mechanical sound analysis in the cochlea of the inner ear, is essential for our understanding of the functioning of the auditory system as a whole. Because sound transmission precedes further auditory sound processing, its sufficient description is required as a basis for research on the remaining, neural part of the system, which is now coming into full swing.

Modern research on auditory sound transmission can be divided into two time periods that coincide roughly with the first and second halves of the 20th century. In the first period, experiments on postmortem preparations and associated theory were its mainstay; in the second, the weight of research shifted to live animals. Both periods brought great surprises to the interested scientific and clinical communities. During the first, G. V. Békésy discovered that sound produced traveling waves in the cochlea, partially contradicting H. Helmholtz's 19th century resonance theory. Nevertheless, he confirmed Helmholtz's prediction of a vibration maximum whose location along the cochlear canal depended on sound frequency and which was regarded by Helmholtz as a place code for pitch. During the second period, B. Johnstone and his associates and W. Rhode showed that the cochlear vibration maximum was much sharper in live animals than postmortem. D. Kemp provided the likely explanation for the difference by concluding from his experiments on sound emissions from the cochlea back to the ear canal that the cochlea supplies metabolic energy to the traveling waves. W. Brownell found experimental evidence for a likely mechanism underlying the energy supply.

Good fortune allowed my own scientific work to be a part of both periods and gave me the opportunity to interact directly with their main players at the cutting edge of research. Because of this circumstance this book has been written to some extent in a historical perspective, and this preface contains biographical notes referring to interactions with other scientists, which in part, shaped my work. They are included in response to urgings by numerous colleagues who caught glimpses of them. To make them meaningful, I first have to mention how I happened to work on sound transmission in the ear.

While studying electrical engineering at the Federal Institute of Technology in Zürich, Switzerland, I discovered that I was much more interested in human behavior and its underlying mechanisms than in engineering that dealt with inert matter. Against the advice of my mentors at the Federal Institute, who promised me a bright future in engineering, I followed my inner calling and took the first opportunity that presented itself for working on the human organism, once I concluded my engineering studies. The opportunity presented itself quite accidentally in the form of a position of a research assistant at the Department of Otolaryngology of the Medical School of Basel. My supervisor, Professor Erhadt Lüscher, who was the chairman of the department and was genuinely interested in scientific research, needed the skills of an applied physicist to complement his surgical, anatomical, and physiological ones. My task consisted of developing audiological diagnostic methods and of building the required electroacoustic equipment. One of the methods even took hold in the clinical world as the Lüscher-Zwislocki Test for monaural detection of loudness recruitment. It is still used in a modified form of the SISI test. The audiological work introduced me to auditory psychophysics and made me learn some English. I still remember plowing through such classics of the auditory literature as *Hearing* of Stevens and Davis (1938), and *Speech and Hearing* of Harvey Fletcher (1929).

Before I had left the Federal Institute, I was encouraged by my advisor, Professor Franz Tank, who happened to be the rector of the institute at that time, to undertake doctoral studies. The rules of the institute allowed me to work on my dissertation in Basel but my audiological work did not appear appropriate for the purpose. Fortunately, related study of the classical literature on hearing brought me to Helmholtz and to the controversy surrounding his resonance theory of pitch perception. To my astonishment, I discovered that the theory, as originally proposed, did not rest upon a solid foundation of physics and that the same was true for many of the arguments against it. Resolution of the situation on the basis of accepted applied physics appeared as a worthy topic for a doctoral dissertation to be submitted to the Federal Institute. My institute advisors agreed with me, so did Professor Lüscher.

The ensuing search of the literature revealed to me two authors who subsequently had a lasting effect on my career—Georg von Békésy and

Otto F. Ranke. Békésy had already performed some of his pioneering experiments on postmortem preparations of human cochleas and had been able to see that, at low frequencies, sound elicited traveling waves in their apical-most parts. He also had undertaken several series of mechanical model experiments in which he had attempted to match what then appeared as the relevant cochlear parameters. In agreement with Helmholtz's prediction, the models had revealed a vibration maximum whose location changed with sound frequency. The model cochlear partition vibrated nearly in one phase up to the vibration maximum. At the maximum, the phase changed by 180° and gave rise to a traveling wave. The vibration pattern seemed to be consistent with Helmholtz's resonance theory up to the maximum but not beyond it, and Békésy thought that he might have discovered a new physical phenomenon.

Ranke proposed two mathematical models that more or less directly attempted to explain the vibration pattern Békésy had seen. He stressed the importance of hydrodynamics in theoretical treatment of cochlear sound propagation. I had a reasonable knowledge of theoretical hydrodynamics from my engineering studies—I even had to design a water turbine; but this knowledge did not include surface waves, the kind one sees on the surface of water and in the cochlea. Fortunately, I discovered the classical work of H. Lamb on the subject, which became my hydrodynamical bible. His analysis suggested to me, somewhat to my surprise, that, formally, sound propagation in the fluid-filled cochlea could be treated mathematically like gravity waves that arise on the surface of oceans. It also convinced me that Ranke's derivations were severely flawed and led me to criticize them in my doctoral dissertation. Although Ranke was able to calculate a wave pattern resembling superficially the wave pattern later observed by Békésy and reproduced by him in one of his chapters, the similarity did not hold up under closer scrutiny. The pattern included a standing wave and a strange wave reflection depending on the depth of cochlear canals, contrary to Békésy's observations.

Combining my newly acquired knowledge of the theory of surface waves with Békésy's measurements of the compliance of the cochlear partition, I was able to derive a differential equation for sound transmission in the cochlea. The structure of the equation was formally the same as of the well-known transmission-line equation. However, it had a variable propagation coefficient because of the compliance of the cochlear partition, which increases toward the cochlear apex. As a consequence, the general solution available for the linear transmission-line equation was not applicable, and I was forced to use approximate solutions. My equation in its first form, together with its derivation and its partial solution, were first published in 1946 in a Swiss journal, Experientia. The solution showed a transversal wave on the cochlear partition, whose wavelength became shorter toward the cochlear apex and whose amplitude increased. I assumed that, in reality, the wave amplitude did not increase indefinitely but, as a result of energy losses due to friction, reached a maximum and decayed beyond it. To verify the theoretical result, I made a mechanical model of the cochlea, similar to those of Békésy's but larger and equipped with an optical system allowing the surface waves to be observed conveniently on a projection screen. The mechanical model agreed with my mathematical theory, showing a traveling wave whose amplitude increased up to a maximum and decayed beyond it. There was no uniphasic motion of the partition up to the maximum, contrary to Békésy's interpretation of his model experiments. The location of the maximum changed with sound frequency in the direction predicted by Helmholtz. I believe that this was the first demonstration of the possibility of a cochlear amplitude maximum not based on Helmholtz-type resonance but occurring in the presence of traveling waves. A year later Békésy published his amplitude and phase measurements in postmortem preparations of human cochleas. To my elation, Békésy's graphs indicated to me that they were consistent with my theoretical conclusions. The relevant 1947 article can be found in Békésy's *Experiments in Hearing* (1960) out of chronological order, ahead of some earlier articles containing descriptions of his measurements of static cochlear parameters, which I used in my numerical calculations.

My doctoral dissertation appeared in 1948 as a Supplement of Acta Oto-Laryngologica, in agreement with the rules of the Federal Technical Institute. The dissertation contains a full mathematical treatment of cochlear waves, including a complete solution of my cochlear differential equation and a theoretical analysis showing that the amplitude maximum of the waves cannot occur at the resonance location of the cochlear partition, contrary to Helmholtz's expectation. The dissertation contains numerous references to Békésy's work and reflects my admiration for it. However, it disagrees with two of its minor results. Since subsequent research proved me right, I sometimes used these disagreements to show to my students that experimental results are not always right, and theories are not always wrong. Thinking that, on the whole, my dissertation would please Békésy, I sent him a copy. His thank-you note, which arrived several months later, was congratulatory but disappointingly noncommittal.

The enormous effect the publication of my dissertation would subsequently have on my life was entirely unforeseen by me. First of all, the Technical Institute in Zürich received a letter from Ranke in which he accused me of plagiarism. The institute forwarded the letter to me with a request for an answer. I must confess, I wrote it with some relish because the accusation appeared to be entirely unjustified and only gave me an additional opportunity to criticize Ranke's theories preceding my own work. My approach was entirely different from Ranke's since it was based on a long-wave approximation and a transmission line differential equation, whereas he approached cochlear sound propagation as a boundary-value problem in which the mechanical properties of the cochlear partition were excluded from the differential equations describing fluid motion. Ranke emphasized short waves by comparison to the depth of the cochlear canals. Because of this emphasis and my long-wave approximation, the controversy between Ranke and myself was later called the short-wave–long-wave controversy. As is so often true in science, neither side was entirely correct or entirely incorrect. It is now clear from numerous experiments that the cochlear waves are relatively long before they approach their amplitude maximum but are not so in the vicinity of that maximum. For this reason, I subsequently modified my equation to include waves of any length. The modification affects somewhat the calculated wavelength and amplitude but not the fundamental insights obtained with the long-wave approximation. The latter is still used by many authors as an admissible simplification. I should add here that my criticism of Ranke's theories was not directed at the length of the waves but, rather, at the inadequacy of his mathematical derivations.

On April 21, 1950, I received a somewhat enigmatic telegram from the United States of America, which started a new phase of my life. The telegram stated: "Mass Institute Technology holding Speech Communication Conference May 31 through June third eager have you come at our expense and speak to us about your work on cochlea we can arrange boat transportation and stay here please cable letter follows + = Lock Chairman+" Having made certain with the help of some friends that the telegram actually meant what it seemed to, I telegraphed Lüscher, who was away, requesting his permission to go. He answered in French the next day: "Felicitations d'accord+" A few days after telegraphing my acceptance to Chairman Lock, I received an extensive letter from him describing the details of the trip arrangements, instructing me that I would be expected to make a presentation of 30 to 40 minutes , and apologizing for the lateness of the invitation—it was due to a delay in the approval of the necessary funds. The signature of the letter finally let me know that Lock was William N. Lock, Head of the Department of Modern Languages at M.I.T. Enclosed with the letter was a preliminary program of the conference. Of particular interest to me was the morning session of the third day, entitled: "Perception of Speech." The first paper was to be given by Georg von Békésy, then at Harvard University, the second by me, and the third by Norman R. French of Bell Telephone Laboratories. The program committee of the conference consisted of J. B. Wiesner, J. C. R. Licklider, and L. L. Beranek, in addition to the Chairman, W. N. Lock. The conference was sponsored by the Acoustical Society of America, the Carnegie Project for Scientific Aids to Learning at the Massachusetts Institute of Technology and the Psycho-Acoustic Laboratory of Harvard University.

While my travel arrangements were being finalized, Professor Lock sent me a more advanced program. Going through it, I discovered to my great surprise that Békésy's name was eliminated and replaced by that of Ranke. I was shattered. Instead of being on the same program with the famous scientist I admired, I would have to follow in the program my enemy. It was hardly possible to find a rational explanation for the change despite endless speculations. Of course, I did not change my travel plans.

I wrote my paper in German, and my English teacher translated it. Because he was British rather than American, the paper ended up by being in English rather than American English. I did not know the difference then. My English was very rudimentary—sufficient for reading my paper, which I learned almost by heart, but not sufficient for the anticipated discussion. The organizing committee kindly provided me with an interpreter for this purpose.

Because of time constraints I had to fly to the United States instead of taking a boat. I landed in New York, where I had an uncle, and took a train to Boston. Two members of the organizing committee, Lickleider and Beranek, picked me up at the railroad station and drove me to a graduate M.I.T. dormitory that was to be my residence for the duration of the Conference. A few hours later, they brought me to the residence of Professor Lock, where I and one other conference participant from Europe were invited to have dinner with the committee members and their spouses. After dinner, during which I attempted valiantly to participate in an English conversation, I found myself surrounded by all the committee members, except Lock, in what appeared to be a small living room. Sitting in comfortable chairs, they started first discretely, then less so, to ask me questions about my theory of cochlear waves. It soon became obvious that they were worried about my paper, not being certain that my theory rested on a solid scientific foundation. I suspect that the doubt arose from Ranke's paper in which he criticized my theory severely, as I was to learn subsequently. Jerry Wiesner and Leo Beranek, who knew well the kind of mathematics I was using, seemed to play the role of chief inquisitors. Somehow, using multilingual communication, I was able to dispel their concerns, and the situation ended up by turning against Ranke. In the process, I learned that the committee was unable to defray Ranke's travel expenses, and that Ranke decided not to come, sending his manuscript instead. Licklider was assigned the task of reading it. At the end of our conversation, he was not sure if he wanted to go through with it but, of course, had little choice left since the conference was to start in one day.

Trying to be fair to me, Licklider gave me a copy of Ranke's paper for study and for preparing a defense against his criticisms. I spent part of the next day doing just that with the help of an interpreter assigned to me, who was most helpful, and I regret very much not remembering his name. During the late morning, I was brought to Harvard to meet Békésy. We met in the conference room of the Psycho-Acoustic Laboratory. Békésy entered soon after we had arrived. He was a small man in a gray lab coat, rather bold, with a large nose separated from a small mouth by a small, graying mustache. He was slightly hunched over and looking down rather than up. He greeted me amiably in German and inquired about my trip and Switzerland, where he used to live in his youth. He made some courteous remarks concerning my doctoral dissertation and asked me some innocuous questions. He was evasive about the conference but thought that Ranke's theoretical work fitted it better than his experimental one. The very pleasant encounter with Békésy made his withdrawal from the conference less painful for me.

There seemed to be an evening party every day of the conference, and I gained the impression that alcohol improved considerably my command of English. In any event, it brought some interesting revelations my way. The most important provided me with an explanation of Békésy's withdrawal from the conference. Apparently, he was irked by the decision of the conference committee to invite me rather than Ranke who seems to have been a friend of his for many years. Perhaps he was also unhappy with my rather strong criticisms of Ranke's theories and my criticisms of two of his own experiments, although they were very mild. In any event, he wanted Ranke to be invited and made space for him on the program by withdrawing from it. I think that his withdrawal meant a great loss for the conference.

According to all indications, my paper was an unmitigated success, aided perhaps by my youth—I was only 28 years old and looked younger. Békésy, who came to hear it, invited me for lunch, and I received two prestigious job offers the next day—one from Harvey Fletcher of the Bell Telephone Laboratories and one from S. S. Stevens, the director of the famous Psychoacoustic Laboratory at Harvard. I admired both men enormously and regarded them as the fathers of American auditory psychophysics. I will never forget Fletcher's telephone call at 8 a.m. telling me that he had a research job for me, if I were interested, but that he suspected Stevens would make me an even better offer. He did in the evening of the same day, and I became a Research Fellow at the Psychoacoustic Laboratory. On my request, motivated by my Swiss obligations, the contract was for 1 year only but, then, became extended to 3 and eventually 6 years.

My stay at the Harvard's Psychoacoustic Laboratory was probably the most exciting period of my professional life. Békésy was the Senior Research Fellow there, and he came to my office almost every day in the afternoon for a chat in German. We spoke about everything, except science. Our scientific interactions were only sporadic but highly meaningful. They occurred mainly when Békésy felt attacked by what he called in his Experiments in Hearing "his three best enemies." I think it is a permissible indiscretion at this point in time to reveal two of them whose names are well known in auditory science—Hallowel Davis and Glen Wever. I do not know who the third one was. The attacks I remember resulted from misunderstanding the physics of the cochlea, and I felt compelled to come to Békésy's support by means of short articles in which I attempted to explain with the help of mathematics and applied physics the actual situation. Invariably, I found Békésy to be right and the "enemies" wrong. I do not mean this facetiously. On one occasion concerning sound transmission to the inner ear through bone conduction, Békésy and I disagreed. This led to an exchange of letters to the editor in which I had to apologize for misinterpreting Békésy's conclusions, and Békésy reformulated his conclusions. Stevens was the arbitrator. From the perspective of time, it appears to me that I was scientifically correct but procedurally wrong—Békésy was a senior scientist. The relationship between us was smoothed out soon after the incident, and I was welcome whenever I ventured into his office to admire his new art acquisitions. Art collection was his avocation.

The Psychoacoustic Laboratory was part of the Department of Experimental Psychology, and the Research Fellows of the laboratory were invited to participate in the Faculty meetings of the department. We also had informal daily lunches together, sitting around a large oval table. Some fascinating scientific and philosophical discussions took place around that table. The department was at its peak, having as its members such luminaries as E. G. Boring, S. S. Stevens, and B. F. Skinner. Some of the most unforgettable arguments took place between Stevens and Skinner on the subject of nature versus nurture. In those days, Skinner's behavioristic point of view was much more popular than Stevens's genetic one. The balance has been steadily changing ever since, however. Békésy rarely participated in the discussions but, sometimes, I seemed to detect a faint contemptuous smile in his expression. Perhaps some of them did not fit rationally in his world of physics. Most of the time, I shared Békésy's silence, soaking up new information coming to me from fields with which I was not familiar.

Undoubtedly, Stevens, with whom I interacted the most outside of Békésy, had the strongest influence on me, and I learned from him a lot about the scientific method, theory of measurement, psychophysics, and the structure of science and its organization. But these matters are for the most part outside the subject matter of this book, and I intend to discuss them elsewhere. Stevens did not guide my work, however. I was my own man, perhaps somewhat to his disappointment. Soon after I had arrived, he appeared in my office and, after exchanging some pleasantries, inquired about my plans. I had some very definite scientific plans and told him that I best worked by myself. He said that this was fine, but I detected a tinge of disappointment in his response. Only later did I realize that he might have hoped for me to work with him on some of his projects. Nevertheless, that is how things remained, and I was free to pursue my research within the means of the laboratory in which ever direction I wished to go. Stevens never put any pressure on me to become involved in his work, although we discussed it from time to time. Some of the projects I decided to pursue dealt with the theory of cochlear waves and some with outer ear acoustics, especially in its application to earphones and ear protectors against noise. Because work at the Psychoacoustic Laboratory was funded by the Office of Naval Research, some applied research was welcome.

Near the end of my career at Harvard, I tackled the middle ear through acoustic impedance measurements at the tympanic membrane. By performing these measurements on both healthy and pathological middle ears, I was able to analyze the middle ear function and determine the contributions of its various anatomical parts. This provided a scientific foundation for clinical diagnostic methods that are now in general use. To do the measurements, I had to invent new instruments and methods. One acoustic principle developed then has been adapted for routine clinical measurements. The work was greatly facilitated by the Eye and Ear Infirmary of the Harvard Medical School, letting me perform some of it within its facilities and supplying me with patients. I am grateful to Alan S. Feldman, who worked there at that time, for opening for me the door to the Infirmary and for helping me with the audiological evaluation of the patients.

The work on the middle ear continued when I moved to Syracuse University to organize a research laboratory within the Gordon D. Hoople Hearing and Speech Center. The laboratory was called Bioacoustic Laboratory. In part because of my ONR connection at Harvard, I was able to secure ONR funds, and the work at the laboratory took off at high speed. Soon, I had several coworkers and visiting faculty from several countries. Eberhard Zwicker from Germany, one of the most prominent members of the community of auditory scientists was one of them. I first met him at Harvard, and his stay in Syracuse coined a lasting friendship between us and our families. Our research work moved in parallel but rarely overlapping channels, however, and the literature reveals few references to each other's results. We also had different approaches. His was closer to that of an engineer, mine was strongly influenced by the medicophysiological environment at the Medical School of Basel and the psychological world at the Psychoacoustic Laboratory.

The success of the Bioacoustic Laboratory allowed it to expand beyond auditory psychophysics and the physics of the ear. Soon, we added research on the senses of touch and vision and became highly interdisciplinary, when we included physiology and anatomy. As it expanded, the laboratory changed its name twice to better reflect its activity. It first became the Laboratory of Sensory Communication, then the Institute for Sensory Research. The disciplinary diversification allowed me to include the added disciplines in my own research that now ranged beyond auditory psychophysics and applied physics and mathematics to neurophysiological experiments. Outside of psychophysics, it initially covered the sound transmitting parts of the ear, including the outer, middle and inner parts of the ear but, finally, expanded to electrophysiological recording of the responses of the cochlear sensory cells, the hair cells. The present book is the culmination of this work.

At the end of this preface I wish to gratefully acknowledge a number of people who made decisive direct or indirect contributions to my work. Unfortunately, some of them are no longer alive.

First of all, I should mention my grandfather, Ignacy Moscicki, who was a prominent physical chemist at the end of the 19th and the beginning of the 20th centuries. He instilled in me the interest in science and the moral principles according to which I attempted to live. Next, I should mention Professor Franz Tank, Rector of the Federal Technical Institute, who supervised my thesis required by the institute for an engineering diploma and guided me all the way through my doctoral work, although he was not my official advisor, my doctoral dissertation having been outside his field of specialization. Professor Erhard Luescher, for whom I worked at the University of Basel as a research assistant, introduced me to the world of biology and medicine, especially to the auditory system and diagnostic hearing testing. He taught me how to write scientific articles and insisted that sentences do not have to be long even in German. He himself was known to be an excellent writer in both German and English. Luesher was an exemplary supervisor, asking sharp questions and providing useful criticism without unnecessarily constraining the scope of my research. Professor S. Stevens launched me on the almost boundless waters of American and international science. He provided the means for me to work on as many projects as my imagination, energy, and time allowed, provided valuable philosophical and scientific guidance, and made sure I understood the importance of good writing. His future wife, Geraldine Stone, then administrative secretary, was my first and very patient editor of my English articles. Because my English was very poor at first, I saw more red ink on my manuscript pages than ever before or afterward. She also was very helpful to me in bridging the social gap between Europe and America. Ski trips with her and Smitty Stevens (S. S. Stevens) provided badly needed relaxation. A special place must be reserved in my heart for my former wife, Sunny Zwislocki Goldman, whom I met soon after my arrival in Cambridge, Massachusetts. She introduced me to America and made it my country of choice. She was my main English teacher and a reviewer of many of my articles. She shared my life for 40 years before she died on July 17, 1992, 1 week after our wedding anniversary. She made my private life exciting without letting it interfere with my work.

In Syracuse, Gordon D. Hoople, a prominent otologist and Chairman of the Board of Trustees of Syracuse University, supported me in many ways, most importantly, by introducing me to the world of otology, locally and nationally, and by bringing me to the attention of Chancellor Tolley. He was an important reason for me to come to Syracuse. Louis DiCarlo, a well known audiologist, brought me and Syracuse University together and did most of the necessary leg work. He supported me at every possible opportunity ever since. Chancellor Tolley gave me and my associates our first independent building, where the Laboratory of Sensory Communication took shape and later developed to the Institute for Sensory Research. Wilbur LePage, as Chairman of the Department of Electrical Engineering, invited me to join his department on a dual appointment, which became the main appointment, and supported the development of our laboratory with great effectiveness. At a crucial time, Ralph Galbraith, Dean of the college allowed us to become an Institute at the departmental level and gave us effective administrative support. Vice Chancellor John Prucha allowed the institute to double its size and to become one of the best housed institutions of its kind in the Country by assigning to it a substantial portion of a first rate research building. Chancellor Melvin Eggers doubled the size of the institute again by constructing for it an additional wing of first-rate laboratory space.

The Syracuse people I mentioned above were my administrative superiors. But I must also thank my numerous coworkers without whom the success of my work would have been substantially diminished. I have to single out three individuals who made a particular difference in my scientific career. Ronald T. Verrillo and Robert B. Barlow should be considered as the two pioneers who helped me in creating the Institute for Sensory Research. Ron organized and made famous the tactile branch of the institute, Bob added vision research and connected it to the activities of the Marine Biological Laboratories at Woodshole, Massachusetts. His work has become a successor to H. K. Hartline's (Nobel, 1967) invertebrate vision research at The Rockefeller University. A very special thanks go to Earl Kletsky who was my Assistant Director during the formative years of the Laboratory of Sensory Communication and later during the organization of the Institute for Sensory Research. He eventually became Administrative Director of the institute and, subsequently, Assistant Dean of the College of Engineering. In addition, he provided effective mathematical and electronic support for my theoretical work on the cochlea. His contributions are not nearly adequately reflected in published articles. I also should thank Robert L. Smith, who came to the institute as a graduate student and ended up by becoming its current director. He taught me animal surgery, more specifically, the surgery of the peripheral auditory system of Mongolian gerbil. I should also mention Gisle Djupesland, who joined me for a time to add his otological skills to my engineering ones in making acoustic measurements in the outer ear. I also received invaluable help in anatomy from Norma B. Slepecky and, in photomicroscopy, from Steven C. Chamberlain and his assistant, William P. Dossert. I should also thank several graduate and undergraduate assistants who contributed meaningfully to my research work and expanded it significantly. Their names appear in the text and in the references at the end of almost every chapter.

Finally, my deep appreciation goes to several technicians without whose help much of the experimental work would have been impossible. First of all, there was Bernhard Klock who manufactured the acoustic bridge for acoustic impedance measurement in the ear and also the prototype of the ear-like coupler for earphone calibration. He was followed by Michael W. Serafini, who made the string instrument for calibration of the stiffness of micropipettes used to measure the tectorial membrane compliance, and by Richard B. Mitchell, who made the precision drill for making accurate holes in the cochlear capsule. Of course, each one of them contributed other pieces of mechanical equipment, but the mentioned ones became the most important. On the electronics side, I first enjoyed the help of Robert Gardinier who made the hardware model of both the middle ear and the simplified cochlea. Next, I have to mention the outstanding work of Arthur J. Wixson who made the hardware model of the live cochlea and manufactured several pieces of electronic equipment, among them, the frequency discriminator that allowed us to produce efficiently transfer functions of various cochlear potentials. He has been followed by Dean J. Arpajian who wrote the final computer program for the cochlear model on the basis of the original program written by Earl J. Kletsky, having been aided at first by John F. Bruno and Michael S. Schechter. I should also thank Christos G. Stathatos Jr. for installing my computers and all around help with their use.

I also thank most sincerely Nicole M. Sanpetrino for her dedicated help with the manuscript, especially the graphic work, but also the final proofreading and formatting.

Last, but certainly not least, I thank my current wife, J. (Jagoda) Marie Zwislocki for her encouragement and constructive criticism coupled with the great happiness she has brought me.

Work associated with this monograph was supported in part by the National Institute on Deafness and Other Communication Disorders and in part by Syracuse University.

chapter

The Path of Sound

The substantive material of this introductory chapter should appear familiar to the reader of this monograph. It is included as a point of departure for the rest of the text in an effort to prepare the reader for what it contains and how it is organized. Nevertheless, certain statements may contain new information for any particular reader, depending on his or her background.

My research on sound transmission in the ear began with the inner ear, then moved to the middle ear, finally, to the outer ear for a short period of time. This order is contrary to the direction of sound propagation. But, as every electrical engineer who deals with networks and transmission lines knows, this is the correct order because the performance of every preceding stage depends on the input properties of the following one.

To be accurate, I should mention that my early research on the cochlea concerned only mathematical theory of postmortem preparations. No necessary empirical information was available for the live cochlea. My experimental and theoretical work on the latter did not begin until the 1970s, after my research on the middle and outer parts of the ear had been completed. This work was made necessary by new evidence indicating that not all the conclusions based on postmortem cochlear preparations were valid for the mechanics of the live organ.

Since the input properties of the sound transmitting parts of the ear have become known by now, it is possible to follow in this book the direction of sound propagation. This is done beginning with the outer ear all the way to the hair cells in the cochlea of the inner ear which, acting as microscopic microphones, transduce the mechanical vibration associated with sound into electrochemical processes culminating in nerve action potentials. The path does not stop at a dead end, however, because part of the generated electrochemical energy, enhanced by metabolic energy, is returned to the mechanical vibration through a positive electromechanical feedback. The first part of the path can be visualized best by looking at a longitudinal section of the outer and middle ear, as sketched in Fig. 1.1. The sketch also includes the fluid filled canals of the inner ear with a clearly visible spiral of the cochlea, its auditory part. Because in reality the human inner ear canals are embedded in hard bone, the artist had to mentally chip it away to show their course. The cochlear canal has been opened partially to show that it is divided longitudinally by a partition. The partition determines the mode of sound propagation in the cochlea. As shown in Fig. 1.2, it is not a simple structure but consists of a multilayered plate, called the basilar membrane,

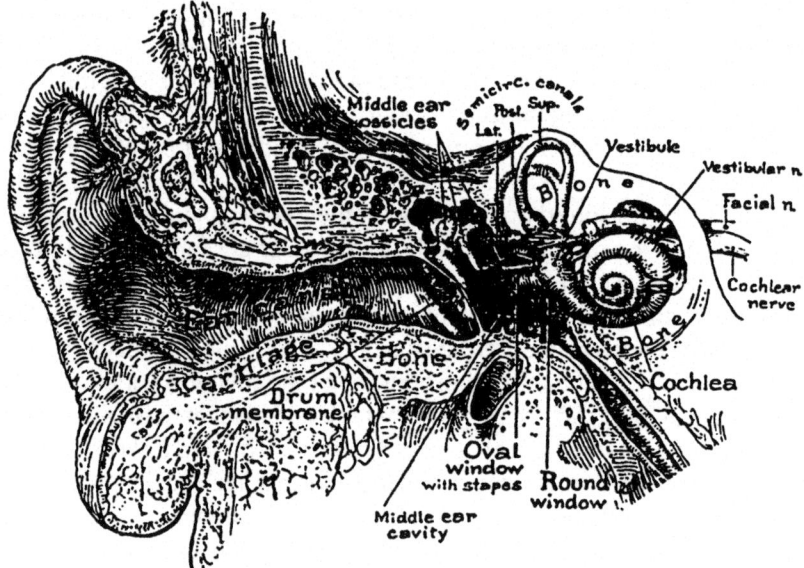

FIG. 1.1. Sketch of the longitudinal section of the outer and middle ear with visualized cochlea, which is opened at one location to indicate the inner partitions—the basilar and Reissner's membranes. (Modified from Brödel, 1946; cit. Zwislocki, 1984). Wever, E. G., & Lawrence, M: *Physiological Acoustics*. Copyright © 1954 by Princeton University Press. Reprinted by permission of Princeton University Press.

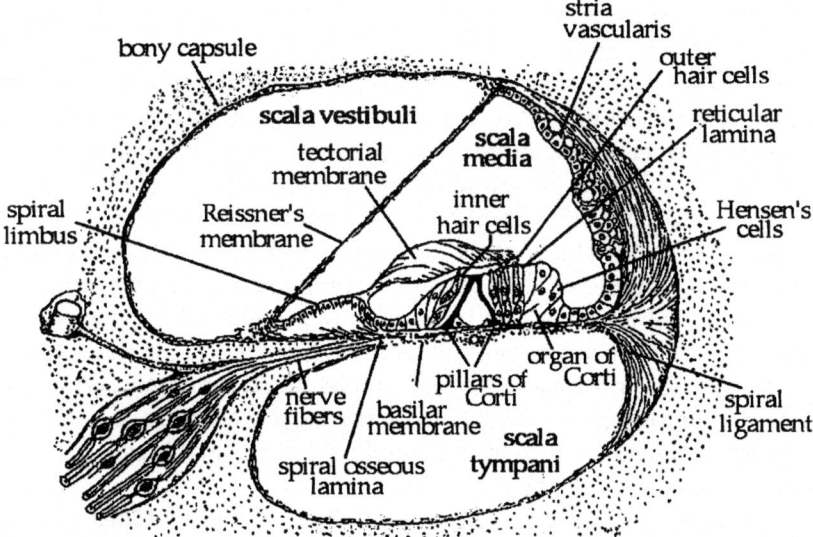

FIG. 1.2. Sketch of the cross section of the cochlear bony canal. (Modified from Rasmussen, 1943; cit. Zwislocki, 1984).

which supports a complex cell mass containing the sensory hair cells stimulated by its vibration. The partition with the hair cells is essential for hearing, and the analysis of its structure and function occupies a substantial part of this monograph.

When sound strikes the human head, most of its energy is reflected but some of it enters the auricle and the ear canal and is led to the tympanic membrane where, in the speech frequency range, about half of the incident energy is transformed into the vibration of the membrane and half is reflected again. From the point of view of auditory sensitivity, the two reflections and an added one in the concha of the auricle may appear as a waste of sound energy, but the reflections are used well by the nature. They serve to enhance the auditory sensitivity in certain frequency regions important in auditory communication. In addition, the reflections at the head and the auricle depend on the direction of incident sound and produce intensity and time differences between the two ears, enabling us to localize the source of sound.

The vibration of the tympanic membrane is transmitted to the three ossicles of the middle ear—the malleus, incus, and stapes, which connect the tympanic membrane to the oval window of the inner ear. The long process of the malleus, the manubrium, is embedded in the tissue of the tympanic membrane and increases its stiffness, improving in this way sound transmission to the ossicles. The malleus is connected to the incus through a massive joint that can be considered as practically rigid from the point of view of sound transmission, except at very high sound frequencies. When the manubrium is entrained by the vibration of the tympanic membrane, the first two ossicles rock around an axis determined in part by the ligaments that hold them in place and in part by their center of gravity, the latter becoming particularly important at high sound frequencies.

The long process of the incus is attached by a small cartilaginous joint to the stapes, which is the smallest bone in the human body. The rigidity of the incudo-stapedial joint appears to vary among mammalian species. According to indirect measurements, it seems to be practically rigid in guinea pigs, quite flexible in Mongolian gerbils and semirigid in humans. When the incus rocks, its long process pushes the stapedial footplate in and out of the oval window, where it is held by the annular ligament. In this way, sound is transmitted to the inner ear.

The advantage of the elaborate system of the ossicles and their rocking motion, as compared to a simple rod-like columella encountered in birds and amphibians, did not become clear until Békésy (1949) demonstrated that it is more stable in sound transmission and prevents certain distortions. It also acts as part of a mechano-acoustic transformer enhancing the sound pressure at the entrance of the inner ear relative to the sound pressure at the tympanic membrane, as was already pointed out by Helmholtz (1877) in mid-19th century.

It is not always clear that the air-filled cavities of the middle ear, which are in communication with the large volume of air in the pneumatic cells of the mastoid bone, play an important role in auditory sound transmission. In fact, they do not only provide a cushion of air necessary for an unimpeded vibration of the tympanic membrane but, as a result of their complicated geometry, also affect the dependence of auditory sensitivity on sound frequency. As I was able to demonstrate (Zwislocki, 1975), they combine their effects with those of wave reflections in the outer ear and those of ossicular mechanics to provide a surprisingly uniform sound transmission in the range of speech frequencies.

Sound is propagated in the outer ear in the form of compressional waves and, in the middle ear, through ossicular vibration. In the inner ear, the mode of sound propagation changes again and takes the form of transversal waves that run along the cochlear partition, somewhat like waves on the surface of water, as illustrated schematically in Fig. 1.3. It should be pointed out that the transversal waves are made possible by the round window of the inner ear, which is situated on the opposite side of the cochlear partition from the oval window. In the absence of the round window, vibration of the stapes would simply compress and decompress the inner ear fluid, producing little motion because of small compressibility of the fluid. The flexible membrane of the round window provides an easy release for the alternating pressure by allowing the cochlear fluid to oscillate between it and the oval window. Because the two windows are on its opposite sides, the cochlear partition is forced to participate in this oscillation. It is true that the helicotrema opening at the apical end of the cochlear canal provides a fluid passage between the two sides of the partition, but the cochlear waves do not reach it, except at very low sound frequencies, as illustrated. The wave pattern shown in the figure is consistent with a large number of measurements. It shows that the wave amplitude increases up to a maximum as the wave progresses toward the apex, then decays rapidly, whereas the wave length decreases continuously.

It has been established experimentally that the location of the maximum depends on sound frequency. For high frequencies it is near the oval window and moves away from it, toward the end of the cochlear canal, as sound frequency is decreased. Following a suggestion of Helmholtz's (1877, 1954), who predicted the existence of the maximum, its location has been believed to be the physiological code for the subjective pitch of sounds. Indeed, some measurements appeared to show a similarity between the ways the location and the pitch depend on sound frequency.

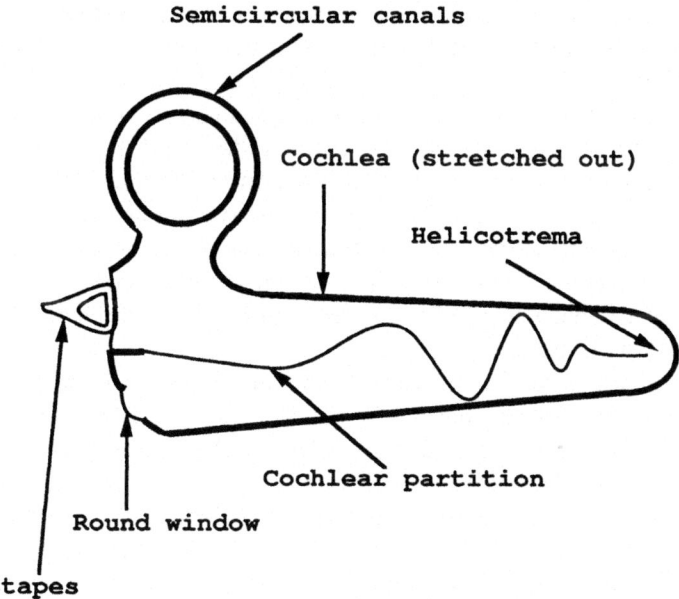

FIG. 1.3. Sketch of transversal waves on the cochlear partition, whose amplitudes are magnified.

The cochlear waves were first observed by Békésy in 1928 on postmortem preparations of human temporal bones and were investigated by him in greater detail in 1947. I explained their nature mathematically in terms of established physical laws, the first publication appearing in 1946, a more extensive one in 1948. These early discoveries are still valid in principle, but subsequent experiments on live animal preparations and associated theory have introduced important modifications. They revealed that the vibration maximum in a live cochlea is much sharper than after death, and that its location depends on sound intensity, so that it cannot be a direct code for pitch that remains practically invariant. They also revealed that the relationship between the vibration of the cochlear partition and the stimulation of the sensory cells is much more complex than originally assumed.

As the waves are propagated along the cochlea, the cochlear partition is deflected transversally back and forth at each location. The pattern of this deflection in the width direction approximates a rocking motion of the basilar membrane around an axis situated near the inner pillars of Corti. According to the classical theory, this motion produces a shear motion between the top of the organ of Corti, the reticular lamina that holds the hair cells, and the tectorial membrane that rotates around a different axis, the ridge of the spiral limbus. The shear motion, in turn, produces deflection of the hairs, or stereocilia, of the hair cells, leading to a depolarization of the hair cells and excitation of the nerve fibers that end on them.

In the following chapters, I analyze separately sound processing in each of the main parts of the sound-transmitting system of the ear. The analysis is based on my past work and, in part, constitutes its review and synthesis. However, much unpublished material is added, and the analysis is updated to coincide with the cutting edge of current research. This is particularly true for the cochlea whose function is so complicated that it defies comprehensive description in a journal article. Accordingly, I gave up on publishing exhaustively the results of my current research on the mechanics and electromechanics of the cochlea in journal articles and have reserved it for this book.

The outer and middle ear each occupy one chapter, but the cochlea occupies three, not only because of the volume of research it has commanded but also because it appeared to me that its complex function can best be explained in steps. Therefore, the first chapter in the sequence concerns the postmortem cochlea that is greatly simplified by comparison to the live cochlea. Once the principles of the simplified function are understood, it is easier to comprehend those of the more complex one.

The analysis of each part of the ear includes the following five aspects not necessarily in the order listed here: description of its structure, measurement of its dynamic characteristics, independent determination of its physical constants, construction of a mathematical, network—or even physical model and, finally, comparison of the model's characteristics with those measured on the natural system. The comparison has a double purpose—validation of the model, which is always a simplified cartoon of the real system, and determination of the effects of individual elements of the system. The extent to which the model characteristics agree with the natural ones may be regarded as a measure of our understanding of the system. Knowledge of the effects of the system's elements is not of purely academic value but can have applications to medical diagnostics, as became clear to me in particular on the occasion of my analysis of the middle ear function.

REFERENCES

Békésy, G. v. (1928). Zur theorie des hörens: Die Schwingungsform der Basilar membrane [On the theory of hearing: The form of vibration of the basilar membrane]. *Physikalische. Zeitschrift. 29*, 793–810.

Békésy, G. v. (1947). The variation of phase along the basilar membrane with sinusoidal vibrations. *Journal of the Acoustical Society of America, 19*, 452–460.

Békésy, G. v. (1949). The structure of the middle ear and the hearing of one's own voice by bone conduction. *Journal of the Acoustical Society of America, 21*, 217–232.

Brödel, M. (1946). *Three unpublished drawings of the anatomy of the human ear.* Philadelphia: Saunders.

Helmholtz, H. L. F. (1877). *On the Sensations of Tone as a Physiological Basis for the Theory of Music* (reprinted second English edition by Henry Margenau (1954), Dover Publications Inc., New York).

Rasmussen, H. T. (1943). *Outlines of Macro-anatomy.* Dubuque, IA: Brown.

Zwislocki, J. J. (1946). Über die mechanische klanganalyse des Ohrs [On the mechanical transient analysis by the ear]. *Experientia, 2*, 10–18.

Zwislocki, J. J. (1948). Theorie der Schneckenmechanik: qualitative und quantitative Analyse [Theory of the mechanics of the cochlea: Qualitative and quantitative analysis]. *Acta Oto-Laryngologica Supplement, 72*, 1–76.

Zwislocki, J. J. (1975). The role of the external and middle ear in sound transmission. In E. L. Eagles (Ed.), *The Nervous System, Vol. 3.* New York: Raven.

Zwislocki J. J. (1984). Biophysics of the mammalian ear. In W. W. Dawson & J. M. Enoch (Eds.), *Foundation of Sensory Science.* Berlin, Germany: Springer.

chapter

The Outer Ear

*P*robably the most extensive investigations of outer ear acoustics were performed by E. A. G. Shaw (e.g., 1974) and his associates. My own work on the subject had a limited and practical purpose of developing a better acoustic coupler for earphone calibration. A working group of CHABA (Committee on Hearing Bioacoustics and Biomechanics of the National Research Council), which I chaired, decided that the available couplers were not entirely satisfactory and provided guidelines for the development of one that would be more up to date (Zwislocki et al., 1967). Since no laboratory picked up the challenge for a year or so and I needed a universal coupler that would allow me to calibrate various types of earphones with the same reference, I undertook the task myself. The development work extended on and off over 4 years, beginning in 1969. It required several kinds of measurements on the outer ear, which ultimately became useful in their own right. They complemented my earlier studies of the cochlea and the middle ear and helped me in accounting for the known dependence of auditory sensitivity on sound frequency. Here, I review these measurements and their analysis, which showed that the complicated geometry of the outer ear can be drastically simplified without appreciably changing its acoustic characteristics. This insight allowed me to design a coupler that lent itself to rigorous specification and was reasonably easy to manufacture.

MEASUREMENT OF SOUND PRESSURE TRANSFORMATION

To sufficiently specify the acoustic properties of the outer ear for our purposes, it was necessary to measure the sound pressure transformation between the tympanic membrane and several points along the ear canal and the concha. Pressure distribution across the ear canal was not investigated because, within the frequency range of our measurements, it was possible to assume that it was uniform. The assumption is permissible when the wavelength of sound is much greater than the linear cross sectional dimensions. At 10 kHz, the highest frequency used, the wavelength in the ear canal amounts to 3.52 cm, almost 5 times the canal's average diameter of about 0.75 cm.

Similar measurements were performed in the past, especially in the classical study of Wiener and Ross (1946), but we wanted to verify the older results and to be able to use the same instrumentation for comparative measurements on the natural ear and the coupler we were to develop. Our work was greatly facilitated by three circumstances—the presence of Dr. G. Djupesland, an internationally known otologist who was spending his sabbatical leave with me, the help of a clever machinist, and the availability of a mechanical adjustable arm for rigidly holding various instruments in the ear, which I had developed at Harvard and was allowed to take with me to Syracuse.

The measurements were described in two reports from my laboratory (Zwislocki,1970, 1971) and were summarized in two papers (Djupesland & Zwislocki, 1972, 1973). They were performed on 7 subjects—3 women and 4 men, whose ages ranged from 16 to 44 years old and who had normal hearing and no history of middle ear disorders. The appearance of their tympanic membranes and ear canals was normal by visual inspection.

A ½" probe tube microphone of Brüel and Kjaer with a probe tube of 5 cm length and 0.1 cm outer diameter, much smaller than the diameter of the ear canal, was used. The microphone was connected to associated Brüel and Kjaer electronics and held in the ear by means of the adjustable arm. It was attached to the arm through a calibrated micrometer-like screw arrangement that permitted precise insertion of the probe tube into the ear canal. The arm is described in greater detail in the next chapter. The business end of the setup is shown in Fig. 2.1, with the probe tube partially inserted into a subject's ear. At the beginning of an experiment, the microphone was positioned by hand so that the probe tube was over the entrance to the ear canal and oriented toward the tip of the manubrium of the malleus. The mechanical arm was made rigid by tightening its ball joints, and the probe tube was advanced until it gently touched the tympanic membrane. It was then lifted by 1 mm, and sound

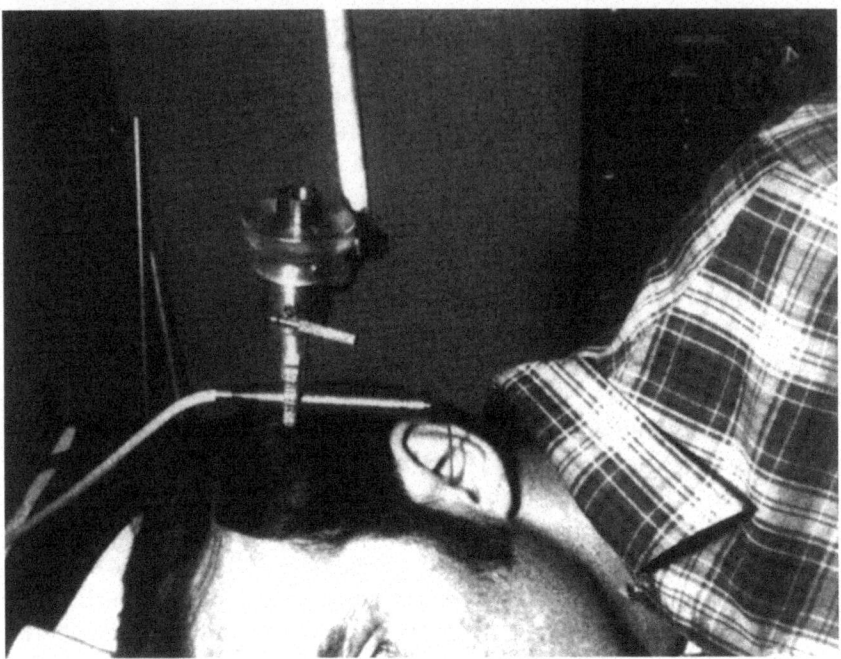

FIG. 2.1. Measuring microphone mounted on the adjustable holder with the probe tube at the entrance to the ear canal of a subject (Zwislocki, 1970).

pressure measurements over a frequency range from 0.2 to 20 kHz were performed. Each subject reclined on a small examination table, with his or her head turned sideways so that the right ear faced upward. The head was supported by a pillow filled loosely with cork chips and tied to the table by an adjustable belt. The arrangement steadied the head quite effectively. The experiments took place in a soundproofed booth with the inner-wall surfaces consisting of perforated plates backed up by sound absorbing material. The sound consisted of sinusoids produced by an Altec 405 loudspeaker mounted on a square baffle hanging over the subject's head. The distance between the entrance of the ear canal and the loudspeaker cone amounted to 110 cm. Because the measurements were comparative in nature, no absolute calibration of the sound generating and measuring equipment was necessary. Nevertheless, we ascertained that neither the sound field at the ear nor the probe tube had any pronounced maxima or minima in their frequency characteristics. In addition, we ascertained that the sound pressure measured by the microphone was transmitted from the tip of the probe tube by closing the

THE OUTER EAR   11

tube opening with wax and measuring the resulting sound attenuation. The residual sound transmission was negligible.

Relative sound pressure was measured at three locations: (1) at a distance of 0.1 cm from the tympanic membrane at the tip of the umbo; (2) inside the ear canal, 1 cm from its entrance; (3) at the entrance; (4) at the tip of the tragus; and (5) 1 cm above it. The locations are illustrated schematically in Fig. 2.2. Sound pressure at the tympanic membrane served as reference. Because the relative sound pressure is quite sensitive to the location of measurement, a substantial effort was made to place the probe tip as precisely as possible. This was achieved by monitoring the placement with the help of a Zeiss surgical microscope. The probe tube was advanced within the ear canal by operating the screw arrangement of the mechanical arm until it reached the desired location. The first set of measurements was performed at 0.1 cm above the tip of the umbo. After its completion, the probe tube was withdrawn to make its tip coincide with the entrance of the ear canal. Subsequently, the tip was placed at 1 cm from the entrance inside the ear canal, then outside the ear canal at the tip of the tragus and, finally, 1 cm above it. Every measurement series was repeated 6 times in separate sessions, so that the probe tube was

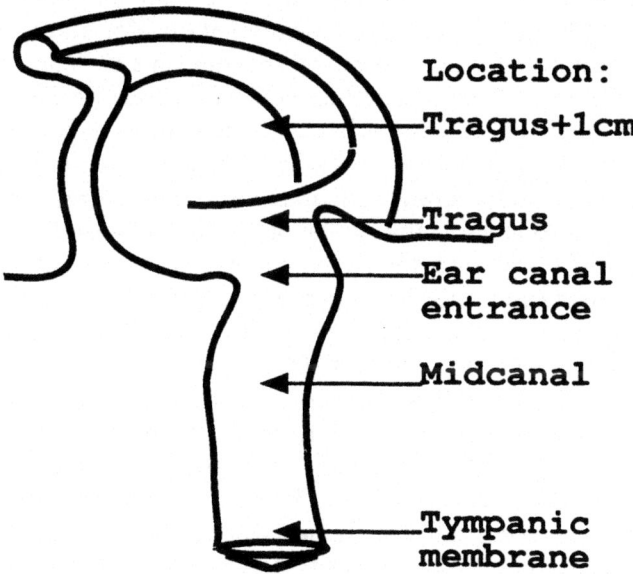

FIG. 2.2. Schematic of the acoustically crucial parts of the outer ear (after Shaw, 1974). From the External Ear by E. A. G. Shaw, 1974. In *Handbook of Sensory Physiology*, edited by W. D. Keidel and William E. Neff, New York: Springer-Verlag. Copyright © 1974 by Springer-Verlag. Reprinted with permission.

placed anew for every measurement. This procedure should have practically removed the effects of any error in the probe tube placement and in expected intra subject variability.

The careful placement of the probe tube tip at both ends of the ear canal allowed us to measure accurately the length of the ear canals of our experimental subjects. This length was defined as the distance between the tip of the umbo and the floor of the concha. The length was the same in all 3 female subjects, measuring 2.2 cm. It was more variable in the male subjects, extending from 2.2 to 2.5 cm, with a median of 2.4 cm and a mean of 2.38 cm. Thus, the average ear canal seems to be a little longer in men than in women. The average of the male and female medians comes to 2.25 cm and the overall mean, to 2.29 cm, which can be rounded to 2.3 cm. This is in agreement with previous results of Wiener and Ross (1946) and of Teranishi and Shaw (1968).

The group results of the sound pressure measurements are plotted in Fig. 2.3 as ratios (SPL differences) between the sound pressures measured at the tympanic membrane and at the other locations. Medians and interquartile ranges are used rather than means and standard deviations in an effort at better approximating typical values rather than average values. The latter may be appropriate for purposes of standardization, but the averaging procedure often eliminates characteristic details from measured functional relationships and can lead to misinterpretations of their underlying mechanisms by smoothing out relative maxima and minima of individual functions. In Fig. 2.3a are plotted the sound pressure ratios between the tympanic membrane and a location in the ear canal 1 cm away from its entrance, about 1.25 cm from the tympanic membrane. The maximum ratio corresponds to the quarter wave resonance of the ear canal portion between the tympanic membrane and the plane of measurement. Note the substantial variability reflected in the interquartile ranges, especially around the maximum. It stems from interindividual differences in the sound pressure ratios and in the frequency location of the maximum.

The sound pressure ratios between the tympanic membrane and the entrance of the ear canal are shown in Fig. 2.3b. The data are in reasonable agreement with the older data of Wiener and Ross (1946; see Djupesland & Zwislocki, 1972, 1973). The maximum is shifted to a lower frequency as a result of a greater distance between the tympanic membrane and the plane of measurement. Note again the very large variability around the maximum, particularly at its high frequency skirt, and the skewed distribution of the data in that region. Both are caused by the deviation of a small number of individual data from the median with respect to magnitude as well as to frequency. The frequency vari-

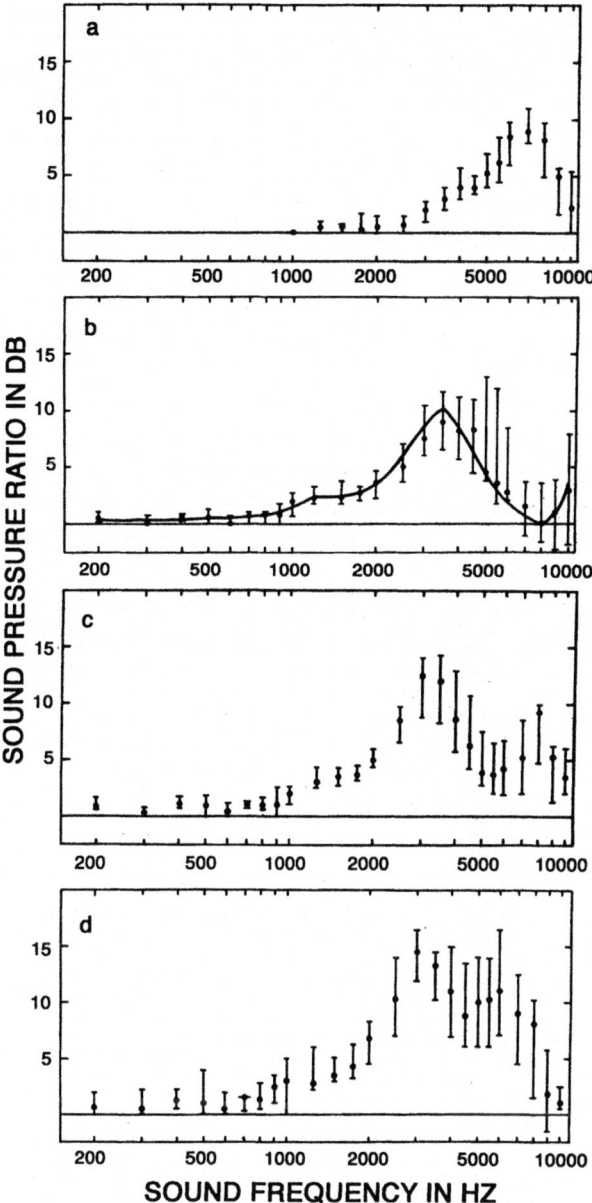

FIG. 2.3. Sound pressure transformation measured between the umbo location of the tympanic membrane and several locations in the outer ear: (a) ear canal, 1 cm from its entrance, (b) entrance of the ear canal, (c) the tragus, (d) 1 cm above it. Closed circles indicate the population medians, and the vertical lines the interquartile ranges. The continuous line in (b) indicates medians obtained after normalization of all sound frequencies to the maximum response.

ability must have had the effect of broadening the apparent width of the maximum so that the maximum does not match any typical pattern. To gain a measure of the effect, the median frequency of the peak was determined, and the individual data normalized with respect to it. The medians of these normalized data are plotted by means of the smooth curve. A slight shift of the maximum toward a lower frequency and a slight increase in its size appear to be the main effects of the normalization. The expected increased sharpness of the maximum is not clearly apparent because it is reflected mainly in the increased height of the maximum.

The sound pressure ratios between the tympanic membrane and the two locations outside the ear canal—at the tragus and 1 cm beyond it, are shown in Fig. 2.3c and Fig. 2.3d. A second maximum at a higher frequency has been added to the first, and the first has been shifted a little more toward lower frequencies. Both phenomena can be ascribed to a further increased distance of the measuring plane from the tympanic membrane, which lowered the frequency of the quarter-wave resonance and made a three-quarter wave resonance possible around 8 kHz. Also, the quarter-wave resonance of the concha additionally increased the sound pressure ratio in the vicinity of 4.5 kHz (Teranishi & Shaw, 1968). According to Fig. 2.3d, the various resonances appreciably enhance the sound pressure at the tympanic membrane relative to the sound pressure just outside the ear in the broad frequency range between about 2 and 8 kHz. This has a beneficial effect on auditory sensitivity.

In many natural listening conditions, the source of sound is located in front of the listener's head at a sufficient distance so that conditions of a free sound field are approximated. In this situation, sound reflection and diffraction at the listener's body and head contribute to sound pressure transformation between the sound source and the tympanic membrane. The total sound pressure transformation is somewhat awkward to measure directly, but it can be obtained by adding the SPL transformations between the source and the entrance to the ear canal and to that in the ear canal. The results of two different computations of the total transformation are compared in Fig. 2.4. One (thick solid line) was obtained by combining Djupesland's and my measurements of the transformation in the ear canal (thin dashed line) with the measurements of the body and head effects (thin solid line) performed by Wiener and Ross (1946). The other (thick dashed line) was provided by Shaw (1974) as a weighted average of data contributed by several investigators. Because the difference between the two resulting curves does not exceed 2 dB, we may be confident that the total transformation is reasonably well established.

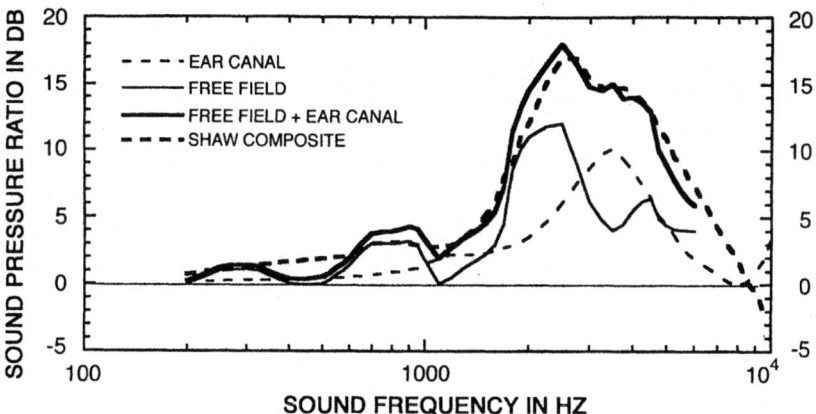

FIG. 2.4. Sound pressure transformation between the free sound field and the tympanic membrane. The thick solid line is a composite of the ear canal measurements of Djupesland and Zwislocki (1972, 1973) and the body and head effects measured by Wiener and Ross (1946). The thick dashed line was obtained by Shaw (1974) as a weighted average of several sets of measurements by several authors.

MODELING THE OUTER EAR ACOUSTICS

One purpose of modeling the acoustics of the outer ear is to find out how far the complicated shape of the outer ear can be simplified without intolerably affecting its acoustic characteristics. My modeling efforts were performed in two steps. First, the pressure distribution in the ear canal was modeled with the help of an electrical network analog (Zwislocki, 1965, 1970); second, the acoustic characteristics of both the ear canal and the auricle were simulated with the help of a mechanical model. The latter effort led to the development of a coupler for earphone calibration (Zwislocki, 1970, 1971, 1980) called *ear-like coupler* or *ear simulator*.

For the modeling purposes described here it is necessary to introduce acoustic variables that are analogs of electrical network variables. This set of variables can be found perhaps the most conveniently in H. F. Olson's *Dynamical Analogies* (1958). According to the theory of electroacoustic analogies of the first kind, the acoustical analog of an electrical capacitance, C_e, is an acoustical compliance, C_a; the acoustical analog of an electrical inductance, L, is an acoustical inertance, M; and the acoustical analog of an electrical resistance, R_e, is an acoustical Resistance, R_a. Furthermore, sound pressure, p, is an analog of the electrical voltage, v, and the volume velocity, $\dot{V}$ (displacement of a volume of fluid per unit of time), an analog of electrical current, i. The amplitudes of the latter variables, which in general depend on time, are denoted by capital letters—P, $\dot{V}$, V and I. Finally, the acoustical input impedance, $Z_a = P/\dot{V}$, is an analog of the electrical input impedance, $Z_e = V/I$. Additional variables are introduced in further text as needed.

To find connections between the acoustical variables introduced above and the geometrical dimensions of an acoustical system, we must introduce appropriate mathematical formulas. The connections arise from the theory of acoustical waves and are given here without mathematical derivations, which can be tedious. The acoustical compliance is defined as $C_a = V/\rho_o c^2$, where V is the static volume of fluid in an enclosure; ρ_o, the density of the fluid (e.g. air) at rest; and c, the velocity of wave propagation in the fluid. The effect of an acoustical inertance is produced by constrictions through which the fluid must flow, or relatively narrow tubes. The defining formula is $M = \rho_o l/\pi r^2$, where ρ_o has already been defined; l is the effective length of the tube; π has the usual meaning; and r is the effective radius of the tube. It should be noted that the effective length, l, of a tube is somewhat larger than the geometric length and that, in the case of a tube with a circular cross section, the effective radius, r, is equal to the geometrical one.

The next step in modeling the acoustics of the outer ear is to determine the relevant geometrical dimensions. The most obvious one is the length of the ear canal. As already mentioned, this length was determined automatically by placing the probe tube with the help of a calibrated micrometer-like screw. A group of 7 subjects was involved, 4 male and 3 female. The length of the male ear canals ranged from 2.3 to 2.5 cm with a median of 2.3 cm. The female ear canals were all 2.2 cm long. Although the population sample was small, and a definite conclusion cannot be reached, there appears to be a disparity between the two populations. As its result, and because of similar disparities in other dimensions, models of a typical male ear should have different parameter values from models of a typical female ear. Strictly speaking, every individual ear requires different parameter values, and a model based on typical dimensions may not represent any natural ear at all. Nevertheless, the approximation is close enough to be useful in many measurements and for an understanding of the fundamental processes. With this justification in mind, I have averaged the dimensions of the male and female ears. For the length of the ear canal, the average of the male and female medians is 2.25 cm. This figure is in excellent agreement with Wiener and Ross's (1946) results and with the value accepted by Teranishi and Shaw (1968) on the basis of acoustic measurements.

We need two additional dimensions of the ear canal—its volume and its average cross-sectional area. Because of the irregular shape of the ear canal the latter is not easy to determine directly. A partial volume of the ear canal was determined on the occasion of acoustic impedance measurements at the tympanic membrane (e.g., Zwislocki, 1957a, 1957b). Perhaps the most accurate measurements were achieved during measurements with the acoustic bridge (e.g., Zwislocki, 1963; Zwislocki & Feldman, 1970). Because of the configuration of the speculum in which the bridge was held and the shape of its sealing tip, the insertion depth remained approximately constant in all ear canals. It amounted to approximately 0.8 cm. This left a residual length of 1.45 cm in a median ear canal. The volume of air associated with this length was measured by filling it with alcohol by means of a calibrated syringe (e.g., Zwislocki, 1957a, 1957b). The particular results used here were obtained on 10 male and 12 female subjects (Zwislocki & Feldman, 1970). The median residual volume came out to be 0.70 cc for the male ears and 0.58 cc for the female ears. The mean of these two median values is 0.64 cc and does not deviate substantially from the grand mean of the population, which was found to be 0.67cc. The median cross-sectional area of the ear canal can be obtained by dividing the median residual volume by the residual length. A value of 0.44 cm^2 is found. It leads to an average median diameter of 0.748 cm, a value only 7% larger than the value estimated by Békésy and Rosenblith (1951) and accepted by Teranishi and Shaw (1968). Extrapolating to the full length of the ear canal, the volume becomes 0.99 cc, practically, 1 cc, again, in excellent agreement with Békésy and Rosenblith.

For modeling the acoustics of the concha of the auricle we need dimensions analogous to those determined for the ear canal (Zwislocki, 1970). As derived from Delany's (1964) acoustic measurements and his network modeling, the air volume of the concha approximates 4.28 cc and the equivalent volume of the ear canal and the middle ear, 1.81 cc, together, 6.09 cc. The latter figure is very close to that assumed for certain standard couplers for earphone calibration (e.g., Beranek, 1988), which is 6.0 cc. Accepting that the 6 cc figure is correct, we can subtract from it the 1 cc volume of the ear canal and the equivalent volume of the middle ear of 0.65 cc (e.g., Zwislocki & Feldman, 1970) to obtain for the concha 4.35 cc, in reasonable agreement with the value derived from Delany's work. We measured the depth of the concha directly on 6 people and found a rather constant value of 0.9 cm. Approximating the concha by a cylinder, the two values can be used to calculate its average diameter. The resulting value is 2.48 cm, which agrees almost exactly with the outer rim of the coupler for earphone calibration suggested by Delany and his coworkers (Delaney, Whittle, Cook, & Scott, 1967).

The ear canal can be modeled according to the theory of electroacoustic analogies in terms of a lumped element electrical transmission line. Such a line is shown in Fig. 2.5. It contains two sections, one corresponding approximately to the outer part of the ear canal, which accommodates various ear devices, the second remaining free. The network acts as a low pass filter, and its elements have to be chosen so that the cut-off frequency remains outside the range of interest. Four equations are available to achieve this and to match the electrical characteristics of the line to the acoustical characteristics of the ear canal. The first two define the characteristic impedance of the ear canal and the corresponding characteristic impedance of the electrical transmission line:

$$Z_a = \rho_o c/A \quad Z_e = (L/C_e)^{1/2} \qquad 2.1$$

The symbols have already been defined, except for A, which means the average cross-sectional area of the ear canal and C_e, which means the electrical capacitance. The third equation defines the cutoff frequency:

$$f_c = 1/\pi (LC_e)^{1/2} \qquad 2.2$$

Finally, the fourth gives the total acoustical capacitance of the ear canal, corresponding to the total canal volume:

$$C_{at} = V_t/\rho_o c^2 \qquad 2.3$$

Combining these equations we obtain the defining equations for the electrical network elements:

$$C_e = 1/\pi f_c Z_e \qquad 2.4$$

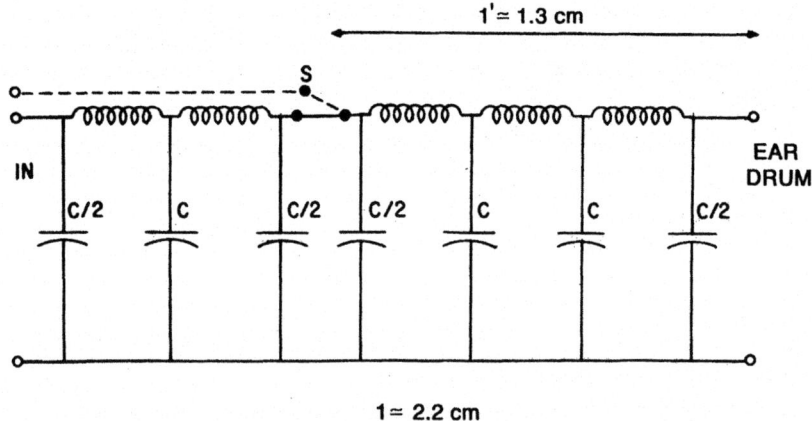

FIG. 2.5. Network analog of the ear canal. The switch, S, marks the location of the tips of most ear inserts (Zwislocki, 1970).

$$L = C_e Z_e^2 \qquad 2.5$$

$$n = C_{at}/C_a \qquad 2.6$$

Note that, if $Z_e = Z_a$, then numerically, $C_a = C_e$.

The numerical values of the constants C_e, L and n can be calculated as follows. For a temperature of about 34° C prevailing in the ear canal (Lilly, 1970, personal communication), the density of air is $\rho_o = 1.15 \times 10^{-3}$ g/cm^3 and the sound velocity, $c = 3.52 \times 10^4$ cm/sec. Together with the cross-sectional area of the ear canal, $A = 0.44$ cm^2, these numerical values lead to a characteristic acoustic impedance of the ear canal, $Z_a = 92$ acoustic Ohms (dyne sec/cm^5). In the MKS system, these values become: $\rho_o = 1.15$ Kg/m^3; $c = 3.52 \times 10^2$ m/sec; $A = 0.44 \times 10^{-4}$ m^2; and $Z_a = 9.2 \times 10^6$ Newton sec/m^5, respectively. On the basis of preliminary calculations, we select for the cut-off frequency $f_c = 24.5$ KHz. Together with the Z_e value already determined, this leads to a $C_e = 0.14$ µF and an $L = 1.2$ mH. The number of sections becomes $n = 5$. These numerical values specify entirely the electrical network analog of the ear canal.

To test the adequacy of the analog, we can measure the voltage transformation between its input and its load at the other end. The voltage transformation should agree with the corresponding sound pressure transformation in the ear canal. In the latter, the acoustic impedance at the tympanic membrane defines the load. As a consequence, the network analog has to be terminated by an electrical analog of this impedance. Such an analog was obtained by developing a network analog of the middle ear, as described in chapter 3. If, after loading the ear canal analog with the analog of the middle ear, a correct voltage transformation is obtained, we can conclude that both analogs, that of the ear canal and that of the input impedance of the middle ear, are satisfactory. The voltage transformation is compared with the sound pressure transformation in Fig. 2.6. The former, plotted as a function of sound frequency, is shown by the solid line, the latter by the closed circles. The crosses indicate the voltage transformation when the ear canal analog is loaded by a simple resistance of 300 Ohms. Note that the peak transformation is about the same for both the middle ear load and the resistive load but that the width of the peak is greater with the former. This seems to be due to the reactive component of the middle ear impedance. As is shown in chapter 3, the middle ear impedance is purely resistive at the location of the peak of the sound pressure transformation but is capacitive (negative) at lower frequencies and inductive (positive) at the higher ones. Capacitive impedance has the effect of lowering the frequency of the resonance peak, inductive impedance, of increasing it. As a consequence, an impedance that is capacitive below the frequency of

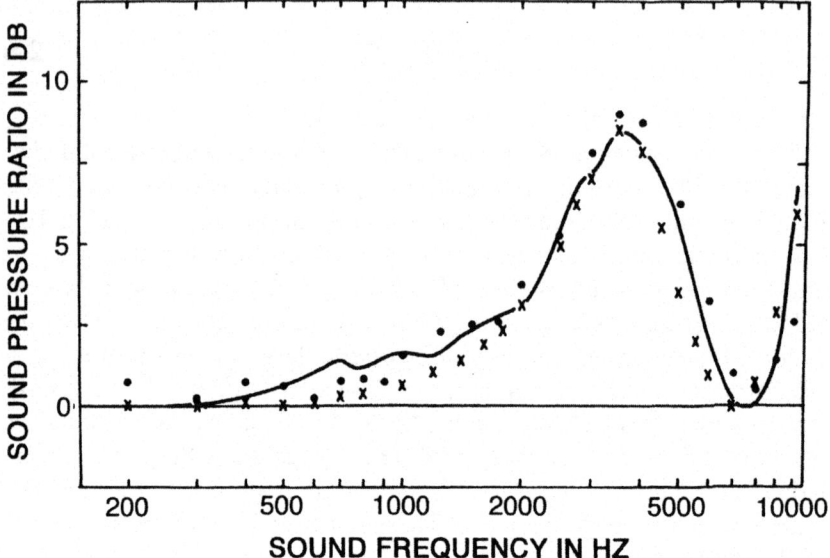

FIG. 2.6. Voltage transformation in the network model of the ear canal, when terminated by the network model of the middle ear (solid line) and by a 300 Ohm resistance (crosses), compared to sound pressure transformation measured in the ear canal between the tympanic membrane and the canal entrance (Zwislocki, 1970).

the peak and inductive above it has the effect of broadening the peak. A broader peak may be expected to be beneficial for the sensitivity of hearing, and the match between the length of the ear canal, which determines the frequency of the peak, and the nature of the middle ear impedance in its vicinity suggest an exquisite piece of evolutionary adaptation.

One additional way of testing the model results is to calculate the sound frequency of the peak from wave theory, more specifically, from the theory of wave reflections. The ear canal is modeled by a uniform tube closed at one end but open at the other. According to the theory, the lowest wave resonance occurs when the effective length of the tube is equal to a quarter wave length, so that

$$\lambda/4 = l + 0.411D \qquad 2.7$$

where λ is the wave length, l, the geometric length of the tube, and D, the diameter of the tube. Because l and D are known, λ can be calculated and from it the frequency of the resonance peak. This frequency is simply f = c/λ. With the numerical constants already given, the predicted frequency of the lowest resonance peak becomes f = 3.441 KHz, in excellent agreement with the model data. It should be pointed out that, if the impedance of the middle ear deviated from pure resistance at the resonance frequency, this frequency would have been shifted and no close agreement would have been found.

Because, according to Fig. 2.6., an electrical transmission line based on a uniform, straight tube adequately simulates the sound pressure transformation in the ear canal, it appears permissible to model the rather convoluted ear canal with such a tube. This leads to great simplification in constructing an acoustic ear simulator, and the straight tube approximation has been attempted for the concha as well. The resulting configuration is shown in Fig. 2.7. It includes, in addition, a damped resonance system simulating the mechanical properties of the auricle.

The ear simulator shown in the figure consists of two main parts—an upper part containing simulators of the auricle, concha, and the outer part of the ear canal, and a lower part simulating the inner part of the ear canal

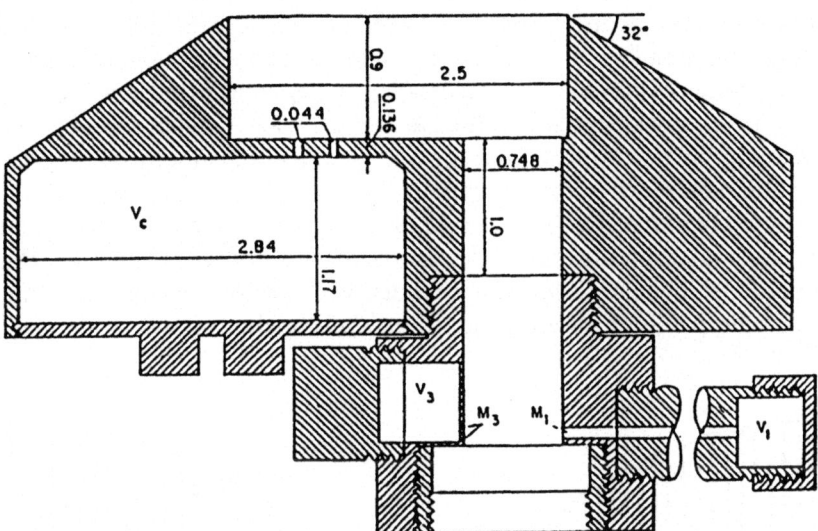

FIG. 2.7. Technical drawing of the ear simulator. From top to bottom, the cavities represent the concha cavity with a large cavity coupled to it through two narrow tubes, the ensemble mimicking the impedance of the auricle, and the ear canal. The enlargement at the bottom fits the measuring microphone. The tubes M_1 and M_3 leading to the cavities V_1 and V_3 constitute 2 of the 4 resonators mimicking the impedance of the middle ear. The bottom part of the simulator can be detached at the horizontal line through the ear canal cavity for calibration of insert phones and other insert devices. This part has been standardized under the designation of *occluded ear simulator* (ANSI S3.25-1979; Zwislocki, 1970).

and the acoustic impedance of the middle ear (Zwislocki, 1970, 1971). This part also includes at the bottom a receptacle for the Brüel and Kjaer 0.5 inch microphone. The dimensions of the concha and ear canal cavities have been already given previously. However, the length of the ear canal had to be adjusted to take into account the temperature difference between the natural ear canal and its simulation and also the equivalent volume of the microphone. The length was reduced from 2.25 cm to 2.15 cm. The circular outer shape of the coupler at the top was made to match the shape proposed by Delany and his coworkers (1967). The large cavity under the concha cavity, marked V_c, and the two small openings coupling it to the concha cavity were designed to simulate the acoustic impedance of the auricle according to measurements of Ithell, Johnson, and Yates (1965). This impedance affects the sound pressure generated by an earphone in the ear canal only at sound frequencies below 0.5 KHz. Note that the ear canal opens into the concha cavity near its side wall, like in the natural ear, rather than in the middle. This is essential because of the transversal resonance modes of the concha. The lowest such mode is illustrated in Fig. 2.8. The concha and the entrance of the ear canal are schematized by the thick lines, the sound pressure distribution, by the shaded areas. Note that the sound pressure has a null in the middle of the simplified concha and a maximum at the location of the ear canal indicated by the opening in the concha floor. Thus, the eccentric location of the ear canal avoids a null in sound transmission to the tympanic membrane and produces maximum sound pressure at the entrance to the ear canal.

The lower part of the simulator contains four side branches situated right above the microphone receptacle. Two of these branches are visible in the longitudinal median section of the simulator. The other two are at right angles to the plane of the drawing. The side branches consist of damped

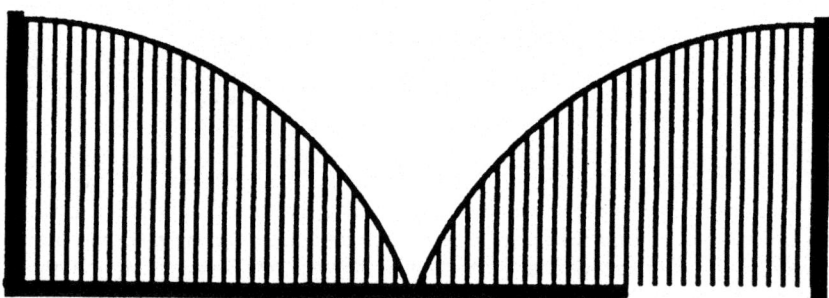

FIG. 2.8. Schematic representation of the first transversal mode of sound pressure distribution across the concha. The heavy lines indicate the walls of the simplified concha; the gap between them at the bottom indicates the ear canal entrance.

Helmholtz resonators, each tuned to a different frequency. Together, they simulate the acoustic input impedance of the middle ear.

The lower part of the simulator has been standardized in the United States under the designation of the *occluded ear simulator* (ANSI S3.25-1979, revised in 1989) and is used mainly for calibration of insert phones, especially, for hearing aids. It also provided the basis for international standardization.

To test the adequacy of the ear simulator, sound pressure transformation was measured at several of its sites with the same instrumentation and acoustic conditions as for real ears. The location of the microphone membrane, which corresponds to the location of the tympanic membrane, served as the reference point. The measurements were made at a distance of 0.9 cm from the entrance to the ear canal model, at the entrance to the ear canal model, at the upper rim of the concha, corresponding to the tragus location, and 1 cm above that location. A measurement was also made at the center of the concha model. The results are shown in Fig. 2.9. In all panels, the solid lines indicate the simulator results obtained inside and above the ear canal model; the dashed line in panel c corresponds to the central location of the concha model. Median data obtained on natural ears at corresponding locations are shown by means of closed circles. Differences between the natural ear and simulator data do not exceed 3 dB, except for the central concha location in the simulator. The agreement would have been even better were the natural ear data normalized with respect to median frequencies of the peaks, as this was done in Fig. 2.3b. The frequency scatter of the individual peaks tends to flatten the median ones. Note the sharp peak in pressure transformation between the center of the model concha and the tympanic membrane model, as shown by the dashed line. The peak results from the pressure null at the concha center, as illustrated in Fig. 2.8. This can be shown by calculating the sound frequency of the lowest transversal mode of the concha. The difference in sound pressure between the ear canal location and the center of the concha can be obtained by determining the distance between the dashed and solid lines at the location of the peak. It amounts to about 11 dB. This is the amount by which the sound pressure at the tympanic membrane is increased by placing the ear canal eccentrically in the concha. Of course, under natural conditions, the difference may be somewhat smaller because of the complicated shape of the concha, which is likely to decrease the resonance peaks. This effect would be easy to implement by small modifications. For example, the diameter of the concha model could be made slightly variable as a function of the distance from the concha floor. Such a solution was already implemented in the past in a standard coupler (e.g., Beranek, 1988).

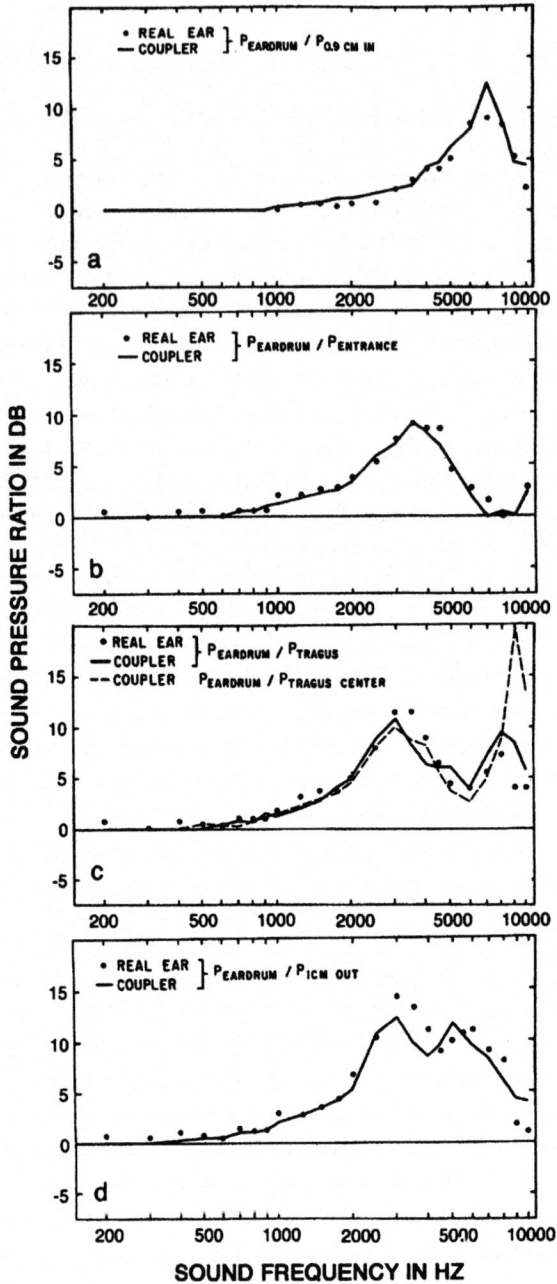

FIG. 2.9. Sound pressure transformations in the simulator (solid lines), as compared to those in the outer ear (closed circles). The dashed line shows the transformation between the ear drum and the center of the concha. The peaks in the real ear appear somewhat broader than in the simulator, in part because of averaging over several subjects.

Agreement between the empirical data obtained on natural ears and on severely simplified network or physical models justifies the simplifications and shows that satisfactory simulation of outer-ear acoustics can be achieved with geometrically simple and, therefore, easy to specify and manufacture devices. Further improvements could be achieved by small changes in the dimensions of the simulator. Perhaps it is regrettable that, in the past, more conservative approaches prevailed and slowed down development of more advanced devices for earphone calibration and sound recording.

REFERENCES

American National Standard for an Occluded Ear Simulator; ANSI S3.25-1979(R 1989), American National Standards Institute, Washington, DC.

Békésy, G. v., & Rosenblith, W. A. (1951). The mechanical properties of the ear. In S. S. Stevens (Ed.), *Handbook of experimental psychology* (pp. 1075–1115). New York: John Wiley.

Beranek, L. L. (1988). *Acoustical measurements* (rev. ed.). Woodbury, NY: American Institute of Physics for the Acoustical Society of America.

Delany, M. E. (1964). The acoustical impedance of human ears. *Sound and Vibration, 1*, 455–467.

Delany, M. E., Whittle, L. S., Cook, J. P., & Scott, V. (1967). Performance studies on a new artificial ear. *Acoustica 18*, 231–237.

Djupesland, G., & Zwislocki, J. J. (1972). Sound pressure distribution in the outer ear. Scand. *Audiology, 1*, 197–203.

Djupesland, G., & Zwislocki, J. J. (1973). Sound pressure in the outer ear. *Acta Oto-Laryngologica. 75*, 350–352.

Ithell, A. H., Johnson, E. G. T., & Yates, R. F. (1965). The acoustical impedance of human ears and a new artificial ear. *Acoustica, 15*, 109–116.

Olson, H. F. (1958). *Dynamical Analogies* (2nd ed.). Princeton, NJ: Van Nostrand.

Shaw, E. A. G. (1974). The external ear. In W. D. Keidel & W. D. Neff (Eds.), Handbook of Sensory Physiology (pp. 455–490). Heidelberg, Germany: Springer.

Teranishi, R., & Shaw, E. A. G. (1968). External-ear acoustic models with simple geometry. *Journal of the Acoustical Society of America, 44*, 257–263.

Wiener, F. M., & Ross, D. A. (1946). The pressure distribution in the auditory canal in a progressive sound field. *Journal of the Acoustical Society of America, 18*, 401–408.

Zwislocki, J. J. (1957a). Some measurements of the impedance at the eardrum. *Journal of the Acoustical Society of America, 29*, 349–356.

Zwislocki, J. J. (1957b). Some impedance measurements on normal and pathological ears. *Journal of the Acoustical Society of America, 29*, 1312–1317.

Zwislocki, J. J. (1963). An acoustical method for clinical examination of the ear. *Journal of Speech and Hearing Research, 6*, 303–314.

Zwislocki, J. J. (1965). Analysis of some auditory characteristics. In R. D. Luce, R. R. Bush, & E. Galanter (Eds.), *Handbook of mathematical psychology*. New York: Wiley.

Zwislocki, J. J. (1970). *An acoustic coupler for earphone calibration;* Special Report LSC-S-7. Syracuse, NY: Institute for Sensory Research, Syracuse University.

Zwislocki, J. J. (1971). *An ear-like coupler for earphone calibration;* Special Report LSC-S-9. Syracuse, NY: Institute for Sensory Research, Syracuse University.

Zwislocki, J. J. (1980). An ear simulator for acoustic measurements: Rationale, principles and limitations. In G. A. Studebaker & I. Hockberg (Eds.), *Acoustical factors affecting hearing aid performance* (pp. 127–147). Baltimore: University Park Press.

Zwislocki, J. J., Benson, R. W., Grason, R. L., Niemoeller, A. F., Rudmose, W. F., & Shaw, E. A. G. (1967). *Couplers for calibration of earphones;* (U). NAS-NRC Committee on Hearing, Bioacoustics, and Biomechanics, Working Group 48.

Zwislocki, J. J., & Feldman, A. S. (1970). *Acoustic impedance of pathological ears*, ASHA Monographs No. 15. Washington, DC: American Speech and Hearing Association.

chapter

The Middle Ear

In 1950s the function of the cochlea seemed to be understood, and it appeared logical to reexamine the function of the middle ear that controls directly the input to the cochlea. I suspected that the middle ear contributed substantially to the shape of the overall transfer function of the ear, which must affect the auditory sensitivity as a function of sound frequency. Békésy's (1960) measurements on this subject were ambiguous. As in every other transmission system, the function of the middle ear depends on its load which, in its case, is provided by the input impedance of the cochlea. Békésy (1960) measured it on cadaver ears, and I calculated it with the help of my theory of cochlear mechanics (Zwislocki-Moscicki, 1946, 1948). I also reinterpreted Békésy's empirical results that appeared to contain a methodological flaw. After the reinterpretation, his results became consistent with the theoretical ones. This set the stage for a comprehensive study of the middle ear.

Such a study appeared of more than purely academic interest since the middle ear is subject to numerous malfunctions whose differential diagnostics was rudimentary then and had not changed much since the 19th century. I hoped that my work would lead to more sophisticated methods. In addition, the performance of the outer ear and of many devices that are placed on the auricle or inserted into the ear canal depends on the input impedance of the middle ear. Probably, the final trigger for my work was provided by Metz's (1946) comprehensive clinical study in which he attempted to measure the acoustic impedance at the tympanic membranes of normal and pathological middle ears with the help of a Schuster bridge. The rather large instrument was cumbersome and difficult to calibrate, which led to a substantial scatter of the data. Accordingly, Metz was not hopeful for clinical applications of his method, except for measurement of impedance changes, such as those produced by middle ear muscle contractions or changes in the pressure differential across the tympanic membrane. It was a challenge for me to improve the method or find new ones, so that the impedance measurements could be used more universally in middle ear diagnostics. In personal conversations, Békésy expressed some doubt that I would succeed because of a small, highly flexible part of the tympanic membrane, called parce flaccida, that, he thought, would effectively shunt the effects of other middle ear structures. I was not to be deterred. Much later, Rosowski, Teoh, and Flandermeyer (1996) were able to demonstrate that the role of parce flaccida varies among animal species and is negligible in humans, except at very low sound frequencies.

Both the fundamental goal of determining the transfer function of the live human middle ear and the various possible applications of research on the middle ear led me to extensive measurements of its acoustic impedance. To study the middle ear transfer function in a nonhuman animal, it is possible to open the bone containing the middle ear and study the motions of its various parts directly. This is not possible for a normal human ear, and the only accessible measurement is that of the acoustic impedance at the tympanic membrane. Because of shunt elements, such as the parce flaccida and the somewhat flexible incudo-stapedial joint, the transfer function cannot be derived directly from the input impedance. For this reason, I introduced a perturbation method based on middle ear pathologies. When the affected parts can be identified and their changes specified, as this is often the case during surgery, the changes can be related to changes in the acoustic impedance at the tympanic membrane and, in this way, the effects of these parts determined. In addition, pathology tends to simplify the middle ear system. For example, a missing incus eliminates the effects of the incudo-stapedial joint, the stapes, and the cochlea. The use of pathology had the corollary benefit of relating various pathologies to characteristic changes in the impedance function. Once the relations were established, it became possible to use the impedance measurements for differential diagnosis of middle ear disorders.

MEASUREMENTS OF THE ACOUSTIC IMPEDANCE AT THE TYMPANIC MEMBRANE

Introduction

Several methods for acoustic impedance measurement at the tympanic membrane, including that of Metz, had been suggested but appeared unsatisfactory for my purposes (see Metz, 1946; Zwislocki, 1957a), and it became necessary for me to develop additional ones. In total, I developed four methods—two entirely physical and two partially psychophysical (Zwislocki, 1957a, 1961, 1970). Although perhaps ingenuous, the psychophysical ones do not seem to have much practical value because of the time they consume and because they include subjective judgment. They were described sufficiently in the past (Zwislocki, 1957a), and I am omitting them here. Of the two physical methods I describe, one is based on an "infinite" resistance source (Zwislocki, 1957a, 1957b) the other, on a highly modified acoustic bridge of Schuster type (Zwislocki, 1961, 1963, 1968; Zwislocki & Feldman, 1970). Although the second method is inherently more accurate and allows the measurements to be made at higher sound frequencies, it proved to be too cumbersome for routine applications. Current clinical impedance or admittance measurements and tympanometry have adopted the first method.

Measurements With the "Infinite" Resistance Source

The method is based on the assumption that the linear dimensions of the source are small compared to the wavelength of sound. It was possible to obtain a sound source having a high internal resistance and obeying this assumption by utilizing the acoustic properties of narrow tubes. On the basis of several practical considerations, a dynamic earphone connected to the ear canal by a metal tube 40 mm in length and 0.4 mm in internal diameter was selected, as shown schematically in Fig. 3.1 (Zwislocki, 1957a). The tube length was acceptable up to almost 2000 Hz. The acoustic impedance of this arrangement was controlled almost entirely by the impedance of the tube, which had an acoustic resistance on the order of 8000 dyne sec/cm^5 ($8*10^8$ Newton sec/m^5) and an acoustic inertance of about 0.6 dyne sec^2/cm^5 ($6*10^4$ Newton sec^2/m^5) within the frequency range of measurements, which extended from 100 to 2000 Hz. It should be noted that these parameters depend somewhat on sound frequency, particularly in higher frequency ranges, and a careful selection of their numerical values is necessary. The unavoidable inertance produced a positive reactance

30 CHAPTER 3

whose numerical value grew with sound frequency and reached that of the resistance slightly above 2000 Hz. A severe transmission loss resulted above that frequency, which limited the range of measurements. The high-frequency cutoff is an inherent disadvantage of the method.

As was found during the experiments, the negative reactance measured in the ear canal interacted with the inertance of the tube to produce a resonance frequency around 225 Hz. The latter had no effect on the measurements, however, because of the strong damping exerted by the tube resistance in that frequency range.

The sound pressure generated in the ear canal by the high impedance source was measured with the help of a condenser microphone via a probe tube with dimensions similar to those of the source tube. Both magnitude and phase of the sound pressure were measured in terms of the output voltage of the microphone. The phase was determined with the help of a nulling phase shifter and a monitoring oscilloscope.

The two tubes were secured in the ear canal by means of a flexible plug consisting of a viscous core surrounded by an elastic skin, as shown in Fig. 3.1. The plug was lubricated with petroleum jelly to facilitate its insertion and improve the seal. At low sound frequencies, the acoustic impedance measured in the ear canal is relatively high and any air leaks around the

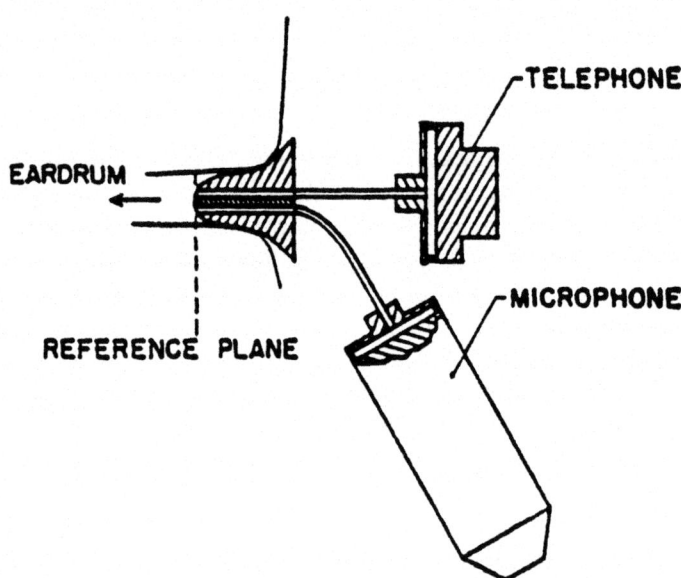

FIG. 3.1. Electroacoustic setup for acoustic impedance measurement in the ear canal with the "infinite" impedance source. Reprinted with permission from Zwislocki, J. J. (1957a). Some measurements of the impedance at the eardrum. *Journal of the Acoustical Society of America, 29,* 349–356.

THE MIDDLE EAR

plug may effectively shunt it. They must be strenuously avoided. When inserting the plug, care must also be taken not to allow the walls of the rather tortuous ear canal to obstruct the tubes. Obviously, this would affect the generated sound pressure and invalidate the results.

To stabilize the measurements, the source and microphone assembly was held by an adjustable arm, the same as described in chapter 2, and the head of the subject was pressed against a support. Various kinds of support were tried, but a cushion filled with small cork chips seemed to fulfill the requirements the best. The cork chips acted nearly like sand but were much lighter and easier to contain. The cushion adapted itself readily to the contour of the head without creating uncomfortable pressure points. Once adapted, it did not change its shape and held the head in a stationary position.

The acoustic impedance in the ear canal was measured by comparison with a known acoustic impedance produced by a cavity with rigid walls. To minimize phase errors, the dimensions of the cavity were made similar to those of the portion of the ear canal remaining between the tip of the plug and the tympanic membrane. The resulting cavity had a volume of 1.1 cc. The plug with the tube assembly was introduced alternately into the cavity and into the ear canal, and the sound pressure was measured in both by the same method, keeping the source input constant. The impedance prevailing in the ear canal at the tip of the earplug was calculated from the sound pressure measurements with the help of the following two equations. If the complex amplitude of the sound pressure generated at the source tube input is called P_i and the corresponding sound pressure in the standard calibration cavity, P_s, then, the ratio between the two sound pressures is given by

$$P_s / P_i = Z_s / (Z_i + Z_s) \qquad 3.1$$

where Z_i is the acoustic impedance of the tube and Z_s, that of the standard cavity. Similarly, we have for the ear canal

$$P_o / P_i = Z_o / (Z_i + Z_o) \qquad 3.2$$

where P_o is the sound pressure and Z_o the acoustic impedance prevailing in the ear canal. Division of the second equation by the first leads to

$$P_o / P_s = (Z_o / Z_s)(Z_i + Z_s)/(Z_i + Z_o) \qquad 3.3$$

If $Z_i >> Z_s$ and also $Z_i >> Z_o$, Eq. 3.3 can be simplified and reduced to

$$P_o / P_s = Z_o / Z_s \qquad 3.4$$

where Z_i is eliminated. Of course, it must be realized that the impedances are composed of real and imaginary parts. In other words, each can be expressed as the vector sum of a resistance and a reactance, $Z = R + jX$.

How well the inequalities were satisfied could be tested only after P_o had been measured and Z_o calculated. In the worst case, in the vicinity of 100 Hz, the reactive component of Z_o, jX_o, had an average magnitude of about $1.9*10^3$ dyne sec/cm^5 ($1.9*10^8$ Newton sec/m^5), only about 4 times smaller than the predominating resistive component of Z_i, R_i. The resistive component of Z_o was negligible. The situation was remedied by choosing $Z_s \approx Z_o$ ($X_s \approx X_o$) on the basis of preliminary measurements, so that $(Z_i + Z_s)/(Z_i + Z_o)$ became almost unity. A cavity volume of 1.1 cc produces a reactance of $2*10^3$ dyne sec/cm^5 ($2*10^8$ Newton sec/m^5) at 100Hz, only 5% larger than the average reactance in the ear canal. To calculate the error committed by assuming that $(Z_i + Z_s)/(Z_i + Z_o)$ equals unity, we must write the expressions in the parentheses in their approximate component forms, $(R_i^2 + X_s^2)^{½}$ and $(R_i^2 + X_o^2)^{½}$. In so doing, the safe assumption is made that X_i, R_s and R_o are negligible at 100 Hz. When the numerical values are introduced, we find that the error amounts to only 0.3%. At higher frequencies, the situation is even better.

From Eq. 3.4, we obtain for Z_o

$$Z_o = jX_s xP_o / P_s \qquad 3.5$$

taking into account the fact that R_s in Z_s is negligible. Since, for a cavity of known volume, the reactance is given as $|X_s| = 1/\omega C_s$, C_s obeying the formula $C_s = V/\rho_o c^2$, and the sound pressures P_o and P_s are measured, Z_o can be determined uniquely.

The median empirical results for Z_o obtained on 9 subjects with normal hearing and no history of middle ear infections are plotted in Fig. 3.2 in terms of the reactive and resistive components. At low sound frequencies the magnitude of the reactive component predominates. It approaches the impedance of a volume of air of 1.36 cc. Part of this volume is contained in the ear canal but another part is contributed by the middle ear. The total size of the equivalent volume is relevant for calibration of insert earphones. Unfortunately, the standardized coupler cavity for such earphones has a volume of 2.0 cc, which is 47% too large. But, since it has been standardized, it is resistant to change. Scientists and engineers appear to be quite conservative professionally even when they are liberal politically. I should add that, even if the standard cavity were reduced to 1.36 cc, it still would not be satisfactory in mimicking the acoustic conditions in the ear. According to Fig. 3.2, the resistive component of the impedance is as important as the reactive one at all but the lowest sound frequencies. Such a component cannot be produced in a simple cavity with rigid walls. As a consequence, sound pressure generated in the standard cavity is not the same as in the ear (e.g., Beranek, 1988). The situation is aggravated by the circumstance that the difference depends on the earphone system, so that a universal correction factor cannot be applied.

Because of a substantial interindividual variation in ear canal volume, the acoustic impedance measured in the ear canal cannot be applied directly to an investigation of the middle ear. The acoustic impedance of the volume shunts its acoustic impedance. When the infinite impedance method is used, the effect has to be eliminated analytically for every individual. To do so, the volume has to be measured independently. Several methods have been suggested over the years, none appear to be flawless. A simple method I used throughout relies on filling the volume with liquid—70% alcohol proved to be particularly appropriate. Because of high surface tension, water tends to leave air pockets. The measurement of the volume has to be made as accurately as possible, because the calculated impedance at the tympanic membrane is sensitive to it.

I used several methodological variants in filling the ear canal with alcohol. In the first, the earplug holding the input and output tubes was exchanged for an identically shaped ear plug, but with a single relatively wide

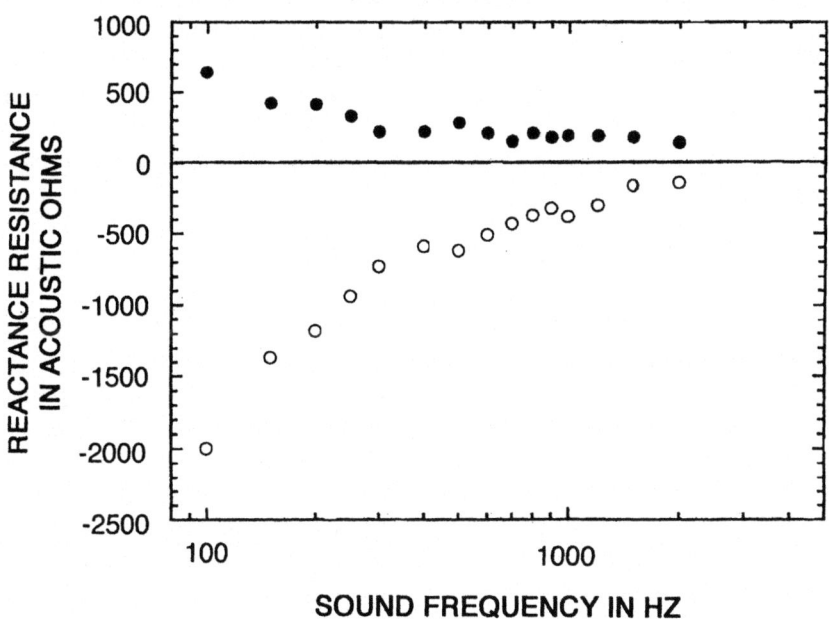

FIG. 3.2. Median acoustic resistance and reactance measured in the ear canals of 9 subjects with no known middle ear defects.

tube in the middle. The alcohol was injected through it by means of a calibrated syringe and a blunt hypodermic needle. When the alcohol became visible in the tube, the injected amount was determined with the help of the syringe scale. Repeated measurements showed that the method provided only a 15% accuracy for a single measurement. Although efforts were made subsequently to improve the accuracy, the volume measurement remained a problem and contributed to data variability, especially with respect to the resistance component of the middle ear impedance. Nevertheless, the median value of the ear canal volume of the series of measurements discussed above agrees well with estimates obtained by others using quite different methods (e.g., v. Gierke & Warren, 1953; Rabinowitz, 1981). It seems that the average residual volume left in the ear canal after insertion of an ear plug, or ear insert, varies from about 0.55 to 0.6 cc, depending on the shape of the plug.

Once the residual volume of an ear canal has been determined, its impedance can be included in the expression for the acoustic impedance measured in the ear canal at the tip of the microphone probe tube. Neglecting sound absorption, the expression for the volume impedance is reduced to a reactance, $Z_v = jX_v = -j/\omega C_v$, where $C_v = V_v/\rho_o c^2$, and V_v means the residual volume. The other symbols have already been introduced. Now, the impedance measured at the tip of the probe tube can be expressed in terms of the impedance of the residual volume and the impedance prevailing at the tympanic membrane, connected in parallel,

$$Z_o = Z_v Z_d /(Z_v + Z_d) \qquad 3.6$$

The latter equation can be solved for the impedance at the tympanic membrane, Z_d,

$$Z_d = Z_o Z_v /(Z_v - Z_o) \qquad 3.7$$

Because both Z_o and Z_v are known, Z_d can be calculated after expressing the impedances in terms of their real and imaginary components.

The method for calculating the acoustic impedance at the tympanic membrane described above has two apparent weaknesses. First, it is based on the assumption that the residual volume of the ear canal can be regarded as a lumped element, in other words, that the sound pressure has a constant phase throughout the volume. This approximation is good at sufficiently low sound frequencies associated with wave lengths that are much larger than the linear dimensions of the residual ear canal, in particular, its length. A more accurate method would take into account phase differences between the probe tube and the tympanic membrane, which result from wave propagation. In this method the ear canal is approximated by a straight tube of constant cross section. I have already demonstrated in the preceding chapter that the approximation is acceptable. However, the distance between the tip of the probe tube and the tympanic membrane has to be known. Measurement of this distance is more difficult than the measurement of the residual volume, so that the theoretical advantage of the method over the lumped element method is difficult to achieve in practice. As a consequence, I used the lumped element method, but with a variant that minimized the phase error. As already mentioned, the length of the calibration cavity was approximately equal to the average length of the residual ear canal. Under these conditions, the phase errors became small of the second order.

The second weakness of the lumped element method I used results from neglecting sound absorption in the walls of the ear canal. This sound absorption, although expected to be small, is not known. There are indications that the simplification introduces errors in impedance derivations at high sound frequencies, particularly in the resistance values (e.g., Onchi, 1961).

The resistance and reactance values at the tympanic membrane obtained in two separate series of experiments performed with the infinite impedance method, on 9 subjects each, are shown in Fig. 3.3 (Zwislocki, 1957b). The points indicated by circles have been derived from the impedance data shown in Fig. 3.2. The crosses belong to a second series in which the method of measuring the ear canal volume was slightly modified. The resulting difference between the two sets of data in Fig. 3.3 stems in part from this modification. Note that the differences in the reactance are very small, except above 1000 Hz where the reactance values are close to zero and are particularly vulnerable to inaccuracies in measurements of the ear canal volume. The differences between the two sets of resistance values are more substantial and persistent above 400 Hz. In particular, the smaller values of the second series suggest that the modified method of volume measurements tended to underestimate the ear canal volume, and the residual volume acted in the calculations as a small shunt capacitance. The resistance measured in the second series at 2000 Hz is very small, considerably smaller than can be inferred from the standing wave ratio that is found in the ear canal. As discussed in chapter 2, this ratio suggests a value around 300 acoustic Ohms. Several other sets of impedance measurements imply similar underestimation of resistance values in the midfrequency range of the ear (see reviews by Zwislocki, 1962 and Rabinowitz, 1981). Rabinowitz's own data did not contain such an underestimation. However, he estimated the residual ear canal volume in an indirect way, and possibly, overestimated it for his conditions. This would lead to higher resistance values. It is also possible that he avoided the underestimation by

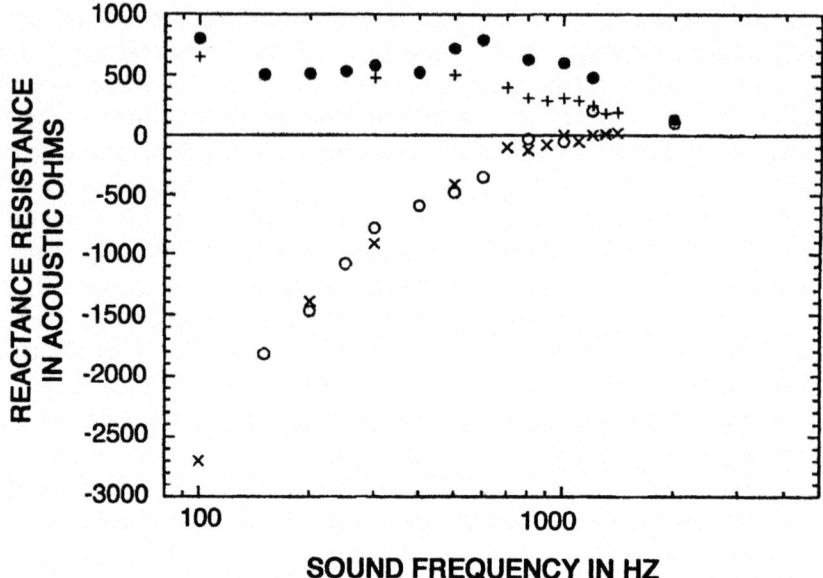

FIG. 3.3. Median acoustic resistances and reactances at the tympanic membrane determined on two groups of 9 subjects with the "infinite" impedance method. The method of determining the ear canal volume differed slightly between the two groups.

taking the phase difference between the tip of the probe tube and the tympanic membrane into account. Sound absorption in the walls of the ear canal may have played a role also.

For purposes of middle ear analysis and diagnostics, the second series of impedance measurements included 15 patients with otosclerosis, in addition to the 9 subjects with normal middle ears (Zwislocki, 1957b). One of the patients had an interrupted incudo-stapedial joint, which disconnected the ankylosed stapes from the tympanic membrane. The obtained median otosclerotic reactance and resistance values referred to the tympanic membrane are compared to the corresponding normal median values in Fig. 3.4 and Fig. 3.5, respectively. The results obtained on the patient with the interrupted incudo-stapedial joint are included as a separate set. All the reactances are negative at low frequencies, indicating that the middle ear system is stiffness controlled. Note that the reactance magnitude of the otosclerotic patients (filled circles) is about twice as large as the normal reactance magnitude (unfilled circles) at low sound frequencies but converges on it above 1000 Hz. This suggests that the mechanical coupling between the stapes and the tympanic membrane is strong at low sound

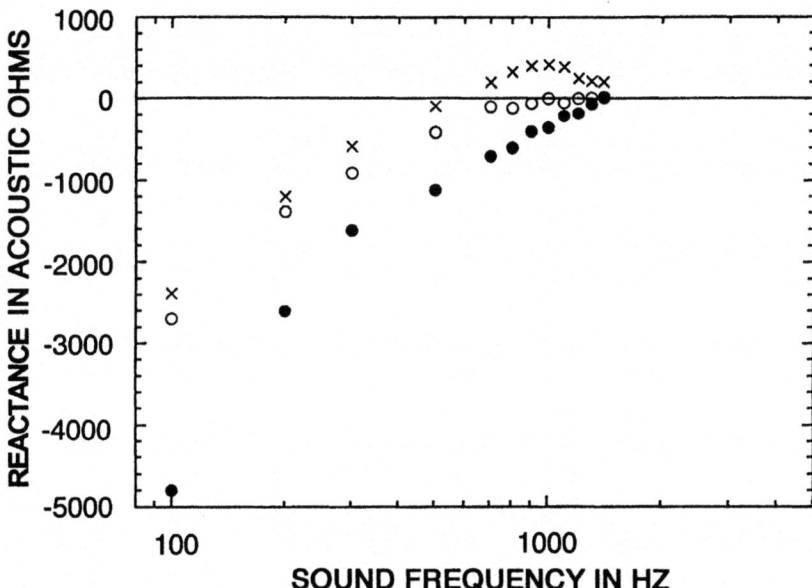

FIG. 3.4. Median acoustic reactance at the tympanic membrane determined with the "infinite" impedance method on 15 otosclerotic patients (filled circles), 9 subjects with normal middle ears (open circles), and one patient with an interrupted incudo-stapedial joint (crosses).

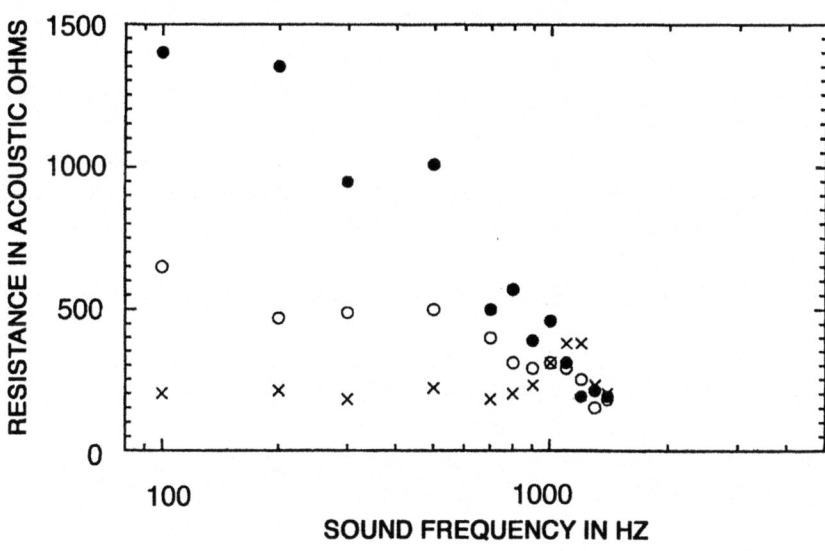

FIG. 3.5. Median acoustic resistances corresponding to the reactances of Fig. 3.4.

frequencies but becomes weak above 1000 Hz. The gradual deterioration of the coupling, as sound frequency is increased, must be caused by flexible shunts, such as the incudo-stapedial joint and the highly flexible parts of the tympanic membrane. It has been shown, however, that pars flaccida plays a minor role in this effect (Rosowski et al., 1996). The reactance magnitude of the patient without incus (crosses) is lower than normal, presumably, because the stiffness of the stapes attachments is disconnected. This is consistent with the increased reactance magnitude produced by stapes ankylosis in otosclerotic patients. In subsequent measurements on different patients with ossicular discontinuity greater reductions were found. The series resonance of the middle ear system should coincide with the point at which the reactance curve crosses zero ordinate. However, the crossing point is not well defined in Fig. 3.4. For the normal population it seems to lie around 1000 Hz. It is around 1400 Hz for the otosclerotic population, and around 550 Hz for the patient with the ossicular discontinuity. Whereas, the shift of the resonance to a higher frequency in the presence of stapes ankylosis can be explained entirely by the increased stiffness, this is not true for the missing incus. Here, the stiffness, which is directly proportional to the reactance magnitude, is decreased by only 15%, the change in the resonance frequency amounts to about 80%. This suggests that interruption of the incudo-stapedial joint increased the mass effect of the middle ear ossicles, perhaps through a change in the mode of ossicular motion (Møller, 1961).

The acoustic resistance values of the three populations, shown in Fig. 3.5, seem to be consistent with the reactance values of Fig. 3.4. The otosclerotic resistance is higher than normal, and the resistance of the patient with ossicular discontinuity is lower. This is particularly true at low sound frequencies. Above 1000 Hz all three sets of data converge, indicating again that the coupling between the tympanic membrane and the rest of the middle ear is much stronger at low than at high sound frequencies. Of some interest is the increased resistance of the ear with ossicular discontinuity slightly above 1000 Hz. It could indicate an antiresonance, also called parallel resonance, arising from an interaction of the inertance of the ossicular chain and the shunt compliance of the tympanic membrane. It is shown in chapter 4 that most of the acoustic resistance measured at the tympanic membrane in normal ears is contributed by the input impedance of the cochlea. This is not true, of course, for the ear with ossicular discontinuity—its resistance is contributed entirely by the middle ear.

Measurements With the Acoustic Bridge

The bridge has been described in several articles (Zwislocki, 1961, 1963, 1968; Zwislocki & Feldman, 1970) and, in greater detail, in a laboratory report (Zwislocki, 1970). I originally developed it as a Schuster type bridge by radically modifying the models that had become known in the past, especially, Metz's (1946) model. The main advantages of the bridge are: an extended frequency range, reaching 7000 Hz for reactance measurements and 4000 Hz for resistance measurements, and direct reference of the measurements to the plane of the tympanic membrane without the need for any calculations. The bridge was manufactured by Grason-Stadler Co., mainly for clinical use. However, it has been superseded by less cumbersome instruments that allow direct reading of the measured variables and do not require nulling adjustments. These instruments are based on the "infinite" impedance source (Zwislocki, 1957a) and are usually applied in connection with tympanometry (e.g., Feldman & Wilber, 1976, for review). They provide only relative impedance measures, however. Absolute impedance values that can be derived from them are known to be inaccurate (e.g., Rabinowitz, 1981; Shanks, Wilson, & Cambron, 1993).

The main parts of the bridge are schematized in Fig. 3.6. They include two main tubes, A and B, arranged symmetrically with respect to an electrodynamic transducer, E, that radiates sound into them in phase opposition. The two tubes are connected symmetrically through a Y tube, Y, whose external outlet can be connected to a microphone or to the ears of the experimenter through a stethoscope-like tube arrangement. During measurement, tube A is inserted into the ear canal via a rigid speculum with a soft sealing tip. The arrangement, when properly used, prevents air leaks to the ambient air space, which can falsify the measurements. Tube B terminates in a variable acoustic impedance arrangement that consists of three adjustable elements: a first tube with an adjustable volume of air, V_1, a second tube with an adjustable volume of air, V_2, and an adjustable, conically shaped slit between them, designed to produce a variable acoustic resistance, R_A. When a slit is narrow enough, it produces an almost purely resistive impedance. In the bridge, we have been able to achieve such an impedance up to about 2,000 Hz. In previous models of the Schuster bridge, the resistance was produced with the help of sound absorbing materials. For reasons of stability, such materials have been avoided in my bridge, which consists entirely of metal parts.

The purpose of the tube containing volume V_1 is to mimic the residual portion of the ear canal contained between the tympanic membrane and the tip of tube A. For this reason, the diameter of the tube has been matched approximately to the median effective diameter of the ear canal. As a result, when volume V_1 is matched to the residual volume of an individual ear canal, the distance between tube B and the resistance element, R_A, approximates the distance between the tip of tube A and the tympanic membrane. In this way phase errors due to travel time of acoustic waves in

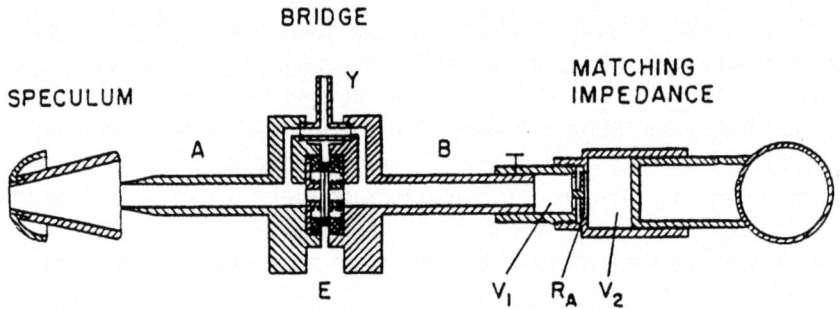

FIG. 3.6. Acoustic bridge with the speculum to seal it in the ear canal. The volume V_1 is adjusted to match the residual volume of the ear canal, the acoustic resistance R_A and Volume V_2 are varied to match the resistance and reactance components of the acoustic impedance at the tympanic membrane (Zwislocki, 1970).

the ear canal are minimized. Volume V_1 is adjusted prior to impedance measurements on the basis of independent measurements of the ear canal volume. Similarly to the measurements with the "infinite" impedance source, the volume is measured by filling the residual ear canal volume with 70% alcohol. For this purpose, the speculum of the bridge with its sealing tip is inserted into the ear canal, as for impedance measurement, and the ear canal is filled with the alcohol up to the tip of the speculum by means of a calibrated syringe. The injected volume is read on the syringe scale before withdrawing the alcohol and letting the ear canal dry for several minutes.

Volume V_2 and resistance R_A serve for matching the reactive and resistive components of the acoustic impedance at the tympanic membrane. They are adjusted alternately so as to obtain minimum sound output in the middle branch of the Y tube. When this is achieved, wave reflections at the tympanic membrane and the variable bridge impedance (R_A and V_2) are the same, which means that the impedances are the same. The numerical values of V_2 can be read directly on a scale calibrated in cubic centimeters. Unfortunately, this is not possible for the resistance, R_A, which has only a position scale. The resistance values must be found on a calibration curve shown in Fig. 3.7. The abscissa axis corresponds to the scale divisions on a rotating dial, the ordinate axis to corresponding acoustic resistance values. These values remain approximately independent of sound frequency up to about 2,000 Hz. Between this frequency and 4,000 Hz, the resistance fluctuates somewhat up and down and increases rapidly at higher sound frequencies. The resistance as a function of sound frequency is shown in Fig. 3.8 for three settings of the resistance dial. As had to be expected, the resis-

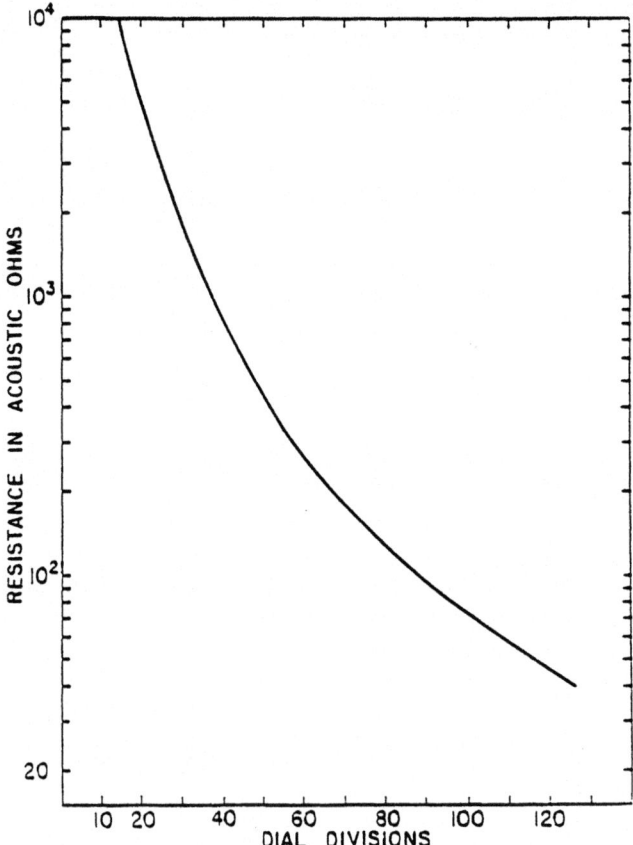

FIG. 3.7. Calibration curve for the variable resistance element, R_A, of the acoustic bridge (Zwislocki, 1970).

tance element is not entirely free of a positive reactive component. This component increases with sound frequency and becomes of the same order of magnitude as the resistance component above 1,000 Hz, as shown in Fig. 3.9 for three dial settings of the resistance element. The reactive component makes it necessary to correct the V_2 values read on the V_2 scale by taking the difference between this component and the reactance associated with R_A.

In some ears, the reactance at the tympanic membrane around 1,000 Hz and above is positive with values too large to be matched by the combination of volume V_2 and the reactive component of the resistance element. To measure these values, the piston controlling V_2 is exchanged for a tube and piston assembly connected to the resistance element through a small orifice. The assembly is shown in Fig. 3.10. The orifice and the variable

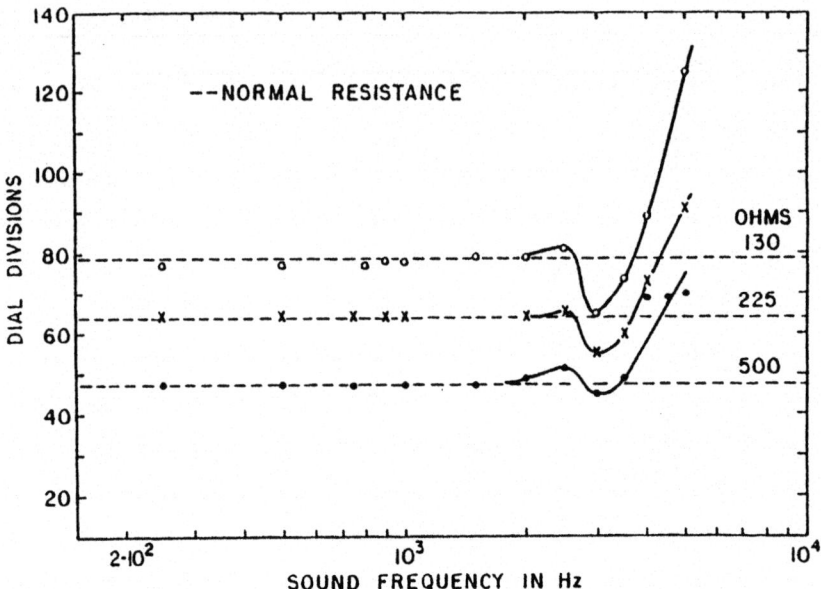

FIG. 3.8. Acoustic resistance values of the resistance element, R_A, as functions of sound frequency for three settings of the resistance dial (Zwislocki, 1970).

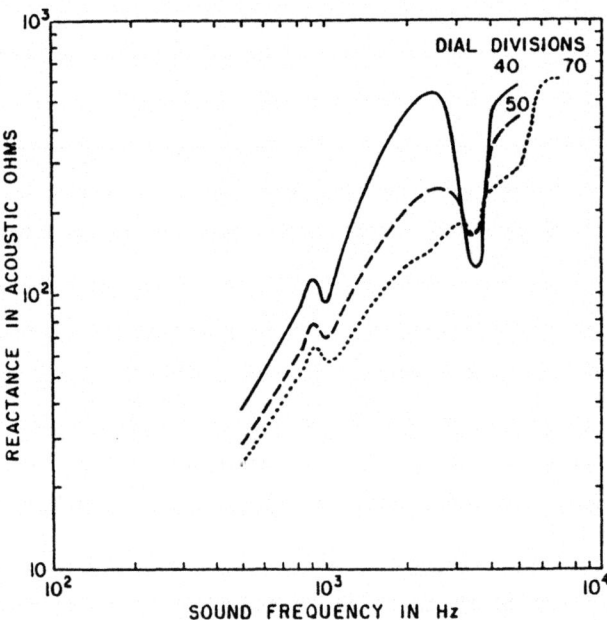

FIG. 3.9. Acoustic reactance values of the resistance element, R_A, as functions of sound frequency for three settings of the resistance dial.

THE MIDDLE EAR 43

new V_2 volume constitute a series resonator (Helmholtz resonator), which makes it possible to obtain both negative and positive reactance values over a wide range at sound frequencies of interest. Unfortunately, these values change with the frequency, and the measurement requires calibration charts.

A photograph of the bridge positioned in a person's ear is shown in Fig. 3.11, together with its holding arm. Note three adjustable joints in the latter. The top and bottom joints are ball joints, the middle joint is axial. The arm can be repositioned as a whole by sliding it along a rigid beam shown at the top of the photograph and by rotating it at the top ball joint. The angular position in this joint can be fixed by means of the small lever to the left of the joint. The levers serving for tightening the remaining two joints are placed in such a way that finger pressure can be exerted on them relative to the movable arm, so that no net force is exerted on the arm. This means that the joints can be tightened with one hand without displacing the arm, while the other arm is holding the speculum in the ear. The impedance elements of the bridge can also be adjusted with one hand because of a sy-

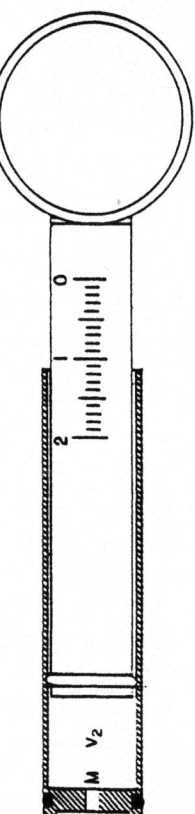

FIG. 3.10. Variable bridge attachment for measurement of both negative and positive reactances. M indicates acoustic inertance. (Zwislocki, 1970).

44 CHAPTER 3

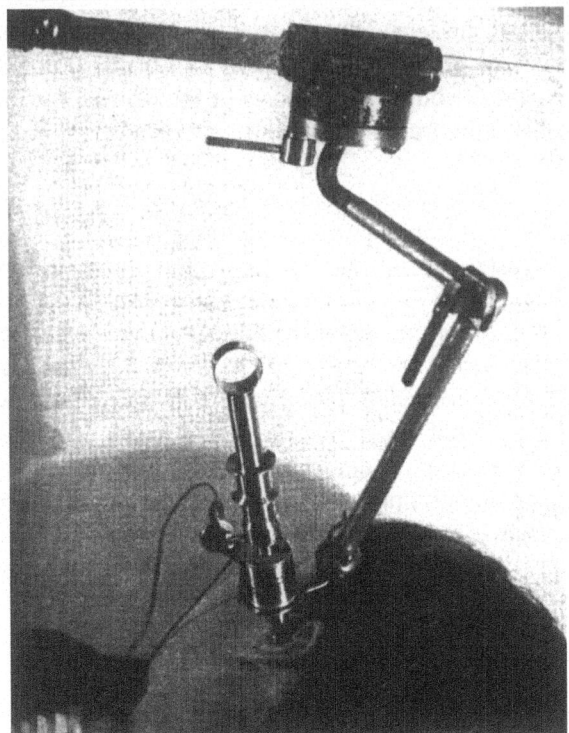

FIG. 3.11. The acoustic bridge held in the ear canal of a subject by the adjustable arm (Zwislocki, 1970).

ringe like configuration of the bridge. The reactive component is adjusted with the thumb inserted in the ring at the top of the bridge by sliding the piston controlling volume V_2 up and down. The resistive component is adjusted by holding the flanged part of the bridge cylinder between the index and middle fingers and rotating it.

The most detailed data on the middle ear impedance obtained with the bridge are shown in Figs. 3.12 through 3.15 (Zwislocki, 1970, 1975). Figures 3.12 and 3.13 show the median reactance and resistance values and the interquartile ranges of individual data for 12 female ears. Only one ear was used per subject. The next two figures include the medians for 10 male ears in addition to the female median results of Figs. 3.12 and 3.13. The interquartile ranges for the male ears have been omitted since they are nearly the same as for the female ears. In addition to the laboratory results indicated by the closed circles and crosses, the open circles in Figs. 3.14 and 3.15 show more limited results obtained in a clinical study on a mixed population of 33 subjects (Feldman, 1967). The agreement between the laboratory and clinical results is remarkable for the reactance but less so for the resistance.

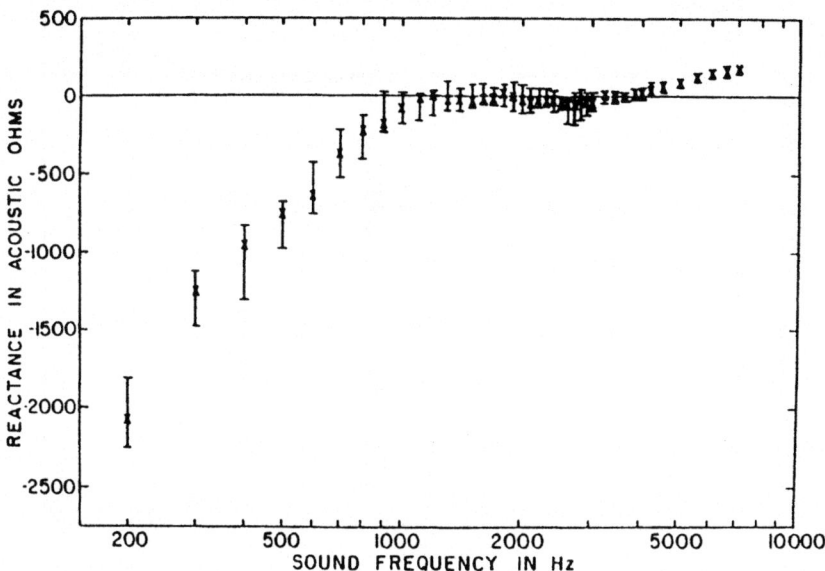

FIG. 3.12. Median acoustic reactance at the tympanic membrane measured with the bridge on 12 female ears, one per subject. Vertical lines indicate the intersubject interquartile ranges (Zwislocki, 1970).

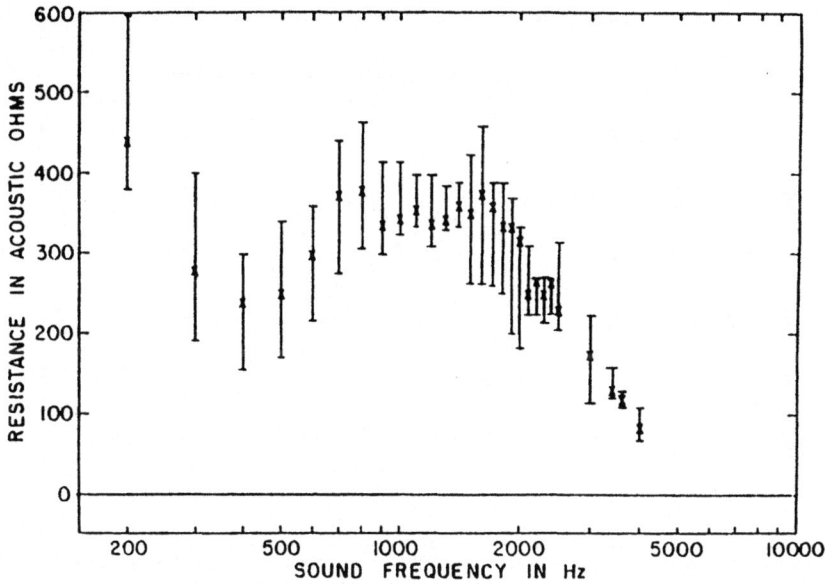

FIG. 3.13. Median acoustic resistance values corresponding to the reactance values of Fig. 3.12 (Zwislocki, 1970).

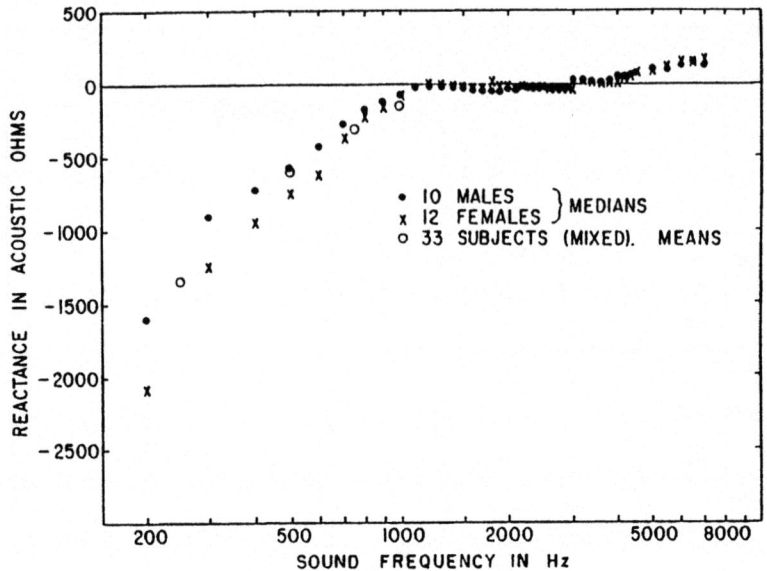

FIG. 3.14. Median acoustic reactance at the tympanic membrane measured with the bridge on 10 normal male ears (closed circle), compared to the female reactance values of Fig. 3.12 and to the reactance values determined on a mixed group of 33 subjects without known middle-ear defects (Zwislocki, 1970).

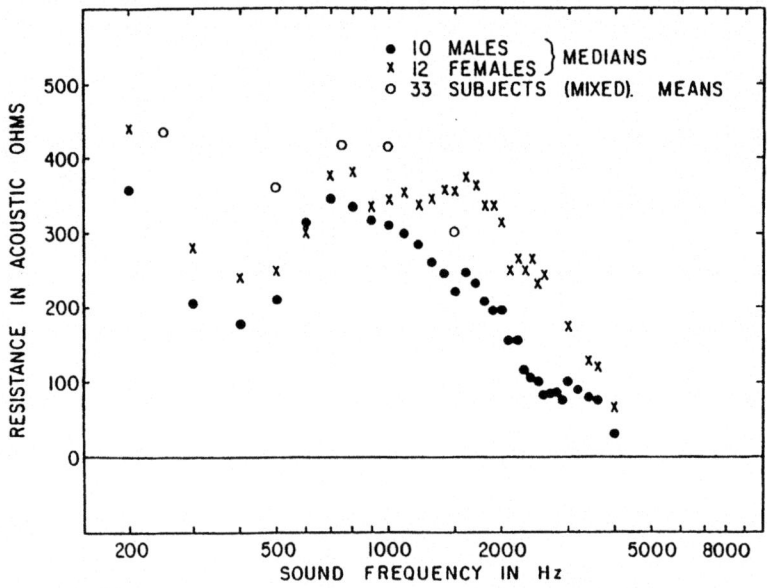

FIG. 3.15. Same as Fig. 3.14, for the resistance (Zwislocki, 1970).

Several features of the data in Figs. 3.12 through 3.15 should be noted. One is that the relative variability of the resistance data is much larger than of the reactance data, particularly, at low sound frequencies and around 2,000 Hz. The large variability at the low frequencies may be due simply to measurement inaccuracy. In this frequency range, the resistance component is a small fraction of the total impedance magnitude, and a small error in measurement may be reflected as a relatively large error in the measured resistance. Such a situation does not prevail around 2,000 Hz, where the resistance is a major component of the impedance. Here, the relatively large variability may be due to interindividual differences in the frequency location and sharpness of a parallel resonance in the middle ear, which can be shown to take place in this frequency region.

Another important feature is that the reactance is negative and is inversely proportional to sound frequency at low sound frequencies and weaves its way around the zero line between 1,000 Hz and 4,000 Hz, as shown in Figs. 3.12 and 3.14. These characteristics mean that the reactance is controlled by compliance, or its reciprocal—stiffness, at low sound frequencies but reveals a complicated mechanism at higher sound frequencies, which cannot be modeled as a simple resonator, as this was suggested in the past. The complicated course of the resistance curves in Figs. 3.13 and 3.15 is consistent with this conclusion. However, the decay of the resistance to nearly zero toward high sound frequencies must be artifactual, as is mentioned in chapter 2 on the outer ear. The measured low resistance is in conflict with the rather high standing wave ratio measured in the ear canal at these frequencies. Rabinowitz (1981) suggested that the artifact arises when the phase difference between the measuring plane and the tympanic membrane is not taken into account. He did include an appropriate phase correction and avoided the artifact. But, the geometry of the bridge and of calibration cavities I used should have taken care of the phase correction automatically. Onchi (1961) avoided the problem in his experiments on cadaver ears by measuring sound pressure inside the tympanic cavity. He suggested that the artifact may be due to sound absorption in the walls of the ear canal, but it is not clear why such sound absorption should not affect the standing wave ratio also. His impedance values are much higher than obtained in vivo, so that the relevance of his measurements to live middle ears is questionable. It is also possible that the artifact was produced by a bias in the measurements of the ear canal volume. Such possibility is invited in particular by the fact that some measurements performed with the bridge lead to an approximately flat resistance curve hovering around 400 acoustic Ohms between 250 and at least 1500 Hz (Feldman & Zwislocki, 1965). Further research concerning this problem would be welcome. Meanwhile, the resistance at high sound frequencies can be obtained from sound pressure distribution along the ear canal, as was suggested in the past (Zwislocki, 1970; Shaw, 1974).

To ascertain the reliability of the data of Figs. 3.12 to 3.15 obtained with the bridge, they are compared with some of my earlier data (Fig. 3.3) and to data of Rabinowitz (1981) in Figs. 3.16 and 3.17, both sets obtained with the "infinite" impedance source. All the data are for male subjects. The bridge data are indicated by the solid line, two sets of my older data by the circles, and Rabinowitz's data by the crosses. The agreement among the four sets is good for the reactance but not for the resistance. On the average, the bridge produced the lowest resistance values and my earliest measurements with an "infinite" resistance source the highest. All the resistance measurements shown, except those of Rabinowitz, produced decreasing values at high sound frequencies. Rabinowitz's results show a maximum around 2,000 Hz and an asymptote of about 300 Ohms at high sound frequencies. They agree the best with the resistance magnitudes derived from the standing wave ratio (Zwislocki, 1970) and with theoretical considerations discussed below. Møller (1961) contributed another important set of data on the human middle ear impedance, which were omitted in Figs. 3.16 and 3.17 in order to avoid crowding. His reactance magnitudes were rather lower at low sound frequencies than those shown in Fig. 3.16 and remained negative be-

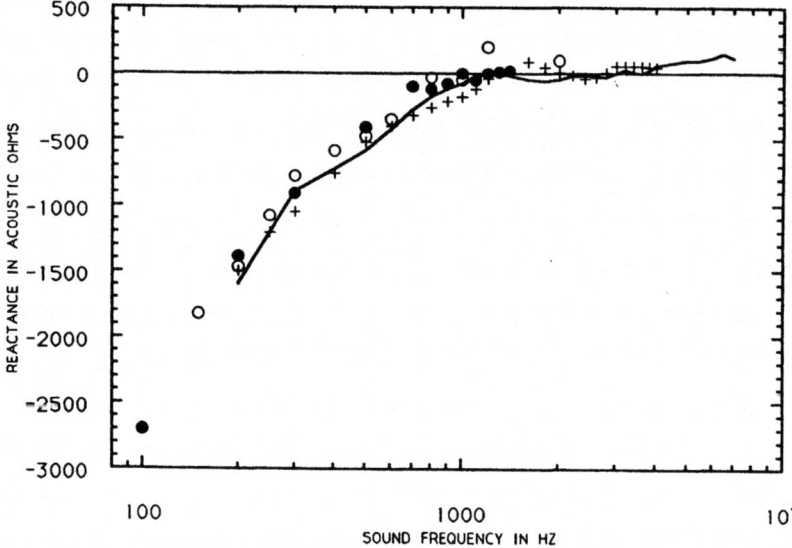

FIG. 3.16. Comparison of the male reactance data of Fig. 3.14 obtained with the acoustic bridge (solid line) to three sets of male data obtained with the "infinite impedance source." The two sets reproduced from Fig. 3.3 are indicated by circles, the data obtained by Rabinowitz, by crosses.

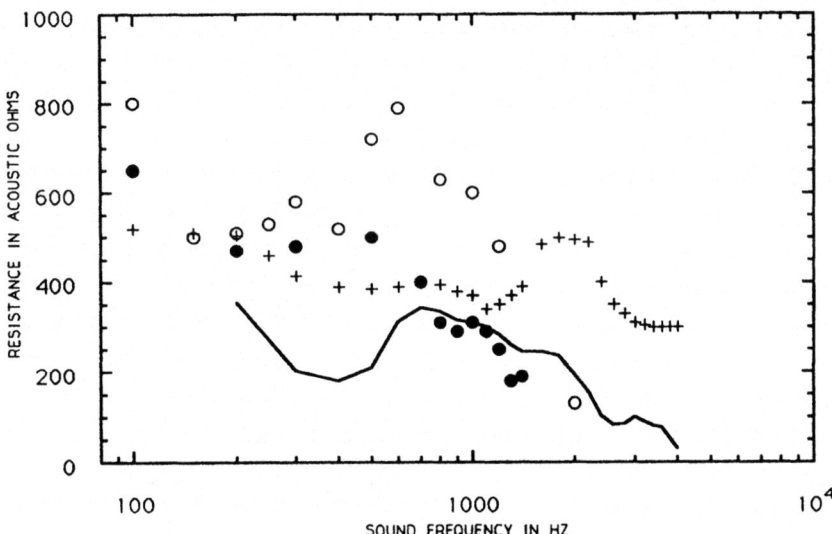

FIG. 3.17. Acoustic resistance data corresponding to the acoustic reactance data of Fig. 3.16.

yond 1,000 Hz. His resistance values remained roughly frequency independent, lying somewhat below those of Rabinowitz.

The bridge was used extensively on patients with middle ear disorders. This was done for two reasons. One aimed at developing a clinical diagnostic method, the other, at an analysis of the middle ear function. As mentioned above, the complexity of the middle ear structure prevents an unambiguous analysis of its function on the basis of impedance results obtained solely on normal ears. Pathology tends to simplify the system by making some of its parts inoperative and perturbing the function of others. Results obtained on 33 normal ears, 24 otosclerotic ears, and 4 ears with ossicular separation are shown in Fig. 3.18 (Zwislocki, 1968). Instead of the reactance, which is negative at low sound frequencies and varies strongly with the frequency, the compliance in equivalent volume of air is plotted. One advantage of doing so is that, from the theoretical point of view, the compliance should remain constant at low sound frequencies, and departures from constancy indicate directly a disagreement between the experimental results and the theoretical expectations. All the compliance curves in the upper panel of Fig. 3.18 have horizontal asymptotes at low sound frequencies, in agreement with theoretical expectations. Note that the compliance is the smallest in otosclerotic ears and the largest in ears with ossicular separation. This order agrees, of course, with theoretical expectations and gross clinical observations. All three compliance curves have vertical asymptotes toward higher frequencies, indicating infinite compliance values. An infinite compliance value signifies a series reso-

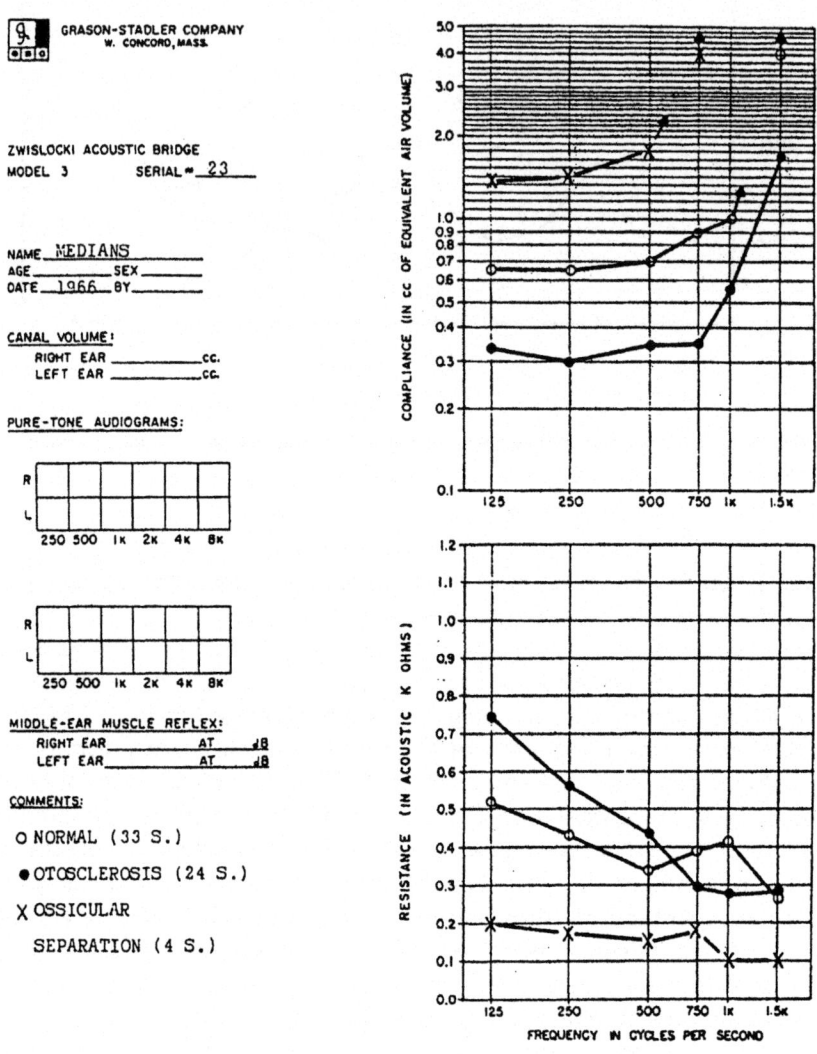

FIG. 3.18. Median acoustic reactance and resistance data obtained with the acoustic bridge on the group of 33 subjects with normal middle ears (Figs. 3.14 and 3.15), on a group of 24 otosclerotic ears, and on a group of 4 ears with ossicular separation. The data are plotted on a clinical form, and the reactance ones are expressed as equivalent volumes of air (Zwislocki, 1968). From "On Acoustic Research and Its Clinical Application," by J. J. Zwislocki, 1968, *Acta Oto-Laryngologica, 65,* pp. 86–96. Copyright © 1968 by Taylor & Francis. Reprinted with permission.

nance that comes about through interaction of the compliance of the system with its inertance. It is known from the theory of simple resonators that the resonance frequency is inversely proportional to the square root of the compliance when the inertance is kept constant. This is nearly true for the compliance curves of Fig. 3.18, so that the two pathologies included do not seem to affect appreciably the effective inertance of the middle ear system.

The resistance curves plotted in the lower panel of Fig. 3.18 seem to be intuitively consistent with the compliance curves. The highest values are associated with the smallest compliance values and the lowest with the largest compliance values. There is a departure from this pattern around 1,000 Hz, however, where the otosclerotic ears produce a smaller resistance than normal ears, although the latter have a larger compliance. As discussed below, the departure probably originates in a parallel resonance that increases the resistance in normal ears but is less pronounced in otosclerotic ears. A slight relative resistance maximum is also detectable in the curve associated with ossicular separation, although at a somewhat lower frequency. The mechanism underlying the maximum is probably the same as in normal ears, and the frequency shift is likely to be due to a lowering of the parallel resonance frequency by the separation.

Individual results for a patient with confirmed massive adhesions to the large ossicles are shown in Fig. 3.19. The compliance is very small by comparison to the normal range indicated by two thin lines and a shaded area between them, and the resistance is very large. This is not surprising since the ossicular chain has been fixed almost completely, especially in the left ear, and the compliance and resistance are controlled practically completely by the tympanic membrane vibration relative to a fixed umbo. The low compliance and high resistance values measured in the presence of a fixed ossicular chain are crucial for the understanding of the middle ear function. They show directly, that the tympanic membrane does not shunt much energy away from the ossicular chain in spite of some determinations of the vibration pattern of the tympanic membrane, which suggest great flexibility (e.g., Tonndorf & Khanna, 1971; Tonndorf & Khanna, 1972).

Bridge measurements were also used to assess how well some middle ear prostheses restore normal sound transmission properties of the middle ear (Zwislocki, 1968). Impedance results obtained on 8 patients with wire-fat prostheses, 15 with Teflon prostheses and 11 with stainless steel prostheses are shown in Fig. 3.20. Since the investigation was performed in 1968, the data cannot be regarded as representative of later more improved techniques. Nevertheless, they point to some characteristic features indicative of the prosthetic quality. It is clear from the upper panel that the compliance associated with the wire-fat and teflon prostheses is sub-

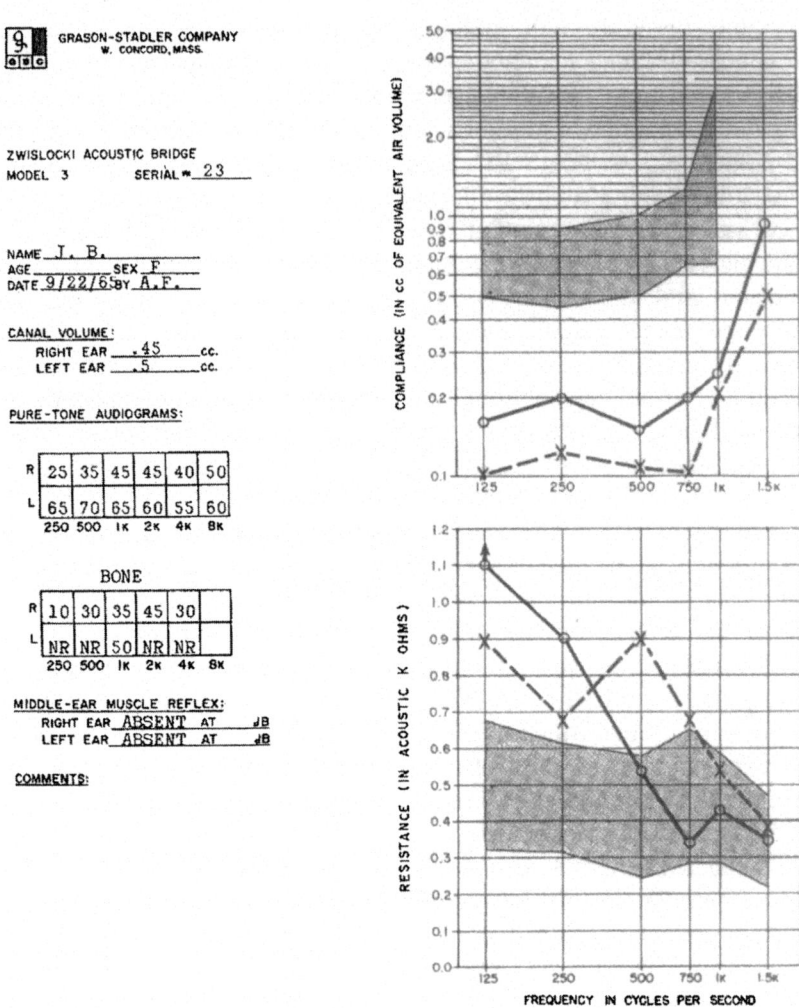

FIG. 3.19. Bilateral acoustic reactance and resistance data obtained with the acoustic bridge on a patient with confirmed massive adhesions to the large ossicles. The thin lines indicate the accepted ranges of normal values (Zwislocki, 1968). From "On Acoustic Research and Its Clinical Application," by J. J. Zwislocki, 1968, *Acta Oto-Laryngologica, 65,* pp. 86–96. Copyright © 1968 by Taylor & Francis. Reprinted with permission.

stantially larger than normal, suggesting some loss in sound transmission. The somewhat lowered resistance associated with the two kinds of prostheses and shown in the lower panel is consistent with this conclusion. On the other hand, the stainless steel prosthesis appears to have restored normal acoustic conditions, as is evident from the data indicated by crosses.

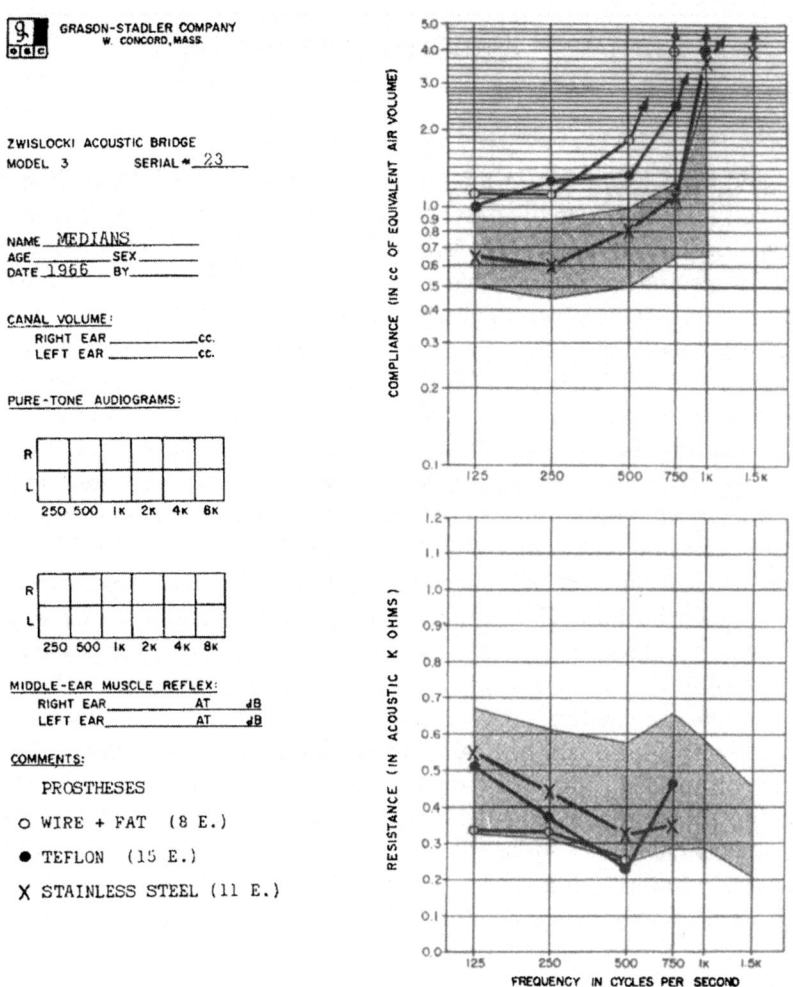

FIG. 3.20. Performance of three types of middle ear prostheses tested by means of reactance and resistance measurements at the tympanic membrane with the bridge. The thin lines indicate the normal ranges (Zwislocki, 1968). From "On Acoustic Research and Its Clinical Application," by J. J. Zwislocki, 1968, *Acta Oto-Laryngologica, 65*, pp. 86–96. Copyright © 1968 by Taylor & Francis. Reprinted with permission.

ANALYSIS OF THE MIDDLE EAR FUNCTION

Introduction

Judging from references in the literature, the analysis of the middle ear function I performed in the past (Zwislocki, 1962) is still regarded as valid, except for small modifications. As a consequence, I review it here. But, I do introduce some changes relative to my earlier publications to take advantage of subsequent experimental information.

Historically, I devised my first highly simplified model of the middle ear function in 1957 (Zwislocki, 1957a). Onchi devised an incomplete mechanical model earlier (Onchi, 1949) but was unable to solve the differential equations associated with it. He completed his work later (Onchi, 1961) by converting the mechanical model to an electrical network analog and by experimentally determining the numerical values of its elements. The main difference between the final model of Onchi's and my 1957 model is an explicit realization of the complex effects of the middle ear cavities by him. Unfortunately, Onchi's measurements were performed on cadaver ears, and his impedance values depart severely from those obtained on live humans. He also obtained an effective volume of the middle ear cavities, which is much smaller than that determined repeatedly in subsequent experiments. My latest model (Zwislocki, 1962) incorporated the elements of my first model and additional elements of Onchi's (1961) model with some modifications, especially with respect to the numerical values. Møller (1961) devised a different model based on different assumptions and his own experimental results. He tried to perturb the dynamic characteristics of the middle ear by coating the tympanic membrane with various substances, but the success of this procedure has appeared to me to be uncertain.

There are two fundamentally different classes of models of biological systems. In one, a mathematical set of equations or equivalent network or computer models are developed that are capable of reproducing isolated characteristics of a natural system. The elements of these models are not in one-to-one correspondence with the physical entities, so that it is not possible to investigate the effects these entities may have on the characteristics of interest by varying the numerical values of the models' elements. Such models are appropriate and, indeed, the only ones possible when the structure of a natural system is largely unknown. Because biological systems tend to be nonlinear and often do not follow the principle of parsimony, being products of evolutionary patch work, their structures are not unambiguously derivable from relationships between the inputs imposed on them and their corresponding outputs. The usefulness of these models becomes doubtful when the structure of a natural system is known sufficiently well to model it, physical element by element. This leads to a second class of models in which every element is a model of a natural element. Under these conditions, it is possible to investigate theoretically the effects the natural elements have on the system's characteristics of interest. Reciprocally, it is often possible to pinpoint changes in any particular element of the natural system on the basis of changes in the measured system's characteristics. This possibility is of crucial diagnostic value in medicine and, more generally, in biology as a whole.

In analyzing the middle ear function, I was fortunate enough to be able to use models of the second class. As a result, the effect of every relevant part of the middle ear on the impedance measured at the tympanic membrane and on the transfer function of the middle ear could be identified. By reversing the process, measured changes in the impedance became indicative of pathological changes in specific parts of the middle ear. It is possible that this recognition has contributed significantly to current middle ear diagnostics.

I found it convenient to model the middle ear in terms of an electrical network analog of the first kind (e.g., Zwislocki, 1957b, 1962) that can be derived from a mechanical network analog (Zwislocki, 1963) on the basis of the theory of electroacoustical analogies. Because the middle ear system is quite complicated and includes acoustical elements in addition to mechanical ones, it proved convenient to express all its elements in terms of their electroacoustical rather than electromechanical analogs. These analogs have already been introduced in chapter 2. To recapitulate, an electrical capacitance, C_e, is an analog of an acoustical compliance defined as $C_a = V/\rho_o c^2$, with V—an enclosed volume of air, ρ_o—the static density of air, and c—the corresponding sound velocity. An electrical inductance, L, is an analog of an acoustical inertance defined as $M_a = l\rho_o/\pi r^2$, with l—the effective length of a tube, and r—its radius. Finally, an electrical resistance, R_e, is an analog of an acoustical resistance R_a associated with losses of acoustical energy due to friction, heat transfer and radiation of sound away from the system under consideration. Usually, maximum resistance effects are associated with narrow passages in an acoustical system. It is a basic property of electroacoustical analogies of the first kind that the electrical impedance, $Z_e = V/I$, with V—voltage magnitude and I—current magnitude, is an analog of the acoustical impedance, $Z_a = P/\sqrt{}$, with P—sound pressure magnitude and $\sqrt{}$—volume velocity magnitude.

In electromechanical analogies of the first kind, the electrical impedance, Z_e, is an analog of the mechanical impedance, Z_m. The latter is obtained from the acoustical impedance by multiplying it by the square of the area exposed to sound pressure: $Z_m = Z_a A^2$. Since $Z_a = P/\sqrt{}$, the definition of Z_m can be interpreted as $Z_m = AP/\sqrt{}/A$, where $AP = F$ is the total force exerted on the area A by sound pressure P, and $(\sqrt{}/A) = U$ is the average linear velocity associated with the volume velocity through the same area. Multiplication of the acoustical impedance by a constant means that all its elements have to be multiplied by the same constant. If, for instance,

$$Z_a = R_a + j(\omega M_a - 1/\omega C_a) \qquad 3.8$$

then

$$Z_m = A^2 R_a + j(\omega A^2 M_a - A^2/\omega C_a) \qquad 3.9$$

Therefore, the mechanical circuit elements are obtained from the acoustical ones as follows:

$$R_m = A^2 R_a \,;\, M_m = A^2 M_a \,;\, C_m = C_a / A^2 \qquad 3.10$$

The equations given in the last two paragraphs are fundamental for the analysis of the middle ear function described below.

General Structure of the Model

The structure of a network analog of the middle ear can be derived almost entirely on the basis of the available anatomical information. It is helpful to begin with a block diagram that shows the connections among the various functional units of the natural system. Such a block diagram can be derived in a reasonably straightforward manner from the schematic middle ear representation of Fig. 3.21 (Zwislocki, 1962), which includes the tympanic membrane (ear drum) with the three ossicles—malleus, incus and stapes, attached to it, the ligaments holding the ossicles in place, and the middle ear cavities. The latter consist of the tympanic cavity situated directly in the back of the tympanic membrane, the epitympanum filled almost completely by parts of the large ossicles and the antrum into which opens the system of pneumatic cells of the mastoid bone. The last ossicle, the stapes, is held in the oval window of the inner ear by the annular ligament and is exposed directly to the perilymph of the inner ear. The input impedance of the inner ear has a powerful effect on middle ear mechanics.

There are two muscles in the middle ear—tensor tympani that is attached to the malleus, and stapedius that is attached to the stapes. Whereas, in humans the role of the tensor tympani is not clear, the stapedius muscle, when contracted, partially immobilizes the stapes and reduces sound transmission to the inner ear. The most extensive research on the two muscles, especially the stapedius muscle, was performed by

THE MIDDLE EAR   57

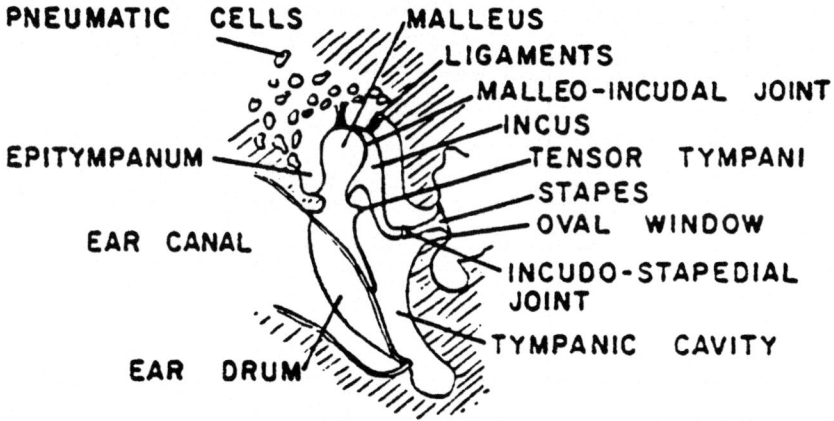

FIG. 3.21. Schematic representation of the middle ear (Zwislocki, 1962). Reprinted with permission from Zwislocki, J. J. (1962). Analysis of middle ear function. Part I. Input impedance. *Journal of the Acoustical Society of America, 34,* 1514–1523. Copyright © 1962, Acoustical Society of America.

Møller and Borg (e. g., Møller, 1983, for review). Because my work on the acoustic effects of the two muscles was rather limited, it is not included in this book.

The block diagram is shown in Fig. 3.22. Its general structure is such that the series blocks are associated with acoustic energy flow from the ear canal to the inner ear and the shunt blocks, with energy leakage. It may appear surprising that the first block, *middle ear cavities,* refers to the middle ear cavities rather than to the tympanic membrane which, in the ear, is more distal. The reversal of the order is necessary because vibrations of all parts of the tympanic membrane are transmitted to the air contained in the cavities but not to the ossicular chain included in the following blocks. In terms of the electrical analog, all the electrical current must flow through the cavities but not through the ossicles coupled only to the central part of the tympanic membrane. Because of the flexible coupling of the more peripheral parts of the tympanic membrane to the central part, some of the current is shunted away from the ossicles through these parts. The shunting effect is embodied in the shunt block *T. Membrane Shunt.* The central part of the tympanic membrane is assumed to be practically rigidly coupled to the malleus, and the incudomaleolar joint is considered to be practically rigid, so that all these elements can be grouped in one block, *T. Membrane—Malleus—Incus.* There is sufficient evidence for these assumptions in the literature. For example, Békésy (1960) found that the middle ear sys-

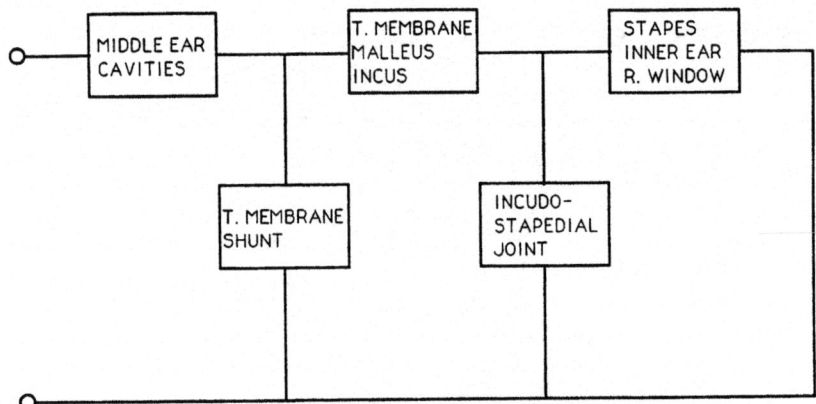

FIG. 3.22. The block diagram of the mechanical middle ear network.

tem acted as if about 65% of the tympanic membrane was rigidly coupled to the malleus. Clinical observations and animal experiments indicate that the malleus and incus vibrate as one body up to sound frequencies at least as high as 10,000 Hz (Rhode, 1971). There is solid experimental evidence that the incudo-stapedial joint is not entirely rigid. Its relative flexibility is the reason why total ankylosis of the stapes increases the stiffness of the system measured at the tympanic membrane by only a factor of about two. When the more distal ossicles are immobilized, the stiffness is increased by as much as 5 times. As a consequence, the incudo-stapedial joint must act as a shunt and is placed accordingly in the block diagram as the block, *Incudo-stapedial Joint*. The last block, *Stapes—Inner Ear—R. Window*, indicates that all these parts are rigidly coupled, and no energy loss occurs among them. Because the inner ear fluids can be considered as incompressible within the frequency range of interest here (e.g., Békésy, 1960; Nedzelnitsky, 1980), the volume of fluid displaced by the stapes footplate must equal the volume displacement of the round window membrane. This is shown in the diagram as a common current path—a series connection. The inner ear, lying between the stapes and the round window, must be included in the same path. Note that, in contradistinction to some previous work, I refer in the last block to the inner ear rather than to the cochlea. This is more accurate since the oval window in which the stapes footplate is embedded does not open directly into the cochlea but rather into the vestibule that is shared with the semicircular canals of the vestibular system. Such an arrangement interposes a column of fluid between the stapes and the cochlea proper.

To understand how the middle ear works, the specific network of every block of Fig. 3.22 has to be defined. This can be done on the basis of the anatomy and known physical properties of the relevant anatomical parts but, first, the transformer effect of the middle ear has to be introduced (Zwislocki, 1962).

Middle Ear as a Mechanical Transformer

Since Helmholtz (1864), the middle ear has been regarded as a mechanical transformer that adapts the high acoustical impedance of the liquid filled inner ear to the low acoustical impedance of air in the ear canal. The adaptation increases the transfer of acoustic energy to the inner ear by almost three orders of magnitude. Three different modes of impedance transformation were proposed by Helmholtz, one resulting from the curvature of the roughly conical tympanic membrane, one from the lever ratio of the middle ear ossicles, and one from the ratio of surface areas of the tympanic membrane and the stapes footplate. The first mode never found any clear experimental confirmation, but Békésy (1960) was able to confirm the last two.

The following analysis replicates basically analyses I performed on earlier occasions (e.g., Zwislocki, 1962, 1975). Somewhat different analyses were performed by others. As is evident in Fig. 3.21, the levers are embodied by the manubrium of the malleus, which is attached to the tympanic membrane, and the long process of the incus, whose tip is attached to the head of the stapes. With respect to the axis of ossicular rotation, the manubrium has a lever length that has been estimated by Békésy to be about 1.3 times greater than that of the incudal process. Békésy also estimated the effective area of the tympanic membrane to be on the order of 0.55 cm^2 and that of the stapes footplate, on the order of 0.032 cm^2. For the lever system to be in equilibrium, the force acting on the tympanic membrane multiplied by the malleal lever length must be equal to the force acting on the stapedial footplate multiplied by the lever length of the incus. We must have, therefore,

$$A_t \, P_t \, l_m = A_s \, P_c \, l_i \qquad 3.11$$

with A_t and A_s—effective areas of the tympanic membrane and the stapes footplate, respectively, P_t and P_c—the sound pressures at the tympanic membrane and in the inner ear near the stapes, and l_m and l_i—the effective lever lengths of the malleus and incus. By rewriting the equation we can obtain

$$(P_c / P_t) = (A_t \, l_m / A_s \, l_i) \qquad 3.12$$

the sound pressure ratio between the inner ear and the tympanic membrane. With the constants given above, it amounts to 22.4, which is equivalent to 27 dB. This ratio holds for an ideal transformer. Békésy obtained about 3 dB less on postmortem preparations, probably, because of losses in the middle ear.

To determine the impedance transformation, it is necessary to calculate the velocity transformation in addition to the pressure transformation. For the acoustic impedance, we need the volume velocity rather than the linear point velocity. For the tympanic membrane, we have $V_t = U_t A_t$, and for the stapedial footplate, $V_s = U_s A_s$, where V_t and V_s stand for the volume velocities of the tympanic membrane and the stapes footplate, respectively, and U_t and U_s, for their point velocities averaged over their areas. When the malleus and incus rotate around their common axis, their velocities U_t and U_s are directly proportional to the lengths of their effective levers, l_m and l_i, so that $(U_s/U_t) = (l_i/l_m)$ and

$$(V_s / V_t) = (A_s l_i / A_t l_m) \qquad 3.13$$

Division of equation 3.6 by equation 3.5 produces

$$Z_t / Z_c = (P_t / V_t)(V_s / P_c) = (A_s l_i / A_t l_m)^2 \qquad 3.14$$

Introducing the numerical values of the constants on the rightmost side of the equations 3.14, we see that $Z_t/Z_c \approx 1/500$, so that the impedance of the inner ear appears reduced by a factor of about 500 at the tympanic membrane. In the mid audible frequency range the inner ear impedance is determined mainly by the cochlea. If the input impedance of the latter in humans is estimated at about $0.4 * 10^6$ dyne sec cm^{-5} (but see chap. 5, this volume), the reduction would make it appear at the tympanic membrane as 800 dyne sec cm^{-5}, except that it is further reduced through shunt elements in the middle ear, as is shown by the network analysis described below.

The significance of the transformer action of the middle ear becomes clear when the ratio between the acoustic energy transmitted to the ear, E_t, and the acoustic energy impinging on the tympanic membrane, E_i, is calculated. The ratio obeys the equation (e.g., Kinsler & Frey, 1950)

$$(E_t / E_i) = 4 Z_t Z_a / (Z_t + Z_a)^2 \qquad 3.15$$

If we introduce for Z_t the cochlear input impedance, the ratio becomes $0.92 *10^{-3}$, if we introduce the transformed cochlear impedance, the ratio increases to 0.37. In the first instance, only about 0.1% of the incident energy would be transmitted to the ear, in the second, 37%, more than 1/3. Of course, the latter is approximated only in the vicinity of 1,000 Hz, at the maximum of the middle ear transfer function, which shows bandpass characteristics, as is shown further below.

In the following analysis, the numerical values of all the middle ear parameters are referred to the tympanic membrane, tacitly taking the transformer effect into account.

The Middle Ear Cavities

The electrical network analog of the block *Middle Ear Cavities* shown in Fig. 3.23 is determined by the geometry of the cavity system. When drawing it, it is necessary to realize that, in the electroacoustical analogies of the first kind, acoustical compliances (cavities) connected in parallel lead to electrical capacitances connected in parallel. The situation for the inertances is somewhat more complicated. When an inertance (constriction) precedes a compliance (cavity), this is reflected as a series connection between an in-

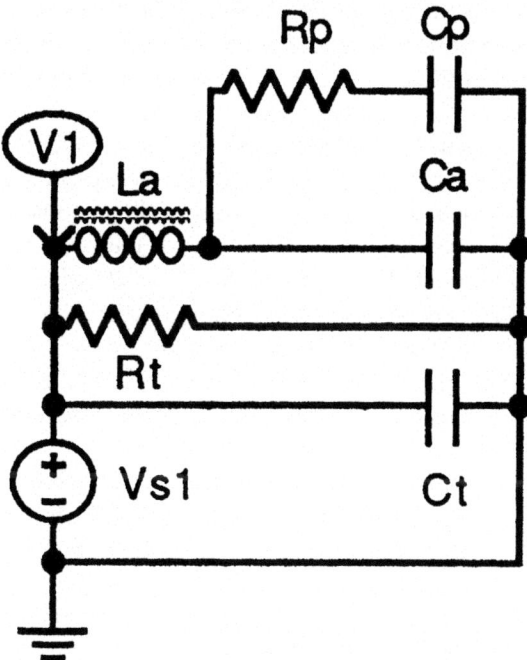

FIG. 3.23. Analog electrical network of the middle ear cavities. The symbols C_p, C_a, and C_t indicate the volume compliances of the pneumatic cells, the antrum and the tympanic cavity, respectively; L_a indicates the inductance associated with a narrow passage to the antrum, R_p and R_t, resistances associated with sound absorption in the pneumatic cells and the wall of the tympanic cavity.

ductance and a capacitance in the electrical analog. When, on the other hand, the order is reversed, the series connection must be replaced by a parallel connection. In general, acoustical compliances preceding other network elements always enter in a shunt configuration. For example, sound absorption in the walls of a cavity enters the electrical analog in a branch parallel to the branch of the capacitance modeling the cavity. On the other hand, a resistance associated with a constriction preceding a cavity enters the analog in series with the compliance modeling the cavity.

At very low sound frequencies at which the effects of series inertances and resistances can be neglected, the middle ear cavities are all in parallel, and the resulting total compliance is a simple sum of the compliances of the tympanic cavity, the antrum and the pneumatic cells. In the electrical analog, all the corresponding capacitances—C_t, C_a and C_p are in parallel. Sound absorption in the walls of the tympanic cavity is modeled by the resistance, R_t, in parallel with C_t. The constriction separating the tympanic cavity from the antrum is modeled as the inductance L_a, the relatively narrow passages to the pneumatic cells, as the resistance R_p. The latter resistance is a rough approximation of the actual situation because of multiple connections among the pneumatic cells. I felt that such an approximation was justified because of the relatively small estimated value of the resistance.

The numerical values of the electrical elements have been calculated on the assumption that, when expressed in Farads, Henrys, and Ohms, they are equal to the numerical values of the corresponding acoustical elements in the cgs system. The capacitances have been calculated from the associated volumes of air with the help of the formula $C_a = V/\rho_o c^2$, as defined above, setting $C_e = C_a$. For the assumed body temperature of 37°C and an atmospheric pressure of 760 mm Hg, the density of air is $\rho_o = 1.139 * 10^{-3}$ g/cm³, and the speed of sound, $c = 3.532 * 10^4$ cm/sec. With these numerical values of the physical constants, the following capacitance values are obtained:

$V_t = 0.5$ cc $C_t = 0.35$ μF
$V_a = 0.5$ cc $C_a = 0.35$ μF
$V_p \approx 7.0$ cc $C_p = 5.0$ μF

The volumes of air V_t, V_a, and V_p were obtained by me in the past as follows (Zwislocki, 1962). The first two volumes are enclosed in reasonably simply shaped cavities and were estimated on four temporal bone preparations by filling the cavities with colored water. The results were found subsequently to be in good agreement with those of Onchi (1961). The situation was much more complicated for the volume, V_p, belonging to the pneumatic cells. That volume was determined with the help of three methods. In the first, the temporal bones were sealed on the outside with wax. Then, they were attached to a vacuum pump to evacuate the air they contained. Finally, the air was replaced with colored water. The volumes so obtained amounted to 2.43, 4.1, 8.6 and 10.5, respectively, with a mean of 6.4 cc, twice as large as given by Onchi on the basis of one specimen. The volumes included both the tympanic cavity and the antrum. Similar results were obtained by other investigators (e.g., Molvaer et al., 1978). In particular, Molvaer performed his measurements on 55 temporal bones and found an average volume of pneumatic cells of 6.5 cc. My results with water filling were checked by measuring the acoustic impedance at the entrance to the bony ear canal of the temporal bones. The volumes calculated from these measurements amounted to 2.1, 3.4, 11.7 and 17.44 cc, respectively, with a mean of 8.6 cc. Finally, impedance measurements were performed with the same acoustic bridge on 6 patients during stapedectomy. During one part of the surgery, a portion of the tympanic membrane is reflected allowing direct access to the middle ear cavities. The volumes obtained on the patients ranged from 5.5 to 14 cc, with a mean of 8.7 cc. The mean of my three sets of measurements amounts to 7.9 cc. After subtraction of the volumes of the tympanic cavity and the antrum, this leaves 6.9 cc for the pneumatic cells.

Once the analog capacitances of the middle ear cavities were determined on the anatomical basis, it became possible to find the numerical value of the analog inductance, L_a, resulting from the constriction in the epitympanum. This was done by matching the zero and pole frequencies in the input impedance curves of Onchi's (1961), apparently the only ones available for the middle ear cavities. The frequencies were on the order of 500 and 2,000 Hz, leading to $L_a = 20$ mH. This value is 43% greater than in my 1962 paper because the cavities network has been modified somewhat to more faithfully reflect the structure of the cavity system.

Finally, the resistances of the cavities' network have been chosen so as to match the characteristic features of Onchi's impedance curves. These were: a low resistance decreasing with sound frequency in the low frequency region, a resistance maximum in the vicinity of 2,000 Hz, and a resistance approaching zero at high frequencies. In addition, a moderate negative reactance at low frequencies, becoming positive above 500 Hz, then, almost symmetrically negative above 2,000 Hz. No attempt was made to match Onchi's absolute values because the ones given by him for the cavity volumes were much smaller than those found by others, and his impedance values were generally higher. The accepted resistance values are: $R_t = 300$ Ohms and $R_p = 20$ Ohms. Again, they are somewhat different than in my 1962 paper because of a slight network modification aiming at a more direct structural relationship to the cavity system. The significance of each resistance has already been explained.

The network of the middle ear cavities of Fig. 3.23 with the numerical values introduced above produces the input impedance characteristics shown in Fig. 3.24. They follow the same pattern as determined by Onchi but with smaller magnitudes, more compatible with my own anatomical date and those obtained by others. Note in particular the negative reactance above 2,000 Hz, which counteracts to some extent the inertia of the middle ear ossicles and extends the passband of the middle ear, as is shown further in this chapter.

The Tympanic Membrane Shunt

As already mentioned, the acoustic effect of the tympanic membrane has been approximated by dividing the membrane in two parts—one considered to be rigidly coupled to the malleus and constituting Békésy's effective area of 0.55 cm^2 and a residual one elastically coupled to the former and capable of moving relative to the malleus. The latter produces a shunt effect diverting part of the sound energy from the ossicular chain. Its acoustic properties can be studied independent of the ossicular chain and the cochlea by fixing the malleus. This immobilizes the central part of the tym-

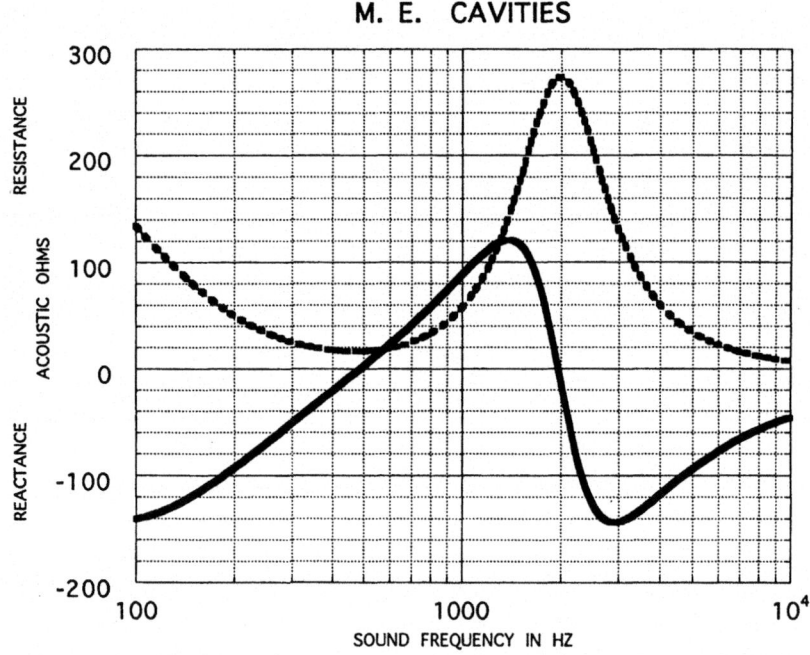

FIG. 3.24. Calculated resistance and reactance components of the acoustic impedance of the middle-ear cavities.

panic membrane and restricts the tympanic membrane motion to the residual part. Such an immobilization is produced by massive adhesions to the large ossicles, which occur in some patients with hearing loss. Unfortunately, only one such patient with surgically confirmed massive adhesions to the large ossicles was available to me. The adhesions were bilateral. Acoustic impedance measurements performed on this patient produced the compliance and resistance data shown in Fig. 3.19 (Zwislocki, 1968). They reveal a very low compliance, as expressed in equivalent volume of air, and a very high resistance, both bilaterally. Clinical observations were consistent with these results.

The assumed network analog of the tympanic membrane shunt is shown in Fig. 3.25 in series with the network analog of the middle ear cavities. The rest of the middle ear system is considered to be practically disconnected as a result of the ossicular fixation. The shunt branch is represented as a second order system with one resonance mode, the higher modes being assumed to be strongly damped. This network is somewhat at variance with the network published previously, which was based on less extensive empirical data. It consists of an inductance, L_d, between two capacitances, C_{d1} and C_{d2}, all connected in series. They represent the mass of the pertinent part of the tympanic membrane suspended elastically between the malleus and the bony annulus supporting the tympanic membrane. The resistance R_{d1} is assumed to result from the internal viscosity of the membrane, and the resistance R_{d2}, from sound absorption in the tympanic ring. The numerical values of the network elements have been determined on the basis of the data of Fig. 3.19 averaged over both ears. They are: $L_d = 15$ mH, $C_{d1} = 0.12$ µF, $C_{d2} = 0.35$ µF, $R_{d1} = 300$ Ohms and $R_{d2} = 600$ Ohms.

Available anatomical date make it possible to check if the derived inductance value is realistic (Zwislocki, 1962). Because of the units used, it should be numerically equal to the acoustic inertance of the pertinent part of the tympanic membrane. The inertance value multiplied by the effective area of the tympanic membrane squared (0.55 cm^2) should be equal to the mass of the shunt part of the tympanic membrane. The multiplication produces a value of 4.5 mg. Wever and Lawrence (1954) gave 14 mg for the total mass of the tympanic membrane. According to Békésy, the total area of the tympanic membrane is on the order of 0.85 cm^2. This means that the effective area of 0.55 cm^2 is about 0.65 of the total area, and the remaining area is 0.35 of the total. Multiplying the total mass of the tympanic membrane by 0.35 we obtain 4.9 mg. From Békésy's (1960) data a somewhat smaller mass results, amounting to 3.3 mg (Zwislocki, 1962). The value of 4.5 mg derived from the impedance data with the help of the network analog lies between the two extremes and, therefore, appears to be realistic.

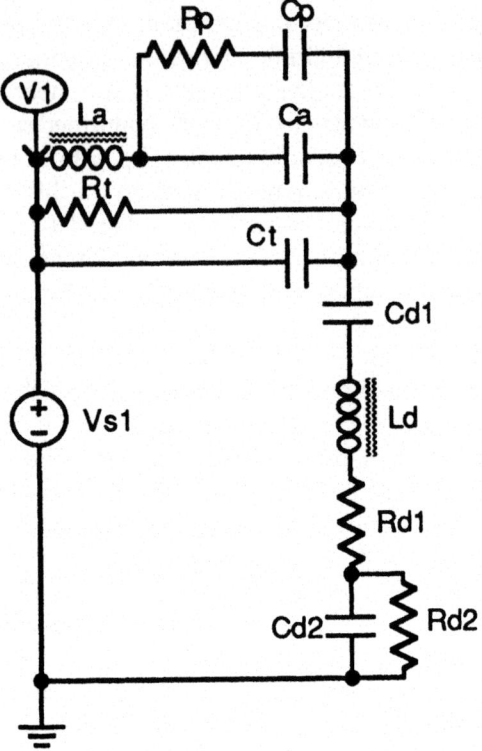

FIG. 3.25. Network analog of the shunt portion of the tympanic membrane added to the network analog of the middle ear cavities.

The theoretical impedance characteristics obtained with the network analog of the ear with fixed ossicles are compared in Fig. 3.26 to the empirical ones derived from the data of Fig. 3.19. Within the frequency range of the measurements, the agreement for both the reactance and resistance curves is reasonable, although it was possible to obtain the empirical data on only one patient.

Middle Ear Without Incus

When the incus is removed, the cochlea becomes disconnected from the tympanic membrane, and the contributions of the inertance of the malleus and of the compliance of its ligamental attachments to the bony walls of the middle ear as well as the inertance and the compliance of the tympanic membrane portion that moves together with the malleus can be deter-

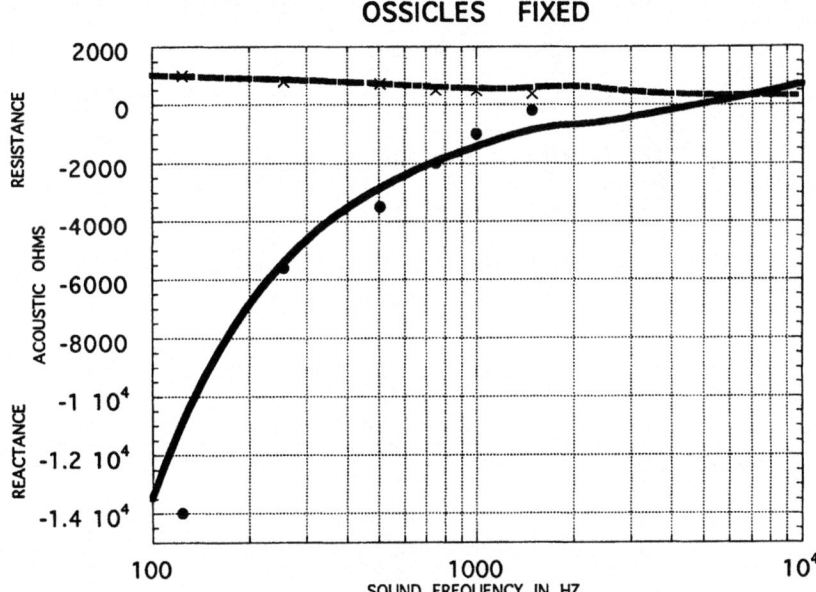

FIG. 3.26. Theoretical impedance characteristics calculated for an ear with fixed ossicles on the basis of the network of Fig. 3.25 (solid lines) compared to measured values of Fig. 3.19 (circles and crosses).

mined (Zwislocki, 1957b, 1962). Unfortunately, neither the inertance nor the compliance contributions can be separated analytically. The same is true for the associated resistances. This is indicated in the block diagram (Fig. 3.22) by lumping all these elements together in one block. The corresponding network diagram is shown in Fig. 3.27 with all these elements connected in series.

The numerical values of the tympanic membrane and malleal complex have been determined from typical impedance data obtained on two subjects (Zwislocki, 1962) in the same way as in the past (Zwislocki 1957b, 1962). Once the network elements of the middle ear cavities and the shunt part of the tympanic membrane are specified, the capacitance of the complex can be derived from reactance values at low sound frequencies, and the inductance, from the zero crossing of the reactance curve. The resistance value has been chosen so as to match the resistance data. The following values have been obtained: $C_o = 1.1\ \mu F$; $L_o = 40$ mH; $R_o = 85$ Ohms. The fit of the theoretical impedance curves computed with these values to the empirical data of the two subjects is shown in Fig. 3.28. It must be considered good within the frequency range of measurements. Unfortunately, no empirical data were available above 1,200 Hz, so that the network ana-

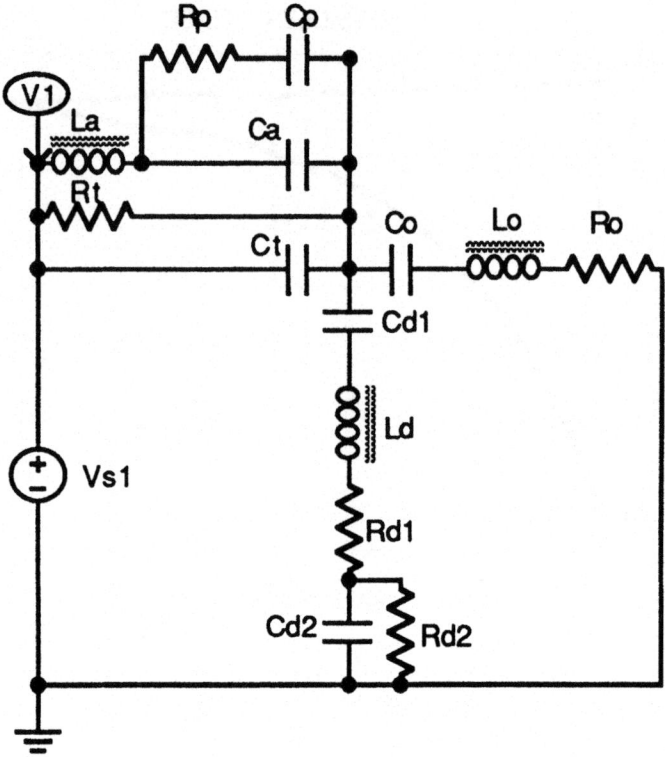

FIG. 3.27. Electrical analog of a middle ear with an interrupted ossicular chain.

log for the middle ear without the incus is valid at the higher frequencies only by extrapolation.

As explained above, the mechanical mass corresponding to the acoustical inertance can be calculated by multiplying the inertance value by the square of the effective area of the tympanic membrane. A value of about 12 mg is obtained for the effective mass of the malleus plus that of the tympanic membrane within its effective area. Békésy and Rosenblith (1951) gave a mass of 23 mg for the malleus, and a mass of 9.1 mg can be derived for the effective area of the tympanic membrane from the measurements of Wever and Lawrence (1954) and of 5.1 from those of Békésy. With these values, the total mass, should range between 28.1 and 32.1 mg. This value is considerably larger than the dynamical mass derived from the impedance data with the help of the network analog. However, according to dynamical measurements of Frank (1923), the effective mass of the malleus and incus together is much smaller because the two ossicles rotate around

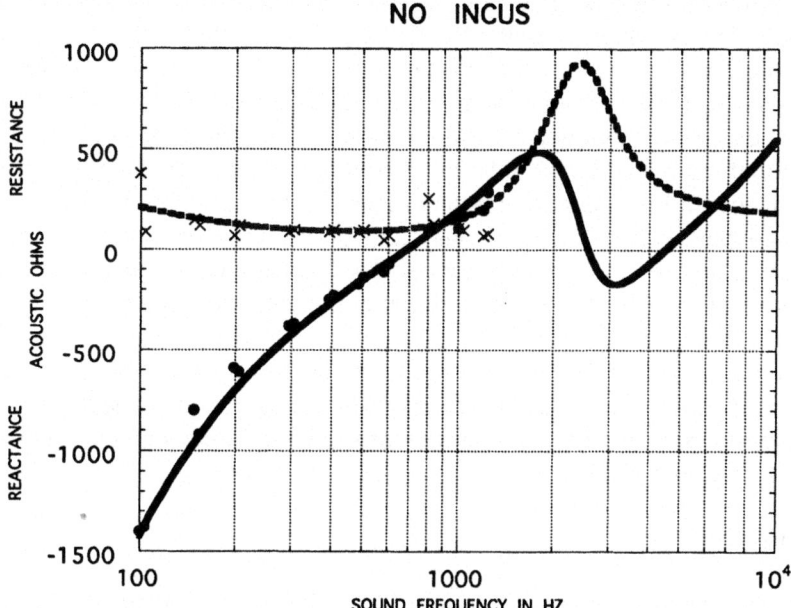

FIG. 3.28. Acoustic resistance and reactance curves calculated on the basis of the network of Fig. 3.27 for an ear with interrupted ossicular chain (dotted and solid lines, respectively), compared to empirical data obtained on two patients (crosses and circles).

their common point of gravity. He obtained an effective mass of 12 mg for both ossicles acting together. The corresponding mass of the malleus should be about half as large—6 mg. If the mass of the effective part of the tympanic membrane is added, we obtain 11.1 to 15.1 mg, close to the value derived from the impedance data. However, the accuracy of Frank's measurements is difficult to estimate. It is possible that the actual effective mass of the large ossicles *in situ* is even smaller because it is doubtful that Frank was able to obtain optimum conditions after extracting the ossicles from the ear, as his experiments required. It is also possible that the malleus without the incus is off balance, and that its effective mass contribution is larger than when balanced with the incus. For all these reasons, it is only possible to conclude that the effective mass of the malleus and tympanic membrane complex, as derived with the help of the network analog from the measured impedance at the tympanic membrane, appears to be of the right order of magnitude.

In the block diagram of Fig. 3.22, the block containing the tympanic membrane and the malleus also includes the incus, more specifically its inertance and the compliance of its ligamental attachments. I was able to obtain data on only one patient who had an interrupted ossicular chain at the level of the incudo-stapedial joint so that the incus was preserved (Zwislocki, 1957b). In this patient, the impedance data indicated a somewhat decreased compliance and increased resistance relative to the two ears without incus. These differences could be ascribed to the presence of the incus and its ligamental attachments but the experimental series involved was an early one and yielded generally somewhat higher impedance values than did subsequent series. Measurements on otosclerotic ears, to be discussed next, suggested that the values obtained without the incus did not change appreciably when the incus was present. A possible explanation is that the presence of the incus improved the ossicular balance by bringing the axis of ossicular rotation closer to the ossicular center of gravity, compensating in this way for the added static mass and added ligamental attachments.

Otosclerotic Middle Ear

From the point of view of this analysis, stapedial fixation practically disconnects the inner ear from the tympanic membrane but leaves the incus and the incudostapedial joint in the circuit. It appears reasonable to assume that the joint is practically free of inertia and its effect can be modeled by a capacitance in series with a resistance, the capacitive effect being produced by cartilaginous tissue, and the resistance, by friction between opposing surfaces of the joint. The network elements of the joint, C_s and R_s, must be connected in series with the tympanic membrane and ossicular elements, C_o, L_o and R_o, as shown in Fig. 3.29, since they must belong to the same current path.

Again, the numerical values of the circuit elements of the joint have been obtained from the empirical impedance data on the assumption that the already determined numerical values of the elements preceding the joint remained unchanged. Several series of such data were obtained in the past with only a fair agreement between them. For the purposes of this chapter, two extensive series of measurements involving 17 ears of 9 patients and 24 ears of 24 patients, respectively, have been used. They were obtained with two different methods (Zwislocki 1962, 1968). The capacitance value has been derived from the reactance data at low sound frequencies, where the inertial effects are negligible. Low to medium sound frequencies have been used to estimate the resistance value because the empirical resistance data obtained thus far at higher frequencies appeared unreliable. The resulting numerical values—$C_s = 0.18\,\mu F$ and $R_s = 1,300\,Ohms$, have led to the analog impedance curves of Fig. 3.30. The reactance curve lies within

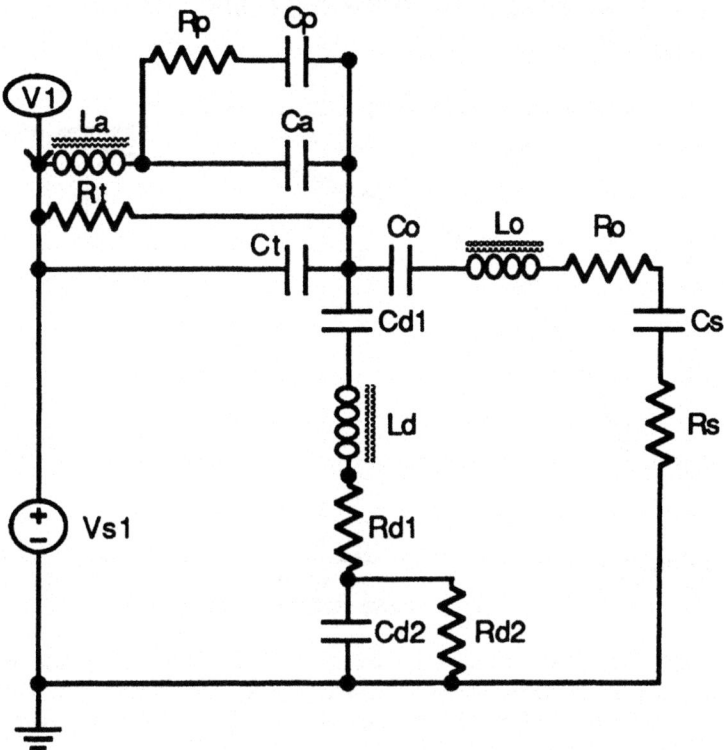

FIG. 3.29. Network analog of the otosclerotic ear with a fixed stapes. The symbols C_s and R_s indicate the compliance and resistance analogs of the incudo-stapedial joint.

the scatter of the averaged empirical data up to about 1000 Hz and may depart from them somewhat around 1500 Hz. The resistance curve agrees approximately with the averaged data up to about 600 Hz but lies clearly above them at higher frequencies. As already mentioned, however, the empirical resistance values obtained in the past are not reliable at these frequencies and tend to be underestimated (Onchi, 1961; Rabinowitz, 1981).

With respect to the derived values of the network elements, the high resistance of the incudo-stapedial joint, $R_s = 1,300$ Ohms, may appear surprising. It is probably due to an unnaturally enhanced motion within the joint, caused by the immobilization of the stapes. The enhanced motion should be related to an unnatural motion of the malleus and incus, which, under these conditions, may not rotate around an axis passing exactly through their center of gravity (Møller, 1961). This could be reflected in a somewhat increased value of the inductance L_o by comparison to a normal

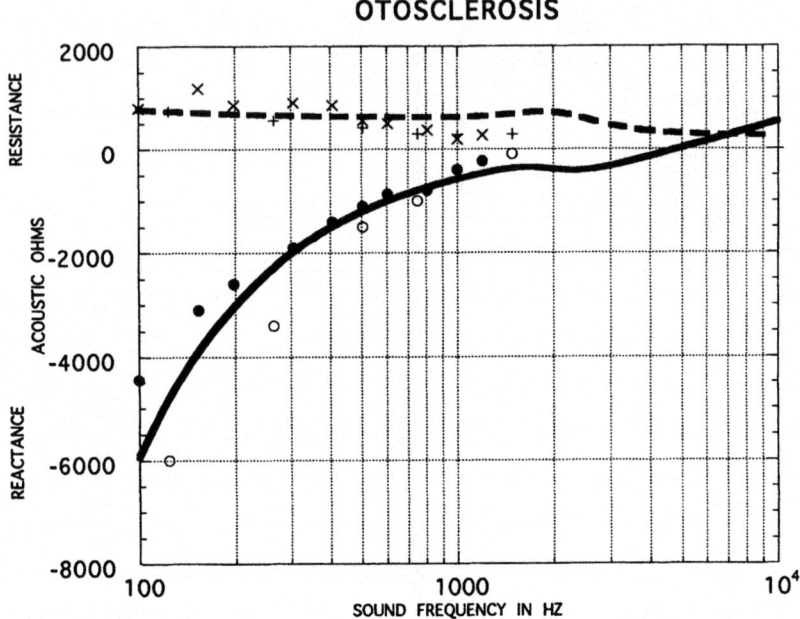

FIG. 3.30. Acoustic resistance and reactance curves calculated on the basis of the network of Fig. 3.29 for the otosclerotic ear (dashed and solid curves, respectively), compared to two sets of measurements obtained on 17 and 24 ears, respectively.

ear. Results obtained on normal middle ears did not prove sensitive enough to ascertain such an effect, however.

Normal Middle Ear

In a normal middle ear, the inner ear is mechanically connected to the tympanic membrane and contributes appreciably to the acoustic impedance measured there. The effect of the cochlea appears to be particularly pronounced, (Zwislocki, 1957b, 1962, 1965, 1975). I showed mathematically a long time ago (Zwislocki-Moscicki, 1948) that the cochlea should have a resistive input impedance, except at very low sound frequencies where an inductive component manifests itself. It was possible to derive a first experimental confirmation of the resistive nature of the impedance from Mundie's (1962) experiments on guinea pig middle ears (Zwislocki, 1963). Subsequently, Lynch, Nedzelnitsky, and Peake (1982) demonstrated it directly on cat cochleas, and Merchant, Ravicz, and Rosowski (1996) on human cochleas.

A first experimental evidence that the cochlear input impedance has an inertial component at very low sound frequencies could be inferred from Dallos's (1970) cochlear microphonics measurements. Later, it was confirmed through direct experiments of Lynch et al. (1982). I have estimated its value for the analog network of the middle ear on the basis of the experimental information that its effect can be felt only below 100 Hz. This is achieved when the inertance is assumed to be on the order of 3 g/cm^4, producing an inductance of 3 H. It has been found subsequently that it has a negligible effect on the acoustic impedance at the tympanic membrane.

The stapes is not located directly at the cochlear input but rather in the vestibule, and a short column of liquid separates it from the cochlea proper (e.g. Zwislocki-Moscicki, 1948; Zwislocki, 1962). Of course, this column of liquid must act as a mass. Its corresponding acoustic inertance can be calculated on the basis of purely anatomical considerations (Zwislocki, 1962). If the distance between the oval window and the cochlea is estimated at 2 mm, and the area of the stapes footplate, at 3.2 mm, and the density of the perilymph is equated approximately to that of water, a column of perilymph with a mass of about 6.4 mg results. This mass has to be augmented by that of the stapes footplate—2.5 mg according to Békésy and Rosenblith (1951), giving in total 8.9 mg. In order to obtain the corresponding acoustic inertance, the mass has to be divided by the square of the stapedial area. This produces a numerical value of 8.7 g/cm^4. To refer the inertance to the tympanic membrane, the value has to be divided by the middle ear transformer ratio of about 22 squared. This leads to a value of 20 mg/cm^4 and to that of 20 mH in the analog network. The effect of this inertance on the impedance at the tympanic membrane has been found here to be negligible in part because of partial decoupling of the tympanic membrane from the cochlea.

The analog network of the cochlea must also include the compliances of the round window membrane and of the annular ligament holding the stapedial footplate in the oval window. Because the two compliances are in series, their values cannot be determined separately on the basis of the impedance measurements at the tympanic membrane. Békésy (1960), and much more recently, Merchant et al. (1996) seem to be the only ones who determined the human compliance of the round window separately from the cochlear compliance as a whole. According to these measurements, the compliance of the oval window appears to be between 2 and 3 times smaller than the compliance of the round window. Unfortunately, both series of experiments had to be performed on postmortem preparations, so that the validity of their results for live ears is somewhat uncertain.

The complete network analog derived here for the normal ear is shown in Fig. 3.31. It differs from that for the otosclerotic ear by the series network consisting of the compliance C_c and the resistance R_c standing for the compliance of the oval and round windows combined and the resistive input impedance of the cochlea, respectively. This network parallels the network representing the incudo-stapedial joint. The numerical values of C_c and R_c have been determined from acoustic impedance measurements at the tympanic membrane with the help of the remaining part of the network analog already determined. Summarizing, the following numerical values have been obtained for the two cochlear network elements: $C_c = 0.52$ µF; $R_c = 800$ Ohms. It should be pointed out that the best fit of the network impedance curves to the empirical impedance data required a modification of the resistance of the incudo-stapedial joint from 1,300 to only 200 Ohms. The change is justified by the very likely alteration of the relative motion between the incudal and stapedial parts of the joint. The logic of such an assumed change has been mentioned above, and was discussed more extensively by Møller (1961). No change in the compliance, C_s, of the joint was necessary.

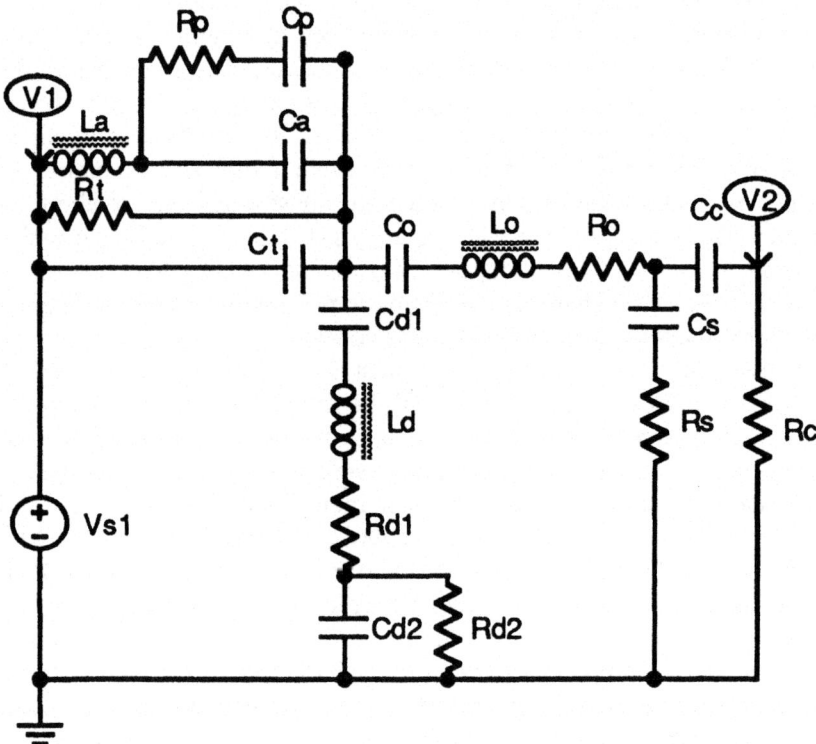

FIG. 3.31. Network analog of the normal ear, where C_c and R_c indicate the combined compliance of the oval and round windows and the resistive input impedance of the cochlea, respectively.

The model impedance curves are compared to two sets of empirical data in Fig. 3.32. The reactance data shown by the open circles were obtain by me with the acoustic bridge (Zwislocki, 1971, 1975), those shown by the closed circles, by Rabinowitz (1981) with the help of an "infinite" impedance source (Zwislocki, 1957a). The resistance data are from Rabinowitz since mine proved to be questionable. Rabinowitz's data are in agreement with the data obtained with the bridge by Feldman and Zwislocki (1965), which, unfortunately, do not extend beyond 1500 Hz. In view of the complexity of the middle ear system and the variability of the impedance data obtained by different investigators with different methods, the agreement between the theoretical curves and the empirical data seems to be satisfactory. The greatest discrepancy occurs in the resistance values above 2000 Hz. It may be due to oversimplifications in the derivation of the analog network, especially, of the part concerning the tympanic membrane shunt where higher modes of motion have been neglected. On the other hand, it may reflect the tendency of the resistance measurements at the tympanic

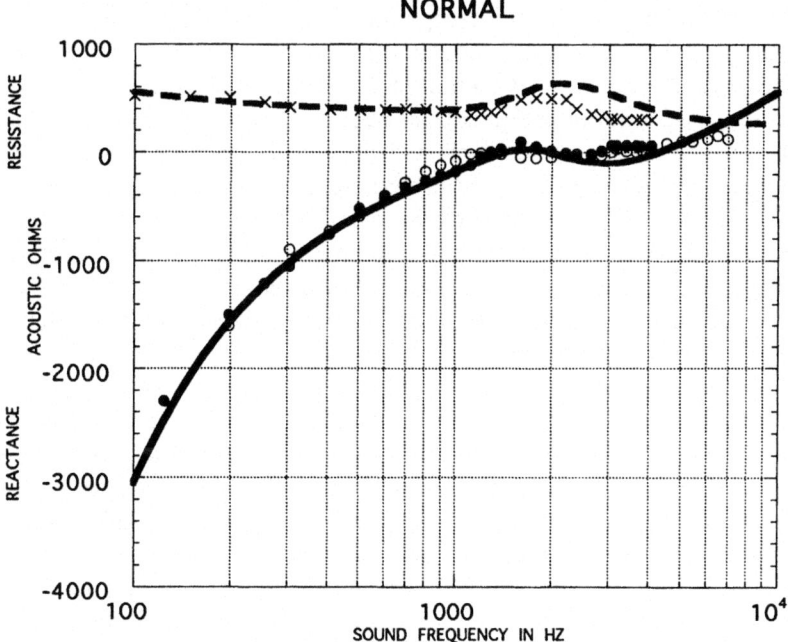

FIG. 3.32. Acoustic resistance and reactance curves calculated on the basis of the network of Fig. 3.31 for the normal ear (dashed and solid curves, respectively), compared to two sets of measurements for the reactance (open circles—Zwislocki, 1971, 1975; closed circles—Rabinowitz, 1981) and one set for the resistance (crosses—Rabinowitz, 1981).

membrane to underestimate the resistance values at higher sound frequencies. In the reactance curves, the greatest pattern difference occurs above 5000 Hz. It may be due to an oversimplified network for the tympanic membrane shunt but also to inaccurate empirical results that, at these frequencies must be considered as tentative.

It is possible to compare the numerical cochlear resistance and compliance values derived from the impedance measurements at the tympanic membrane to corresponding values obtained directly on postmortem human preparations by Békésy (1960). Békésy made his measurements after extraction of the stapes and the round window membrane, so that his results were not contaminated by the impedances of these structures. Although application of data obtained postmortem to in vivo conditions may be questionable, Békésy's data on the compliances of cochlear windows seem to be of the same order of magnitude as corresponding data obtained by Lynch et al. (1982) on live cats. However, his cochlear resistance values appeared to be too small. It was thought that this may have resulted from postmortem deterioration of the basilar membrane, and it seemed on the basis of animal experiments that his postmortem measurements of basilar membrane compliance produced values that were 4 (Zwislocki, 1974) or even 8 times (Schmiedt & Zwislocki, 1977) larger than in vivo. As is discussed in chapter 4, the cochlear input impedance is inversely proportional to the square root of basilar membrane compliance. Therefore, for comparison with in vivo conditions, Békésy's cochlear resistance values were multiplied by a factor of about two or even 2.8 (Zwislocki, 1975). Averaging two sets of Békésy's low frequency data that seemed not to be contaminated by an inadvertent inertia effect, apparent at higher frequencies, a value of approximately $0.2 * 10^6$ dyne sec/cm^5 was obtained. Multiplied by 2 or 2.8, this gave corrected values of $0.4 * 10^6$ or $0.56 * 10^6$ dyne sec/cm^5. The model value of the cochlear input resistance is referred to the tympanic membrane. For comparison with Békésy's data, it has to be referred to the oval window. For this purpose, it has to be multiplied by the transformer ratio of the middle ear squared, which amounts to a factor of about 500. In this way, the model R_c value of 800 Ohms becomes one of $0.4 * 10^6$ dyne sec/cm^5 in agreement with the lower corrected value of Békésy's.

All this is history, however. It is shown in chapter 4 by application of more rigorous mathematical theory to sound propagation in the cochlea that Békésy's measurements from which the cochlear input impedance can be derived need less correction. The theory shows that the numerical value of this impedance must be expected to be smaller at very low sound frequencies than at frequencies above about 500 Hz. Around 1 kHz, the impedance value that can be derived from several series of Békésy's measurements is on the order of $0.5 * 10^6$ dyne sec/cm^5, in reasonable agreement with the much more recent work of Merchant, Ravicz, and Rosowski (1996), which produced values on the order of $0.6 * 10^6$ dyne sec/cm^5. Their results are close to the values derived in chapter 5 for in vivo conditions. However, their determination of the impedance included the stapes and the round window membrane. To obtain the impedance of the cochlea proper, they measured the impedance of the stapes in the presence of an empty cochlea and subtracted the obtained value vectorially from the total. When Békésy measured the combined impedance of the cochlea, the stapes and the round window membrane, he obtained values quite similar to those of Merchant et al. (1996).

The value of the cochlear input impedance obtained by Merchant et al. (1996) postmortem is 50% higher than the value derived here from the network analog of the middle ear. In view of the indirect way the impedance was derived in both cases and the likelihood of some postmortem changes, the difference cannot be considered as significant.

For the cochlear input compliance, it is possible to derive from Békésy's data obtained on one specimen an average value of about $0.3 * 10^{-9}$ cm^5/dyne. The network model suggests a value of $0.52 * 10^{-6}$ cm^5/dyne, as referred to the tympanic membrane. When referred to the oval window through division by the square of the middle ear transformer ratio, this value becomes $1.2 * 10^{-9}$ cm^5/dyne, four times larger than Békésy's value. Merchant et al. (1996) obtained an even lower value than Békésy. According to them, the compliance of the oval window is almost three times smaller than that of the round window after death, so that it dominates the input compliance of the cochlea. Is it possible that death produced hardening of the annular ligament in the oval window, reducing in this way the effective compliance of the window? Was the resistance entering into the measurements of the cochlear input impedance also affected? Such changes should become evident in measurements of the acoustic impedance at the tympanic membrane. According to Merchant et al., death has little effect on this impedance. On the other hand, Onchi's (1961) results obtained postmortem indicate much greater values of the impedance than found in vivo. A similar difference was found by Zwislocki and Feldman (1963). The difficulty in making a decision with respect to the effect of death on the annular ligament is compounded by the possibility of simultaneous changes in stapedial impedance and the shunt impedance of the incudo-stapedial joint. Because both are in parallel, changes in one can be compensated to some extent by changes in the other. This would be true in particular, if tissue shrinkage produced simultaneously increased stiffness of the annular ligament and a partial decoupling in the incudo-stapedial joint. Analysis based on the analog network of Fig. 3.31 demonstrates that, under such conditions, the impedance at the tympanic membrane could remain almost invariant. In Fig. 3.33, the impedance curves of Fig. 3.32, derived for live middle ears and shown as intermittent lines are compared to impedance curves obtained with cochlear impedance parameters derived from postmortem measurements of Merchant et al. (1996). Specifically, the cochlear input resistance was assumed to have a numerical value of $0.6 * 10^6$ dyne sec/cm^5 instead of $0.4 * 10^6$ dyne sec/cm^5 and the compliance, a value of $0.2 * 10^{-9}$ cm^5/dyne instead of $1.2 * 10^{-9}$ cm^5/dyne used in the determination of the normal impedance curves. The new values, when referred to the tympanic membrane, produced the corresponding values of 1,200 dyne sec/cm^5 and $0.18 * 10^{-6}$ cm^5/dyne. To compensate for these changes, the parameters of the incudo-stapedial joint were changed from 200 to 400 dyne sec/cm^5 for the resistance—an unimportant change, and from $0.18 * 10^{-6}$ to $0.52 * 10^{-6}$ cm^5/dyne for the compliance. It is apparent in Fig. 3.33 that the parameter changes produced very small changes in the computed impedance curves referred to the tympanic membrane.

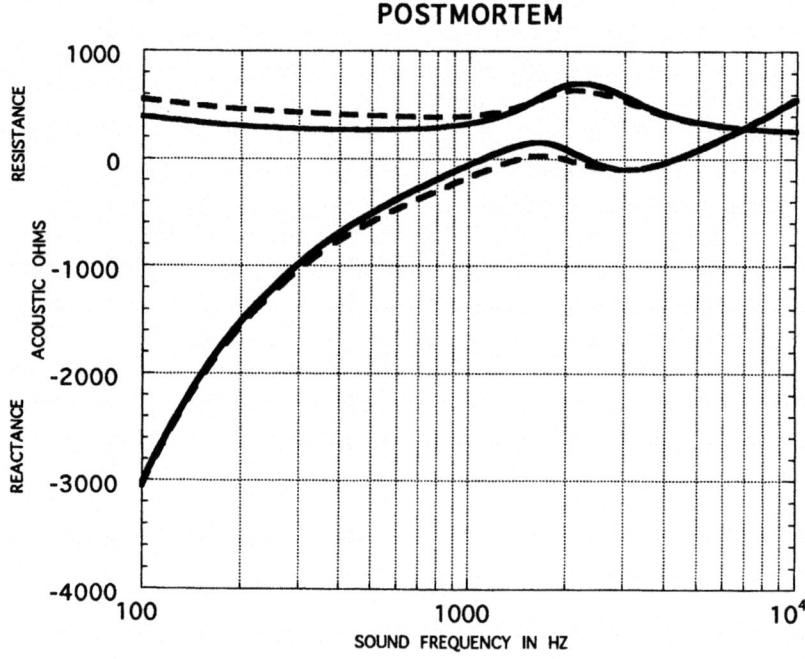

FIG. 3.33. Acoustic resistance and reactance at the tympanic membrane calculated with the cochlear parameters obtained by Merchant et al. (1996) on postmortem preparations (solid lines), compared to the results shown in Fig. 3.32.

THE MIDDLE EAR 79

Perhaps one of the most useful applications of the network model is to an indirect determination of the middle ear transfer function, which cannot be measured directly in live humans. Such a function is plotted in Fig. 3.34 by means of the intermittent line. It is expressed in terms of sound pressure level at the cochlear input versus sound frequency for a constant sound pressure at the tympanic membrane. The function is not normalized but referred to the sound pressure at the tympanic membrane and shows the absolute transmission loss across the middle ear, except for the transformer effect. Inclusion of the effect would move the line upward by 27 dB. The smallest theoretical loss is introduced by the middle ear in the vicinity of 1,000 Hz, and is on the order of 1.5 dB. It decreases the dynamic transformer ratio to about 25.5 dB. Békésy (1960) obtained a transformer ratio of about 25 dB and Onchi (1961), one of about 20 dB on the average on postmortem preparations. Some of Onchi's preparations reached a ratio of 25 dB. For comparison with the theoretical curve, the open and closed circles show corresponding representative data determined on postmortem preparations. The open circles indicate median data obtained by Onchi (1961) on 7 specimens. Like the model data, they are expressed in terms of cochlear sound pressure level for a constant sound pressure at the tympanic membrane. The closed circles indicate the medians of four sets of data reviewed by Goode, Nakamura, Gyo, and Aritomo (1989). These data were determined in terms of stapes displacement as a function of sound frequency for a constant sound pressure at the tympanic membrane, and were converted by me to stapes velocity by multiplication by sound frequency. Because of the resistive character of the cochlear input impedance, the velocity should differ from the sound pressure at the cochlear input only by a multiplicative constant independent of sound frequency. All the empirical values were normalized so as to coincide with the theoretical value at 1000 Hz. There is reasonable agreement between the theoretical curve and the empirical values at sound frequencies higher than 1000 Hz. At lower frequencies, there appears to be a systematic difference. This difference can be decreased by introducing for the cochlear input parameters of the network analog the values derived from Merchant and associates' (1996) measurements on postmortem cochleas, as this was done for the impedance curves in Fig. 3.33. The result is shown by the solid curve. According to the curve, substitution of the postmortem values had a smaller effect on the transfer function above 1000 Hz. The approximate agreement between the theoretical curve derived on the basis of the postmortem impedance data and the postmortem transfer data suggests that both sets of data are mutually consistent. Thus, the difference between the low-frequency portion of the theoretical transfer function derived for in vivo conditions and the postmortem empirical data may be due to postmortem

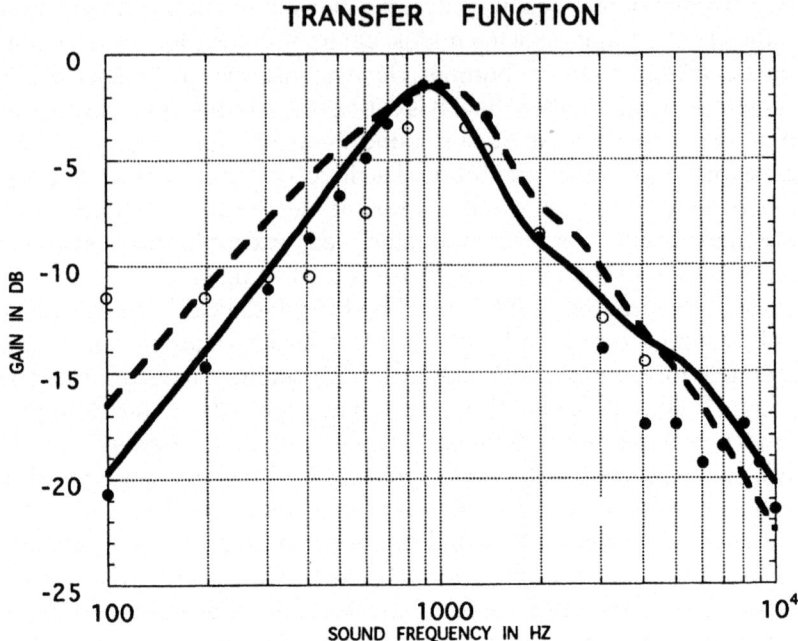

FIG. 3.34. Middle ear transfer function calculated on the basis of the network model of Fig. 3.31 with the cochlear parameter values derived analytically (intermittent line) and those obtained by Merchant et al. (1996) postmortem (solid line), compared to corresponding data obtained postmortem by Onchi (1961; open circles) and Goode et al. (1989; closed circles).

changes, especially, a decreased compliance of the oval window. On the other hand, the erratic differences between the theoretical curves and the empirical data at high sound frequencies may be due to oversimplifications in the network analog, especially in connection with the tympanic membrane shunt.

Recent measurements of umbo response in live humans (Goode, Ball, & Nishihara, 1993) provide a unique opportunity to test the network model against in vivo conditions. A model transfer function based on the velocity of umbo motion (electric current through C_o, L_o, and R_o) is shown by the solid curve in Fig. 3.35. Closed circles indicate corresponding mean data derived from three groups of investigators (Gundersen, 1971; Vlaming & Feenstra, 1986; Gyo et al., 1987) working on cadaver ears, as reviewed by Goode, Nakamura, Gyo, and Aritomo (1989). Open circles have been derived from measurements of umbo displacements in 6 ears of live human subjects by Goode et al. (1993). They represent the means of the data transformed to velocity. Note the agreement between the two sets of experimental data and the theoretical curve at low and medium sound frequencies. At higher frequencies, the empirical data are scattered above and below the theoretical curve, but both sets lie above the curve between 3000 and 6000 Hz. The rapid decay of the postmortem data at high frequencies may have been caused by postmortem changes, but the relatively high gain values obtained on live subjects may be suspect also. The investigators placed a 0.5 to 0.75 mm square piece of 3M reflective tape on the tympanic membrane for light reflection. The vibration of the tape piece may have been affected by vibration of tympanic membrane portions whose amplitudes of displacement were larger than that at the umbo location (e.g., Tonndorf & Khanna, 1972). In any event, the effect of death on the umbo vibration seems to have been moderate, except perhaps at the high frequencies, and the theoretical curve appears to be a realistic representation of the middle ear transfer function measured at the umbo.

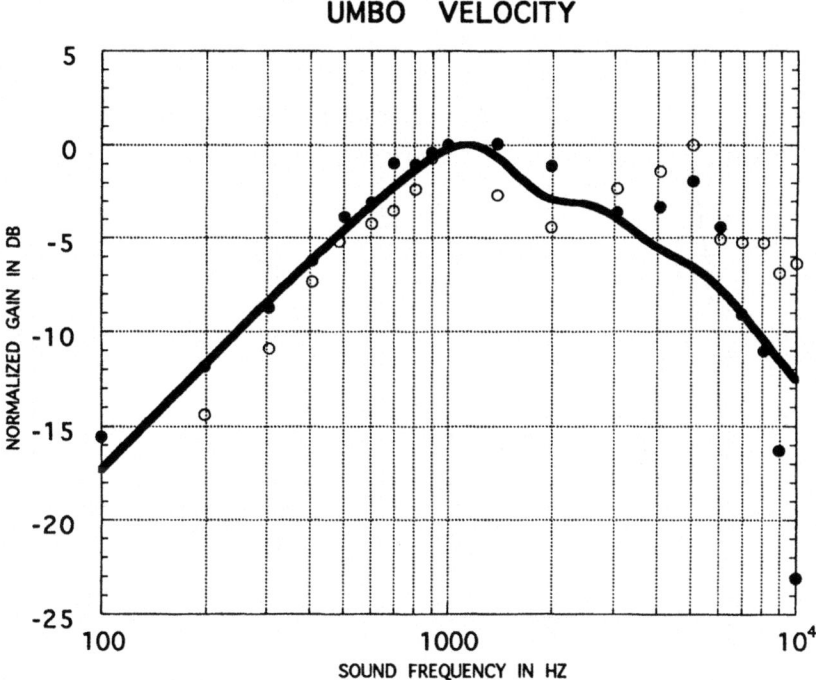

FIG. 3.35. Middle ear transfer function at the umbo derived on the basis of the network model of Fig. 3.31 (solid line), compared to mean empirical data of three groups of investigators obtained on postmortem preparations (closed circles) and by one group using live subjects (open circles).

Another potentially useful application of the network model is to transient stimuli presented at the tympanic membrane of a normal intact ear. The theoretical transient sound pressure at the cochlear input in response to a rectangular condensation pulse of 0.1 msec at the tympanic membrane is shown in Fig. 3.36. Such pulses are used in auditory neurophysiology and psychophysics, except that they are not strictly rectangular because of transformations in the earphone and the ear canal. The model response shows high damping and is practically over within 1 msec. It is somewhat similar in shape and duration to action potentials recorded in the auditory nerve. It suggests suppression of neural activity over a period of about .5 msec, which follows an excitatory phase. The suppression phase must be expected to interact with the phase of neural refractoriness. Because such responses can be measured in intact, live human ears, Fig. 3.37 shows a theoretical velocity transient at the umbo location, produced by the same 0.1 msec pulse. Interestingly, the umbo transient appears to be slightly less damped than the cochlear transient.

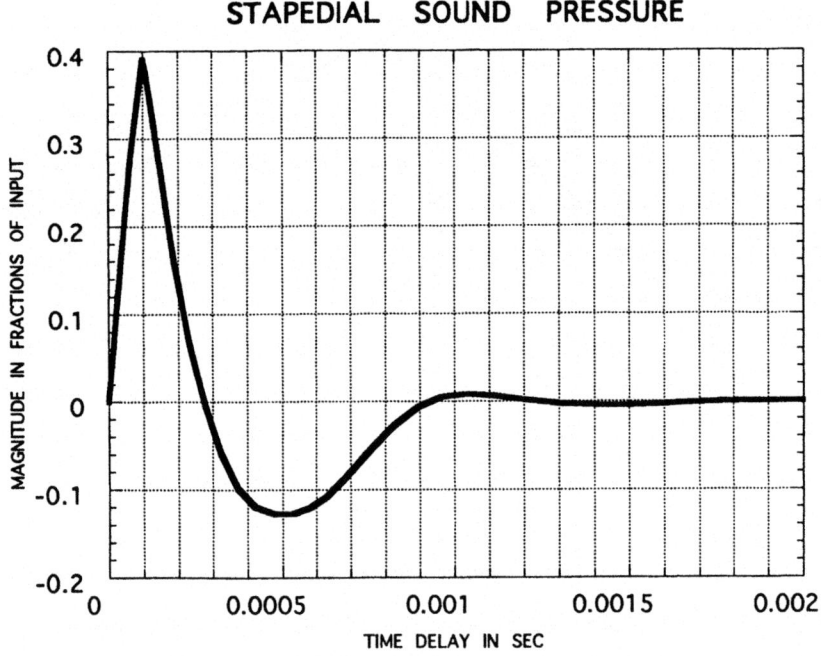

FIG. 3.36. Theoretical transient response to a 0.1 msec pulse at the cochlear input.

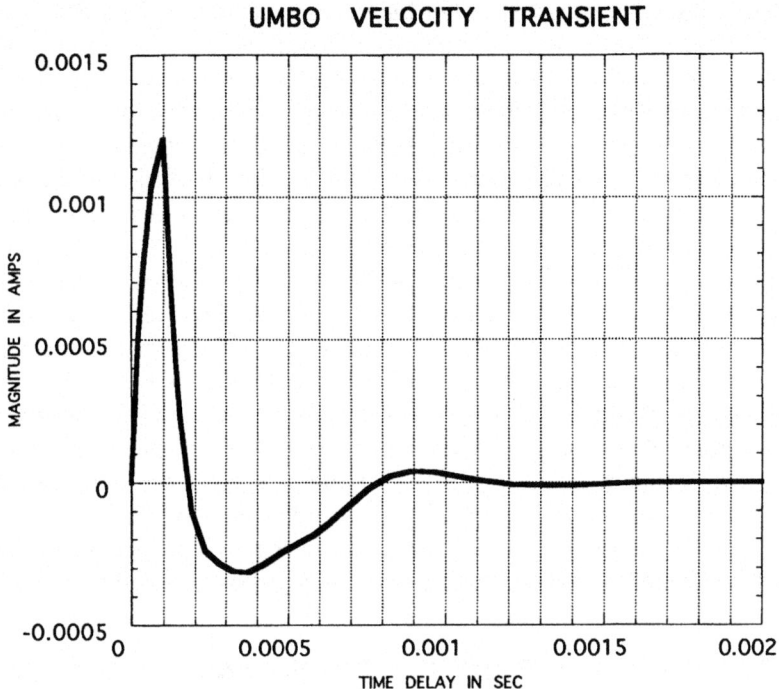

FIG. 3.37. Theoretical transient response to a 0.1 msec pulse at the umbo.

MIDDLE EAR TRANSFER FUNCTION VERSUS THE THRESHOLD OF AUDIBILITY

On the basis of analyses similar to those described in this and the preceding chapters, it was suggested in the past that the human threshold of audibility as a function of sound frequency is determined to a large extent by the combined transfer functions of the outer and middle ear (e.g., Zwislocki, 1965, 1975). The suggestion is explored here further with the help of added experimental evidence and theoretical improvements.

The outer-ear transfer function has been constructed in chapter 2 and displayed in Fig. 2.4. It is based entirely on several sets of empirical data and may be regarded as established to a satisfactory approximation (e.g., Shaw, 1974). To obtain the combined transfer function, I add to it the model transfer function obtained with the help of the network analog for the normal ear, as displayed in Fig. 3.31. This function is shown in Fig. 3.34 by the intermittent line. It differs from the empirical data above 1000 Hz only in detail but not in the main trend. The systematic difference at lower frequencies is ascribed to the effect of death rather than to the failure of the model. This is done because the empirical impedance data obtained in vivo and the anatomical data on which the model is based are the best established in this frequency range. They result from several series of independent measurements that are in mutual agreement. On the other hand, the effective compliance of the oval window measured postmortem appears low in relation to the value of this compliance derived from the acoustic impedance measurements at the tympanic membrane of live humans.

The combined transfer function is shown in Fig. 3.38 by means of the solid curve. The intermittent curve shows the threshold of audibility in a free sound field according to Sivian and White (1933). Many measurements of this threshold have been made, and the data obtained are not entirely consistent with each other, but the classical results of Sivian and White seem to be as representative as any for laboratory conditions. In spite of the somewhat erratic patterns of both curves, a general trend of the difference between them can be discerned. The threshold curve is tilted downward relative to the transfer curve at an angle equivalent to 3 dB per octave. This result is in rough agreement with earlier findings based on different sets of

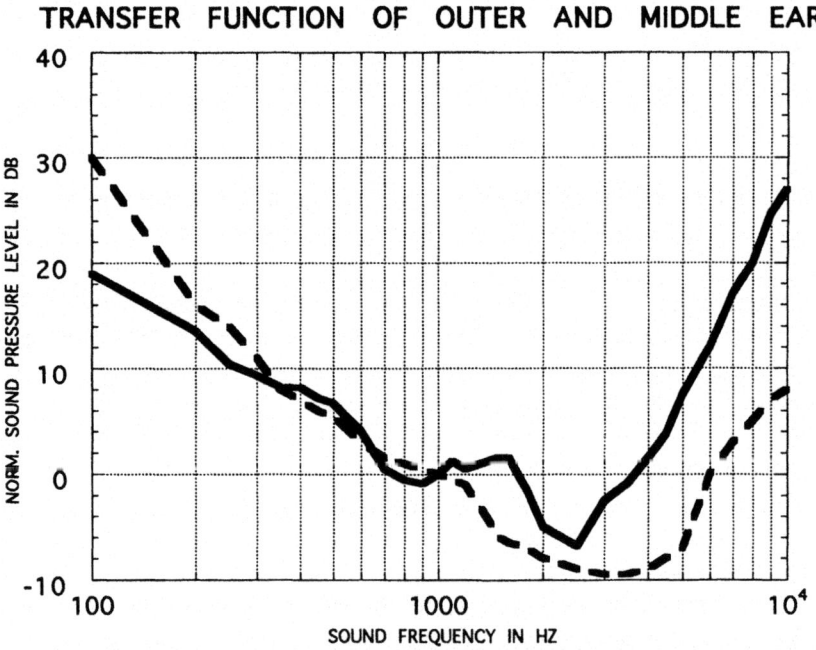

FIG. 3.38. Composite theoretical transfer function of the outer and middle ear derived in this chapter and the preceding one (solid line), compared to the threshold of audibility, as measured by Sivian and White (1933).

data (Zwislocki, 1965, 1975). An attempt was made then to explain the tilt at least partially by neural temporal integration, but other factors may play a role. For example, the vibration maximum in the cochlea could increase with the sound frequency at which it occurs relative to the sound pressure amplitude at the input to the cochlea. Some of Békésy's (1960) measurements speak against it, however. When the relative tilt is compensated for by multiplying the transfer function with a linear function having the 3 dB per octave tilt, a reasonable coincidence with the threshold curve is achieved. This is illustrated in Fig. 3.39. Thus, in agreement with my earlier conclusion (Zwislocki, 1965, 1975), it appears that the combined transfer function of the outer ear and the middle ear does account for the course of the threshold of audibility as a function of sound frequency, except for the 3 dB tilt whose underlying mechanism has yet to be determined. This is contrary to Békésy's (1960) conclusion based on his postmortem measurements indicating a rather flat transfer function. However, his work on the subject was done in early 1940s, and substantial progress was made subsequently in the measurement of acoustic and mechanical events in the ear.

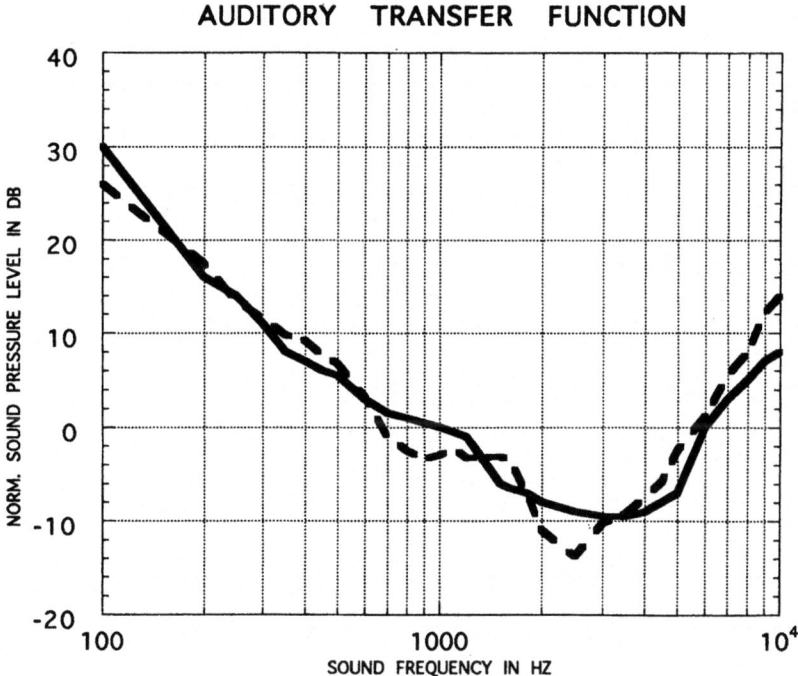

FIG. 3.39. Same as Fig. 3.38 but with the theoretical transfer function tilted toward high sound frequencies by 3 dB/octave.

It seems to me that the network model described in this chapter is a sufficiently accurate representation of the middle ear structure and function to be useful in the study of sound transmission in the human ear. We have used a very similar hardwired model (Zwislocki, 1962) in connection with a hardware model of the cochlea to study the sound transmission in real time. Application to speech sounds proved of particular interest. It was possible to observe, for example, that the amplitude patterns of speech formants in the cochlear model were much better defined in the presence of the middle ear model than in its absence. This probably resulted from high pass properties of the modeled middle ear at low speech frequencies. The model can also be used to study further the relationships between the acoustic impedance at the tympanic membrane and pathological changes in the middle ear. Of course, further refinements are needed. In this respect, it is somewhat unfortunate that currently used clinical methods for measurement of the acoustic impedance and related variables at the tympanic membrane, although rapid and convenient, do not provide accurate information about absolute values of these variables.

REFERENCES

Békésy, G. v. (1960). *Experiments in Hearing*. New York: McGraw Hill.

Békésy, G. v., & Rosenblith, W. A. (1951). The mechanical properties of the ear. In S. S. Stevens (Ed.), *Handbook of Experimental Psychology*. New York: Wiley.

Beranek, L. L. (1988). *Acoustical Measurements* (rev. ed.). Woodbury, NY: American Institute of Physics for the Acoustical Society of America.

Dallos, P. (1970). Low-frequency auditory characteristics: Species dependence. *Journal of the Acoustical Society of America 48*, 489–499.

Feldman, A. S. (1967). Acoustic impedance studies of the normal ear. *Journal of Speech and Hearing Research 10*, 165–176.

Feldman, A. S., & Wilber, L. A. (1976). *Acoustic impedance and admittance: The measurement of middle ear function*. Baltimore, MD: Williams and Williams.

Feldman, A. S., & Zwislocki, J. J. (1965). Effect of the acoustic reflex on the impedance at the eardrum. *Journal of Speech and Hearing Research, 8*, 213–222.

Frank, O. (1923). Die leitung des schalles im ohr [Sound transmission in the ear]. Sitzber. *Math-physikalische Klasse Bayerische Akadamie der Wissenschaften München, 53*, 11.

Goode, R. L., Ball, G., & Nishihara, S. (1993). Measurement of umbo vibration in human subjects—method and possible clinical applications. *American Journal of Otology 14*, 247–151.

Goode, R. L., Nakamura, K., Gyo, K., & Aritomo, H. (1989). Comments on "Acoustic transfer characteristics in human middle ears studied by a SQUID magnetometer method." *Journal of the Acoustical Society of America, 86*, 2446–2449.

Gunderson, T. (1971). *Protheses in the ossicular chain: Experimental and clinical studies*. Baltimore, MD: University Park Press.

Gyo, K., Aritomo, H., & Goode, R. L. (1987). Measurement of the ossicular vibration ratio in human temporal bones by use of a video measuring system. *Acta Oto-Laryn- gologica (Stockholm), 103*, 87–95.

Helmholtz, H. L. F. (1954). *On the sensations of tone as physiological foundation for the theory of music* (E. Higgens, trans.). New York: Dover. (Original work published 1863)

Kinsler, L. E., & Frey, A. R. (1950). *Fundamentals of acoustics*. New York: John Wiley.

Lynch, T. J., Nedzelnitsky, V., & Peake, W. T. (1982). Input impedance of the cochlea in cat. *Journal of the Acoustical Society of America, 72*, 108–130.

Merchant, S. N., Ravicz, M. E., & Rosowski, J. J. (1996). Acoustic input impedance of the stapes and cochlea in human temporal bones. *Hearing Research, 97*, 30–45.

Metz, O. (1946). The acoustic impedance measured on normal and pathological ears. *Acta Oto-Laryngologica* [Suppl.] (Stockholm), *63*.

Møller, A. R. (1961). Network model of the middle ear. *Journal of the Acoustical Society of America, 33*, 168–176.

Møller, A. R. (1983). Frequency selectivity of phase-locking of complex sounds in the auditory nerve of the rat. *Hearing Research, 11*, 267–284.

Molvaer, O. I., Vallersnes, F. M., & Kringlebotn, M. (1978). The size of the middle ear and the mastoid air cell: System measured by an acoustic method. *Acta Oto-Laryngologica, 85*, 24–32.

Mundie, J. R. (1962). *The impedance of the ear—a variable quantity* (Report No. 576). Fort Knox, Kentucky. U.S. Army Medical Research Laboratory.

Nedzelnitsky, V. (1980). Sound pressures in the basal turn of the cat cochlea. *Journal of the Acoustical Society of America, 68*(6), 1676–1689.

Onchi, Y. (1949). A study of the mechanism of the middle ear. *Journal of the Acoustical Society of America, 21*, 404.

Onchi, Y. (1961). Mechanism of the middle ear. *Journal of the Acoustical Society of America, 33*, 794.

Rabinowitz, W. M. (1981). Measurement of the acoustic input immittance of the human ear. *Journal of the Acoustical Society of America, 70*, 1025–1035.

Rhode, W. S. (1971). Observations of the vibration of the basilar membrane in squirrel monkeys using the Mössbauer technique. *Journal of the Acoustical Society of America, 49*, 1218–1229.

Rosowski, J. J., Teoh, S. W., & Flandermeyer, D. T. (1996). The effect of the Pars Flaccida of the tympanic membrane on the ear's sensitivity to sound. In E. R. Lewis, G. R. Long, R. F. Lyon, P. M. Narins, C. R. Steele, & E. Hecht-Poinar (Eds.), *Proceedings of the International Symposium on Diversity in Auditory Mechanics* (pp. 129–135). Berkeley, California: World Scientific.

Schmiedt, R. A,. & Zwislocki, J. J. (1977). Comparison of sound-transmission and cochlear-microphonic characteristics in Mongolian gerbil and guinea pig. *Journal of the Acoustical Society of America, 61*, 133–149.

Shanks, J. E., Wilson, Q. H., & Cambron, N. K. (1993). Multiple frequency tympanometry: Effects of ear canal volume compensation on static acoustic admittance and estimates of middle ear resonance. *Journal of Speech and Hearing Research, 36*, 178–185.

Shaw, E. A. G. (1974). Transformation of sound pressure level from the free field to the eardrum in the horizontal plane. *Journal of the Acoustical Society of America, 56*, 1848–1861.

Sivian, L. J., & White, S. D. (1933). On minimum audible sound fields. *Journal of the Acoustical Society of America, 4*, 288–321.

Tonndorf, J., & Khanna, S. M. (1971). The tympanic membrane as part of the middle ear transformer. *Acta Oto-Laryngologica, 71*, 177–180.

Tonndorf, J., & Khanna, S. M. (1972). Tympanic membrane vibrations in human cadaver ears studied by time-averaged holography. *Journal of the Acoustical Society of America, 52*, 1221–1233.

Vlaming, M. S. M. G., & Feenstra, L. (1986). Studies on the mechanics of the reconstructed human middle ear. *Clinical Otolaryngology, 11*, 411–422.

Von Gierke, H. E., & Warren, D. (1953). *Protection of the ear from noise: Limiting factors* (Benox Report: An Exploratory Study of the Biological Effects of Noise). Chicago, IL: University of Chicago.

Wever, E. G., & Lawrence, M. (1954). *Physiological Acoustics*. Princeton, NJ: Princeton University Press.

Zwislocki, J. J. (1957a). Some measurements of the impedance at the eardrum. *Journal of the Acoustical Society of America, 29*, 349–356.

Zwislocki, J. J. (1957b). In search of bone-conduction threshold in a free sound field. *Journal of the Acoustical Society of America, 29*, 795–804.

Zwislocki, J. J. (1961). Acoustic measurement of the middle ear function. *Annals of Oto Rhino-Laryngology, 70*, 599–606.

Zwislocki, J. J. (1962). Analysis of the middle ear function. Part I. Input impedance. *Journal of the Acoustical Society of America, 34*, 1514–1523.

Zwislocki, J. J. (1963). An acoustic method for clinical examination of the ear. *Journal of Speech and Hearing Research, 6*, 303–314.

Zwislocki, J. J. (1965). Analysis of some auditory characteristics. In R. D. Luce, R. R. Bush, & E. Galanter (Eds.), Handbook of Mathematical Psychology (Vol. III, pp. 1–97). New York: Wiley.

Zwislocki, J. J. (1968). On acoustic research and its clinical application. *Acta Oto-Laryngologica, 65*, 86–96.

Zwislocki, J. J. (1970). *An acoustic coupler for earphone calibration*. (Special Report LSC-S-7). Syracuse, New York: Syracuse University, Laboratory of Sensory Communication.

Zwislocki, J. J. (1971). *An ear-like coupler for earphone calibration*. (Special Report LSC-S-9). Syracuse, New York: Syracuse University, Laboratory of Sensory Communication.

Zwislocki, J. J. (1974). Cochlear waves: Interaction between theory and experiments. *Journal of the Acoustical Society of America, 55*, 578–583.

Zwislocki, J. J. (1975). The role of the external and middle ear in sound transmission. In E. L. Eagles (Ed.), The Nervous System (Vol. 3, pp. 45–55). New York: Raven Press.

Zwislocki, J. J., & Feldman, A. S. (1963). Postmortem acoustic impedance of human ears. *Journal of the Acoustical Society of America, 35*, 856–865.

Zwislocki, J. J., & Feldman, A. S. (1970). Acoustic impedance of pathological ears. *ASHA Monographs, No. 15*. Washington, DC: American Speech and Hearing Association.

Zwislocki-Moscicki, J. J. (1946). Über die mechanische Klanganalysedes Ohrs [On sound analysis in the ear]. *Experientia, 2*, 10–18.

Zwislocki-Moscicki, J. J. (1948). Theorie der Schneckenmechanik: Qualitative und Quantitative Analyse [Theory of the mechanics of the cochlea: Qualitative and quantitative analysis]. *Acta Oto-Laryngologica, Supplement, 72*, 1–76.

chapter

The Cochlea Simplified by Death

*A*s mentioned in the introduction to the book, the function of a normal live cochlea is extraordinarily complicated and difficult to explain. Death simplifies it without abolishing the fundamental mode of its mechanical operation. Analysis of the simplified cochlear mechanics can serve as a first step in the analysis of the normal one in a way analogous to that used in the analysis of the middle ear function.

Beginning with a postmortem cochlea is not only helpful didactically but also correct chronologically. All the early information concerning the cochlear structure and function came from postmortem preparations, and our understanding of the cochlear function advanced gradually as new empirical information came in. At first, only the gross anatomy became known, and the associated function was a matter of speculation. Some of it was formulated mathematically and some was expressed in rather farfetched physical models. These developments were reviewed from different aspects in my doctoral dissertation (Zwislocki-Moscicki, 1948), by Wever (1949), and by Békésy (1960). They served as points of departure for our understanding of the cochlear function. This was particularly true for the work of the great anatomists of the 19th century, who gave their names to various cochlear parts and for Helmholtz (1954/1863) whose psychophysical experiments imposed early limits on cochlear theorizing and whose resonance theory, although not entirely correct, led to the place theory of pitch perception, which endured until now. The controversy surrounding this theory was the trigger for my dissertation work that aimed at resolving it. The controversy concerned the physical incompatibility between a sharp resolution of sound frequencies and good time resolution. As is well known in physics, sharp frequency resolution requires sharply tuned resonators that have long transients at the onset and termination of sounds and exclude sharp time analysis. Paradoxically, the ear is able to reconcile sharp frequency resolution with sharp time resolution.

Although, as is now clear to me, my doctoral dissertation failed to solve adequately the frequency resolution–time resolution paradox, it seems to have produced results that subsequently proved to be of value. It introduced the first useful mathematical description of sound propagation in the cochlea, as expressed mainly in a differential equation of the transmission-line type (Zwislocki-Moscicki, 1946, 1948; Zwislocki, 1950, 1965). The equation is still at the base of much cochlear theorizing, and I like to view it as the fundamental equation of the cochlea. Its approximate solution revealed theoretically the mode of cochlear sound propagation, including the mechanism underlying the local vibration maximum, very different from that suggested by Helmholtz. It also made it possible to determine theoretically the acoustic input impedance of the cochlea (Zwislocki, 1948, 1965, 1975), the only cochlear variable that can be measured in live humans, albeit indirectly, through the measurement of the acoustic middle ear impedance (e.g., Zwislocki, 1957, 1962, 1963, 1975). The input impedance is of crucial importance for both the middle ear and cochlear functions, and its measurement provides one of the few opportunities for checking the theory.

The modern era of cochlear research began with the experimental work of Békésy who, in the first half of the 20th century, was able to make the first microscopic observations of sound waves in human cochlear preparations (for a short review see Zwislocki, 1984). They were transversal waves on the cochlear partition, running toward the cochlear apex. Békésy was not able to determine their extent in his early experiments since, for technical reasons, his observations were limited to the apical region. He attempted to overcome this deficiency with the help of dimensional mechanical models based on a strict model theory. However, he was forced to make some quantitative assumptions based on insufficient empirical knowledge of relevant physical constants, which led him partially astray, as subsequent research has demonstrated. He seems to have attempted to reconcile the resonance mechanism proposed by Helmholtz with traveling waves he saw as originating at the resonance location and running toward the cochlear apex. His models did not appear to indicate any wave propagation up to the maximum, and he wrote (Békésy, 1960, translation from a German, 1928 article): "The portion of the membrane from the stapes to the place of resonance vibrates almost completely in the same phase. At the place of resonance there appears a phase reversal, and from here on waves arise that with a uniform rubber membrane become progressively smaller in wavelength" (pp. 421–422). The inherent misconception caused substantial confusion later on and still does among scientists and clinicians whose backgrounds do not include relevant parts of applied physics.

Békésy's work had by far the greatest influence on my doctoral dissertation, and I thought at times that the dissertation looked like a theory for his experiments. Most important, his measurements provided some key empirical values for my theoretical calculations and allowed for their validation. They also attracted my attention to some phenomena I would not have anticipated otherwise. On the whole, my theory agreed with Békésy's empirical results not only qualitatively but even quantitatively. There were some points of divergence, however. Perhaps the most important, which I did not understand entirely while writing my dissertation, concerned the mechanism of the vibration maximum. According to my theory, the maximum was not due to resonance but arose in the presence of traveling waves that extended on both sides of the maximum (Zwislocki-Moscicki, 1946, 1948). The point of resonance was located beyond the maximum, toward the apex, in a region where the vibration amplitude was small. Beyond this point, no wave propagation took place, in direct contradiction of Békésy's description. Another point of divergence concerned the cochlear input impedance. According to the theory, it was resistive, except at the lowest audible sound frequencies (Zwislocki-Moscicki, 1948; Zwislocki, 1965). According to Békésy, it had a prevailing mass component at all but the lowest frequencies.

Although my theoretical conclusions were based on conditions prevailing in postmortem cochleas, subsequently, they have been found to apply to some fundamental aspects of the live ones. It is a matter of great satisfaction to me that they have been validated by the work of others, most of the time tacitly but sometimes explicitly (e.g., Rhode, 1971; Dallos, 1973; Lynch, Nedzelnitsky, & Peake, 1982; Merchant, Ravicz, & Rosowski, 1996; Olson, 2001) although with a time delay ranging from 20 to 50 years.

In this chapter, experimental results and theory are not presented in separate sections but are intermingled, except in one subsection dedicated entirely to the derivation of the cochlear differential equation. The intermingling has been found necessary because the cochlear anatomy and physical constants have had to be expressed and interpreted mathematically for inclusion in the theory. Subsequently, the theoretical results have had to be validated by comparing them to the empirical results one by one. The intermingling also reflects to some extent the historical chronology, because some of the theoretical insights preceded their empirical validation.

FUNCTIONAL ANATOMY AND PHYSICAL CONSTANTS

To understand sound propagation in the cochlea it is necessary to first have a sufficient knowledge of the cochlear functional anatomy. As is well known and has been stated in the introduction to this volume, the cochlea consists of a rather long, spirally wound canal embedded in hard bone. The canal is divided longitudinally by a double partition, as shown in Fig. 4.1 in cross section. The upper part of the partition consists of a very thin membrane, Reissner's membrane, which serves as a chemical barrier and is assumed to have a negligible mechanical effect because of its great compliance. The lower partition consists in part of bone, the osseous spiral lamina, and in part of a flexible plate, called inaccurately the basilar mem-

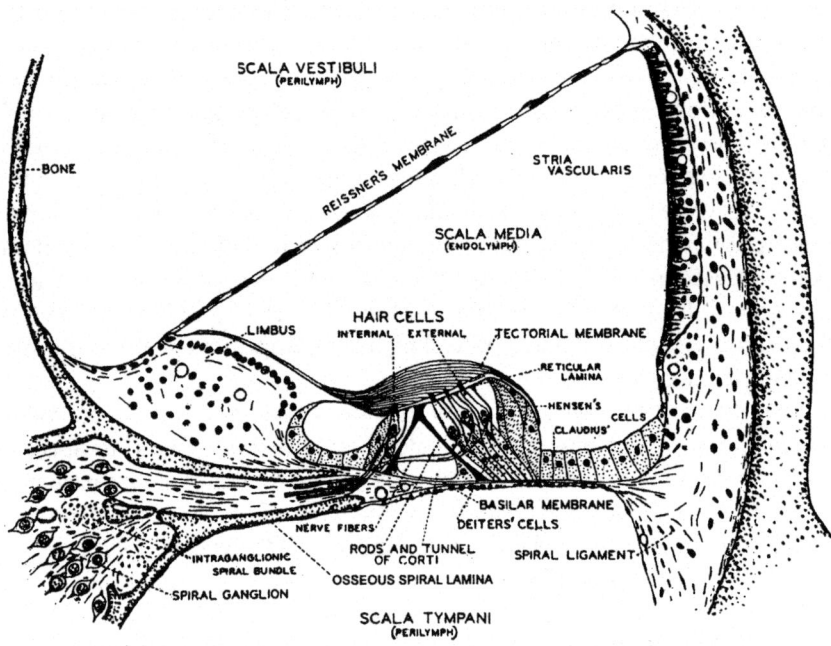

FIG. 4.1. Partial cross section of cochlear canal (modified from Davis and Associates, 1953). Reprinted with permission from H. Davis et al. (1953). Acoustical trauma in the guinea pig. *Journal of the Acoustical Society of America, 25,* 1180–1189. Copyright © 1953, Acoustical Society of America.

brane. The latter has been considered since Békésy's early work in the 1920s as the structure determining the type of sound waves propagated in the cochlea.

The upper chamber in the drawing, called scala vestibuli, is connected to the middle ear via the stapes in the oval window. The lower chamber, scala tympani, communicates acoustically with the air space of the middle ear via the flexible membrane of the round window (Figs. 1.1 and 1.2). All the chambers, or canals, including scala media within the double partition, are filled with liquid resembling water in its physical properties. However, the liquid contained in scala vestibuli and tympani, called the perilymph, has a different chemical composition than the endolymph contained in scala media. Békésy (1960 for English translation) found the endolymph to have the consistency of jelly after death, but the finding has remained controversial and certainly does not apply to live cochleas. What is important for sound propagation in the cochlea is the fact that the cochlear liquids can be considered as incompressible (e.g., Geisler & Hubbard, 1972). Any pressure change produced by the displacement of the stapes in the oval window is equalized by a practically equal volume displacement of the round window membrane.

To describe the cochlear wave propagation mathematically it is necessary to define the cochlear geometry and determine the dimensions associated with it. For this purpose, Fig 4.2 shows the basal coil of the cochlea, together with one canal of the vestibular system and the vestibule, which is part of both the vestibular and cochlear systems, in a schematic representation. The endolymphatic spaces, in particular scala media, are indicated by the crosshatched areas. The middle ear and the outer ear are also shown for orientation. Note that, when sound is transmitted to the cochlea through the oval window, it does not enter the cochlea directly but first travels through the fluid spaces of the vestibule. This affects the acoustic input impedance of the inner ear, which loads the stapes.

To better visualize the configuration of the human cochlear canal, I constructed a wax model of one specimen magnified 25 times (Zwislocki, 1948). Such a model is fashioned by projecting serial sections of a cochlea on sheets of paper impermeable to wax, tracing the projected contours of the cochlear canal on them and spreading wax on them so that plates of uniform, precisely controlled thickness are produced. The plates are trimmed along the tracing and stacked along appropriate landmarks. Then, with a hot knife, the plates are fused on the outside and polished. The model is shown in Fig. 4.3 in an approximately apical and a side view. The side view makes a split produced by the cochlear partition visible. The apical view shows the enlargement due to the vestibule. The beautiful curve traced in three-dimensional space by the outer edge of the basilar

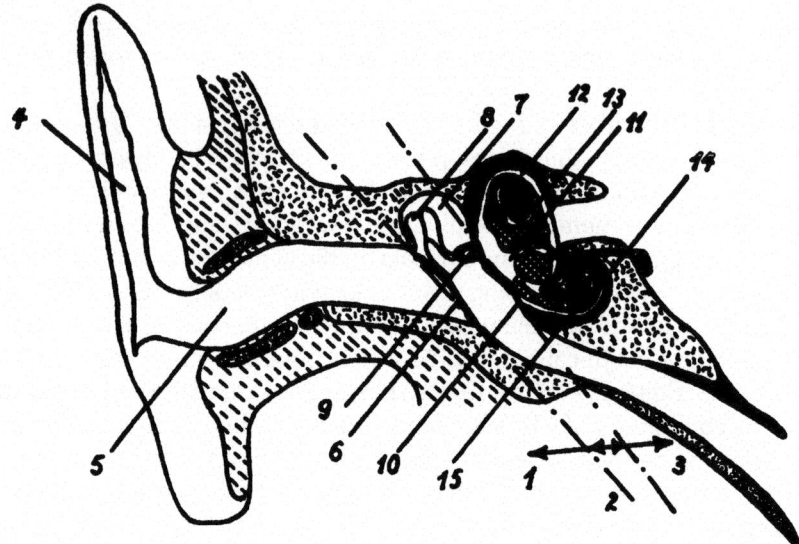

FIG. 4.2. Longitudinal section of the outer, middle, and inner ear (modified from Zwislocki-Moscicki, 1948). From *Theorie der Schneckenmechanik: Qualitative und Quantitative Analyse [Theory of Cochlear Mechanics: Qualitative and Quantitative Analysis]*, by J. J. Zwislocki-Moscicki, 1948, Acta Oto-Laryngologica, Suppl. 72, 1–76. Copyright © 1948 by Taylor & Francis. Reprinted with permission.

membrane is reproduced in Fig. 4.4, as projected on two planes, one parallel and one perpendicular to the cochlear axis. The latter projection follows approximately a logarithmic spiral obeying the equation

$$r = 0.31 \exp(-0.13\phi) \text{ (cm)} \qquad 4.1$$

where r means the variable radius of the spiral, and ϕ, the angle from the outer end. The exact shape of the spiral has probably no effect of interest on sound transmission in the cochlea and is reproduced here mainly for reasons of esthetics.

The model allowed me to determine rather precisely the dimensions of the cochlear canal. They are shown graphically in Fig. 4.5, for scala vestibuli in the upper panel and for scala tympani in the lower one. In each panel, the uppermost drawing refers to the plane of the basilar membrane, the middle one, to a plane perpendicular to it; the lowermost drawing refers to the cross sectional area of the scala. Because the dimensions in Fig. 4.5 are plotted along a straightened out cochlea, it was necessary to decide on the reference distance from the cochlear axis for the length measurement. The distance of the outer edge of the basilar membrane was selected, so that the basilar membrane appears approximately in its natural length. The

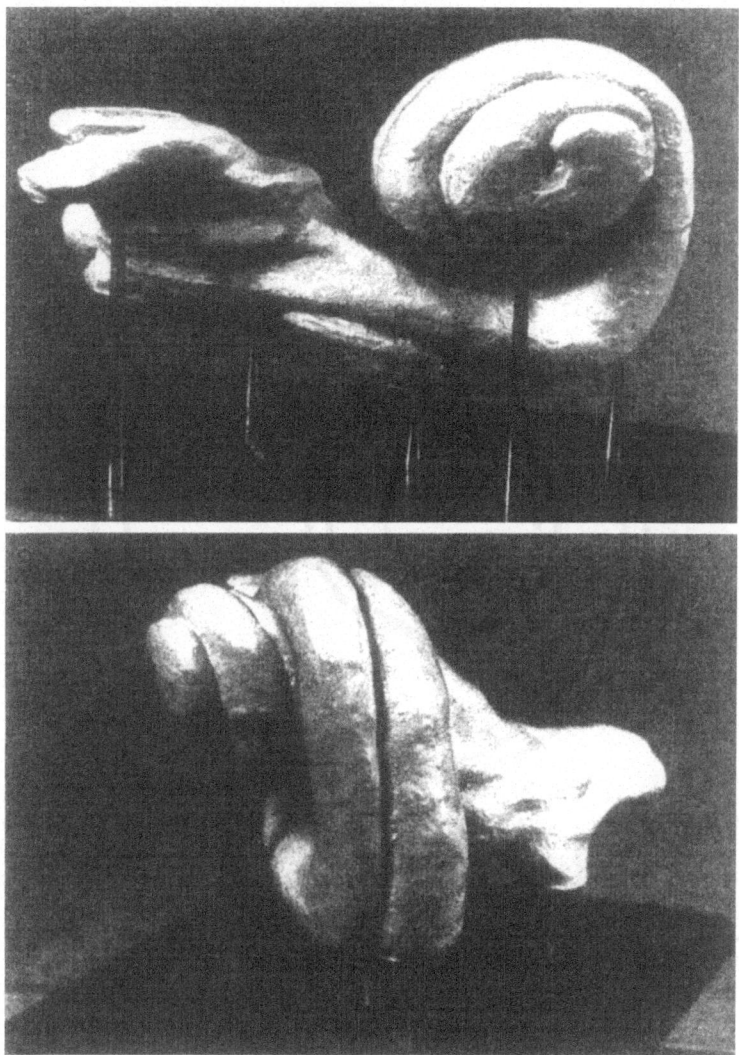

FIG. 4.3. Wax model of the human cochlea photographed from two nearly orthogonal angles (Zwislocki-Moscicki, 1948). From *Theorie der Schneckenmechanik: Qualitative und Quantitative Analyse [Theory of Cochlear Mechanics: Qualitative and Quantitative Analysis]*, by J. J. Zwislocki-Moscicki, 1948, Acta Oto-Laryngologica, Suppl. 72, 1–76. Copyright © 1948 by Taylor & Francis. Reprinted with permission.

distances are referred to its basal end. They are indicated here in percent. This way of reporting the distance is the most useful since it allows generalization across individual cochleas, which have been shown to vary appreciably in length (e.g., Ulehlova, Voldrich, & Janisch, 1987). The basilar

96 CHAPTER 4

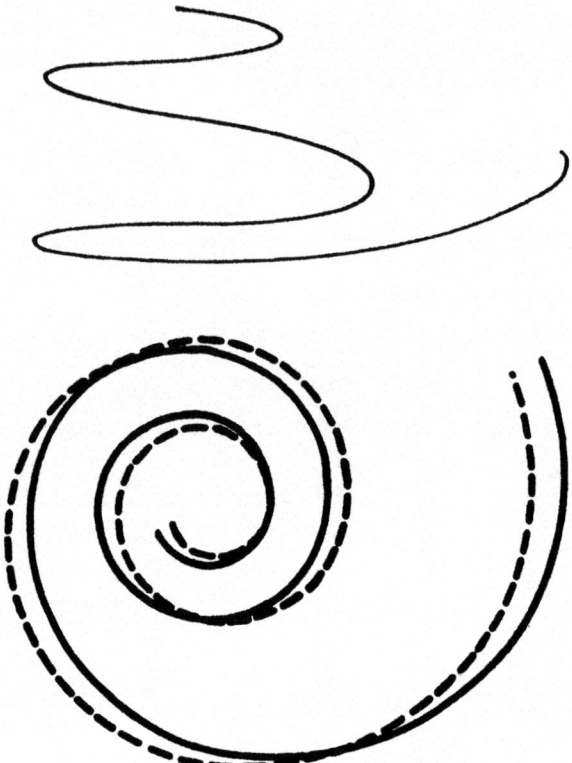

FIG. 4.4. Cochlear spiral along the outer edge of the basilar membrane—top and side views. The intermittent line follows a logarithmic spiral (Zwislocki-Moscicki, 1948). From *Theorie der Schneckenmechanik: Qualitative und Quantitative Analyse [Theory of Cochlear Mechanics: Qualitative and Quantitative Analysis]*, by J. J. Zwislocki-Moscicki, 1948, Acta Oto-Laryngological, Suppl. 72, 1–76. Copyright © 1948 by Taylor & Francis. Reprinted with permission.

membrane is indicated in the figure by the hatched area for orientation purposes only. The width of the area does not represent accurately the width of the membrane. Note that, in the human cochlea, the cross sectional dimensions of both scalae are similar and vary somewhat irregularly with the distance from the cochlear base but, on the whole decrease.

To validate the cross sectional dimensions derived from one specimen, I was able to compare them to measurements made by Wever (1949) on an additional three cochleas (Zwislocki, 1965). The comparison was made on the basis of the total cross sectional area of the cochlear canal. This appeared sufficient since, according to both mine and Wever's data, the dimensions of scalae vestibuli and tympani are similar in humans, ex-

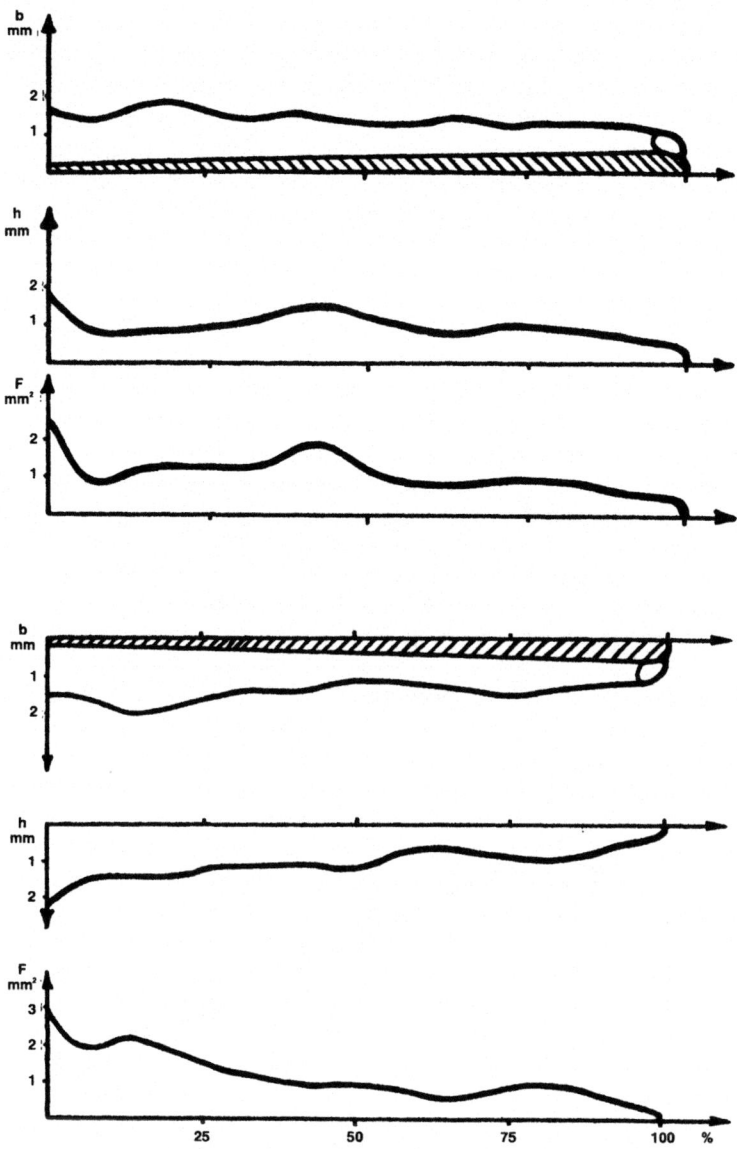

FIG. 4.5. Dimensions of a human cochlear specimen as functions of cochlear distance in percent. From top to bottom: width of scala vestibuli parallel to the basilar membrane which is indicated by the shaded area; height of scala vestibuli; cross sectional area of scala vestibuli, including scala media; width of scala tympani parallel to the basilar membrane; height of scala tympani; cross sectional area of scala tympani (Zwislocki-Moscicki, 1948). From *Theorie der Schneckenmechanik: Qualitative und Quantitative Analyse [Theory of Cochlear Mechanics: Qualitative and Quantitative Analysis]*, by J. J. Zwislocki-Moscicki, 1948, Acta Oto-Laryngologica, Suppl. 72, 1–76. Copyright © 1948 by Taylor & Francis. Reprinted with permission.

cept at the location of the vestibule, where the area on the side of scala vestibuli becomes relatively large. As can be seen from the scatter of the points in Fig. 4.6, the agreement among all the data is reasonably good. For application in subsequent numerical calculations, the data of Fig. 4.6 have been approximated with the help of the least squares method by the exponential function

$$Q_{tot} = Q_V + Q_T = 4.64 * 10^{-2} * e^{-0.0157x} (cm^2) \qquad 4.2$$

where the numerical constants have been rounded off, and the distance, x, is expressed in percent of the total length. The fit is only moderately good but the exponential function greatly facilitates mathematical manipulations, and the effect of the variability of the cross sectional area on the cochlear wave pattern is reduced by a square root relationship.

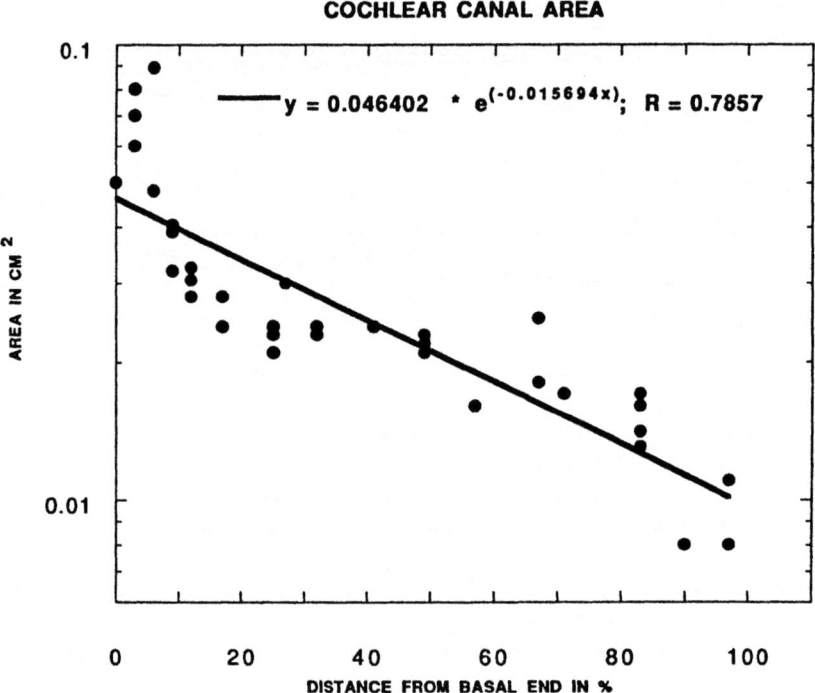

FIG. 4.6. Cross sectional area of the human cochlear canal as a function of distance from cochlear base in percent. The solid line indicates an exponential approximation of the experimental points (modified from Zwislocki, 1965). From Analysis of Some Auditory Characteristics, by J. J. Zwislocki, 1965. In Handbook of Mathematical Psychology, Vol. III, edited by R. D. Luce, R. R. Bush, & E. Galanter (pp. 1–97). Copyright © 1965 by the author. Reprinted with permission.

When sound is propagated in the cochlea, it has to overcome fricative forces resulting from the viscosity of cochlear fluids, the perilymph and endolymph. Békésy (1928) obtained for the viscosity coefficient of the perilymph approximately, $\eta = 2*10^{-2}$ (dyne sec/cm^2). A similar value was found much later by Rauch (1964) for the endolymph. The effect of fluid viscosity depends on the cross sectional areas of scalae vestibuli and tympani. Since it is known to be small, we can assume that $Q_v = Q_t = Q_{tot}/2$. Then, for an approximately cylindrical shape, the specific resistance of each scala can be calculated from Poiseuille's law according to the equation (Zwislocki-Moscicki, 1948)

$$\mathbf{R_{FV}} = \mathbf{R_{FT}} = \frac{16\pi\eta}{Q} \qquad 4.3$$

or, with the numerical value of η and equation 4.2

$$\mathbf{R_{FV\,T}} = \mathbf{7.57} * \mathbf{e^{0.0157x}} \left(\frac{\text{dyne sec}}{\text{cm}^4}\right) \qquad 4.4$$

This produces a resistance value of 7.57 dyne sec/cm^4 at the cochlear base and of 36.32 dyne sec/cm^4 at the cochlear apex. A somewhat larger constant value of 100 dyne sec/cm^4 was assumed in my doctoral dissertation (Zwislocki-Moscicki, 1948).

As was already mentioned, the wave pattern depends critically on the mechanical properties of the basilar membrane, especially its compliance. Békésy (1941, 1947, 1960) was the first to measure it and probably the only one to measure it along the whole cochlear length in several subhuman species and in humans. His measurements were performed by two fundamentally different methods. One consisted of pressing calibrated hairs against its surface and the other of loading it with water under small pressure. Comparison of his graphs suggests that the two methods produced what appears to be mutually disparate results, the first yielding values about one order of magnitude larger than the second. The latter method may be more difficult to implement but is the most relevant for the cochlear wave propagation during which fluid pressure is exerted on the basilar membrane over its entire width. The values obtained with it are consistent with subsequently determined dynamical parameters of the cochlea, such as the speed of wave propagation and input impedance.

Békésy reported the displacement of the basilar membrane produced by a uniform pressure of approximately 1000 dyne/cm^2 (water column of 1 cm) as a function of cochlear distance in a graph with two ordinate scales, one for the volume displacement and one for the maximum displacement in the radial plane. The two scales differ from each other only by a multiplicative constant, and according to Békésy, both variables follow the same curve. Since the width of the basilar membrane increases with cochlear distance by a factor of about 4, and the relation between the volume displacement and the maximum displacement depends on the width, this cannot be accurate. The curve can hold only for one or the other variable but not for both. Since estimates of maximum displacement must be easier than estimates of volume displacement, I assume that the curve applies to maximum displacement. This is not the assumption I made in my earlier publications (e.g., Zwislocki-Moscicki, 1948; Zwislocki, 1965) where I accepted Békésy's results at their face value.

Békésy's (1941) results on the radial maximum of basilar membrane displacement have been expressed here in terms of specific compliance defined as the quotient of displacement and sound pressure (cm^3/dyne) and are plotted in Fig 4.7 by means of the circles tracing his interpolating curve. The data are approximated according to the least squares method by an exponential function. The fit appears to be quite good. Note that the compliance increases by more than two orders of magnitude from about 10^{-7} cm^3/dyne at the cochlear base to about 10^{-5} cm^3/dyne at the apex.

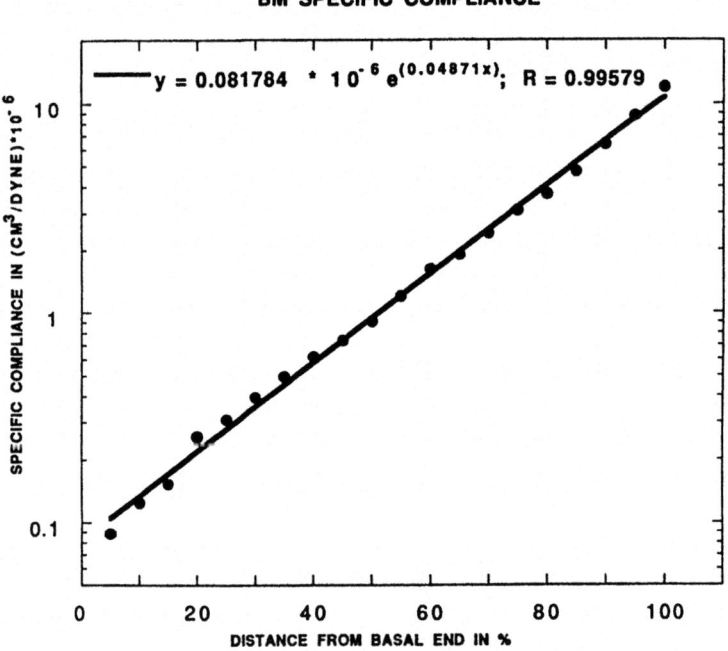

FIG. 4.7. Specific compliance of the human basilar membrane according to Békésy's measurements (circles), and their exponential approximation (solid line).

To find out how reliable are Békésy's data for the basilar membrane compliance, which is a key parameter in cochlear dynamics, we can investigate if the results he obtained with the two methods—hair probes and water pressure, are compatible with each other. Superficially, this does not appear to be true. However, the first method yields a mechanical compliance, the second, a specific compliance per unit length. Before the results of the two methods can be compared, they have to be expressed in the same dimensions. Since the specific impedance must be multiplied by the area exposed to sound pressure to obtain the mechanical impedance, and the compliance is inversely related to the impedance, the specific compliance of the basilar membrane per unit length must be divided by the effective width of the basilar membrane to obtain the mechanical compliance per unit length. As is shown in the following paragraphs, the effective width of the basilar membrane may be estimated to be half the geometric width. Békésy made his measurements with the probe hairs at three cochlear locations, 10, 20, and 30 mm from the stapes. In percent of basilar membrane length, these distances are approximately equal to 30%, 60%, and 90%. According to Fig. 4.7, the corresponding specific compliances are: $C_s = 3.0*10^{-7}, 2*10^{-6}$ and $7*10^{-6}$ cm^3/dyne, and, according to Fig. 4.8, the corresponding basilar membrane widths, w: $2*10^{-2}, 3*10^{-2}$, and $4.7*10^{-2}$ cm. The effective widths w_e are then: $1*10^{-2}, 1.5*10^{-2}$, and $2.35*10^{-2}$ cm, and the mechanical compliances per unit length, C'_m: $3*10^{-5}, 1.33*10^{-4}$, and $3*10^{-4}$ cm^2/dyne. With the probe hairs Békésy obtained at the three locations the following values of mechanical compliance, C_m: $7*10^{-5}, 8*10^{-4}$, and $3*10^{-3}$ cm/dyne. Measurements with probe hairs do not yield mechanical compliances per unit length but involve an unspecified coupling lengths over which the force exerted by the probe is spread. Let us call this length, 2 l (l on each side of the probe). To obtain mechanical compliance values equivalent to those obtained with probe hairs, the compliance values per unit length have to be divided by 2 l, so that $C_m = C'_m/2$ l. This relationship allows us to determine l—l = $C'_m/2$ C_m. We obtain for the three cochlear locations respectively: 0.215, 0.085, and 0.05 cm. If both sets of Békésy's measurements are correct, these values indicate a coupling length substantially greater than the width of the basilar membrane at the cochlear base and about equal to this width near the apex. A coupling length on the order of 1 mm appears to be quite believable because it is greater than basilar membrane width by less than an order of magnitude and is smaller than the lengths of the shortest cochlear waves, as is discussed further. I am inclined to conclude, therefore, that Békésy's values for the basilar membrane compliance are representative for the postmortem conditions.

More recent measurements of basilar membrane compliance performed with methods related to Békésy's experiments with calibrated hairs led to

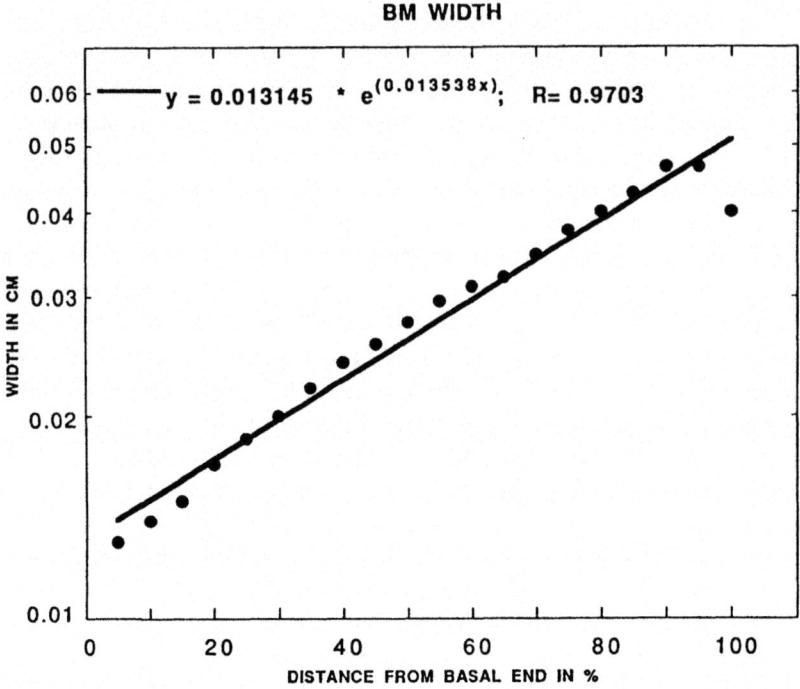

FIG. 4.8. Width of the basilar membrane, as measured by Wever (1949) on 25 specimens of human cochleae (circles), and an exponential approximation of the mean data (solid line).

greater values than he had obtained with this method (Gummer et al., 1981; Miller, 1985). Miller gives for a basal location in guinea pig cochleas a value on the order of $1*10^{-3}$ cm/dyne (1 m/N), more than one order of magnitude larger than Békésy's value at a corresponding location. Difference in animal species is difficult to blame for the discrepancy since Békésy found about the same compliance in humans and guinea pigs in the cochlear base. Perhaps, it can be accounted for by a substantially smaller diameter of the probes used in the more recent experiments.

In cochlear wave propagation, fluid flow interacts with the volume displacement of the basilar membrane. The volume displacement can be calculated from the maximum point displacement if the radial displacement pattern is known. Békésy (1941) sketched the pattern on the basis of his microscope observations, and C. Miller (1985) measured the point compliance as a function of the radial distance. The basilar membrane displacement pattern produced by uniform fluid pressure should follow a proportional function, and it is not surprising that the compliance pattern found by Miller agrees with the pattern sketched by Békésy, although the compliance values obtained by them differ. As an example, Miller's results obtained on one guinea pig specimen as a function of the radial distance are displayed in Fig. 4.9 by means of filled circles. The results may be considered as typical. The two straight line segments fitted to the data by means of the least squares method show that the radial displacement pattern can be approximated reasonably well by a triangle. If this is so, the volume displacement per unit length of the basilar membrane produced by uniform fluid pressure should be approximately equal to half the product of the maximum displacement and the width of the basilar membrane,

$$V_B = \frac{\Psi_{max} w}{2} \qquad 4.5$$

The width of the basilar membrane, as measured by Wever (1949) on 25 specimens of human cochleas, is plotted in Fig. 4.8 as a function of the distance from the basal end by means of the closed circles. The solid line

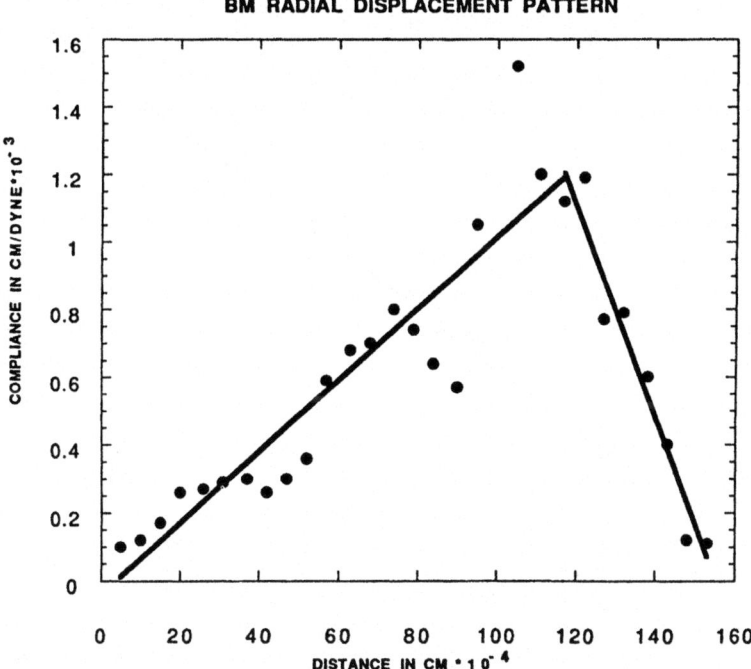

FIG. 4.9. Displacement pattern of the basilar membrane, as measured by Miller (1985) on guinea pigs (circles) and its approximation by a triangle (solid lines).

shows the least squares exponential fit to the data, which is quite good, except for the basal and apical extremes. The acoustic compliance of the basilar membrane computed as half the product of the specific compliance (Fig. 4.7) and the corresponding width given by the curve of Fig. 4.8 is plotted in Fig. 4.10 as a function of the distance from the basal end. The solid line shows the exponential least squares fit to the computed data. The fit appears to be quite good and suggests that the compliance varies from about $6*10^{-9}$ cm^4/dyne at the cochlear base to about $3*10^{-7}$ cm^4/dyne at the apex, a ratio of 500.

Propagation of transversal waves in the cochlea depends not only on the compliance of the basilar membrane but also on its effective mass. Since the basilar membrane can be considered as a thin plate, most of its mass load must be provided by the aggregate of cells attached to it, especially the cells of the organ of Corti. Although this is not often considered, the mass of the tectorial membrane attached to the organ of Corti by the stereocilia of

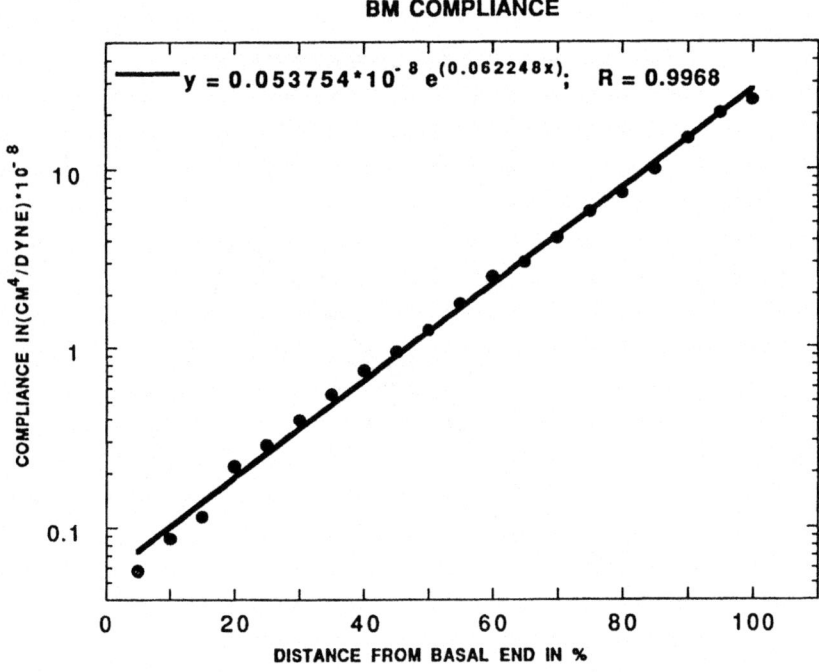

FIG. 4.10. Acoustic compliance of the basilar membrane as a function of the distance from the cochlear base in percent. The values indicated by the circles have been calculated from the experimental data mentioned in the text, the solid line shows their exponential approximation.

the outer hair cells should also be included. It must be pointed out that the effective mass is determined not only by the size and the density of the organ of Corti but also by its location in the width, or radial direction of the basilar membrane (Zwislocki-Moscicki, 1948; Zwislocki, 1980). According to Fig. 4.9, the vibration amplitude increases with the distance from the spiral osseous lamina. Because the mass effect increases with the relative amplitude, the parts of the organ of Corti located near the lamina contribute less to the mass load than the parts further removed. Following Békésy's observation that the endolymph was jelly-like, I assumed in my doctoral dissertation that it could not move in the longitudinal direction and that the mass load of the basilar membrane was produced by its total volume in scala media (Zwislocki-Moscicki, 1948). Since the observation is now questioned, I do not maintain this assumption here. Instead, I calculate the effective mass from the static mass of the organ of Corti and its radial location on the basilar membrane (Zwislocki, 1980). The mass of the tectorial membrane is added as a correction factor.

The static mass of the organ of Corti can be calculated from the organ's cross-sectional area measured in the radial cochlear plane by multiplying it with the average density of the cells. The area of the cells resting on the basilar membrane, which is dominated by the area of the organ of Corti, was given by Wever (1949) based on his measurements on three cochleas. A scatter plot of his data together with their exponential least squares fit are displayed in Fig. 4.11 as a function of cochlear distance. In view of the small change of the area with the distance by comparison to the change in basilar membrane compliance, the fit can be considered as satisfactory. The effective mass can be calculated from the torque generated by the inertia force of the static mass (Zwislocki, 1980). In my doctoral dissertation (Zwislocki-Moscicki, 1948), I calculated it from the kinetic energy that produced an analogous result. The derivation is illustrated in Fig. 4.12. It is assumed for simplicity that all the mass is concentrated in the organ of Corti and in the tectorial membrane overlying it, and that its distribution is roughly rectangular. Comparison of Fig. 4.12 with Fig. 4.1 depicting reasonably faithfully the anatomy of the structures resting on the basilar membrane, suggests that such assumptions are reasonable. In addition, it is assumed that the displacement amplitude of the basilar membrane underlying the organ of Corti increases uniformly with the distance from the spiral lamina, as shown by the displacement pattern in Fig. 4.9. With these assumptions, the incremental torque can be expressed as

$$dT = \omega U_{max} \frac{z}{kw} \rho hz dz \qquad 4.6$$

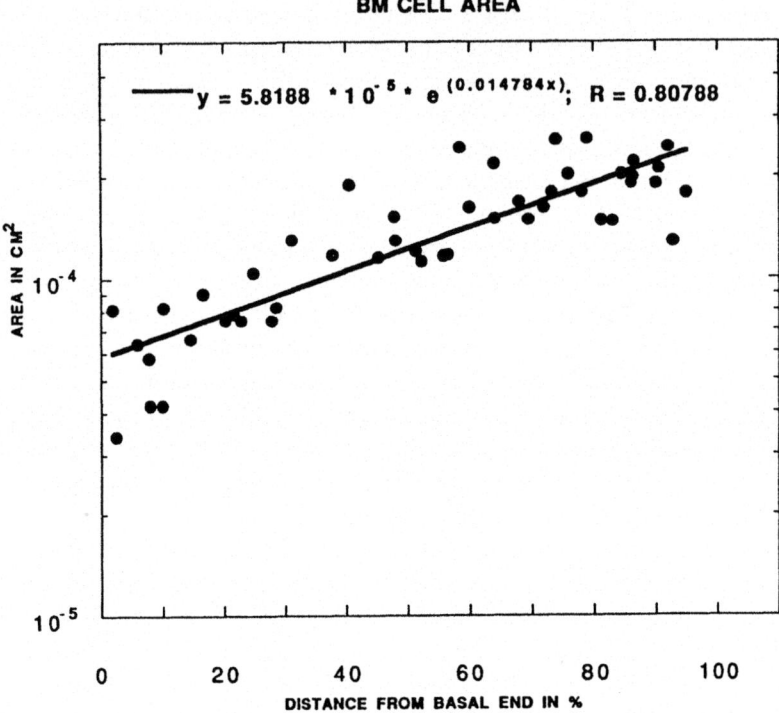

FIG. 4.11. Cross sectional area of the cell mass supported by the basilar membrane, as measured by Wever (1949) on individual human specimens (circles), and their exponential approximation as a function of the distance from the cochlear base in percent.

where T—torque; $\omega = 2\pi f$—angular velocity; Umax—transversal velocity at kw; kw—distance of the margin of the organ of Corti from the osseous spiral lamina along the radial coordinate z; z—space coordinate in the radial plane; ρ—average density of the organ of Corti. The torque is then equal to

$$T = \frac{\omega U_{max} \rho h}{kw} \int_0^{kw} z^2 dz \qquad 4.7$$

After integration and simplification, it becomes

$$T = \frac{kw}{3} \omega U_{max} kwh\rho \qquad 4.8$$

The product kwh = A is the cross sectional area of the simplified organ of Corti, approximately the cross sectional cell area measured by Wever and plotted in Fig. 4.11 as a function of the cochlear distance. When the point of gravity of the mass is known, the torque can be applied to this point by calculating the effective mass concentrated there. Because of the symmetry of the simplified organ of Corti shown in Fig. 4.12, the point of gravity must be at kw/2. The displacement velocity at this distance from the spiral lamina is $U_{max}/2$, so that the inertial torque becomes

$$T = \frac{kw}{4} \omega U_{max} M_e \qquad 4.9$$

By equating the two torque equations, 4.8 and 4.9, we can calculate the effective mass,

$$M_e = \frac{4}{3} A\rho \qquad 4.10$$

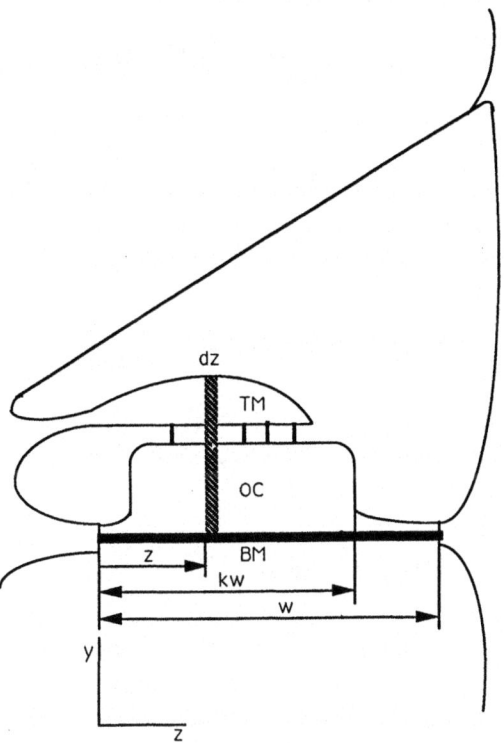

FIG. 4.12. Schematic representation of the basilar membrane, organ of Corti and tectorial membrane in cross section. For further explanation, see text (Zwislocki, 1980).

For convenient calculation of the cochlear wave pattern, we need the acoustic rather than the mechanical mass. The latter is obtained by dividing the mechanical mass by the square of the effective area exposed to sound pressure. The effective area per unit of length of the basilar membrane can be calculated by relating the volume displacement amplitude defined in equation 4.5 to the displacement amplitude at the point of gravity, which is $\Psi_{max}/2$. For this displacement amplitude, the effective width is equal to the geometric width, w. Accordingly, the effective acoustic mass per unit length becomes

$$M_{ea} = \frac{4}{3}\frac{A}{w^2}\rho \qquad 4.11$$

Because the area, A, and the width, w, vary with cochlear distance, the same can be expected to be true for the mass, and it is necessary to write $A(x)$, $w(x)$, $M_{ae}(x)$, where x is the length coordinate. The functions $w(x)$ and $A(x)$ have been defined in Figs. 4.8 and 4.11 by least squares fits to empirical data. The corresponding equations are

$$w(x) = 1.3 * 10^{-2} * e^{0.0135x} \text{ (cm)} \qquad 4.12$$

$$A(x) = 5.82 * 10^{-5} * e^{0.0148x} \text{ (cm}^2\text{)} \qquad 4.13$$

The effective acoustic mass as a function of the cochlear distance in percent is obtained by dividing equation 4.13 by the square of equation 4.12 and multiplying the result by the density. The result is

$$M_{ea} = 0.343 * \rho * e^{-0.0122x} \left(\frac{g}{cm^3}\right) \qquad 4.14$$

According to Percoll density gradients, the density of the organ of Corti is between 1.025 and 1.048 g/cm³ (Holley, 1988). Taking the mean of these values, which is 1.037, we obtain

$$M_{ea} = 0.356 e^{-0.0122x} \left(\frac{g}{cm^3}\right) \qquad 4.15$$

This expression does not take the mass of the tectorial membrane into account. In a cochlea one hour or more postmortem, the tectorial membrane is severely shrunken and its mass is probably on the order of 10% of the mass of the organ of Corti. The shrinking of the membrane was observed in my own preparations (Zwislocki, Chamberlain, & Slepecky, 1988) and is evident in many published histological sections. If a 10% fraction is added to the constant in equation 4.15, the constant can be rounded off to 0.4 g/cm³. With this correction, the effective acoustic mass becomes 0.4 g/cm³ at the basal end of the basilar membrane and 0.12 g/cm³ at its apical end. In a preceding article (Zwislocki, 1980) an effective mass of 0.5 g/cm³ was obtained with an unrealistically high density of 1.2 g/cm³, and was assumed to be constant throughout the cochlea. Because the exact configuration of the organ of Corti—tectorial membrane complex and the radial displacement of the basilar membrane are not well known as functions of cochlear distance, it is difficult to decide if the mass remains practically constant or decreases somewhat, as derived here.

Fletcher (1951) seems to have arrived at higher mass values, especially for the cochlear apex, by postulating a radiation mass and, thus, including part of the mass of the fluid surrounding the basilar membrane. Radiation mass is not included in the fundamental theory of surface waves in incompressible, inviscid fluids (e.g., Lamb, 1945; Lindsay, 1960), so that justification of Fletcher's assumption appears to be tenuous.

The last physical parameter necessary to specify the dynamics of the cochlear partition is its resistive impedance component, which stems from viscous forces. Unfortunately, the resistance cannot be determined reliably from empirical measurements until the mechanism of cochlear sound propagation is specified mathematically, so that the resistance becomes a free parameter in the mathematical derivations. Nevertheless, Békésy measured the logarithmic decrement of basilar membrane vibration, and a mathematical formula makes it possible to calculate the resistance from the decrement for a simple resonator. Although the value the formula provides cannot be accurate for the basilar membrane, which is a distributed structure, it is considered here for completeness. It is shown in the next section of this chapter that the resistance fitting the empirical amplitude distributions must be much smaller and hardly affects such cochlear functions as the phase distribution and the input impedance over wide frequency ranges.

The formula based on the logarithmic decrement can be written as

$$R_{Ba} = 2\sqrt{\frac{M_{Ba}}{C_{Ba}}} \sqrt{\frac{1}{1+\frac{4\pi}{D^2}}} \left(\frac{dynesec}{cm^4}\right) \qquad 4.16$$

where R_{ba} is the acoustic resistance of the basilar membrane per unit length and D, the logarithmic decrement. According to Békésy (1960), the decrement varies from about 1.4 at the cochlear base to 1.8 at the apex. On the assumption that the resistance varies exponentially and in view of the equations for the compliance, C_{Ba}, and mass, M_{Ba}, equation 4.15 can be approximated by

$$R_{Ba} = 3*10^4 e^{-0.0452x} \left(\frac{dynesec}{cm^4}\right) \quad \textbf{4.17}$$

MATHEMATICAL ANALYSIS

In this section, I first calculate the location of the basilar membrane resonance as a function of sound frequency to see if it accounts for the location of maximum basilar membrane vibration, as measured by Békésy. Second, I derive the differential equation for long waves in the cochlea, then generalize it to any wavelength. The derivations are made in the most specific way possible to avoid mathematical abstraction as much as possible. Next, I solve the differential equation for the long waves, neglecting the basilar membrane mass and the resistance elements and show that the solution leads to standing waves in the cochlea. The solution makes it possible to calculate the velocity of cochlear wave propagation at relatively low audible frequencies and the cochlear input impedance for all audible frequencies, except the highest ones. Next, I introduce the effects of the fluid and basilar membrane resistances and also of basilar membrane mass. The latter two sections are focused on the amplitude distribution of basilar membrane vibration.

The mathematical derivations are restricted to simple harmonic, or sinusoidal, motions of the form $y = Y*e^{j\omega t}$, where Y is the complex amplitude of the oscillation, j, the imaginary unit, ω, the angular frequency and t, time. It must be noted, however, that any waveform can be represented by superposition of sinusoidal oscillations of harmonically related frequencies.

For convenience of the reader, most of the symbols used in the mathematical formulations are listed below.

x—distance from the basal end of the basilar membrane

x_B—distance to the amplitude maximum

y—depth coordinate perpendicular to the basilar membrane

f—sound frequency

f_r—resonance frequency

f_B—best frequency (for amplitude maximum)

ω—angular frequency ($\omega = 2\pi f$)

$Q_V(x) = Q_V$—cross sectional area of scala vestibuli

$Q_T(x) = Q_T$—cross sectional area of scala tympani

$Q(x) = Q$—effective cross sectional area of the cochlear canal
Q_o—effective cross sectional area of the cochlear canal at $x = 0$
$\rho = \rho_o$—fluid density (at rest)
$C_{Ba}(x) = C_{Ba}$—acoustic compliance of the basilar membrane
$Z_{Ba}(x) = Z_{Ba}$—acoustic impedance of the basilar membrane
$M_{Ba}(x) = M_{Ba}$—acoustic mass of the basilar membrane
$R_{Ba}(x) = R_{Ba}$—acoustic resistance of the basilar membrane
$R_F(x) = R_F$—effective specific resistance of cochlear fluid
$p_v(x,t) = p_v$—sound pressure in scala vestibuli
$p_t(x,t) = p_t$—sound pressure in scala tympani
$P_{v,t}(x) = $ amplitudes of the sound pressures
$p(x,t) = $ sound pressure difference across the basilar membrane
$P(x) = P$—amplitude of the sound pressure difference
$\xi(x,t) = \xi$—cochlear fluid displacement in the x direction
$\Xi(x) = \Xi$—amplitude of the displacement in the x direction
$u_v(x,t) = u_v$—fluid velocity in scala vestibuli in the x direction
$u_t(x,t) = u_t$—fluid velocity in scala tympani in the x direction
$U_{v,t}(x) = U_{v,t}$—amplitudes of the fluid velocities
$\psi(x,t) = \psi$—cochlear fluid displacement in the y direction
$\Psi(x) = \Psi$—amplitude of the displacement in the y direction
$\psi_B(x,t) = \psi$—displacement of the basilar membrane
$\Psi_B(x) = \Psi$—amplitude of basilar membrane displacement
$v_B(x,t) = v_B$—volume displacement of the basilar membrane (or velocity)
$V_B(x) = V_B$—amplitude of the volume displacement (or velocity)

It may also be useful to list here all the cochlear anatomical and physical parameters arrived at as functions of the distance, x.

$$Q = Q_o e^{\theta x}$$

with $Q_O = 1.16*10^{-2}$ cm^2 and $\theta = 0.015$ for x in % and 0.43 cm^{-1} for x in cm ($x_{max} = 3.5$ cm).

$$\rho = 1 \text{ g/cm}^3$$

$$C_{Ba} = C_{Bao} e^{\gamma x}$$

with $C_{Bao} = 0.538*10^{-9}$ cm^4/dyne and $\gamma = 0.0622$ for x in % and $\gamma = 1.79$ cm^{-1} for x in cm

$$M_{Ba} = M_{Bao} e^{-\mu x}$$

with $M_{Bao} = 0.4$ g/cm^3 and $\mu = 0.0122$ for x in % and $\mu = 0.35$ cm^{-1} for x in cm

$$R_{Ba} = R_{Bao} e^{-\alpha x}$$

with $R_{Bao} = 3*10^4$ dyne sec/cm^4 and $\alpha = 0.0452$ for x in % and $\alpha = 1.3$ cm^{-1} for x in cm, superseded by $R_{Ba} = 1.2*10^3$ dyne sec/cm^4 = constant.

Resonance of the Basilar Membrane

According to modern interpretation of Helmholtz's (1954; original publication in 1863) resonance theory and Békésy's (1928, 1947, 1960) interpretation of his early model experiments as well as his later experiments on postmortem cochleas, the maximum of basilar membrane vibration should occur at the location of basilar membrane resonance. Such a resonance must take place where the reactive part of the basilar membrane acoustic impedance becomes zero. The impedance is defined by the equation

$$Z_{Ba} = R_{Ba} + j\left(\omega M_{Ba} - \frac{1}{\omega C_{Ba}}\right) \qquad 4.18$$

At resonance,

$$\omega M_{Ba} - \frac{1}{\omega C_{Ba}} = 0 \qquad 4.19$$

This occurs for

$$f_r = \frac{1}{2\pi} \frac{1}{\sqrt{(M_{Ba} C_{Ba})}} \qquad 4.20$$

When the numerical values are introduced for the Mass, M_{Ba}, and compliance, C_{Ba}, from equation 4.11 for the first, and Fig. 4.10, for the second, the relationship between the resonance frequency and the distance from the basal end of the basilar membrane can be obtained. The relationship is shown in Fig. 4.13 by the upper solid line. The solid circles and the lower line indicate the corresponding location of the maximum vibration of the basilar membrane, as determined by Békésy (1942, 1960). Clearly, there is a substantial discrepancy between this location and the location of basilar membrane resonance. For sound frequencies above 1000 Hz, the discrepancy exceeds 20% of cochlear length; for lower frequencies, the discrepancy is even larger. A cochlear distance of 20% corresponds to one octave in frequency terms.

The theoretical curve of Fig. 4.13 indicates that no resonance of the basilar membrane takes place below a frequency of 1000 Hz. This is in agreement with my earlier calculations (Zwislocki-Moscicki, 1948) and an experiment of Békésy (1942) in which he measured the phase of basilar membrane vibration in the absence of cochlear fluids. He found that the basilar membrane vibrated in one phase over its entire length for frequen-

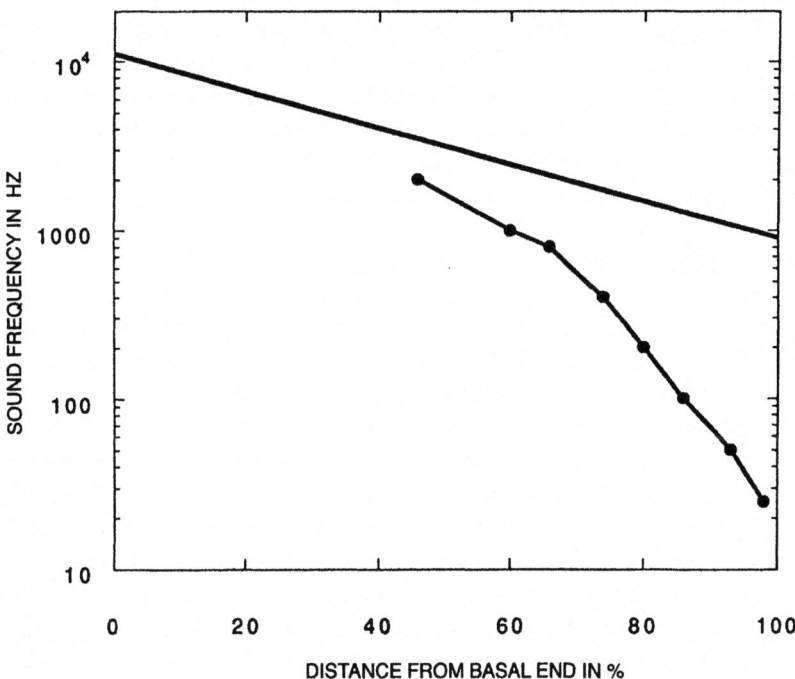

FIG. 4.13. Calculated cochlear location of the basilar membrane resonance (straight line) as compared to the location of maximum vibration measured by Békésy (1960) as functions of the distance from the cochlear base.

cies not exceeding 1000 Hz. A resonance would have produced a phase shift of 180°.

Peterson and Bogert (1950) and Fletcher (1951) obtained a somewhat better agreement of the resonance place with the location of the vibration maximum, but both used values Békésy had found with the calibrated hairs for the compliance of the basilar membrane, which are too large, as has been already explained. In addition, Fletcher seems to have used unjustifiably high mass values.

The discrepancy between the location of the vibration maximum and the location of basilar membrane resonance is much too great to be accounted for by a measurement error, especially since the location of the resonance depends only on the square root of the product of the basilar membrane compliance and mass. I have shown already that Békésy's measurements of the compliance performed by two entirely different methods are consistent with each other and I show subsequently that they are also consistent with the measurements of cochlear wave velocity and input impedance. Measurement of the cross sectional area of the cell mass attached to the basilar membrane, which effectively determines the mass load of the membrane, is a relatively simple matter, and an error in excess of 20% or more can hardly be expected. Therefore, resonance appears as an unlikely mechanism for the local vibration maximum of the basilar membrane. I show in further sections of this chapter, in agreement with my doctoral dissertation (Zwislocki-Moscicki, 1948), that coincidence of the maximum with the resonance could occur only in the complete absence of viscous resistance. This, of course, is not physically possible.

The Cochlear Differential Equation for Long Waves

To keep things as simple as possible mathematically but still have useful results, I first derive a simplified differential equation for waves that are long compared to the depths of scalae vestibuli and tympani (Zwislocki-Moscicki, 1946, 1948; Zwislocki, 1950, 1965). As in practically all mathematical theories of cochlear waves, the cochlear fluids are considered to be incompressible. This is justified for two reasons. First and foremost, a relatively high compliance of the round window membrane allows pressure changes in the cochlea to be practically completely equalized (Fletcher, 1951). In terms of electroacoustic analogies of the first kind, the round window acts as an effective ground. Second, the wavelengths of compressional waves in cochlear fluids can be considered as large compared to the length of the cochlea at least up to a sound frequency of 7000 Hz (Geisler & Hubbard, 1972).

Two additional simplifying assumptions are made. According to one, the curvature of the cochlear canal has a negligible effect on wave propa-

THE COCHLEA SIMPLIFIED BY DEATH   115

gation in the cochlea, as was demonstrate in two very different ways in the past (Zwislocki-Moscicki, 1948; Viergever, 1978). As a consequence, the cochlea can be straightened out for mathematical analysis. According to the other, the effect of the longitudinal elastic coupling within the basilar membrane is negligible. This is so for two reasons. For one, at all cochlear locations where the wave amplitude is measurable, the half wave length is substantially greater than the width of the basilar membrane. This means that the deformation curvature of the oscillating basilar membrane is substantially greater in the width direction than in the length direction. According to the theory of plate oscillation (e.g., Kinsler & Frey, 1950), and the basilar membrane must be considered as a plate (Békésy, 1947), the elastic restoring force is directly proportional to the second derivative of the curvature. As a consequence, the restoring force produced by the curvature in the width direction must be much greater than that produced by the curvature in the length direction, and the longitudinal elastic coupling becomes negligible. In addition, Voldrich (1978) found that, in live cochleas at least, the basilar membrane is stiffer in the width direction than in the length direction.

When, due to a positive phase of sound pressure (i.e., compression), the stapes is pressed into the oval window, the perilymph is displaced inward, and the displacement in scala vestibuli must produce a fluid flow in part through the helicotrema but in part also through the elastic basilar membrane. It is well known that any significant flow through the helicotrema occurs only at the lowest audible frequencies. For higher frequencies, fluid motion produced by sound ceases well before the helicotrema is reached. In any event, for incompressible fluids, the displacement of the stapes must produce an equally large displacement of the round window membrane.

Let us now consider an infinitesimally small longitudinal element of space, dx, in scala tympani, as indicated in Fig. 4.14. Because of the incompressibility of the fluid, the same amount must flow out of the element as flows in. In other words, the difference in fluid flow in the x direction must be compensated for by the deflection of the basilar membrane associated with fluid flow in the y direction. According to Fig. 4.14, this can be expressed mathematically as follows:

$$\int_{QT} u_T dQ_T - \int_{QT} [u_T dQ_T + \frac{\partial \int_{QT} u_T dQ_T}{\partial x} dx] - v_B dx = 0 \qquad 4.21$$

The positive sign indicates fluid inflow, the negative signs, fluid outflow. The equation can be simplified to

$$\frac{\vartheta \int_{QT} u_T dQ_T}{\vartheta x} dx = -v_B dx \qquad 4.22$$

The negative sign means that, when the flow gradient is positive in the x direction and more fluid flows out of the dx element than flows in, there must be an inflow along the negative y direction.

For relatively long waves, a uniform fluid velocity can be assumed over the whole cross section, Q_T, so that

$$\int_{QT} u_T dQ_T = Q_T u_T \qquad 4.23$$

and

$$\frac{\vartheta (Q_T u_T)}{\vartheta x} dx = -v_B dx \qquad 4.24$$

By expanding the left side of equation 4.24 and dividing both sides by dx we obtain:

$$\frac{\vartheta Q_T}{\vartheta x} u_T + Q_T \frac{\vartheta u_T}{\vartheta x} = -v_B \qquad 4.25$$

Because Q_T varies only slowly with x and u_T is small at useable sound intensities, the first term on the left side of this equation is small of the second order and can be neglected, so that a simple relationship between the volume

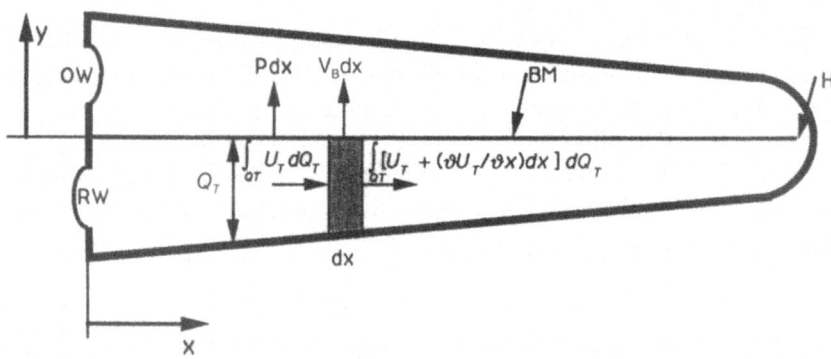

FIG. 4.14. Schematic representation of a straightened out cochlea with indicated mathematical formulations of the dynamical relationships. See text for further explanations.

velocity of the basilar membrane and the gradient of fluid flow in the longitudinal cochlear direction can be obtained,

$$Q_T \frac{\partial u_T}{\partial x} = -v_B \qquad 4.26$$

Because scalae tympani and vestibuli are antisymmetric relative to the basilar membrane, we have by analogy for scala vestibuli:

$$Q_V \frac{\partial u_V}{\partial x} = v_B \qquad 4.27$$

According to Bernulli's equation for small amplitudes of fluid velocity, where quadratic terms can be neglected, the fluid motion in the x direction is produced be the negative pressure gradient in this direction. Accordingly, we can write for scalae vestibuli and tympani, respectively:

$$-Q_V \frac{\partial p_V}{\partial x} = \rho Q_V \frac{\partial u_V}{\partial t} + R_{FV} Q_V u_V \qquad 4.28a$$

$$-Q_T \frac{\partial p_T}{\partial x} = \rho Q_T \frac{\partial u_T}{\partial t} + R_{FT} Q_T u_T \qquad 4.28b$$

The first term on the left of each equation indicates the force produced by the pressure gradient, the first term on the right, the inertia force due to fluid acceleration, and the second term, the fricative force. To obtain a differential equation exclusively for sound pressure, it is first necessary to differentiate the latter two equations with respect to x. This gives:

$$Q_V \frac{\partial^2 p_V}{\partial x^2} = -\rho Q_V \frac{\partial^2 u_V}{\partial x \partial t} - R_{FV} Q_V \frac{\partial u_V}{\partial x} \qquad 4.29a$$

$$Q_T \frac{\partial^2 p_T}{\partial x^2} = -\rho Q_T \frac{\partial^2 u_T}{\partial x \partial t} - R_{FT} Q_T \frac{\partial u_T}{\partial x} \qquad 4.29b$$

It is possible to eliminate the $\frac{\partial u_{V,T}}{\partial x}$ terms by substitution from equations 4.26 and 4.27,

$$Q_T \frac{\partial^2 p_T}{\partial x^2} = +\rho \frac{\partial v_B}{\partial t} + R_{FT} v_B \qquad 4.30a$$

$$Q_V \frac{\partial^2 p_V}{\partial x^2} = -\rho \frac{\partial v_B}{\partial t} - R_{FV} v_B \qquad 4.30b$$

The pressure difference across the basilar membrane is:

$$p = p_T - p_V \qquad 4.31$$

By differentiating the latter equation twice with respect to x, we obtain

$$\frac{\partial^2 p}{\partial x^2} = \frac{\partial^2 p_T}{\partial x^2} - \frac{\partial^2 p_V}{\partial x^2} \qquad 4.32$$

When the right hand expressions of equations 4.30a and 4.30b are substituted for the p_V and p_T terms in equation 4.32 after division of both sides of these equations by Q_V and Q_T, respectively, the p_T and p_V terms are eliminated, and only the term containing their difference remains,

$$\frac{\partial^2 p}{\partial x^2} = \left(\frac{1}{Q_V} + \frac{1}{Q_T}\right) \rho \frac{\partial v_B}{\partial t} + \left(\frac{R_{FV}}{Q_V} + \frac{R_{FT}}{Q_T}\right) v_B \qquad 4.33$$

To simplify this equation, it is convenient to substitute for the expression in the first parenthesis

$$\frac{1}{Q} = \frac{1}{Q_V} + \frac{1}{Q_T} \qquad 4.34$$

where Q can be regarded as an effective cross sectional area of the cochlear canal. Similarly, we can substitute for the expression in the second parenthesis

$$\frac{R_F}{Q} + \frac{R_{FV}}{Q_V} + \frac{R_{FT}}{Q_T} \qquad 4.35$$

With these substitutions, equation 4.33 becomes

$$\frac{\partial^2 p}{\partial x^2} = \frac{\rho}{Q} \frac{\partial v_B}{\partial t} + \frac{R_F}{Q} v_B \qquad 4.36$$

For simple harmonic oscillations, the time derivative of v is simply jωv, so that equation 4.36 can be written in the form

$$\frac{\partial^2 p}{\partial x^2} = \frac{R_F + j\omega\rho}{Q} v \qquad 4.37$$

When we eliminate the time function, which is identical for p and v, basilar membrane volume velocity, v, can be replaced by sound pressure difference across the membrane by introducing the membrane impedance, Z_{Ba} = P/V. We obtain

$$\frac{\partial^2 P}{\partial x^2} = \frac{R_F + j\omega\rho}{QZ_{Ba}} P \qquad 4.38$$

This is the cochlear differential equation for long waves as expressed in terms of the pressure difference across the basilar membrane (for its early derivations see Zwislocki-Moscicki, 1946, 1948; Zwislocki, 1950, 1965). Significantly, it is valid for any acoustic impedance of the basilar membrane. Because R_F is small compared to $\omega\rho$, for most purposes, the equation can be simplified to

$$\frac{\partial^2 P}{\partial x^2} = \frac{j\omega\rho}{QZ_{Ba}} P \qquad 4.39$$

It has been established in the general wave theory that waves occur only when the impedance term is negative. When it is positive, a gradual uniphasic decay of magnitude with the distance occurs. As is shown further, impedance negativity in the cochlea occurs when the reactance term of the impedance is dominated by stiffness. This is true up to the location of the basilar membrane resonance. Beyond the resonance location, the impedance becomes positive, and wave propagation ceases (Zwislocki-Moscicki, 1948).

For sound frequencies and locations well below the basilar membrane resonance, the impedance, Z_{Ba}, is dominated by the compliance term, so that the impedance expression, as defined in equation 4.18a, can be simplified to $-j/\omega C_{Ba}$. Under these conditions, equation 4.39 becomes

$$\frac{\partial^2 P}{\partial x} = \frac{\rho C_{Ba}}{Q} \omega^2 P \qquad 4.40$$

Cochlear Wave Velocity and Related Variables—A First Approximation

The approximation holds best at sound frequencies substantially below the resonance frequency of the basilar membrane. It should be remembered, however, that the compliance, C_{Ba}, and the cross sectional area, Q, depend on the location in the cochlea, x. As a consequence, the following derivations hold within a satisfactory approximation only for conditions under which C_{Ba} and Q vary slowly with x by comparison to the length of cochlear waves. Such conditions prevail generally except for very low sound frequencies.

Since $-\omega^2 p$ is the second time derivative of p, equation 4.40 can also be written in the form

$$\frac{\partial^2 p}{\partial x^2} = \frac{\rho C_{Ba}}{Q} \frac{\partial^2 p}{\partial t^2} \qquad 4.41$$

when the time function is included. From general wave theory it is known that the term $\rho C_{Ba}/Q$ is equal to the square of the inverse wave, or phase velocity. Accordingly, the phase velocity, c_L, of cochlear waves at locations basal to the resonance place is simply

$$c_L = \sqrt{\frac{Q}{\rho C_{Ba}}} \qquad 4.42$$

The expression is in agreement with those derived previously (Zwislocki-Moscicki, 1948; Zwislocki, 1965). Because, at locations basal to the resonance place, the waves tend to be long relative to the canal depth, and the theory of long waves applies, equation 4.42 should hold for most postmortem as well as in vivo conditions. When the numerical values for Q and C_{Ba} are introduced from their defining equations, taking into account equation 4.34 and Fig. 4.10, the following numerical equation is obtained for the velocity:

$$c_L = 46.5 e^{-0.039x} (m/sec) \qquad 4.43$$

The function it describes is shown graphically on linear coordinates in Fig. 4.15. The resulting curve decays exponentially from 46.5 m/sec at the basal end of the basilar membrane to 0.94 m/sec at the apex. This is in reasonable agreement with the results I obtained earlier (Zwislocki-Moscicki, 1948; Zwislocki, 1965).

To compare the theoretical result with Békésy's (1947, 1960) empirical ones it is necessary to convert the velocity to the phase of basilar membrane vibration, which he measured for low frequency sinusoids. This can be done by first calculating the wave delay time at a particular cochlear location. To do this, the cochlear distance scale has to be expressed in centimeters rather than percent. To be compatible with Békésy, I assume the

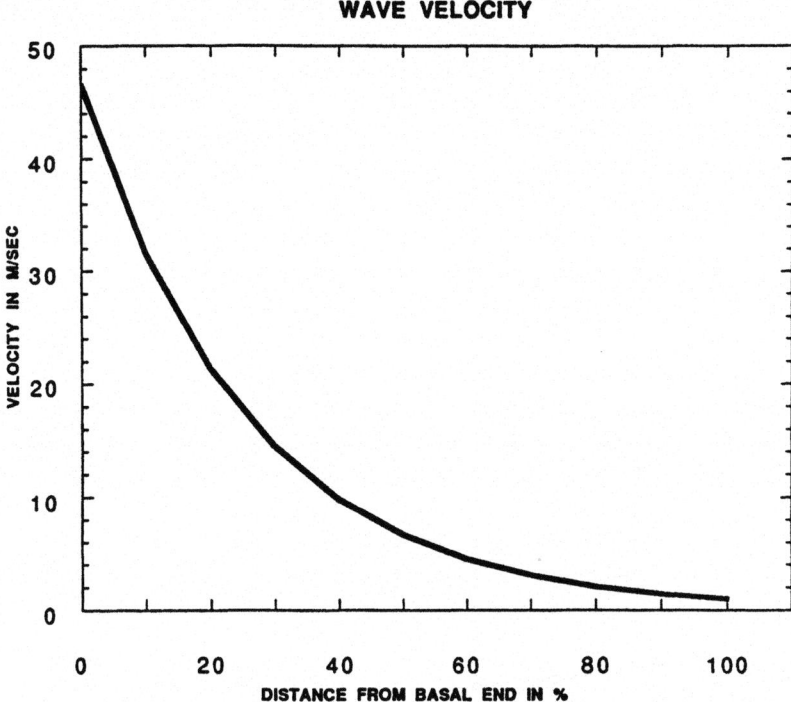

FIG. 4.15. Theoretical cochlear wave velocity for long waves as a function of the distance from the cochlear base in percent—first approximation based on Békésy's postmortem measurements of basilar membrane compliance.

total effective cochlear length to be 3.5 cm, so that 100% is equivalent to 3.5 cm. On this basis, the velocity equation (4.43) becomes

$$c_L = 46.5 * 10^2 e^{-1.11k} (cm/sec) \quad 4.44$$

Since the velocity is defined as an increment in distance per a corresponding increment in time, $c = dx/dt$, the incremental time is given by $dt = dx/c$ and the total delay time by

$$t = \int_0^x \frac{1}{c_L} dx \quad 4.45$$

By executing the integral operation and introducing the numerical values for c_L, we obtain

$$t = 0.194(e^{1.11k} - 1) \text{ (msec)} \qquad 4.46$$

The phase of a sinusoidal oscillation beginning at time $t = 0$ is simply $\phi = 2\pi ft$. If the phase is measured at two frequencies f_1 and f_2, we also have $\Delta\phi = \phi_2 - \phi_1 = 2\pi t(f_2 - f_1)$. The latter formula has the advantage of being independent of the initial conditions, that is, of the phase at $t = 0$, provided it is the same for both frequencies. The phase calculated for two frequencies, 200 and 300 Hz, on the basis of the delay time defined in equation 4.46, is plotted in Fig. 4.16 as a function of the cochlear distance. The data points have been derived from Békésy's direct measurements at the same frequencies. Unfortunately, Békésy did not specify the initial phase unambiguously. The approximate agreement between the theoretical and empirical results shown in Fig. 4.16 has been obtained by adding to Békésy's raw data a phase angle of $\pi/2$. There is good independent justification for this correction. Békésy is likely to have used sta-

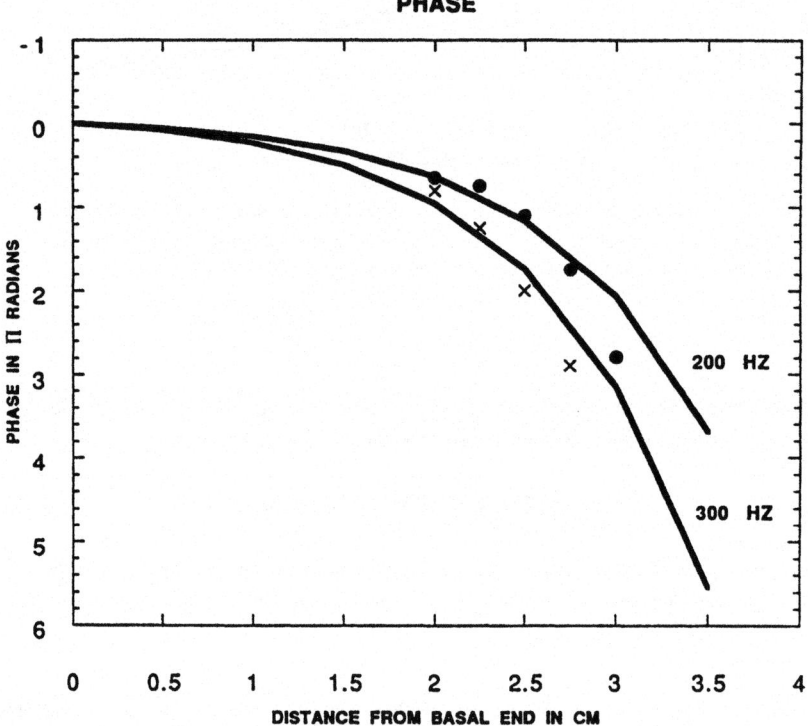

FIG. 4.16. Phase of basilar membrane displacement as a function of the distance from the cochlear base in cm, referred to the phase at the base, for two sound frequencies. Circles and crosses indicate experimental data of Békésy's, the curves are theoretical.

pes displacement for phase reference. It is shown in the next subsection, in agreement with previous calculations (Zwislocki-Moscicki, 1948), that the basilar membrane displacement at the cochlear base must lead the stapes displacement by such a phase angle. The need for the correction was pointed out independently by Flanagan (1962) on the basis of the theory of minimum phase filters. A direct empirical confirmation of these theoretical results was provided by Rhode (1971). It is shown in the next paragraph that the correction is consistent with delay, or wave travel time computations.

The travel time can be calculated from Békésy's phase data for 200 and 300 Hz and also from the phase difference between the 300 and 200 Hz phase curves of Fig. 4.16, as mentioned above. The latter calculation is independent of the initial phase. The results of the three way calculations are compared with each other and with the theoretically obtained travel time from equation 4.46 in Fig. 4.17. The phase correction of $\pi/2$ has been included in the single frequency calculations. Clearly, the cor-

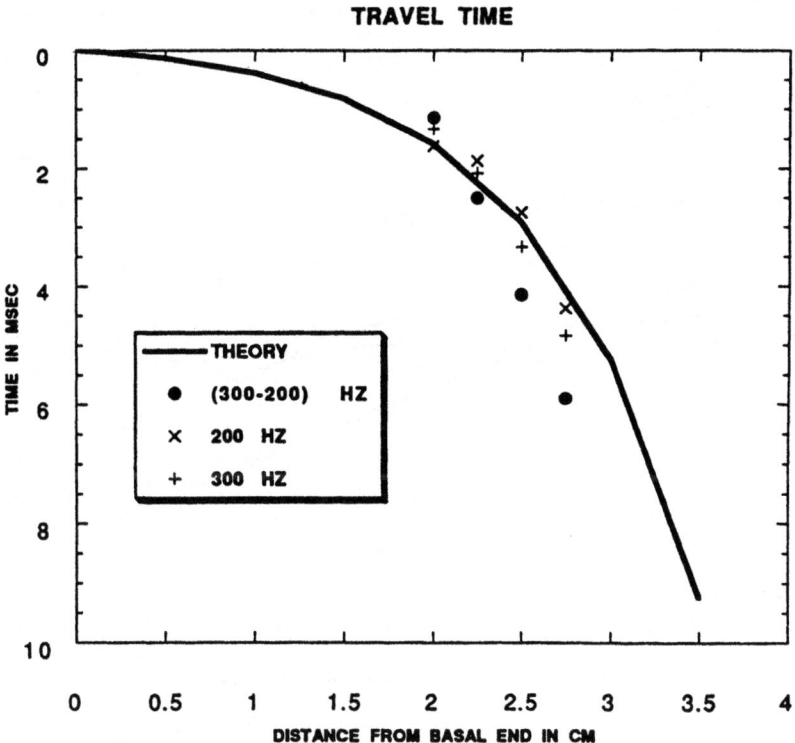

FIG. 4.17. Cochlear wave travel time as a function of the distance from the base in cm. The curve is theoretical, the points are derived from Békésy's phase data.

rection makes the travel time calculated from single frequency phases compatible with the travel time calculated from their differences and with the theoretical results. The differentially determined data depart the most from the others, probably, because of error summation between the two single frequency phase measurements.

In the derivations made thus far, long waves relative to canal depth have been assumed. On the basis of the already calculated wave velocity, it is possible to calculate the wavelength and test this assumption. The wavelength is equal simply to the velocity divided by the frequency of oscillation, $\lambda = c/f$. The numerical results for three frequencies, 200, 800 and 3,200 Hz are plotted in Fig. 4.18 by means of thick lines. The crosses mark the locations of maximum vibration for these frequencies, as found by Békésy (1942, 1960). The location of the cross for 3,200 Hz has been obtained from his data by extrapolation. The thin line with a smaller slope follows the effective canal depth multiplied by 2π. The effective depth has

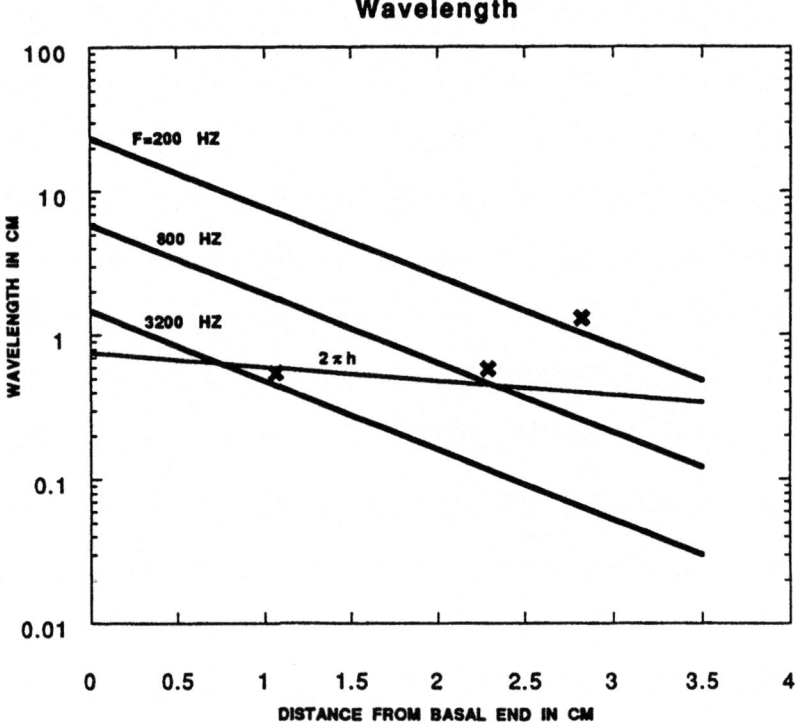

FIG. 4.18. The wave length of cochlear waves as a function of the distance from the cochlear base in cm for three sound frequencies. The crosses mark the locations of maximum vibration amplitude for these frequencies, as determined by Békésy, and the thin line marks the effective canal depth multiplied by 2π.

been obtained on the assumption that the cochlear canal has a roughly circular cross section and that scalae vestibuli and tympani occupy two equal halves of this cross section. With these assumptions, the cross sectional area of the canal is $Q_{tot} = r^2\pi$, where r is the radius. The effective depth, equal to the radius, can be calculated from $h = r = (Q_{tot}/\pi)^{1/2}$. According to the theory of surface waves, the assumption of long waves is applicable when the wavelength exceeds the canal depth multiplied by 2π. The graph of Fig. 4.18 indicates that the assumption is generally tenable for low sound frequencies and, roughly, up to the location of maximum vibration at the higher ones, as I maintained in my earlier publications (Zwislocki-Moscicki, 1946, 1948; Zwislocki, 1965). It is possible to show that, beyond the vibration maximum, the assumption makes the waves appear too long.

Effect of Basilar Membrane Mass on the Wave Velocity

In the above calculations of cochlear wave velocity and of variables derived from it, the mass of the basilar membrane has been neglected. To estimate over what frequency and distance ranges such an approximation is sufficiently accurate I calculate here the effect of the mass. Because of their small effect on the velocity, except at very low sound frequencies and near the resonance location, the resistance components are neglected. The differential equation 4.39 can serve as a point of departure when the impedance term defined in equation 4.18 is simplified to

$$Z_{Ba} = j(\omega M_{Ba} - \frac{1}{\omega C_{Ba}}) \qquad 4.47$$

When the impedance expression is introduced into equation 4.39, the equation can be written in the form

$$\frac{\partial^2 P}{\partial x^2} = -\omega^2 \frac{\rho}{Q\left(\frac{1}{C_{Ba}} - \omega^2 M_{Ba}\right)} P \qquad 4.48$$

Since $-\omega^2 P$ is the magnitude of the second time derivative of the sound pressure, p, the velocity of wave propagation is according to wave theory

$$c = \sqrt{\frac{Q(1 - \omega^2 M_{Ba} C_{Ba})}{\rho C_{Ba}}} \qquad 4.49$$

in analogy to the derivation without the mass. This equation can be rewritten in the form

$$c = c_L \sqrt{(1 - \omega^2 M_{Ba} C_{Ba})} \qquad 4.50$$

where c_L is the low frequency velocity, as defined in equation 4.42. When the numerical expressions for Q, M_{Ba} and C_{Ba} are introduced respectively from their defining equations and Fig. 4.10 and the terms are appropriately arranged, equation 4.50 becomes

$$c = 0.465 * 10^4 e^{0.039x} \sqrt{(1 - 0.73 * 10^{-8} e^{0.05x} f^2)} \qquad 4.51$$

The values of c for the frequencies of 200, 800 and 3,200 Hz are plotted as functions of the cochlear distance in Fig. 4.19 by means of the solid curves arranged according to the frequency from top to bottom. The crosses indicate the locations of maximum vibration found by Békésy for the corresponding frequencies. It is apparent that the effect of the mass is negligible

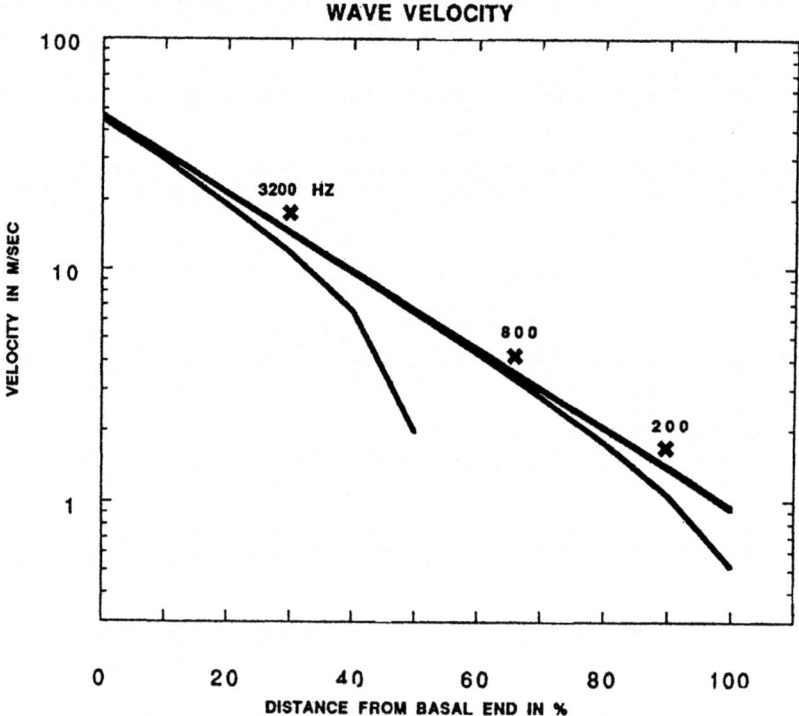

FIG. 4.19. Cochlear wave velocity as a function of the distance from the cochlear base in percent when the mass of the cochlear partition is taken into account. The solid lines show the theoretical velocity values for three sound frequencies; crosses indicate the locations of maximum basilar membrane vibration for the same sound frequencies, as determined by Békésy.

at the low frequencies. For 200 Hz, the curve is practically the same over the entire length of the cochlea as is obtained without the mass. At 3,200 Hz, the effect is small up to the vibration maximum but increases rapidly beyond it. A similar situation prevails at higher sound frequencies.

The Input Impedance

The cochlear input impedance is one of the most important parameters in auditory sound transmission. It affects sound transmission, not only through the middle ear, but also through the outer ear by contributing to the acoustic impedance at the tympanic membrane (Zwislocki, e.g., 1957, 1962). It affects the transfer function of the inner ear with respect to both magnitude and phase (Zwislocki, 1965). It is also important from the experimental point of view, since it is the only cochlear parameter that can be measured in a live human, albeit indirectly, through the measurement of the acoustic impedance at the tympanic membrane. As is shown in the preceding chapter, most of the resistive component of the latter stems from the inner ear. The cochlear input impedance can be measured relatively easily postmortem on intact cochleas. Since it depends directly on the compliance of the basilar membrane, it can serve for verification of the much more difficult compliance measurements.

Once the cochlear wave velocity and the effective cross section of the cochlear canal are known, the characteristic acoustic impedance of the cochlea can be calculated quite easily. According to the general theory of acoustic transmission lines (e.g., Kinsler & Frey, 1950), the characteristic impedance is related to the wave velocity by the equation

$$Z_{Ca} = \frac{r}{Q} c_L \qquad 4.52$$

The low frequency wave velocity is used since the effect of the basilar membrane mass is negligible near the cochlear input for all but the highest sound frequencies. For numerical values of Q and c_L prevailing at the cochlear input, the characteristic impedance becomes the input impedance. It is shown in a further section that equation 4.51 has to be modified for very low sound frequencies where partial wave reflection takes place.

By introducing for c_L its defining formula from equation 4.42, we obtain

$$Z_{Ca} = \sqrt{\frac{\rho}{QC_{Ba}}} \qquad 4.53$$

Because the right side of the equation is real, the input impedance must be resistive in character. This theoretical outcome, when first obtained

(Zwislocki-Moscicki, 1948), seemed quite counterintuitive. It partially contradicted Békésy's (1942) experimental data and, probably for both reasons, was disregarded in the literature for many years. Its first experimental confirmation, although not a direct one, came with a delay of 10 years. It resulted from the analysis of the middle ear function based on acoustic impedance measurements at the tympanic membrane (Zwislocki, 1957, 1962, 1963; Mundie, 1963; Møller, 1965). Dallos (1973) accepted the resistive nature of the cochlear input impedance on theoretical grounds and also on the basis of cochlear microphonic measurements (Dallos, 1970) in his classical book on the Auditory Periphery, but a definitive experimental confirmation based on direct measurements did not occur until 20 years later (Nedzelnitsky, 1980; Lynch et al., 1982; Merchant, Ravicz, & Rosowski, 1992, 1996).

The resistive nature of the cochlear input impedance has important consequences. First of all, it means that all sound energy entering the cochlea is absorbed in it (Zwislocki-Moscicki, 1948; Zwislocki, 1965). Second, it means that basilar membrane displacement at its basal end precedes the stapes displacement by 90° (Zwislocki, 1965). This counterintuitive and long misunderstood phase relationship arises in the following way. Since acoustic impedance is defined as the ratio between sound pressure and volume velocity, $Z_a = P/\dot{v}$, a resistive impedance means that sound pressure and volume velocity are in phase. Because volume velocity precedes volume displacement by 90°, the same must be true for sound pressure. In the cochlea, stapes displacement produces an input sound pressure that leads the displacement by 90°. This sound pressure drives the basilar membrane. Because, at its basal end, the basilar membrane impedance is controlled by compliance for practically all sound frequencies, its displacement must be in phase with the sound pressure, this means, it must lead the stapes displacement by 90°. This relationship was not accepted in the literature until it became verified by Rhode's (1971) direct measurements of basilar membrane vibration, and even then, its consequences were rarely taken into consideration. The 90° lead is the likely explanation of the apparent discrepancy between Békésy's (1947) phase data, as reported by him, and theoretical results obtained in one way by Flanagan (1960) and in another way in this chapter.

After introduction of the numerical value of ρ and the numerical relationships of the variables Q, and C_{Ba} from their defining equations and Fig. 4.10, respectively, equation 4.53 allows us to calculate the numerical values of the cochlear characteristic impedance. The equation becomes

$$Z_{Ca} = 0.4 * 10^6 e^{-0.023x} \qquad 4.54$$

At the cochlear input, $x = 0$, the impedance becomes $0.4*10^6$ dynes sec/cm^5. It should be emphasized that this value constitutes only a rough approximation since the values of the effective cross section of the cochlear canal, Q, and of the compliance of the basilar membrane, C_{Ba}, both of which enter the defining equation of the input impedance, Z_{Ca}, are themselves only approximations of the empirical data, being derived from interpolating functions. A careful consideration of the data near the cochlear base suggests that the actual value of the impedance could be slightly smaller. To be exact, this value is predicted only for audible sound frequencies that are not very low and not very high. It is shown theoretically in a further subsection that smaller impedance values must be expected at low sound frequencies. In addition, the impedance may be complicated by the occurrence of short waves at high frequencies.

The fact that the value of the cochlear input impedance obtained here from cochlear sound transmission theory with the help of anatomical and physical constants of the cochlea agrees with the value derived with the help of a network model of the middle ear from impedance measurements at the tympanic membrane should be significant. The probability of such an agreement occurring by chance in the presence of faulty data or erroneous theoretical formulations is small. It should be noted that the numerical values of the cochlear constants were determined postmortem and the impedance at the tympanic membrane was measured in vivo. This would imply that Békésy's measurements of basilar membrane compliance made postmortem are valid. More recently, Merchant et al. (1992, 1996) determined the cochlear input impedance of human cochleas postmortem using a different method than did Békésy and obtained a somewhat larger value of approximately $0.6*10^6$ dyne sec/cm^5. In view of the many variables involved and likely postmortem changes, this value must be considered as being in reasonable agreement with Békésy's, nevertheless.

The significance of the mutual agreement of the results mentioned in the preceding paragraph becomes less certain in view of interspecies comparisons discussed in chapter 5. These comparisons suggest that the human input impedance of the live cochlea is twice as large as obtained by Békésy postmortem and as inferred from acoustic impedance measurements at the tympanic membrane. It must be realized that the evidence obtained from the latter is indirect, and that the impedance measured at the tympanic membrane is not very sensitive to the conditions at the cochlear input, as has been shown in the preceding chapter. In view of these caveats, it seems that we cannot estimate the magnitude of the cochlear input impedance in vivo better than within a factor of two. It appears to lie between 0.4 and $0.8*10^6$ dyne sec/cm^5.

In past publications (e.g., Zwislocki-Moscicki, 1948; Zwislocki, 1965, 1974, 1975), I had used larger numerical values of basilar membrane compliance and smaller numerical values of cochlear input impedance postmortem than here, based on different interpretations of Békésy's measurements. I had accepted Békésy's data of basilar membrane acoustic compliance directly rather than computing them from his data on maximum basilar membrane displacement (Fig. 4.7), as has been done here (Fig. 4.10). In addition, I had accepted Békésy's direct measurements of cochlear input impedance at low sound frequencies as representative for all sound frequencies. The low frequency data had been used to avoid an artifactual mass effect evident at the higher frequencies. However, the numerical values obtained by him at low sound frequencies are substantially lower than the values obtained at higher sound frequencies, which are compatible with his measurements of the cochlear and the stapedial impedance combined. The values resulting from the latter are consistent with those determined more recently by Merchant et al. (1996).

I had blamed the apparent inconsistency of Békésy's impedance data obtained at low sound frequencies and of his data of basilar membrane compliance with other data on postmortem changes associated with his method of specimen preservation. This appeared especially reasonable in view of some independent evidence for increased compliance of the basilar membrane after death and for the dependence of the change on postmortem conditions (Rhode, 1973). Reinterpretation of Békésy's compliance data, as already discussed, brings them in line with the other data discussed in this chapter. The application of more rigorous mathematical theory to cochlear wave propagation, as introduced below, accounts for the small impedance values obtained by Békésy at low sound frequencies.

Cochlear Waves When the Partition's Resistance and Mass are Negligible

In the cochlear region sufficiently removed from the vibration maximum toward the cochlear base, the impedance of the cochlear partition is dominated by its compliance component, and the mass and resistance effects can be neglected, as has already been mentioned. The cochlear dynamics is adequately defined by equation 4.39. When specific expressions for its parameters are introduced, the equation takes the form

$$\frac{\partial^2 P}{\partial x^2} = -\frac{\rho C_{Bao} e^{\gamma x}}{Q_o e^{-\theta x}} \omega^2 P \qquad 4.55$$

for simple harmonic time functions. The numerical values of the parameters have been introduced in preceding sections. They are repeated here for clarity and convenience.

$\rho = 1$ g/cm^3

$C_{Bao} = 0.538*10^{-9}$ cm^4/dyne

When x in %, $\gamma = 0.0622$

When x in cm ($x_{max} = 3.5$ cm), $\gamma = 1.79$ cm^{-1}

$Q_o = Q_{toto}/4 = 1.16*10^{-2}$

When x in %, $\theta = 0.015$

When x in cm ($x_{max} = 3.5$ cm), $\theta = 0.43$ cm^{-1}

The exponential constants can be lumped together, so that

$$\beta = \gamma + \theta = 0.0772 \text{ or } 2.22 \text{ cm}^{-1}$$

With this simplification, equation 4.55 becomes

$$\frac{\partial^2 P}{\partial x^2} = -\frac{\rho C_{Bao} e^{\beta x}}{Q_o} \omega^2 P \qquad 4.56$$

I showed in the past (Zwislocki-Moscicki, 1946, 1948; Zwislocki, 1965) that this equation has an exact solution in the form of a Bessel function of zeroth order. Several kinds of Bessel functions can be used but, for reasons that should become apparent further in this chapter, the Hankel function of the second kind is found to be the appropriate one (e.g., Jahnke & Emde, 1945). Accordingly,

$$P(x) = P_o H_o^{(2)}(a^{1/2} f e^{\beta x/2}) \qquad 4.57$$

where

$$a^{1/2} = \frac{4\pi}{\beta}\left(\frac{\rho C_{Bao}}{Q_o}\right)^{1/2} \qquad 4.58$$

With the cochlear constants already specified, the constant $a^{1/2}$ is $3.5*10^{-2}$ when the distance, x, is specified in percent, and 1.22, when it is specified in centimeters.

By analogy to the imaginary exponential function, the Hankel function, $H_o^{(2)}$, which is allowed to have an imaginary argument, can be decomposed in two oscillatory components whose mutual relationship depends on the initial phase. In symbols,

$$H_0^{(2)}(x) = J_0(x) - jN_0(x) \qquad 4.59$$

where $J_0(x)$ is the Bessel function of the first kind, popularly called the Bessel Function, and N_0, the Bessel function of the second kind, known as the Neuman Function. Both functions are here of zeroth order. For the purpose of the following calculation of the low frequency wave pattern in the cochlea, it is sufficient to use the Bessel Function, $J_0(x)$. It resembles a sinusoid with a period and an amplitude that decrease with the distance, x. This Bessel function has a maximum for the argument 0 and is close to the maximum for argument values smaller than 1, which prevail at x = 0 for low sound frequencies.

When the cochlear length is in a suitable relationship to the wavelength, standing waves can arise on the cochlear partition. Sound frequencies of the lowest standing wave numbers have been calculated on the assumption that the source impedance at the cochlear input is high so that nearly total wave reflection takes place and that the cochlea is terminated at the helicotrema with an impedance approaching 0. As a consequence, the lowest standing wave should correspond to a quarter wavelength. The following numerical values in Hz are obtained:

$$f_1 = 40; \ f_2 = 93; \ f_3 = 145; \ f_4 = 197; \ f_5 = 250$$

They are lower than those obtained originally (Zwislocki, 1948) on the basis of different assumptions. The resulting wave pattern is shown in Fig. 4.20 for both the pressure differential across the basilar membrane and the basilar membrane volume displacement at the frequency of 200 Hz (near 197 Hz). The pressure wave obeys the equation

$$P(x) = P_0 J_0(a^{1/2} f e^{\beta x/2}) \qquad 4.60$$

and the displacement wave,

$$V_B(x) = P(x) * C_{Ba0} e^{\beta x} \qquad 4.61$$

To obtain the partition displacement, the volume displacement would have to be divided by the width of the basilar membrane, which varies slowly with the distance, x. The wave pattern would be affected little.

The simple relationship between the pressure wave and the displacement wave is obtained only when the mass and resistance components of the cochlear partition impedance are neglected. As has been shown already, the mass component plays a negligible role at the very low sound frequencies,

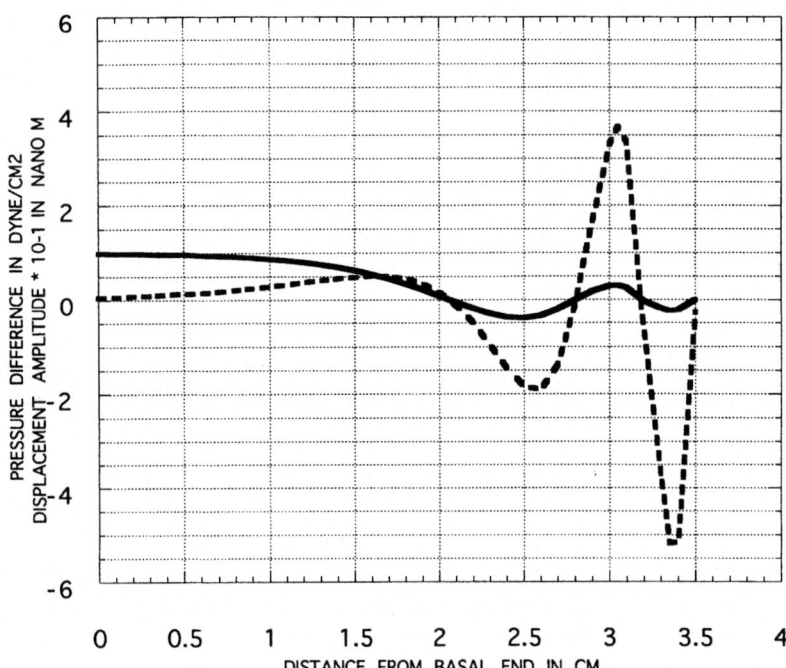

FIG. 4.20. Standing waves in the cochlea calculated for a sound frequency of approximately 200 Hz when the cochlear damping is neglected. The solid line indicates the differential sound pressure across the basilar membrane, the intermittent line, the membrane's displacement.

including 200 Hz. This does not seem to be true for the resistance component, however, as can be deduced from Bekesy's experiments. They show that the displacement amplitude does not increase with the distance from the cochlear base all the way to the helicotrema, as is apparent in Fig. 4.20, but goes through a maximum before the helicotrema is reached and decays beyond it. The resistance and mass effects are considered in further sections. Here, it is mathematically convenient to address the question of the cochlear input impedance at very low sound frequencies.

Cochlear Input Impedance at Low Sound Frequencies

The mass and resistance of the basilar membrane can be neglected near the cochlear input, except for very high sound frequencies, so that equation 4.57 for the pressure differential across the basilar membrane holds. In analogy to electrical transmission lines, the cochlear characteristic impedance can be defined as

$$Z_{Ca}(x) = \frac{P(x)}{Q(x)U_V(x)} \quad 4.62$$

with P(x) meaning the pressure differential across the basilar membrane, Q(x), the effective cross section of the cochlea, and $U_V(x)$ the fluid velocity in scala vestibuli, all expressed as amplitudes. For x = 0, the expression defines the cochlear input impedance. Empirically, the input impedance is usually determined in terms of the sound pressure at the stapes (e.g., Lynch et al., 1982; Merchant et al., 1996), rather than the pressure differential across the basilar membrane, which is approximately equal at the cochlear input to the pressure difference between the stapes and round window locations. Because the sound pressure at the round window is much smaller than at the stapes, except below 50 Hz (Nedzelnitsky, 1980), the pressure differential is fairly well approximated, however.

Since the cochlear input impedance is defined in equation 4.62 in terms of longitudinal fluid motion in scala vestibuli, $U_v(x)$, it may not be clear why the effective cross section of the whole cochlea is used rather than the cross section of scala vestibuli. This is done to account for the fact that motion of the perilymph in scala tympani also contributes to the impedance measured in scala vestibuli, as is made clear by the following analysis.

The pressure differential across the basilar membrane has been defined in equation 4.31 as $p = p_t - p_v$. When both sides of this equation are differentiated with respect to x, equations 4.28a,b make it possible to relate the differential sound pressure to the longitudinal, along the x axis, volume velocity in scala vestibuli by the following set of derivations. We leave out the resistance terms since they have been shown already to be very small compared to the inertia terms. By taking the difference between the two equations, we obtain after trivial transformations

$$\frac{\partial p}{\partial x} = \rho\left(\frac{\partial u_V}{\partial t} - \frac{\partial u_T}{\partial t}\right) \quad 4.63$$

Because of negligible compressibility of the cochlear fluids, we must have equal volume velocities in both scale. This means that

$$Q_T u_T = -Q_V u_V \quad 4.64$$

With the help of this relationship, it is possible to eliminate the term referring to scala tympani in equation 4.63 and obtain a relationship between the pressure differential and the volume velocity in scala vestibuli in the form

$$\frac{\partial p}{\partial x} = \rho \frac{Q_V + Q_T}{Q_V Q_T} Q_V \frac{\partial u_V}{\partial t} \qquad 4.65$$

For simple harmonic motions, equation 4.65 can be rewritten as

$$\frac{\partial P}{\partial x} = -j 2\pi f \rho \frac{Q_V + Q_T}{Q_V Q_T} Q_V U_V \qquad 4.66$$

In this equation, the expression

$$\frac{Q_V + Q_T}{Q_V Q_T} = \left(\frac{1}{Q_V} + \frac{1}{Q_T}\right) \qquad 4.67$$

defines the effective cross section in terms of $1/Q$, so that equation 4.66 can be written in the form

$$\frac{\partial P}{\partial x} = -j 2\pi f \rho \frac{Q_V U_V}{Q} \qquad 4.68$$

The latter equation allows us to calculate $Q_V U_V$ from equation 4.67 for the differential pressure, P. Differentiation of this equation with respect to x produces

$$\frac{\partial P}{\partial x} = -P_0 a^{1/2} f \frac{\beta}{2} e^{\beta x/2} H_1^{(2)}(a^{1/2} f e^{\beta x/2}) \qquad 4.69$$

The volume velocity $Q_V U_V$ can be calculated from equation 4.69 when the pressure derivative on the left side is replaced by its right hand expression of equation 4.68. After a simple transformation, one obtains

$$Q_V(x) U_V(x) = -\frac{j}{4\pi\rho} P_0 a^{1/2} \beta e^{\beta x/2} Q_0 e^{-\theta x} H_1^{(2)}(a^{1/2} f e^{\beta x/2}) \qquad 4.70$$

When equation 4.57 is divided by the latter equation, the desired characteristic acoustic impedance of the cochlea is obtained, in agreement with its definition in equation 4.62. Its expression as a function of distance, x, takes the form

$$Z_{Ca}(x) = \frac{j\ 4\pi\rho}{a^{1/2}\beta e^{\beta x/2} Q_0 e^{-\theta x}} \frac{H_0^{(2)}(a^{1/2}fe^{\beta x/2})}{H_1^{(2)}(a^{1/2}fe^{\beta x/2})} \quad 4.71$$

For sound frequencies above about 1 kHz, the argument of the Bessel functions becomes larger than 1. Then, approximately,

$$H_1^{(2)} = jH_0^{(2)} \quad 4.72$$

and equation 4.71 takes the simple form of

$$Z_{Ca}(x) = \frac{4\pi\rho}{a^{1/2}\beta e^{\beta x/2} Q_0 e^{-\theta x}} \quad 4.73$$

It can be made more explicit by introducing for $a^{1/2}$ its expression from equation 4.58. It then becomes

$$Z_{Ca}(x) = \frac{\rho^{1/2}}{Q_0^{1/2} e^{-qx/2} C_{Ba}^{1/2} e^{\gamma x/2}} \quad 4.74$$

and is equal to the impedance expression of equation 4.53 derived directly from the cochlear differential equation, equation 4.41. The only difference between the two equations is that equation 4.74 has been written in a more explicit form. The expressions indicate that above about 1000 Hz, the characteristic impedance is real, that is, has the character of a resistance. This is no longer true for lower sound frequencies because the relationship of equation 4.72 does not hold, as had been noted previously (Zwislocki-Moscicki, 1948).

When, in equation 4.74, $x = 0$, the equation defines the input impedance of the cochlea. The numerical magnitude values of the impedance have been calculated on the basis of the numerical constants given above and are shown in Fig. 4.21 by the solid curve. According to it, the impedance magnitude is small at low sound frequencies, increases at a rate of about $f^{1/2}$ up to about 1000 Hz and reaches an asymptote of $0.4*10^6$ dyne sec/cm^5 above 1000 Hz, in agreement with the value derived directly from the differential equation. For comparison, the intermittent curve shows magnitudes of the impedance derived from Békésy's (1942) measurements. The agreement is good even in absolute terms between about 160 and 500 Hz, suggesting that his impedance measurements in this frequency region are consistent with his measurements of the basilar membrane compliance on which the impedance calculations have been based. However, above 500 Hz, his measurements indicate numerical values that

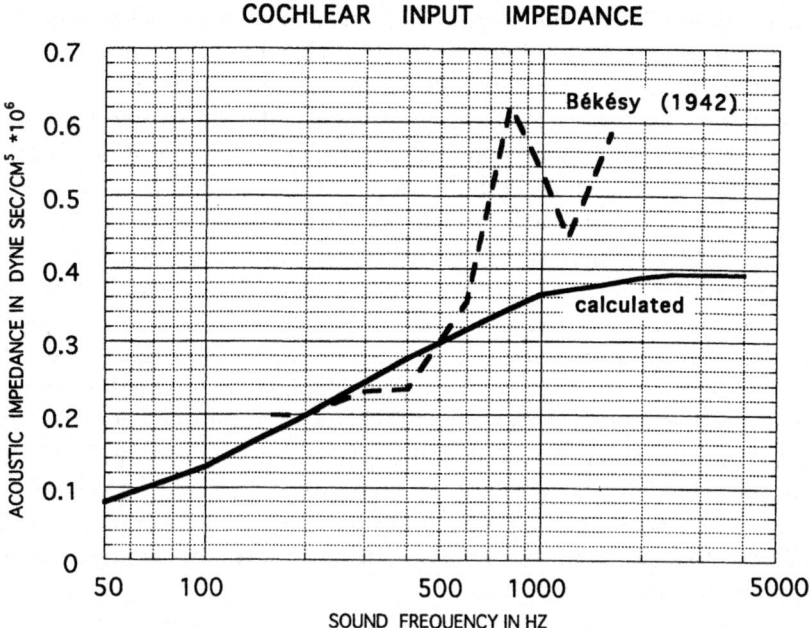

FIG. 4.21. Magnitude of the cochlear input impedance at low sound frequencies. The solid line is theoretical, the intermittent one connects the values derived from Békésy's (1960) measurements.

increase rapidly at first, then, follow an erratic course, in disagreement with the theory. The rapid rise is consistent with Békésy's phase measurements indicating a mass effect not predicted theoretically. From the description of his experimental approach it is possible to conjecture that the mass effect originated in the coupling of a measuring tube to the round window whose membrane was removed. It is likely that, because of capillary forces, some of the cochlear fluid ascended the tube, producing an added fluid column. Such a column would act as a mass connected in series with the cochlea.

A pure acoustic input impedance of the cochlea appears to be difficult to determine empirically. Usually, both the stapes in the oval window and the round window membrane are left in during the measurement and, subsequently, their effects subtracted. Both elements contribute impedance components whose magnitudes can be substantially larger than the magnitude of the cochlear impedance proper. As a consequence, small errors in their determination can lead to substantial errors in the derived cochlear impedance. Some such relationships for the impedance magnitude and phase, measured on human temporal bone preparations, are shown in Figs. 4.22a and 4.22b. The magnitude data have been normalized relative to the values prevailing in the 800 to 1000 Hz region. For reference, the solid curves show the theoretical values. The curves marked "B" have been derived from Békésy's (1942) measurements already mentioned. The curves marked "M" are based on more recent measurements of Merchant's et al. (1996) during which both the stapes and the round window membrane were left in, and only the stapes effect subsequently subtracted. Clearly, Békésy's magnitude curve in Fig. 4.22a has a steeper average slope than the theoretical curve at all but the lowest sound frequencies and his phase curve in Fig 4.22b indicates a greater positive angle, except at these frequencies. Both features are consistent with an artifactual mass effect already mentioned. By contrast, the magnitude curve of Merchant et al. has a negative slope below some 400 Hz, and the phase curve, a negative angle, consistent with a compliance effect. Between 400 and 3000 Hz, both the magnitude and phase curves follow approximately the theoretical curves. The departure of the magnitude curve from the theoretical one at 4000 Hz is difficult to interpret because it is not accompanied by a corresponding departure of the phase curve.

Because of scant experimental data on human cochlear preparations, it may be useful to compare the theoretical results with the experimental impedance data obtained on animals. The most extensive data are available for domestic cats. Because the magnitude of their cochlear input impedance appears to be somewhat larger than in humans, only normalized magnitude data can be compared. This is done in Fig. 4.23a. In addition, corresponding comparisons for the phase data are shown in Fig. 4.23b. In both figures, the solid curves follow the theoretical functions for humans and have been reproduced from Fig. 4.22a and Fig. 4.22b. The other two curves have been taken from an article of Lynch et al. (1982). The curve marked "L" is based on their direct impedance measurements during which the round window membrane was not removed. The curve marked "D" is based on their derivation of the impedance from cochlear microphonic measurements of Dallos (1970) with the help af a cat's middle ear transfer function determined by Guinan and Peake (1967). Whereas the former is referred to sound pressure at the stapes foot plate, the latter is based on the pressure differential across the basilar membrane (Lynch et al., 1982). With respect to the magnitude curves the following points seem to be worth noting. At low sound frequencies, both experimental curves ascend according to the same slope as the theoretical curve. At higher frequencies, all the curves become horizontal. However, the cat experimental curves reach the saturation level at lower frequencies than do human theoretical curves. This difference could originate in a real interspecies variation. The phase

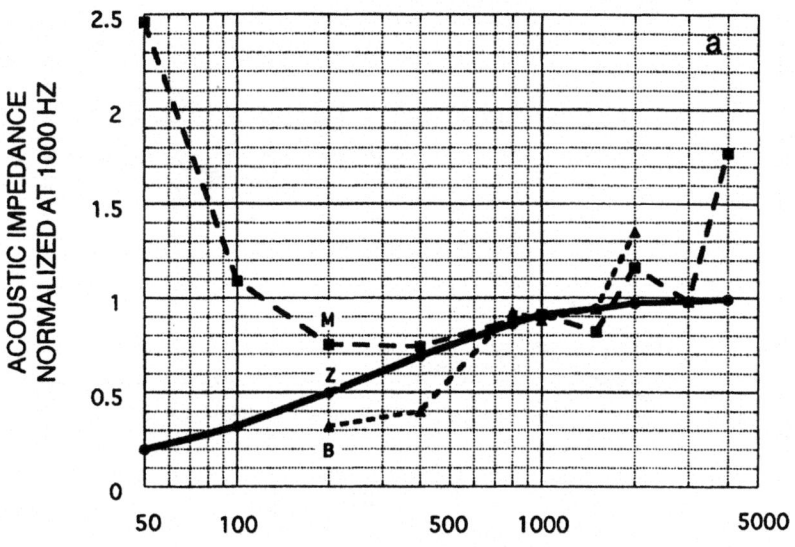
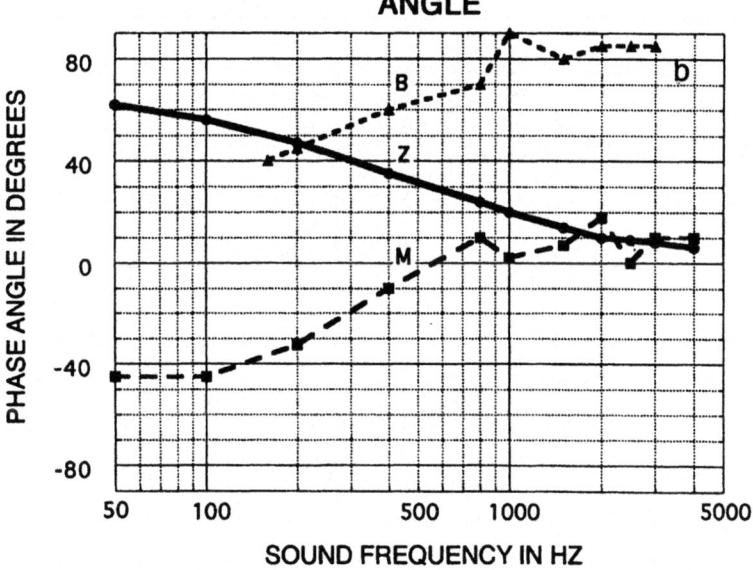

FIG. 4.22. Magnitude and phase angle components of the cochlear input impedance, as calculated from the theory and obtained indirectly experimentally on human postmortem preparations by Békésy (1960) and Merchant et al. (1996). The magnitude values are normalized at 800 Hz.

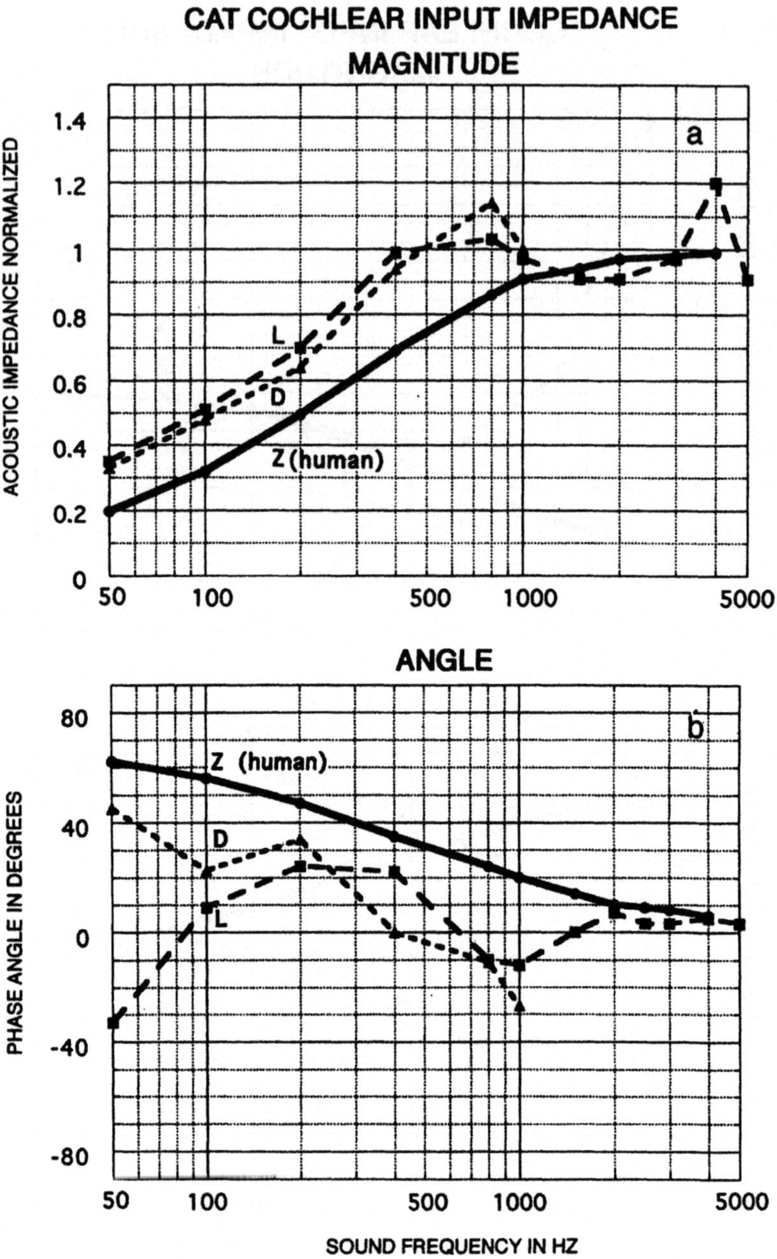

FIG. 4.23. Magnitude and phase angle components of the cochlear input impedance, as calculated from the theory for humans and obtained experimentally on cats. The magnitude data are normalized with respect to the values obtained for high sound frequencies. See text for further explanation.

curves of Fig. 4.23b are roughly consistent with the magnitude curves. They reach 0 angle at a lower frequency than the human curve. In addition, the "L" curve deviates at the lowest frequencies toward negative values, presumably because of the compliance effect of the round window membrane.

On the whole, the experimental results, where artifact free, confirm the theoretical calculations of the cochlear input impedance at low sound frequencies and suggest that the underlying theory is valid. Some of the fine grain of these curves, which is absent in the theoretical curves, may reflect not only experimental artifacts but also wave reflections at the helicotrema, which have been considered in the preceding section of this chapter. There is not enough empirical information to take account of these reflections in the theoretical curves.

What is the physical significance of the decreased cochlear input impedance at low sound frequencies and of its departure from pure resistance toward inertance? A purely resistive input impedance would mean that no sound energy is reflected and all the energy is absorbed in the cochlea. A purely imaginary impedance would mean that no energy is absorbed and all of it is reflected. The phase angle of the cochlear input impedance at low sound frequencies, smaller than half a period but greater than zero, indicates partial wave reflection. Such a reflection is known to take place when the wavelength is not short by comparison to the rate of change of the transmission parameters of a transmission line or a wave guide. In the cochlea, the compliance of the partition varies rapidly with distance. Some variation is also present in the canal cross section.

Effect of the Partition Resistance

As already shown, when the resistance and mass of the cochlear partition are neglected in the calculations, a cochlear wave pattern is obtained according to which the displacement amplitude of the partition increases monotonically all the way to the helicotrema. Such a pattern does not agree with what Békésy (1960) was able to observe. He saw that the amplitude increased up to a maximum at a frequency dependent location and decayed beyond it. I showed in my dissertation (Zwislocki-Moscicki, 1948) that even a small resistance of the cochlear partition was sufficient to produce this kind of a pattern even when the mass was neglected. I now introduce such a resistance, R_{Ba}, and determine its appropriate numerical value. According to preliminary calculations, the value required to match the wave pattern observed by Békésy is much smaller than that derived from his measurements of cochlear transient oscillations. The discrepancy very likely stems from the mechanical coupling among cochlear length elements, dx. Because the cochlear characteristic impedance is essentially resistive, every element, dx, is loaded at both its faces with a resistance. As already calculated the numerical value of such a resistance is relatively high and produces a relatively high damping. The mechanical coupling makes it impossible to determine R_{ba} independently, so that it becomes a so-called free parameter. The theoretician is free to choose its value so as best to fit the data. Fortunately, in the instance of cochlear wave patterns, the choice can be validated because two independent quantitative conditions have to be satisfied. One is the location of the amplitude maximum along the cochlear canal; the other, the sharpness of the maximum and the correlated shape of the amplitude envelope.

Because calculations show post hoc that the absolute value of the resistance component of the cochlear partition impedance is small compared to that of its stiffness component, it can be introduced to a sufficient approximation directly into the solution (equation 4.57) of the simplified cochlear differential equation (equations 4.40 and 4.56). In my original mathematical treatment of the cochlear waves, the resistance was introduced into the differential equation and the equation solved by approximation (Zwislocki-Moscicki, 1948). Both methods lead to a similar result.

The solution, equation 4.57, can be rewritten in the form

$$P(x) = P_0 H_0^2 \left[\frac{2}{\beta} \left(\frac{\rho}{Q} \right)^{1/2} \omega^{1/2} \omega^{1/2} C_{Ba}^{1/2} \right] \qquad 4.75$$

Note that the term $\omega^{1/2} C_{Ba}^{1/2}$ defines the square root of the absolute value of the compliance component of the cochlear partition impedance, so that it is possible to write $\omega C_{Ba} = 1/jZ_c$ and

$$P(x) = P_0 H_0^2 \left[\frac{2}{\beta} \left(\frac{\rho}{Q} \right)^{1/2} \omega^{1/2} \left(\frac{1}{jZ_c} \right)^{1/2} \right] \qquad 4.76$$

By including the resistance component in the impedance term, we obtain

$$Z_{CR} = R_{Ba} + \frac{1}{j\omega C_{Ba}} \qquad 4.77$$

or, after trivial transformation,

$$\frac{1}{jZ_{CR}} = \frac{\omega C_{Ba}}{1 + j\omega C_{Ba} R_{Ba}} \qquad 4.78$$

On the assumption that the resistance component is much smaller than the stiffness component, Z_{CR} can be substituted in equation 4.76 for Z_C. Using equation 4.78, we obtain

$$P(x) = P_0 H_0^2 \left[\frac{2}{\beta} \left(\frac{\rho}{Q} \right)^{1/2} \omega C_{Ba}^{1/2} \left(\frac{1}{1 + j\omega C_{Ba} R_{Ba}} \right)^{1/2} \right] \qquad 4.79$$

The absolute value of the expression $j\omega C_{Ba} R_{Ba}$ being much smaller than unity, the last term on the right side of equation 4.79 can be approximated by $(1 - j\omega C_{Ba} R_{Ba}/2)$, so that equation 4.79 can be rewritten as

$$P(x) = P_0 H_0^2 \left[\frac{2}{\beta} \left(\frac{\rho}{Q} \right)^{1/2} \omega C_{Ba}^{1/2} \left(1 - \frac{1}{2} j\omega C_{Ba} R_{Ba} \right) \right] \qquad 4.80$$

The form of this equation can be simplified by introducing the constant a, as defined by equation 4.58. It becomes

$$P(x) = P_0 H_0^2 \left[a^{1/2} f e^{\frac{\beta x}{2}} \left(1 - \frac{1}{2} j\omega C_{Ba} R_{Ba} \right) \right] \qquad 4.81$$

For numerical values of the argument substantially greater than unity, the Hankel function can be approximated by an exponential function. Then, following the formulation of Jahnke and Emde (1945), we can write

$$P(x) = P_0 \frac{\exp[-a^{1/2} C_{Bao} R_{Bao} \pi f^2 ((\frac{\beta}{2} + \gamma - a)x)]}{[\frac{\pi}{2} a^{1/2} f \exp(\frac{\beta x}{2})]^{1/2}} \exp[(\varphi - a^{1/2} f \exp(\frac{\beta x}{2}))]$$

$$4.82$$

the approximate cochlear wave equation that includes the resistance of the partition. The real part of this equation describes the amplitude, and the imaginary part, the phase. All the constants have already been defined, except the phase constant φ that depends on the initial condition.

The absence of a resistive term in the imaginary part of the equation indicates that the effect of the relatively small partition resistance on the phase is negligible, and the phase relationships depicted in Fig. 4.16 are approximately valid. Only the amplitude of the pressure difference across the partition is calculated here. It is shown in Fig. 4.24 for a resistance $R_{Ba} = 1.2*10^3$ dyne sec/cm^4 that is assumed to be constant along the cochlear length coordinate, x. The amplitude is plotted as a function of the distance, x, from the cochlear input for 6 sound frequencies from 0.2 to 8 kHz. It decays monotonically toward the apex, the faster the higher the frequency.

To obtain the vibration amplitude of the partition, the pressure equation has to be divided by the partition impedance multiplied by $j\omega$. Because the resistance component of the impedance is small compared to the stiffness component, no great error is committed if it is neglected altogether. Then, the impedance expression is reduced to

$$Z_{Ba} = \frac{-j}{\omega C_{Ba}} \qquad 4.83$$

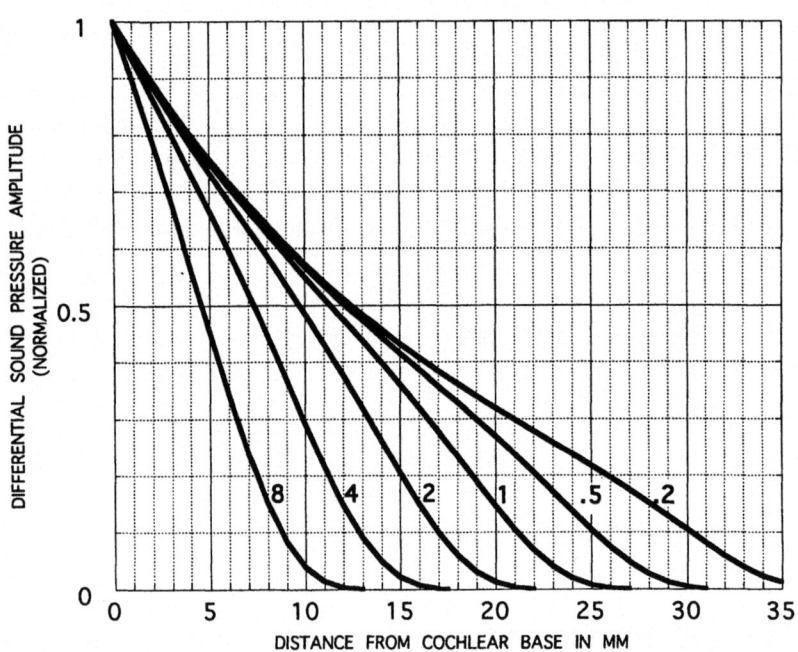

FIG. 4.24. Theoretical differential sound pressure amplitudes across the basilar membrane as functions of the distance from the cochlear base for 6 sound frequencies relative to the amplitude in the cochlear base. The mass loading of the basilar membrane has been neglected.

The expression for the volume displacement of the partition becomes simply a product of the pressure difference and the compliance of the partition,

$$V_B(x) = P(x)C_{Ba} \qquad 4.84$$

The displacement amplitudes calculated according to equation 4.84 are plotted in Fig. 4.25 as functions of cochlear distance for 6 sound frequencies from .2 to 8 kHz. For comparison, a sample of corresponding measurements of Békésy's (Békésy, 1960, Fig. 11–43) is indicated by dots. Since Békésy used a somewhat different frequency grid than used here and measured the amplitudes for a frequency of .8 rather than 1 kHz, his data for this frequency have been translated proportionately along the length axis.

The agreement between Békésy's results and the theoretical calculations must be considered as excellent, especially in view of the measurement diffi-

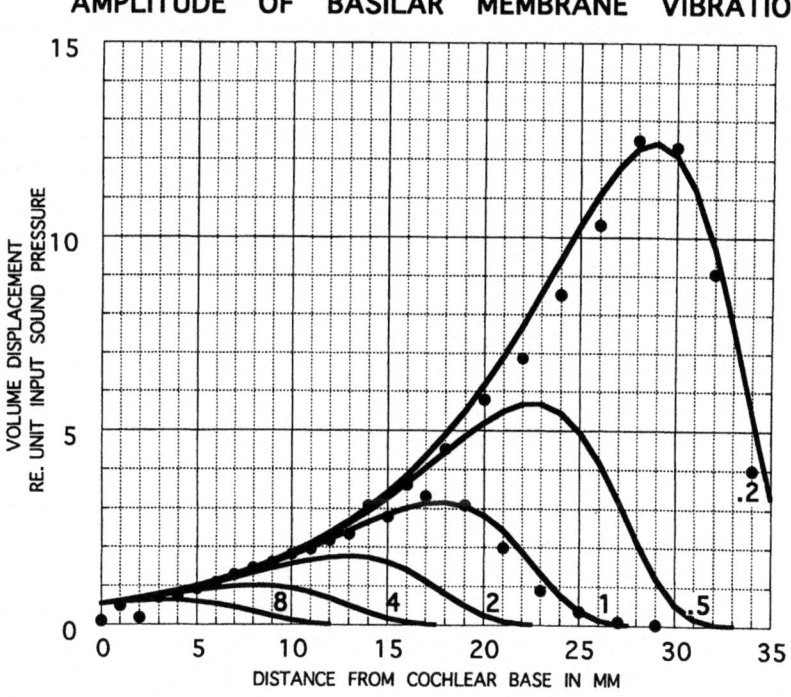

FIG. 4.25. Theoretical amplitudes of basilar membrane displacement as functions of the distance from the cochlear base for 6 sound frequencies. The curves have been calculated on the basis of the differential sound pressure curves of Fig. 4.24. The circles indicate values obtained from Békésy's measurements.

culties and the simplifications made in the theory. The fact that Békésy observed maximum displacement of the partition and volume displacement is used in the theory constitutes a minor correction factor. The agreement indicates that, in Békésy's preparations of postmortem cochleas, the amplitude pattern along the cochlear partition was determined almost entirely by the compliance and the resistance of the partition, the mass having had a negligible effect. I reached this conclusion in my doctoral dissertation of 1948, emphasizing the somewhat surprising discovery that a vibration maximum of the cochlear partition at a frequency dependent location could arise without any resonance, solely through interaction of the locally variable compliance and a more or less constant resistance of the partition. This conclusion flew in the face of the existing dogma that the maximum arose from a mechanical resonance, in which even Békésy seemed to believe. Of course, it met with an almost unanimous skepticism of the scientific community, which, I suspect, still persists in many quarters, although, Dallos used my theory in his book on *The Auditory Periphery* (1973).

How can interaction between the partition compliance and resistance, without any help of the mass, produce a local vibration maximum? The mechanism is rather simple. As the wave is propagated along the cochlea, it encounters an increasing compliance of the partition, which lets its amplitude grow. The larger is the amplitude, however, the faster must the partition move, and the more of the vibration energy is consumed by the partition resistance resulting from friction. Finally, a location is reached at which the effect of the resistance matches, then, outweighs the effect of the compliance—the amplitude of vibration reaches its maximum and begins to decrease.

On the occasion of his visit to my laboratory, Hessel de Vries, a Dutch physics professor, told me how he explained my theory to his students. He compared the mechanism involved to a digestive process in a boa constrictor that swallowed a monkey. As the monkey's body progressed along the boas digestive duct, it encountered increasingly week muscles controlling the snake's girth. As a result, the girth grew. However, the digestive process gradually depleted the monkey's body, so that the snake's girth eventually begun to shrink toward its tail.

A more detailed look at the curves of Fig. 4.25 reveals a systematic tendency for Békésy's results to trace slightly narrower maxima than obtained theoretically. In the next subsection the possibility is investigated that the difference is due to the neglected effect of the mass of the partition.

Effect of the Partition Mass

To test the effect of the partition mass on the cochlear wave pattern observed by Békésy in postmortem preparations, I introduce the mass in equation 4.76 in addition to the resistance. It should be noted that it has been possible to determine the numerical value of the mass independently, as discussed in the first section of this chapter, so that it cannot be adjusted to best fit the data. Its defining equation is, as already stated, $M_{Ba} = M_{Bao}\exp(-\mu x)$, where $M_{Bao} = 0.4$ g/cm³ and $\mu = 0.35$ cm⁻¹. The remaining parameter values are the same as used in the preceding subsection.

The reactance expression for the cochlear partition now becomes

$$Y = -j\frac{1-\omega^2 M_{Ba}C_{Ba}}{\omega C_{Ba}} \qquad 4.85$$

Introduced into equation 4.76 in addition to the resistance, it produces

$$P(x) = P_0 H_0^2 \left[\frac{2}{\beta}\left(\frac{\rho}{Q}\right)^{1/2} \omega^{1/2} \left(\frac{\omega C_{Ba}}{1-\omega^2 M_{Ba}C_{Ba}}\right)^{1/2} \left(\frac{1}{1+j\omega C_{Ba}R_{Ba}}\right)^{1/2}\right] \qquad 4.86$$

Note that, in this and the following two equations, the symbols C_{Ba} and M_{Ba} include the exponential functions of x. On the continued assumption that the mass and resistance terms are considerably smaller than unity and by introducing the constant a, the equation can be simplified to

$$P(x) = P_0 H_0^2 \left[a^{1/2}fe^{\beta x/2}(1+2\pi^2 f^2 M_{Ba}C_{Ba})(1-j\pi f C_{Ba}R_{Ba})\right] \qquad 4.87$$

Its exponential approximation becomes

$$P(x) = P^\circ \frac{\exp\left[-a^{1/2}\pi f^2 C_{Ba}R_{Ba} \; \exp(\beta x/2)(1+2\pi^2 f^2 M_{Ba}C_{Ba})\right]}{\left[a^{1/2}\frac{\pi}{2}f \; \exp(\beta x/2)^{1/2}\right](1+2\pi^2 f^2 M_{Ba}C_{Ba})^{1/2}} \qquad 4.88$$

$$* \exp\left\{j\frac{\pi}{2}\left[\varphi - \frac{2}{\pi}a^{1/2}f \; \exp(\beta x/2)(1+2\pi^2 f^2 M_{Ba}C_{Ba})\right]\right\}$$

With the numerical values given in preceding sections, it generates the differential pressure amplitudes across the cochlear partition plotted in Fig. 4.26 as functions of the cochlear distance, x, for 6 frequencies ranging from 0.2 to 8 kHz. Comparison of these curves with the corresponding curves of Fig. 4.24, calculated without the mass of the partition, demonstrates that the effect of the mass is negligible at low sound frequencies and remains small even at 8 kHz.

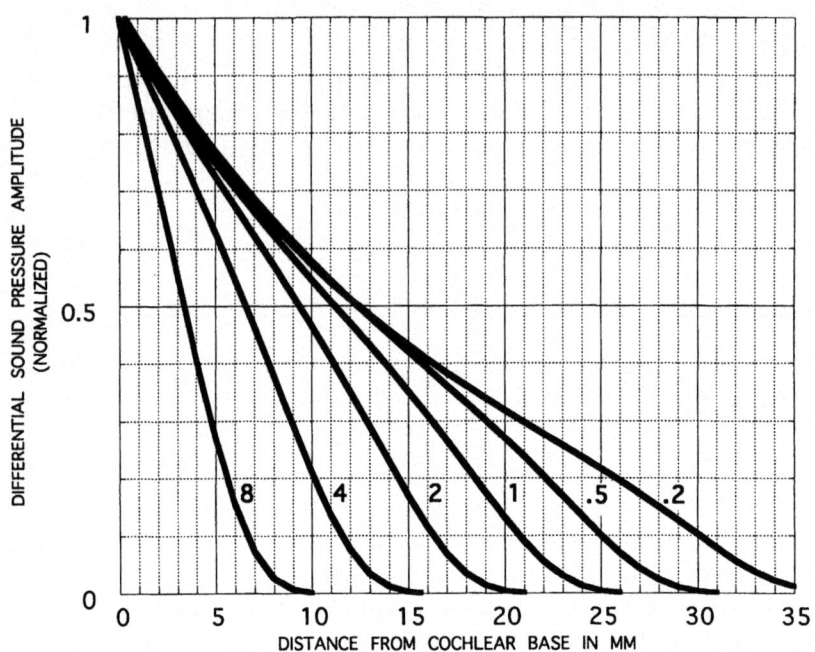

FIG. 4.26. Same as Fig. 4.24, but with the mass loading taken into account.

To obtain the volume displacement amplitudes, the pressure amplitudes have to be divided by the product of the angular frequency, ω, and the impedance of the partition. Because the resistance can be neglected, as already discussed in the preceding subsection, the impedance expression can be reduced to the reactance term defined in equation 4.85. The resulting curves of the displacement amplitude are shown in Fig. 4.27. For comparison, the same results of Békésy's as used in Fig. 4.25 are included. The agreement between the experimental and theoretical data is marginally better than in Fig. 4.25, where the mass effect was omitted. This outcome confirms the conclusion reached in the preceding subsection that the effect of the partition mass on the cochlear amplitude patterns must have been small in Békésy's experiments. Because of the agreement of the data generated in these experiments with the fundamental theory developed here, it seems permissible to extend this conclusion to the cochlea postmortem in general.

To complete the comparisons with Békésy's observations of wave patterns along the cochlea in human postmortem preparations, the phase of

THE COCHLEA SIMPLIFIED BY DEATH   149

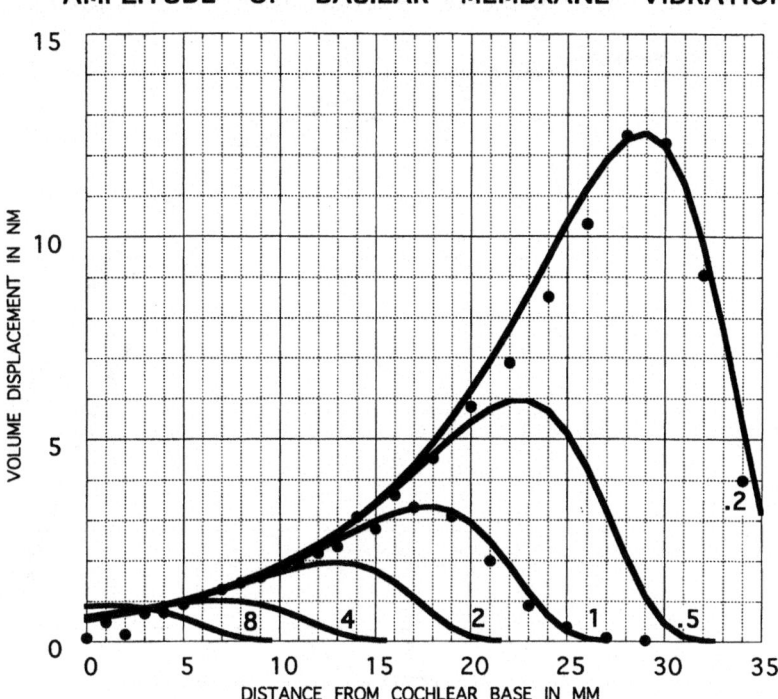

FIG. 4.27.   Analogous to Fig. 4.25, but with the mass loading taken into account.

the partition displacement has been calculated with the help of the phase term of equation 4.87. Probably because of technical difficulties, Békésy measured the phase only at very low sound frequencies. For meaningful comparisons, I take his data for 200 and 300 Hz., the highest frequencies he used. They are shown in Fig. 4.28 by circles and crosses respectively. As before, for reasons already stated, a phase angle of $\pi/2$ has been added to Békésy's original data. The corresponding theoretical results are plotted as the solid and intermittent lines. They indicate phase angles whose absolute numerical values are only slightly larger than those of the phase angles shown in Fig. 4.16, obtained in an earlier section without taking the resistance and mass effects into account. The small difference confirms the earlier conclusion, based on the amplitude results, that the mass of the partition plays a minor role in controlling the wave pattern in postmortem cochleas. It also confirms a purely theoretical corollary according to which the resistance of the partition has only a minor effect on the displacement phase of the partition.

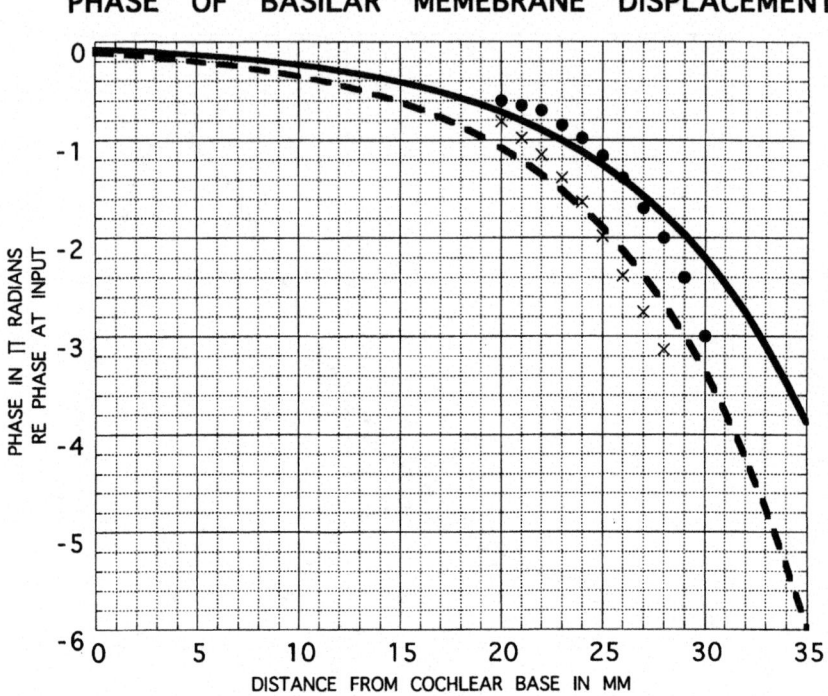

FIG. 4.28. Phase of basilar membrane displacement for two frequencies—200 (solid curve and closed circles) and 300 Hz (intermittent curve and crosses). The curves are theoretical, the points are taken from Békésy's measurements.

The agreement with Békésy's available results is good with respect to the mean overall phases but less so with respect to the slopes of the phase curves. The slope discrepancy is likely to be due to the fact that Békésy's phase data were obtained on a different preparation than the amplitude data used in this chapter. Unfortunately, Békésy did not obtain amplitude data at sufficiently high sound frequencies on the preparation on which he measured the phases to make a comparison with the theoretical results meaningful. In the preparation for which the phase data are available the amplitude peaks were sharper than in the preparation on which the amplitude pattern was measured over an extended frequency range. On theoretical grounds, the sharpness of the peak and the slope of a corresponding phase curve should be correlated.

Perhaps because of technical difficulties, Békésy did not perform his experiments on a sufficient number of preparations to make statistical evaluation and the determination of mean values possible. Because the parameter values used in the theory are kept constant throughout, some differences between the theoretical and empirical results must arise from the variability among individual preparations.

Békésy's results may also be subject to an artifact. To measure the partition vibration in the apical part of the cochlea, he had to open scala vestibuli, increasing in this way the canal depth. It is known in the hydrodynamic theory of surface waves that increasing the canal depth decreases the wavelength of surface waves. This leads to a more rapid change of the vibration phase with distance. The effect of the canal depth is investigated in a separate subsection.

In any event, the secondary discrepancy has no fundamental meaning and certainly does not negate the conclusion that, in the postmortem cochlear preparations, the vibration peak of the partition arises through an interaction between the locally variable compliance and a more or less constant resistance of the partition and not through a resonance effect (Zwislocki-Moscicki, 1948). It should be pointed out in this connection that Békésy's phase measurements indicate wave propagation well beyond the location of maximum vibration (Békésy, 1947, 1960). According to the differential equation of the cochlea (equation 4.38) and the general wave theory, well established in physics, this excludes the possibility of the partition resonance coinciding with its maximum vibration, as has been already stated in connection with the differential equation.

Cochlear Transfer Functions

Transfer functions of a system are obtained when the magnitude and phase of a transmitted sinusoidal signal are measured at its output as a function of oscillation frequency while these variables are kept constant at the input. Applying the definition to the cochlea, one can determine the magnitude and phase of the partition vibration at one location as a function of sound frequency for a constant magnitude and phase of the stapes vibration. Békésy made such measurements (Békésy, 1943) and I accounted for them theoretically (Zwislocki-Moscicki, 1948). Békésy and Rosenblith (1951) reproduced the comparison between the two sets of results in graphic form in their chapter in S. S. Stevens's Handbook of Experimental Psychology. Here, the comparison between the empirical and theoretical results is repeated on the basis of somewhat more exact calculations. Békésy's measurements did not include the phase, so that the calculations performed here are limited to the magnitude.

The necessary equation can be derived from the real part of equation 4.88 by dividing it by the product of angular frequency, ω, and the partition impedance defined by equation 4.85, where the partition resistance is neglected in agreement with preceding calculations. Taking advantage of the prior determination that the mass is relatively small, the inverse impedance, as multiplied by ω, can be approximated by

$$1/j\omega Y = C_{Ba}\left(1 + \omega^2 M_{Ba} C_{Ba}\right) \qquad 4.89$$

According to equations 4.88 and 4.89, the amplitude of the volume displacement of the basilar membrane becomes

$$V(x) = P_o C_{Ba}\left(1 + 4\pi^2 M_{Ba} C_{Ba} f^2\right)$$

$$\frac{\exp\left[-a^{1/2} \pi f^2 C_{Ba} R_{Ba} \quad \exp(\beta x/2)\left(1 + 2\pi^2 f^2 M_{Ba} C_{Ba}\right)\right]}{\left[a^{1/2}\frac{\pi}{2}f \quad \exp(\beta x/2)\right]^{1/2} \left(1 + 2\pi^2 f^2 M_{Ba} C_{Ba}\right)^{1/2}}$$

$$4.90$$

for a constant sound pressure at the cochlear input. When the stapes displacement is kept constant, the sound pressure changes with sound frequency, as has been already explained. It becomes directly proportional to the product of sound frequency and the cochlear input impedance. The term, P_O, in equation 4.90 has to be replaced by

$$P_o(f) = 2\pi f Z_{Ca}(f) D_{St} \qquad 4.91$$

the symbol D_{St} standing for the constant amplitude of stapes vibration. The transfer functions determined by equation 4.90, as modified by equation 4.91, are plotted by means of solid lines in Fig. 4.29. The dashed lines indicate the corresponding results of Békésy (1943, 1960). The general pattern of both sets of curves is similar, and it should be pointed out that both sets were obtained at approximately the same locations along the cochlea, every curve in a set corresponding to a different one. However, Békésy's data appear to reflect considerably sharper filter functions than theoretically calculated. What may be the reasons for the difference? Is it fundamental or contingent upon numerical differences?

One possible factor may result from individual differences among cochlear specimens, as already mentioned. It appears that the specimen that furnished the amplitude distributions as functions of the distance from the cochlear input had broader filter functions than the specimen that furnished the transfer functions. Another factor may result from rather gross theoretical approximations. A more detailed consideration of the location

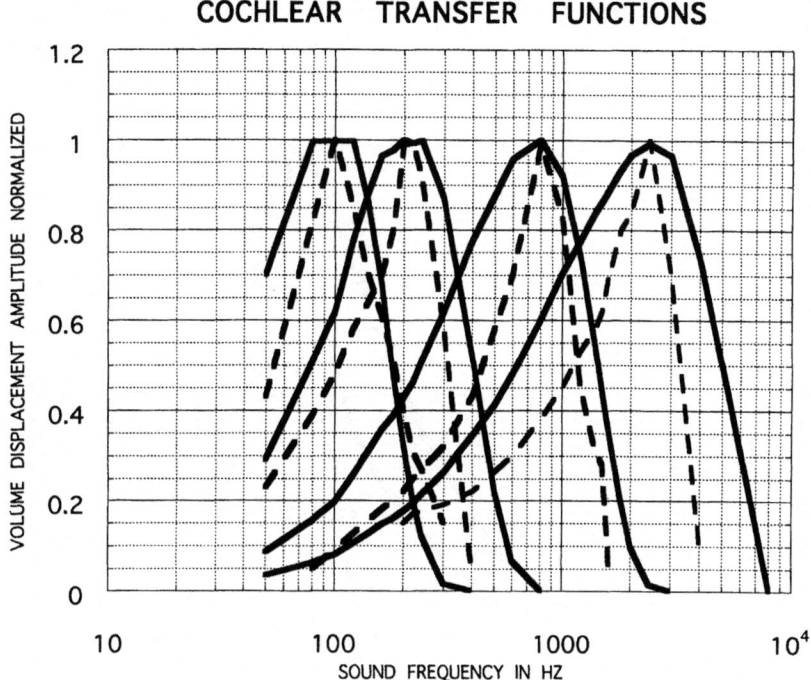

FIG. 4.29. Transfer functions of basilar membrane displacement at several cochlear locations. The solid curves are theoretical (mass load of basilar membrane neglected). The intermittent curves follow Békésy's measurements.

of maximum vibration of the cochlear partition as a function of sound frequency, as described by Békésy (1942, 1960), suggests that the resistance of the partition (R_{Ba} in the theory) did not remain entirely constant along the cochlea, as has been assumed in the theory, but decreased somewhat toward the cochlear input in the apical portion of the cochlea to reach a constant value below the location of the 400 Hz maximum vibration. The numerical value for R_{Ba} of $1.2*10^3$ dyne sec/cm^4 assumed in the theory produced good agreement with Békésy's results concerning the location of the vibration maximum and the amplitude distribution along the cochlea at 200 Hz but may have shifted the location of maximum vibration at higher sound frequencies by about 3 mm toward the cochlear base. Numerical calculations have shown that the shift can be eliminated by making the resistance four times smaller. A transfer function obtained with such a smaller resistance at the location of maximum response for 800 Hz, as determined by Békésy, is shown in Fig. 4.30 by the short dashes. For compar-

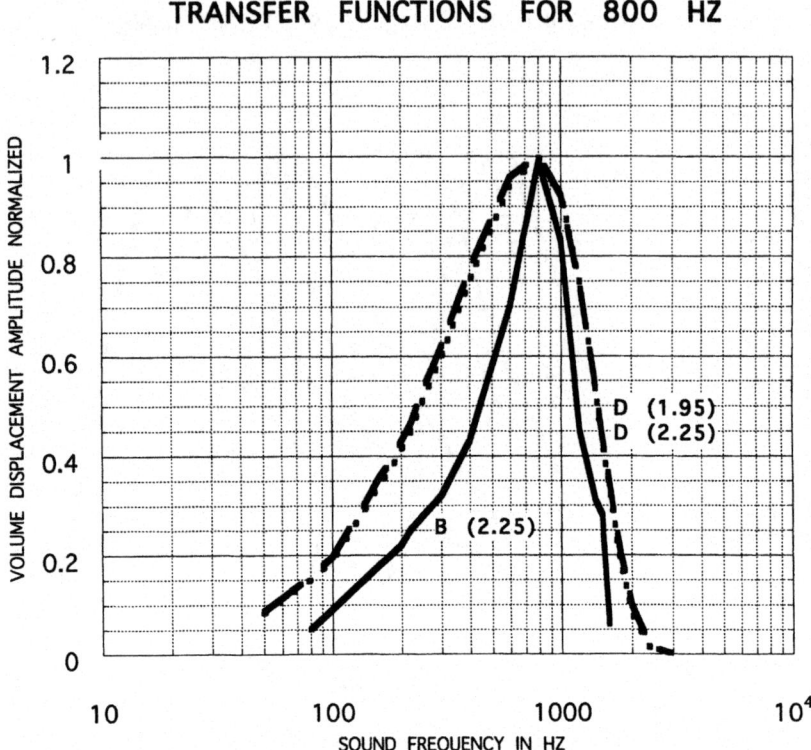

FIG. 4.30. Effect of the basilar membrane resistance on the transfer function of basilar membrane displacement at the location of the 800 Hz displacement maximum. The curve indicated by the long dashes has been reproduced from Fig. 4.29, the curve with the short dashes has been obtained by decreasing the resistance by a factor of 4, the solid curve follows Békésy's measurements.

ison, the curve obtained with the higher resistance is reproduced from Fig. 4.29 by means of long dashes. Békésy's results are also reproduced by the solid curve from that figure. Clearly, the decreased resistance, which made the location of the maximum vibration coincide with that determined by Békésy, did not appreciably affect the filter width of the theoretical transfer function. The reason for this perhaps surprising result is that the decreased resistance shifted the maximum to a cochlear location with a smaller stiffness of the cochlear partition, so that the ratio between this stiffness and the resistance remained constant.

Another possible factor is that the mass of the cochlear partition was contributed not only by the cell mass of the organ of Corti and the tectorial membrane but also by the endolymph of scala media. In the description of his experiments, Békésy makes a remark that can be interpreted to mean that its viscosity was increased to the point where its whole mass vibrated together with the partition. This was assumed in my doctoral dissertation (Zwislocki-Moscicki, 1948). Accordingly, the transfer function of Fig. 4.29 with the maximum at 2400 Hz was recalculated with a mass of the partition increased by a factor of three and also a resistance decreased by a factor of four. The relatively high frequency was chosen to maximize the effect of the mass. The results are compared to the curve reproduced from Fig. 4.29 in Fig. 4.31. This curve is indicated by the solid line. The intermittent curve with the long dashes shows the result of the decreased resistance, that with the short dashes, the cumulative effect of the decreased resistance and in-

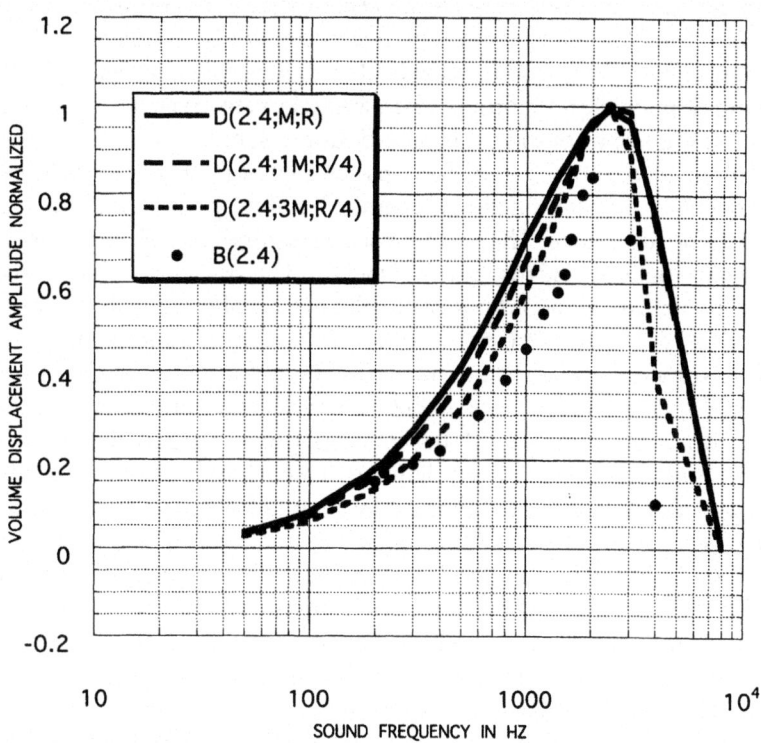

FIG. 4.31. Effects of basilar membrane resistance and mass load on the transfer function of basilar membrane displacement at the location of 2,400 Hz displacement maximum. The solid curve has been reproduced from Fig. 4.29. For the curve with the long dashes, the resistance has been decreased by 4, for the one with short dashes, the mass load has been increased by 3, in addition. The closed circles follow Békésy's measurements.

creased mass. As in Fig. 4.30, the effect of the decreased resistance is small, and the effect of the increased mass is substantially greater, especially at the high frequency cut off. The curve corresponding to the increased mass approaches more closely Békésy's results indicated by the closed circles but is still far from coincidence. Because a further increase of the mass would certainly not be realistic, no reasonable adjustment of the partition parameters appear to be capable of duplicating the relatively narrow cochlear filter widths obtained by Békésy—an experimental artifact must be suspected.

Such a possible artifact may have resulted from opening the cochlear wall for the observation of the partition, which increased effectively the canal depth, so that the waves on the partition ceased to be long by comparison to the canal depth. It is shown in the next subsection theoretically that the greater canal depth should have led to an enhanced maximum of the partition vibration. Here, it should be pointed out that the theoretically calculated transfer functions have the same character as Békésy's, and the difference in the filter width cannot be accounted for by a difference in the resonance effect. Such a difference would produce filter characteristics that have simultaneously both broader peaks and less steep skirts. According to Fig. 4.29, the slopes of the skirts are about the same.

There is still another way of demonstrating that the filter characteristics of the cochlear transfer functions obtained by Békésy could not have been produced by resonance. The way was shown serendipitously by a budding controversy between Békésy and H. Davis and his associates. On the occasion of their classical recordings of cochlear microphonics with differential electrodes, the latter did not see cochlear transfer functions having bandpass characteristics Békésy had found but rather, lowpass characteristics (Tasaki, Davis, & Legouix, 1952). Since their experiments were conducted on live animals and the microphonics were believed to be proportional to the vibration amplitude of the cochlear partition, the difference gave immediate rise to the speculation that Békésy's results obtained postmortem may not be valid in vivo. The controversy was short lived, however. Looking at the methodology employed by Davis and his associates, I discovered that they referred their measurements to a constant response amplitude in the cochlear base, whereas Békésy kept the displacement amplitude of the stapes constant. According to my theory (Zwislocki-Moscicki, 1948), in the presence of a constant displacement amplitude of the stapes, the displacement amplitude of the cochlear partition had to increase with sound frequency according to the product of sound frequency and the cochlear input impedance, as has been already explained. Calculation of the transfer functions under the conditions of Davis and his associates' produced lowpass characteristics agreeing with their results (Zwislocki, 1955).

The condition of a constant displacement amplitude of the cochlear partition (constant cochlear microphonic) in the cochlear base can be realized theoretically with the help of equation 4.89 by first calculating the displacement amplitude of the partition at any desired cochlear location, then, doing the same for the cochlear base, both for a constant sound pressure in the cochlear base and, subsequently, dividing the first result by the second. The outcome is shown in Fig. 4.32 by the intermittent curve. It is clearly consistent with a lowpass characteristic obtained by Davis and his associates. The solid curve, shown for comparison, has been obtained for the same cochlear location and identical parameter values under Békésy's experimental conditions. It is the same curve as the curve with the peak at 2400 Hz, in Fig. 4.29. It has bandpass characteristics consistent with those observed by Békésy. Thus, the theoretical duplication of the relevant methodological difference between Békésy's and Davis and associates' experi-

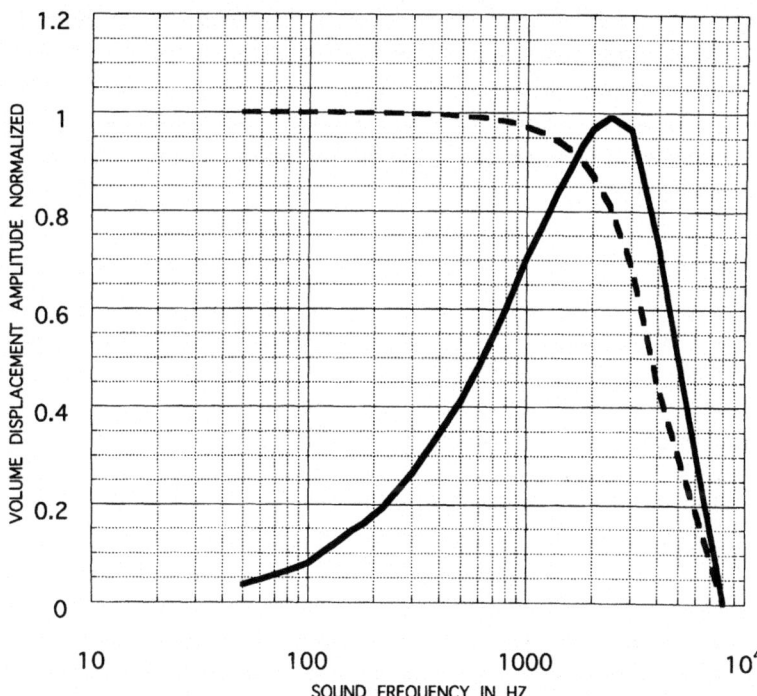

FIG. 4.32. Theoretical transfer functions of basilar membrane displacement at the location of the 2,400 Hz displacement maximum for two input conditions. For the solid curve, the displacement amplitude of the stapes has been kept constant, for the intermittent curve, that of the basilar membrane at the cochlear base.

ments completely accounts for the difference in their results. This means that, instead of disagreeing, Davis's group confirmed for live animals Békésy's observations made on postmortem preparations. They also confirmed the theoretically derived mechanism for the local vibration maximum in the cochlea. If the theory had been wrong, duplication of the different experimental conditions used by Békésy and the Davis group would not have been able to account for the difference in their experimental results. In particular, no lowpass characteristic would be obtained under the conditions used by the Davis group. If Békésy's bandpass characteristics had been due to a partition resonance, these characteristics would have been preserved under the conditions of the Davis group. This follows from equation 4.84 that describes the acoustic reactance of the partition. At resonance, the reactance becomes zero, producing a minimum in the partition impedance. Because the amplitude of the partition vibration is inversely related to the impedance, it must reach a maximum. Therefore, juxtaposition of the results obtained by Békésy and the Davis group furnishes added evidence for the theoretical conclusion that the local maximum of the cochlear partition's vibration was due to an interaction between the partition compliance and resistance and not to a resonance. The conclusion certainly applies to postmortem preparations but, as seems to follow from the Davis' group results, may also be true for live cochleas. It is explained in chapter 5, however, that the latter appears to occur only in deteriorated preparations.

Effect of Canal Depth

The analysis of the cochlear waves described in the preceding subsections is based on the assumption that the waves are appreciably longer than the depths of the cochlear canals, scalae vestibuli and tympani. As successful comparisons with Békésy's empirical results demonstrate, this assumption has been justified at least for postmortem preparations used by him. However, some numerical differences have become apparent with respect to the phase of basilar membrane vibration and the sharpness of the associated magnitude transfer functions. It has been mentioned that the departures may be due in part to Békésy's experimental method requiring opening one of the scalae for visualization of the cochlear partition. The intervation must have produced an effectively infinite canal depth, invalidating the assumption of long waves. In addition, this assumption is tenuous for cochlear regions where the vibration maximum occurs at high sound frequencies and is relatively sharp, suggesting diminished damping. This is particularly true in live cochleae, as discussed in the next two chapters. For all these reasons, it appears neces-

THE COCHLEA SIMPLIFIED BY DEATH   159

sary to be aware of the main effects produced by a relatively deep canal by comparison to a shallow one.

The basic hydrodynamics of long and short surface waves by comparison to canal depth are well known (e.g., Lamb, 1945; Lindsay, 1960; Lighthill, 1978, 1981) and only their specific applications to the cochlea are of interest here. The classical difference between the two kinds of waves resides in their propagation velocities. Whereas the velocity of long waves is independent of the wavelength and the associated oscillation frequency, that of the short waves decreases as the wavelength decreases. In the cochlea, even the long waves can have a variable propagation velocity. This occurs in regions where the mass of the cochlear partition makes itself felt. Its effect on the partition impedance, and through it, on the wave propagation velocity depends on sound frequency. The greater the frequency, the greater is the effect and the smaller the propagation velocity. The effect is added to that of the canal depth.

No attempt is made here at calculating explicitly the cochlear wave pattern for short or intermediate wavelengths. This was done successfully by others, at least in principle, on several occasions (e.g., Siebert, 1974; Steele and Taber, 1979; Viergever, 1980; Lighthill, 1981). Instead, I describe two phenomena associated with short waves, which were predicted mathematically in a preceding article (Zwislocki, 1983). They are germane to the understanding of cochlear phenomenology and to the comparison of empirical and theoretical data.

The analysis begins with the cochlear differential equation, equation 4.39, where the geometrically determined cross sectional area of the cochlear canals, $Q = Q(x)$, is replaced by an effective area, $q = q(x,f)$, that depends on sound frequency and is based on a concept introduced earlier (Zwislocki, 1953). In the presence of long waves, the fluid velocity parallel to the surface is uniformly distributed over the whole cross section of the canal, and the total volume flow is equal to this velocity multiplied by the cross sectional area. In the presence of short waves, on the other hand, the velocity decreases as the distance from the surface increases. As a consequence, the total flow must be calculated by integrating the variable velocity over the area. Under these conditions, an effective cross sectional area can be defined as (Zwislocki, 1983)

$$q(x,f) = \frac{1}{U_B(x,f)} \int_{Q(x)} U(x, y, z, f) dy dz \qquad 4.92$$

where $U_B(x,f)$ denotes the velocity at the cochlear partition in the x direction. With the effective area so defined, the cochlear differential equation can be expressed as

$$\frac{\partial^2 P}{\partial x^2} = \frac{j\omega\rho}{qZ_{Ba}} P \qquad 4.93$$

As has already been stated, wave propagation requires a negative partition impedance, Z_{Ba}, so that the compliance component must be dominant. It has also been shown that, in the region of substantial wave amplitude, certainly up to the vibration maximum, the resistance component must be small. As a consequence, the impedance term in equation 4.93 can be approximated by its reactance component defined in equation 4.85. With this approximation, and a simple transformation, equation 4.93 can be rewritten as

$$\frac{\partial^2 P}{\partial x^2} = -\omega^2 \frac{\rho C_{Ba}}{q(1-\omega^2 M_{Ba}C_{Ba})} P \qquad 4.94$$

According to the wave theory, provided all the parameters vary relatively slowly with x, the fraction on the right side of equation 4.94 approximates the inverse wave propagation velocity squared, so that the velocity can be expressed approximately as

$$c = \left(\frac{q(1-\omega^2 M_{Ba}C_{Ba})}{\rho C_{Ba}} \right)^{1/2} \qquad 4.95$$

Clearly, the velocity must decrease strongly as sound frequency increases in the cochlear region where the term $\omega^2 M_{Ba}C_{Ba}$ gradually approaches unity because of $\omega^2 = (2\pi f)^2$. As is shown below, the effect is enhanced when, in the presence of short waves, the effective cross sectional area decreases as sound frequency increases.

For a demonstration, it is sufficient to approximate the cross section of the cochlear canals by a rectangle of width b(x) divided in two by a partition stretching across the whole width and vibrating with a uniform velocity (Viergever, 1980). Under such conditions the fluid velocity in the x direction, parallel to the partition, is also uniformly distributed over the whole width. In agreement with the theory of surface waves (e.g., Lamb, 1945), it decays approximately exponentially as the distance from the partition increases when the waves are short enough. The approximation is satisfactory already when the canal depth is equal to half the wavelength. Empirical evidence indicates that this can be realistic in the cochlea in the vicinity of the vibration maximum.

The exponential decay function can be written in the form

THE COCHLEA SIMPLIFIED BY DEATH   161

$$U(x, y, \omega) = U_B e^{-k(x, \omega)y} \qquad 4.96$$

where $k(x, \omega) = 2\pi/\lambda(x, \omega) = \omega/c(x, \omega)$ and λ denotes the wave length. Introduced into equation 4.92, this expression generates the desired effective cross sectional area

$$q(x, \omega) = \frac{b(x)}{k(x, \omega)} = \left(\frac{b(x)c(x, \omega)}{\omega}\right) \qquad 4.97$$

Note that it depends on the wavelength contained in the parameter k.

With the effective cross-sectional area according to equation 4.97, the wave propagation velocity specified by equation 4.95 becomes

$$c = \frac{b(1 - \omega^2 M_{Ba} C_{Ba})}{\omega \rho C_{Ba}} \qquad 4.98$$

It should be pointed out that, according to equation 4.98, the propagation velocity can decrease as sound frequency increases even in the cochlear regions where the mass of the partition, M_{Ba}, is negligible but this, of course, only in the presence of short waves. Such waves are unlikely to occur there, however. On the other hand, they are likely to occur where the mass effect is substantial. Here the velocity decays much more rapidly as a function of increasing sound frequency than in the presence of the long waves, as is indicated by comparing equations 4.95 and 4.98.

The effect of the short waves on the velocity of wave propagation and, therefore, on the response phase, may underlie the difference between the phase values calculated on the assumption of long waves and those obtained by Békésy, as shown in Figs. 4.16 and 4.28. This is especially true since Békésy had to open one of the cochlear scalae, in this way increasing the canal depth.

A most interesting relationship is revealed when the cochlear characteristic impedance is determined for the short waves. As has been pointed out in a preceding subsection, a simple relationship exists between the impedance and the wave propagation velocity. According to the theory of transmission lines, as applied to the cochlea, it can be described by the ratio

$$\frac{Z_{Ca}(x, \omega)}{c(x, \omega)} = \frac{\rho}{q(x, \omega)} \qquad 4.99$$

When the effective cross sectional area, $q(x,\omega)$, is specified according to equation 4.97, the ratio becomes

$$\frac{Z_{Ca}(x,\omega)}{c(x,\omega)} = \frac{\omega\rho}{b(x)\,c(x,\omega)} \qquad 4.100a$$

After obvious simplification, it produces

$$Z_{Ca}(x,\omega) = Z_{CS}(x,\omega) = \frac{\rho\omega}{b(x)} \qquad 4.100b$$

This equation indicates that, for short waves, the cochlear characteristic impedance, Z_{Ca}, becomes independent of the locally variable impedance of the cochlear partition. The apparent dependence on the weakly variable canal width is only fictitious and results from the assumption of a rectangular canal cross section. In reality, the cross section is more nearly semicircular, and it should be intuitively clear that, in the presence of such a cross section, even this factor is minimized.

The locally invariant characteristic impedance is crucial for a reflectionless wave propagation. It minimizes wave reflections near the vibration maximum, close to the region of the most variable partition impedance, where short waves are the most likely to occur. This phenomenon seems to have been overlooked prior to my article in 1983, as witnessed for example by the article of Zweig, Lipes, and Pierce (1976) entitled "The cochlear compromise." The authors attempted to calculate the minimum damping allowed for acceptable wave reflections based on the long wave approximation. According to the derivations reproduced here, the compromise is not necessary.

Another effect of the short waves is to enhance the amplitude of the partition vibration. In the absence of empirical data, the effect can be demonstrated theoretically. This may be done in the following simple way. Borrowing from the theory of transmission lines on which my theory has been based, we note that the ratio between the volume velocity of the partition per unit length and the volume velocity along the cochlear canal is equal to the ratio between the cochlear characteristic impedance and the impedance of the partition, also taken per unit length. In symbols

$$\frac{j\omega V_B(x,\omega)}{q(x,\omega)U_B(x,\omega)} = \frac{Z_{Ca}(x,\omega)}{Z_{Ba}(x,\omega)} \qquad 4.101$$

To evaluate the left hand ratio, the characteristic impedances Z_{Ca} and Z_{Ba} have to be specified. By combining equations 4.95 and 4.99, it is possible to obtain a general expression for Z_{Ca}, which becomes an expression for long waves when the cross section $q(x,\omega)$ is replaced by $Q(x)$

$$Z_{CL} = \left(\frac{\rho(1 - \omega^2 M_{Ba} C_{Ba})}{Q C_{Ba}} \right)^{1/2} \qquad 4.102$$

Note that all the parameters on the right side, except ρ, depend on x. For short waves, the characteristic impedance is given directly by equation 4.100b. The partition impedance has been specified in equation 4.85 in terms of its reactance. With these impedance definitions, the volume velocity ratios according to equation 4.101 become

$$\frac{j\omega V_{BL}}{QU_{BL}} = j\omega \left(\frac{\rho C_{Ba}}{Q(1 - \omega^2 M_{Ba} C_{Ba})} \right)^{1/2} \qquad 4.103a$$

for the long waves, and

$$\frac{j\omega V_{BS}}{qU_{BS}} = j\omega \frac{\omega \rho C_{Ba}}{b(1 - \omega^2 M_{Ba} C_{Ba})} \qquad 4.103b$$

for the short ones. The ratio between the two transversal volume displacements can be calculated from the latter two equations for constant volume velocities parallel to the basilar membrane and can be expressed as

$$\frac{V_{BS}}{V_{BL}} = \frac{\omega}{b} \left(\frac{Q \rho C_{Ba}}{1 - \omega^2 M_{Ba} C_{Ba}} \right)^{1/2} \qquad 4.104$$

clearly, the ratio increases with both the sound frequency and the distance along the cochlea, especially in the region where the mass effect makes itself felt, and the product $\omega^2 M_{Ba} C_{Ba}$ approaches unity. Thus, appearance of short waves in the vicinity of the vibration maximum of the basilar membrane should be expected to enhance the maximum.

In connection with the calculation of the cochlear transfer functions, it has become evident that the peak of the transfer functions measured by Békésy tends to be more pronounced than the calculated one on the basis of the long wave approximation. The short wave effect derived here may provide a partial explanation of the difference, especially, since Békésy had to open one of the cochlear scalae, increasing its depth.

Comparison of the volume velocities of the basilar membrane obtained with the long and short waves on the basis of a constant volume velocity along the basilar membrane may not be the most pertinent for the cochlear functioning. Because the short waves increase the cochlear characteristic impedance, they may be expected to decrease the volume velocity along the cochlea. It may be more relevant to calculate the short wave effect on the basilar membrane vibration amplitude on the basis of constant input power. According to the transmission line theory, if the fricative losses are neglected, the power transmitted along the cochlea can be expressed as

$$\Pi = Z_{Ca} q^2(x,\omega) U_B^2(x,\omega) \qquad 4.105$$

Accordingly,

$$q(x,\omega) Q_B(x,\omega) = \left(\frac{\Pi}{Z_{Ca}} \right)^{1/2} \qquad 4.106$$

By introducing this equation into equations 4.103a and 4.103b, respectively, we obtain for the long and short waves in turn

$$j\omega V_{BL} = j\omega \Pi^{1/2} \left(\frac{\rho}{Q} \right)^{1/4} \left(\frac{C_{Ba}}{1 - \omega^2 M_{Ba} C_{Ba}} \right)^{3/4} \qquad 4.107a$$

$$j\omega V_{BS} = j\omega^{3/2} \Pi^{1/2} \left(\frac{\rho}{b} \right)^{1/2} \left(\frac{C_{Ba}}{1 - \omega^2 M_{Ba} C_{Ba}} \right) \qquad 4.107b$$

The ratio of the corresponding volume displacements becomes

$$\frac{V_S}{V_L} = \omega^{1/2} \left(\frac{Q}{b^2} \right)^{1/4} \left(\frac{\rho C_{Ba}}{1 - \omega^2 M_{Ba} C_{Ba}} \right)^{1/4} \qquad 4.108$$

Clearly, the short waves still enhance the amplitude of basilar membrane vibration, but much less than for constant volume flow along the basilar membrane. The amplitude grows somewhat faster than for the long waves with both the sound frequency and the distance along the cochlea. Remember that all the parameters on the right side of the equation are functions of the distance, x. The compliance C_{Ba} grows especially rapidly with it.

Concluding this subsection, it should be pointed out that the significance of the derived short wave effects goes well beyond possibly accounting for the residual differences between Békésy's measurements and the theoretical results described here. They must be taken into consideration in the following chapter in which wave propagation in the live cochlea is considered.

Békésy's Paradoxical Waves

Normally, sound is transmitted to the cochlea through the air in the outer ear, then, the middle ear and the stapes—the so-called *air conduction* in the otological or audiological parlance. It is well known, however, that vibration delivered to the skull bones can also be heard—a phenomenon called *bone conduction*. To be heard, the sound has to be transmitted to the cochlea. Part of this transmission occurs through induced vibration of the ossicles in the middle ear but another part, through compression of the inner ear cavities. Békésy was struck by the observation that, independent of how the sound was delivered to the inner ear, it sounded the same and by the discovery of being able to compensate audible bone conducted sound by air conducted sound so that the combination became inaudible. For these phenomena to happen, the sound had to produce the same vibration pattern on the basilar membrane, whether transmitted through the air or bone conduction. He was able to confirm this conclusion in experiments on mechanical models resembling the cochlea both geometrically and physically, except that the cochlear spiral was straightened out. He found that, no matter at what location sound was delivered to the model's fluid, it generated the same pattern of vibration on the model basilar membrane. The waves were always propagated from the end simulating the cochlear base toward the end simulating the cochlear apex. When the sound source was located near the apical part, the waves appeared to run paradoxically toward it—hence, paradoxical waves.

The discovery that bone conducted sound could be cancelled by air conducted sound gave rise to various speculations concerning the mode of sound transmission in the cochlea. Wever and Lawrence (1952) recorded cochlear microphonics in domestic cats while delivering sound simultaneously to the cochlear apex and through the stapes, as in air conduction. By adjusting the latter in amplitude and phase, they were able to cancel the effect of the former so that no cochlear microphonics were produced. They did not think that such a cancellation could have happened in the presence of traveling waves on the basilar membrane, assuming that the direction of travel would have to depend on the location of the sound source. They ascribed Békésy's findings of traveling waves to an optical illusion. A variant of this experiment was performed by Tasaki, Davis, and Legouix (1952) on guinea pigs. They measured cochlear microphonics at specific locations along the cochlea while delivering sound either through the stapes or an opening in the third cochlear turn. They found their amplitude and phase results to be consistent with traveling waves but also to be independent of the location of the sound source. To reconcile these apparently paradoxical results with each other, the authors concluded that they could not have been produced by true traveling waves and suggested that a traveling wave illusion was created by responses of locally variable elements of the cochlear partition exposed to a uniformly distributed sound pressure.

The incipient controversy surrounding cochlear traveling waves lasted only one year and was nipped in the bud through application of some well-known physical principles and simple mathematics (Zwislocki, 1953). On the basis of the evidence already discussed that the cochlear fluids can be regarded as practically incompressible relative to the compliance of the cochlear windows and the principle of fluid continuity it is possible to derive two equations stating that the total volume of each of the cochlear scalae must be conserved.

$$-\Delta V_L + \Delta V_{St} + \int_0^l V_B dx = 0 \qquad 4.109a$$

$$-\Delta V_T + \Delta V_R + \int_0^l V_B dx = 0 \qquad 4.109b$$

The symbols have the following meanings: $-\Delta V_L$ and $-\Delta V_T$—compression of the inner ear labyrinth, except scala tympani, and of scala tympani, respectively; ΔV_{St} and ΔV_R—volume displacements of the stapes and of the round window membrane; V—volume displacement of the basilar membrane per unit length; and l—the length of the basilar membrane. The minus sign refers to compression; the plus sign, to expansion.

For sinusoidal compressions of the inner ear, the volume displacement of the stapes or the round window membrane must be related to the sound pressure causing the displacement by the respective impedance, so that

$$P_{St} = j\omega Z_{St} \Delta V_{St} \qquad 4.110a$$

$$P_R = j\omega Z_R \Delta V_R \qquad 4.110b$$

The difference between the two pressure variables amounts to

$$P_0 = P_{St} - P_R = j\omega (Z_{St} \Delta V_{St} - Z_R \Delta V_R) \qquad 4.111$$

It establishes the pressure differential across the basilar membrane in the cochlear base. By combining appropriately the latter equation with equations 4.109a and 4.109b, it is possible to replace the volume displacements of the stapes and the round window membrane by the terms expressing the compression of the inner ear

$$P_0 = j\omega \left[Z_{St} \Delta V_L - Z_R \Delta V_T - (Z_{St} + Z_R) \int_0^1 V_B dx \right] \qquad 4.112$$

Within the limits of linearity, the displacement of the basilar membrane must be directly proportional to the pressure gradient across it. Accordingly, it is possible to write

$$V_B = P_0 X \qquad 4.113$$

where X is the displacement of the basilar membrane per unit pressure. With this definition, equation 4.113 can be transformed as follows

$$P_0 = j\omega \frac{Z_{St} \Delta V_L - Z_R \Delta V_T}{1 + j\omega(Z_{St} + Z_R) \int_0^1 X dx} \qquad 4.114$$

In agreement with equation 4.114, compression of the inner ear will always produce a pressure gradient across the basilar membrane in the cochlear base, except when

$$Z_R \Delta V_T = Z_{St} \Delta V_L \qquad 4.115$$

Occurrence of such an equality is very unlikely, however, because the two sets of variables, Z_{St}, Z_R, and ΔV_L, ΔV_T are independent of each other.

According to the equations derived in preceding subsections, generation of a pressure differential across the basilar membrane in the cochlear base gives rise to traveling waves running toward the cochlear apex, in agreement with Békésy's observations. Equation 4.114 says that this must also happen when, instead of being delivered to the stapes, the sound is transmitted to the bone surrounding the cochlea. It should be clear moreover that a similar situation arises when the sound is transmitted through an opening in scala vestibuli near its apical end. Such sound transmission amounts in effect to a periodic change in the volume of scala vestibuli. Since the waves must still travel from the cochlear base toward the apex, as follows from the sound pressure generation across the basilar membrane according to equation 4.114, an apparently paradoxical direction of wave travel results.

It should be clear from these considerations, and especially equation 4.114, that neither Wever and Lawrence's nor Tasaki, Davis and Legouix's experiments succeeded in producing any valid evidence against the existence of traveling waves in the cochlea. Their sound delivery through an artificial opening near the cochlear apex simply produced the same mode of cochlear wave propagation as sound delivery through the stapes. This conclusion is consistent with all of their results.

It should also be realized that equation 4.114 is satisfied independent of the pattern of basilar membrane vibration, as defined by function X. Reciprocally, the pattern must be independent of the way the pressure difference across the basilar membrane, P_O, is generated. In other words, the same pattern occurs independent of whether the pressure difference results from direct driving of the stapes or through bone conduction, as was found by Tasaki et al. (1952).

A caveat concerning generation of the paradoxical waves must be mentioned here. Such waves occur only when sound is delivered to the cochlea through its fluid. When it is transmitted through direct vibration of the basilar membrane, the cochlear waves always travel away from the sound source.

Békésy's Eddies

While observing vibration modes of the basilar membrane in his models of the cochlea, Békésy noticed two eddies in the fluid surrounding the model membrane, one on each side of it. The eddies occurred at the location of maximum vibration of the membrane and were so precisely bound to the maximum that he used them in some experiments to indirectly locate the maximum. To visualize the eddies, Békésy suspended in the fluid some fine particles for light reflection. He observed that the fluid formed eddies in such a way that it flowed in the wave direction near the membrane, away from the membrane a short distance past the maximum and returned along a short path near the canal bottom and back to the membrane on the other side of the maximum.

Many scientists were intrigued by the potential physiological role the eddies could play. It was even suggested that the eddy pressure rather than the vibration of the basilar membrane provided the mechanical stimulus for the hair cells (Ranke, 1931). The suggestion was attractive in view of the known sharp frequency analysis performed by the ear and the impression that the stimulus provided by the eddies was better focused than the maximum of basilar membrane vibration. According to another suggestion, the direct streaming of the fluid along the basilar membrane caused a longitudinal shift in its tissues, leading to a change in pitch. Such a shift, it was thought, could account for the diplacusis known to occur after excessive noise exposure (Rüedi & Furrer, 1947).

To investigate the eddy phenomenon somewhat more systematically, I first endeavored to find out if it was a unique property of the cochlea or a more general property of surface waves. The situation was quickly clarified with the help of some simple experiments (Zwislocki-Moscicki, 1948). Using a bottle with flat side walls, I filled it half way with water in which some fine particles were suspended for motion visualization. When the bottle was held on its side and moved back and forth lengthwise, waves arose on the free water surface. By an appropriate choice of the frequency of back and forth motion, it was possible to generate standing surface waves. A particularly convenient frequency was one that generated a full wavelength, one antinode occurring in the middle and one at each end. The suspended particles clearly indicated direct water streaming associated with the wave motion. Near the surface, water flowed toward the wave antinodes where it was propelled toward the bottom. Along the bottom, it flowed toward the nodes where, it returned to the surface. In this way, loop motions were created resembling eddies. Variation of the canal depth did not have any fundamental effect on the eddy formation, except for changing the eddy shape. On the basis of these experiments, I concluded that eddy formation was not an exclusive property of the cochlea but occurred universally in the presence of surface waves. Of course, rectification processes associated with surface waves are well known (e.g., Lamb, 1945). They are perhaps the most prominent in the formation of breakers on the ocean surface near beaches. The cochlear eddies observed by Békésy are distinct nevertheless in that they occur in the presence of traveling waves at the location of their maximum amplitude.

On the basis of the generality of the eddy phenomenon, it is necessary to conclude that, in addition to the eddies Békésy saw, there must be in the cochlea an eddy motion orthogonal to their plane. This is so because the basilar membrane bulges out during vibration not only along the cochlea but also in the cross sectional direction. Because of the narrowness of the basilar membrane and the relatively long cochlear waves, the resulting curvature in the cross section must actually be greater than along the cochlea. The cross sectional curvature must be present wherever the basilar membrane vibrates noticeably, so that the cross sectional eddy motion must stretch over a substantial part of the cochlear length.

What is the mechanism of eddy formation in the presence of elastic surface waves, such as the waves in the cochlea where the relevant surface, or rather—interface, is provided by the basilar membrane? It has to do with the departure of the surface from being parallel to the surface at rest and with the resulting angle between the inertia force and the elastic force attempting to restore the surface back to its rest position. The inertia force must parallel the motion of the surface and the fluid near it, which is normal to the surface at rest. Only the inertia component parallel to the elastic restoring force, which is normal to the bulged-out surface, leads to the membrane displacement. The vector difference between the inertia force and its component perpendicular to the displaced surface produces a force vector parallel to the surface. This vector must propel the fluid tangentially to the surface, as is illustrated in Fig. 4.33.

Assume that the curved line in Fig. 4.33 schematizes the bulged-out basilar membrane in a cross sectional plane. Vector A represents the inertia force normal to the surface at rest and parallel to the surface motion. Vector B is its component normal to the displaced surface, and vector C, the difference between them. Vector B* represents the elastic restoring force. Note that force vector C is tangent to the bulged out surface. It must produce a tangent fluid flow of a magnitude sufficient to generate through friction and inertia an opposing force vector of equal magnitude and direction. Note also that vector C points toward the maximum of surface displacement, in agreement with the direction of fluid flow observed in the experiment described above. For reasons of symmetry, tangential fluid flow on the other side of the bulge must also be directed toward the maximum of the surface displacement. As can be easily seen by mentally inverting the surface bulge in Fig 4.33, tangential fluid flow in the presence of a negative bulge must still have the same direction. As a result, rectified fluid flow toward the maximum of the surface excursion is maintained during the full cycle of a periodic motion. This produces fluid accumulation near the location of maximum oscillation. Because of its incompressibility, the accumulated fluid must escape away from the surface, as was observed in the bottle experiment. The loop of the rectified flow can only be closed along the canal bottom and its lateral walls.

If fluid circulation of the kind described above actually takes place in the cochlea, as is quite likely, it may have a nonnegligible effect on the cochlear

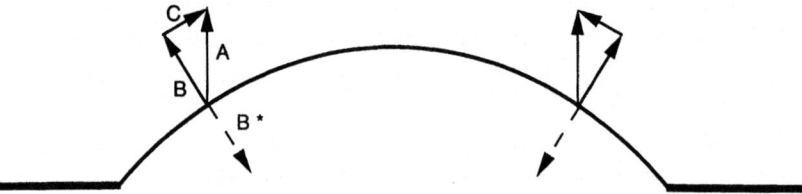

FIG. 4.33. Mechanism of eddy formation in association with surface waves. The curved line represents the deformed liquid surface, vectors A and B, the inertia and pressure forces acting on the surface, respectively. The vector, C, parallel to the liquid surface, is the difference vector between them.

THE COCHLEA SIMPLIFIED BY DEATH

biochemistry by promoting chemical and ionic exchange between the tissues surrounding cochlear scalae, including scala media, and the organ of Corti as well as the tectorial membrane. In this respect, it may be of interest to note that vector C in Fig. 4.33, and with it, the tangential fluid flow increase as the square of the amplitude of the surface—basilar membrane, displacement. This is so because the vector C is proportional to both the magnitude of the inertia vector, A, and the sinus of the angle between the inertia vector and its component, B, normal to the displaced surface.

$$C = A \sin \varphi \qquad 4.116$$

For small angles that must prevail during acoustic stimulation, the sinus is directly proportional to the excursion amplitude of the surface, which in turn, is proportional to the inertia vector, A. Thus, the rapid increase of the tangential flow with the amplitude of the basilar membrane vibration may help increase the metabolic energy supply to the organ of Corti in direct proportion to the acoustic energy that has to be processed.

It is doubtful that the force driving the tangential flow can have any other effects. Even at the highest sound pressure levels that do not damage the ear, it is very small compared to the inertia force acting on the basilar membrane, as can be determined with the help of a very simple calculation. The surface of the bulging basilar membrane produces an average angle with its surface at rest whose tangent is nearly equal to the ratio between the maximum displacement of the membrane in its approximate middle and the half width of the membrane. This tangent is approximately equal to the tangent of the angle substained between the vectors A and B in Fig. 4.33. In turn, the tangent is nearly equal to the sinus of the angle for the small basilar membrane displacements that are encountered. In the middle of the cochlea, the half width of the basilar membrane is about 0.015 cm. The displacement of the basilar membrane near the threshold of audibility has been estimated at around 10^{-8} cm (e.g., Rhode, 1971). At 120 dB SPL, it should be on the order of 10^{-5} cm. Accordingly, the tangent at 120 dB comes out to be on the order of $0.67*10^{-5}$. This means that the tangential force producing the tangential fluid flow is only $0.67*10^{-5}$ fraction of the normal force acting on the basilar membrane.

To understand Békésy's eddies that occur in the direction of wave travel, imagine the surface bulge of Fig. 4.33 moving in one direction, let us say, to the right. At each location through which it travels it produces a rectified fluid flow toward its greatest excursion. As its amplitude increases toward the location of the local maximum, the fluid accumulation traveling with the bulge increases. Past the maximum, where the amplitude of the basilar membrane begins to decrease, the accumulation becomes excessive and has to be dissipated. Thus, the fluid flows away from the membrane. In an incompressible fluid, no compression is possible, and the fluid ejected from the neighborhood of the basilar membrane must flow back to where it came from along the canal bottom either in scala tympani or vestibuli. In this way an eddy is formed.

In my original work (Zwislocki-Moscicki, 1948) I calculated the pressure driving the eddy from the quadratic terms of the surface wave equation (Lamb, 1945). However, the calculation is tedious and hardly justified here in view of the finding that the pressure is too small to play any appreciable role in hair cell stimulation or basilar membrane deformation. Whether the eddy can play a similar metabolic function to that just suggested for the cross-sectional eddy motion is questionable. It seems to be there as an epiphenomenon resulting from the circumstance that all surface waves must be expected to be accompanied by rectified fluid circulation.

REFERENCES

Békésy, G. v. (1928). Zur Theorie des Hörens; Die Schwingungsform der Basilar membrane [On the theory of hearing; The oscillation form of the basilar membrane]. *Physikalische Zeitschrift, 29,* 793–810.

Békésy, G. v. (1941). Ueber die Elastizität der Schneckentrennwand des Ohres [On the elasticity of the cochlear partition in the ear]. *Akustische Zeitschrift, 6,* 265–278.

Békésy, G. v. (1942). Ueber die Schwingungen der Schneckentrennwand beim Präparat und Ohrenmodell [On the oscillations of the cochlear partition in physiological specimen and the ear model]. *Akustische Zeitschrift, 7,* 173.

Békésy, G. v. (1943). Ueber die Resonanzkurve und die Abkingzeit der verschiedenen Stellen der Schneckentrennwand [On the resonance curve and the decay of oscillation at different locations of the cochlear partition]. *Akustische Zeitschrift, 8,* 66.

Békésy, G. v. (1947). The variation of phase along the basilar membrane with sinusoidal vibrations. *Journal of the Acoustical Society of America 19,* 452–460.

Békésy, G. v. (1960). *Experiments in Hearing.* New York: McGraw Hill.

Békésy, G. v., & Rosenblith, W. A. (1951). The mechanical properties of the ear. In S. S. Stevens (Ed.), *Handbook of Experimental Psychology.* New York: Wiley.

Dallos, P. (1970). Low-frequency auditory characteristics: Species dependence. *Journal of the Acoustical Society of America, 48,* 489–499.

Dallos, P. (1973). *The Auditory Periphery.* New York: Academic Press.

Davis, H., & Associates (1953). Acoustic trauma in the guinea pig. *Journal of the Acoustical Society of America, 25,* 1180–1189.

Flanagan, J. L. (1960). Models for approximating basilar membrane displacement. *Bell System Technical Journal, 39,* 1163–1191.

Flanagan, J. L. (1962). Computational model for basilar-membrane displacement. *Journal of the Acoustical Society of America, 34,* 1370–1376.

Fletcher, H. (1951). On the dynamics of the cochlea. *Journal of the Acoustical Society of America, 23,* 637.

Geisler, C. D., & Hubbard, A. E. (1972). New boundary conditions and results for the Peterson–Bogart Model of the Cochlea. *Journal of the Acoustical Society of America, 52,* 1629–1634.

Guinan, J., & Peake, W. T. (1967). Middle ear characteristics of anesthetized cats. *Journal of the Acoustical Society of America, 41,* 1237–1261.

Gummer, A. W., Johnstone, B. M., & Armstrong, N. J. (1981). Direct measurement of basilar membrane stiffness in the guinea pig. *Journal of the Acoustical Society of America, 70,* 1298–1309.
Helmholtz, H. L. F. (1954). *On the Sensations of Tone as a Physiological Basis for the Theory of Music* (2nd English edition by Henry Morgenau). New York: Dover Publications.
Holley, M. C. (1988). Purification of mammalian cochlear hair cells using small volume Percoll density gradients. *Journal of Neuroscience Methods, 27,* 219–224.
Jahnke, E., & Emde, F. (1945). *Tables of Functions with Formulae and Curves.* New York: Dover Publications.
Kinsler, L. E., & Frey, A. R. (1950). *Fundamentals of Acoustics.* New York: John Wiley.
Lamb, H. (1945). *Hydrodynamics.* New York: Dover Publications.
Lighthill, J. (1978). *Waves in Fluids.* Cambridge, UK: Cambridge University Press.
Lighthill, J. (1981). Energy flow in the cochlea. *Journal of Fluid Mechanics, 106,* 149–213.
Lindsay, R. B. (1960). *Mechanical Radiation.* New York: McGraw Hill.
Lynch, T. J., Nedzelnitsky, V., & Peake, W. T. (1982). Input impedance of the cochlea in cat. *Journal of the Acoustical Society of America, 72,* 108–130.
Merchant, S. N., Ravicz, M. E., & Rosowski, J. J. (1992). The acoustic input impedance of the stapes and cochlea in human temporal bones. *Abstract, 15th Midwinter Meeting of the Association for Research in Otolaryngology* (p. 98), St. Petersburg Beach, Florida.
Merchant, S. N., Ravicz, M. E., & Rosowski, J. J. (1996). Acoustic input impedance of the stapes and cochlea in human temporal bones. *Hearing Research, 97,* 30–45.
Miller, C. E. (1985). Structural implications of basilar membrane compliance measurements. *Journal of the Acoustical Society of America, 77,* 1465–1474.
Møller, A. R. (1965). An experimental study of the acoustic impedance of the middle ear and its transmission properties. *Acta Oto-Laryngologica, 60,* 129–149.
Mundie, J. R. (1963). The impedance of the ear—a variable quantity. In J. L. Fletcher (Ed.), *Middle Ear Function Seminar,* Report 576, U.S. Army Medical Research Laboratory, Fort Knox, Kentucky.
Nedzelnitsky, V. (1980). Sound pressures in the basal turn of the cat cochlea. *Journal of the Acoustical Society of America, 68,* 1676–1689.
Olson, E. S. (2001). Intracochlear pressure measurements related to cochlear tuning. *Journal of the Acoustical Society of America, 110,* 349–367.
Peterson, L. C., & Bogert, B. P. (1950). A dynamical theory of the cochlea. *Journal of the Acoustical Society of America, 22,* 369–381.
Ranke, O. F. (1931). Die Gleichrichter-Resonanz-theorie [The Rectifier-Resonance Theory]. München: J. Llehmann.
Rauch, S. (1964). Biochemie des Hörorgans [The Biochemistry of the Organ of Hearing]. Stuttgart: Thieme.
Rhode, W. S. (1971). Observations of the vibration of the basilar membrane in squirrel monkeys using the Mössbauer technique. *Journal of the Acoustical Society of America, 49,* 1218–1229.
Rhode, W. S. (1973). An investigation of postmortem cochlear mechanics using the Mössbauer effect. In A. R. Møller (Ed.), *Basic Mechanism in Hearing* (pp. 49–63). New York: Academic Press.
Rüede, L., & Furrer, W. (1947). Das akustische Trauma. Switzerland: Basel.
Siebert, W. M. (1974). Ranke revisited—a simple short-wave cochlear model. *Journal of the Acoustical Society of America, 56,* 594–600.
Steele, C. R., & Tabor, L. A. (1979). Comparison of WKB calculations and experimental results for three-dimensional cochlear models. *Journal of the Acoustical Society of America, 65,* 1007–1018.
Tasaki, I., Davis, H., & Legouix, J. P. (1952). The space-time pattern of the cochlear microphonics (guinea pig), as recorded by differential electrodes. *Journal of the Acoustical Society of America, 24,* 502.

Ulehlova, L., Voldrich, L., & Janisch, R. (1987). Correlative study of sensory cell density and cochlear length in humans. *Hearing Research, 28,* 149–151.

Viergever, M. A. (1978). On the physical background of the point-impedance characterization of the basilar membrane in cochlear mechanics. *Acustica, 39,* 292–297.

Viergever, M. A. (1980). *Mechanics of the inner ear.* Delft, Netherlands: Delft University Press.

Voldrich, L. (1978). Mechanical properties of the basilar membrane. *Acta Oto-Laryngologica, 86,* 331–335.

Wever, E. G. (1949). *Theory of Hearing.* New York: Wiley.

Wever, E. G., & Lawrence, M. (1952). Sound conduction in cochlea. *Annals Oto-Laryngology, Rhinology, & Larngology, 6,* 824–835.

Zweig, G., Lipes, R., & Pierce, J. R. (1976). The cochlear compromise. *Journal of the Acoustical Society of America, 59,* 975–982.

Zwislocki, J. J. (1950). Theory of the acoustical action of the cochlea. *Journal of the Acoustical Society of America, 22,* 778–784.

Zwislocki, J. J. (1953). Wave motion in the cochlea caused by bone conduction. *Journal of the Acoustical Society of America, 25,* 986–989.

Zwislocki, J. J. (1955). The nature of auditory stimuli and their attenuation. *Symposium on Physiological Psychology,* ONR Symposium Report ACR-1, Office of Naval Research, Washington, DC, 182–190.

Zwislocki, J. J. (1957). Some measurements of the impedance at the eardrum. *Journal of the Acoustical Society of America, 29,* 349–356.

Zwislocki, J. J. (1962). Analysis of middle ear function. Part I. Input impedance. *Journal of the Acoustical Society of America, 34,* 1514–1523.

Zwislocki, J. J. (1963). Analysis of the middle-ear function. Part II. Guinea-pig ear. *Journal of the Acoustical Society of America, 35,* 1023–1030.

Zwislocki, J. J. (1965). Analysis of some auditory characteristics. In R. D. Luce, R. R. Bush, and E. Galanter (Eds.), *Handbook of Mathematical Psychology, Vol. III* (pp. 1–97). New York: John Wiley.

Zwislocki, J. J. (1974). Cochlear waves: Interaction between theory and experiments. *Journal of the Acoustical Society of America, 55,* 578–583.

Zwislocki, J. J. (1975). The role of the external and middle ear in sound transmission. In E. L. Eagles (Ed.), *The Nervous System* (Vol. 3, pp. 45–55). New York: Raven Press.

Zwislocki, J. J. (1980). Two possible mechanisms for the second cochlear filter. In G. van den Brink & F. A. Bilsen (Eds.), *Psychophysical, Physiological and Behavioral Studies in Hearing* (pp. 16–23). The Netherlands: Delft University Press.

Zwislocki, J. J. (1983). Cochlear micromechanics—a model and some of its consequences. In W. R. Webster & L. M. Aitkin (Eds.), *Mechanics of Hearing* (pp. 21–26). Clayton, Victoria, Australia: Monash University Press.

Zwislocki, J. J. (1984). Biophysics of the mammalian ear. In W. W. Dawson & J. M. Enoch (Eds.), *Foundations of Sensory Science* (pp. 109–150). Berlin: Springer-Verlag.

Zwislocki, J. J., Chamberlain, S. C., & Slepecky, N. B. (1988). Tectorial membrane I: Static mechanical properties in vivo. *Hearing Research, 33,* 207–222.

Zwislocki-Moscicki, J. J. (1946). Über die mechanische Klanganalysedes Ohrs [On sound analysis in the ear]. *Experientia, 2,* 10–18.

Zwislocki-Moscicki, J. J. (1948). *Theorie der Schneckenmechanik: Qualitative und Quantitative Analyse [Theory of Cochlear Mechanics: Qualitative and Quantitative Analysis].* Acta Oto-Laryngologica, Suppl. 72, 1–76.

chapter

Live Cochlea: Physical Constants and Fundamental Characteristics

*A*t the end of 1950s, it seemed that cochlear mechanics was reasonably well understood. This feeling began to loose its footing near the end of 1960s when two perplexing discoveries took place, both subsequently verified on numerous occasions. Johnstone and Boyle (1967) found on live guinea pigs that the local maximum of basilar membrane vibration was substantially sharper than seen by Békésy (1960) in postmortem preparations. Their measurements were soon confirmed by Rhode (1971) on squirrel monkeys, who also determined directly on the same preparations the relationship between the in vivo and postmortem vibration maxima (Rhode, 1973). Almost at the same time, Spoendlin (1966, 1970) demonstrated that most of the afferent nerve fibers entering the cochlea ended on IHCs and only 5% to 10% innervated the OHC. The conclusion was inevitable that practically all auditory information reached the auditory nerve through the IHC.

If practically all auditory information goes through the IHCs, what do the OHC do? Before this dilemma could be solved several discoveries had to be made. Kemp (1978) discovered that the cochlea emitted sound when excited by brief acoustic stimuli. He called the emission *oto-acoustic emission* (OAE). The discovery was soon followed by an even more astonishing one that, in many instances, the cochlea could emit sound spontaneously, without any external stimulation (Kemp, 1979; Wilson, 1980; Zurek, 1981). The latter seemed to verify indirectly Gold's (Gold, 1948; Gold & Pumphrey, 1948) suggestion made more than 30 years earlier and appearing quite improbable then that the cochlea performed a sharp frequency analysis due to a positive feedback coupled to a metabolic energy supply. The assumption was in agreement with the psychophysically known sharp sound analysis in hearing but contradicted Békésy's direct observations. Since there were other neural explanations for the sharp analysis found psychophysically, which did not contradict Békésy's findings, as a reviewer, I recommended to the Proceedings of the Royal Society in London rejection of the Gold and Pumphrey's article on the subject, although the feedback suggested by them was known to be at the basis of man-made and natural mechanical and electrical oscillators. The article was published, anyway. Several years elapsed after Kemp's discovery before Brownell and his associates (Brownell, Bader, Bertrand, & Ribaupiere, 1985) discovered a possible mechanism for the acoustic emissions. They found that the OHCs but not the IHCs changed their lengths in response to changes in surrounding electrical field. A few years later, Ashmore (1987) showed that the changes persisted even at high auditory sound frequencies. Santo-Sacchi (1990) pushed the upper frequency limit of the oscillation even further. These findings furnished the answer concerning the role of the OHCs—they most likely provided the electromechanical feedback postulated by Gold.

Another discovery that stunned the world of auditory scientists was made by Russell and Sellick (1977) just before that of Kemp's. They demonstrated with the help of intracellular recordings that the frequency tuning of the IHCs was about as sharp as of auditory nerve fibers. No further filtering beyond the IHCs was necessary. Nevertheless, the question of additional filtering between the basilar membrane and the IHCs has persisted (e.g., Khanna & Leonard, 1982; Russell, Kössl, & Murugasu, 1995; Ruggero, Rich, Narayan, & Robles, 1997). An attempt at answering it is made in this volume.

There were other revelations with less glitter perhaps but with the potential of profoundly affecting our concepts of the cochlear mode of operation. Flock (1977) discovered that the stereocilia of hair cells in the inner ear were stiff, not flexible, as was implied by the classical model of shear motion between the reticular lamina and the tectorial membrane. In addition, the tectorial membrane was found to be soft in a live cochlea, in its natural chemical environment, much softer than the corresponding aggregate of the OHC stereocilia (Zwislocki, 1988, Zwislocki et al., 1988; Zwislocki & Cefaratti, 1989), contrary to Békésy's observations on postmortem preparations. The two findings together reverse the assumption implicit in the classical model of the shear motion according to which the stereocilia are relatively soft and the tectorial membrane relatively stiff. As a result, the dynamics of the shear motion becomes dependent on sound frequency in a complicated manner, and the assumption inherent in the classical model that the hair cells, especially the OHCs, are always depolarized during basilar membrane displacement toward scala vestibuli can no longer be maintained. This conclusion is supported by direct experiments showing that the phase of the hair cell depolarization changes with sound frequency and even sound intensity (Zwislocki & Smith, 1988; Zwislocki, 1990; Szymko, Zwislocki, & Hertig, 1997).

All the insights mentioned above are included in the model of cochlear dynamics in vivo described in chapter 6. In the present chapter, the cochlear structures and physical parameter values determined postmortem and introduced in the preceding chapter are modified when needed to suit the in vivo conditions and supplemented by others determined in more recent years either on live preparations or in vitro under conditions that preserved their in vivo properties. In addition, fundamental dynamical characteristics of the cochlea are introduced, such as the vibration magnitude and phase of the basilar membrane as functions of cochlear location and sound frequency and the magnitude and phase of the shear motion between the reticular lamina and the tectorial membrane as functions of the same independent variables. The latter are derived from recordings of cochlear microphonics (CM) and of intracellular alternating potentials of the OHCs and supporting cells.

Feasibility dictated that practically all anatomical and physiological experiments on live cochleas be performed on selected small mammals, and human cochlear characteristics in vivo be subsequently inferred from them. This does not seem to be a serious drawback because the cochleas of these animals are structurally and physically similar to those of humans. The same intermammalian similarity holds for the dynamical characteristics, so that extrapolation to humans appears to be justified (e.g., Greenwood, 1990). Such extrapolation is further supported by the similarity of psychophysical characteristics.

The numerical transfer from animal cochleas to human can be achieved with the help of determinations of basilar membrane compliance that can be measured directly postmortem and derived from cochlear wave travel times in vivo. Because of structural similarities, it is assumed that the relation between the basilar membrane compliance and the compliance values of other structures, especially of the hair cell stereocilia and the tectorial membrane remain roughly invariant. Of course, anatomical cochlear dimensions of animals and humans can be directly determined and compared. Except for those of the tectorial membrane, they seem to change little after death.

To make the transfer from the animal to the human cochlea possible, it became necessary to determine the compliance of the basilar membrane of our animal model, Mongolian gerbil (*Meriones unguiculatus*). The required procedure and the results are described here for the first time.

No attempt is made in the present chapter or the rest of this book, to discuss the voluminous literature concerning cochlear mechanics. References to the literature are made only as needed to supplement the information provided by the author's and his associates' research so as to achieve a usefully complete picture of cochlear dynamics. Aspects of cochlear mechanics are emphasized in which the author was the most directly involved and that often have not been adequately covered in the literature. One of these aspects concerns the mechanical properties of the tectorial membrane and their effect on the cochlear dynamics as a whole. As is shown in this chapter and, especially, the following one, the effect plays a much greater role than has been assigned to it in the literature thus far and makes it possible to explain some seemingly paradoxical phenomena.

PHYSICAL PARAMETER VALUES OF THE GERBIL TECTORIAL MEMBRANE

In the mathematical treatment of the human cochlea postmortem in chapter 4, the assumption was made that the organ of Corti together with the tectorial membrane are rigidly attached to the basilar membrane and constitute in this way a simple, second order system consisting in terms of lumped elements of a compliance, a mass, and a resistance, taken per unit length. Such an assumption is no longer tenable in a live cochlea. Although the organ of Corti may still be regarded as rigidly attached to the basilar membrane within certain location and frequency ranges of interest, this is no longer possible for the tectorial membrane. It is attached separately to the cochlear bone by means of the spiral limbus and is coupled flexibly to the organ of Corti and, through it, to the basilar membrane via the OHC stereocilia whose stiffness cannot be considered as infinite. According to the measurements of Flock and his associates (Strelioff & Flock, 1984), the stiffness contribution of the IHC stereocilia is negligible.

For the reasons already stated, it is necessary to explore the OHC stereocilia-tectorial membrane complex as a separate entity and study its interaction with the basilar membrane–organ of Corti complex. To do this, the physical parameters of both the tectorial membrane and the stereocilia have to be defined and measured. The predominant parameter of the stereocilia, their stiffness, has been investigated extensively in guinea pigs by Flock's group (Strelioff & Flock, 1984; Strelioff, Flock, & Minser, 1985). The physical parameters of the tectorial membrane in vivo, on the other hand, do not seem to have been determined prior to my work on the dynamics of the live cochlea, and I had to undertake their measurement myself. I did this with the assistance of L. K. Cefaratti. It was possible to determine the mass of the tectorial membrane from its cross-sectional area and its average density but the determination of its effective stiffness required extensive experimentation fraught with surgical and instrumental difficulties. Because the measurements have not yet been repeated by others and still remain unconfirmed, they are described here in some detail to give the reader an opportunity to judge their reliability. It should be pointed out, nevertheless, that, while this manuscript was being written, the method, including the instrumentation, was applied by He and Dallos (1999) to their measurements of the longitudinal, or axial, stiffness of the outer hair cells, and their results were in agreement with those of others using different methods. The agreement should provide an indirect evidence of the method's reliability.

It is somewhat unfortunate that I was unable to measure the stiffness of the tectorial membrane in guinea pigs for direct comparison with the measurements of the stereocilia stiffness performed by Flock and his associates. However, I found the configuration of the guinea pig cochlea unsuitable for the purpose, unlike that of the Mongolian gerbil cochlea. Because of the similarity of cochlear tonotopic maps among mammals (e.g., Greenwood, 1990), it is likely that the stiffness is similar in both. Measurement of CM transfer characteristics made it possible to indirectly verify this assumption (Schmiedt & Zwislocki, 1977).

Within the means and time available to me, I was unable to measure the mechanical resistance of the tectorial membrane, and it remained as a free parameter to be determined by matching the theoretical dynamical characteristics of the OHC stereocilia-tectorial membrane system to the empirical ones.

Elastic Properties of the Tectorial Membrane

It seemed necessary to preserve the natural chemical environment of the tectorial membrane during exploration of its elastic properties. In particular, some earlier investigators (Tonndorf, Duvall, & Reneau, 1962; Kronester-Frei, 1979) had found the membrane to change its appearance under the influence of perilymph. To prevent such a change and a possible associated change in the membrane's stiffness, it appeared necessary to preserve Reissner's membrane separating the perilymph from the endolymph in the cochlea. As a consequence, the tectorial membrane had to be accessed through the lateral wall of scala media. I found the configuration of the Mongolian gerbil cochlea uniquely suitable for this approach, especially, in its second turn (Zwislocki, 1988; Zwislocki, Chamberlain, & Slepecky, 1988; Zwislocki & Cefaratti, 1989). Here, Reissner's membrane forms an unusually wide angle with the basilar membrane which is tilted so that fine rods can be inserted either parallel or almost perpendicularly to the top surface of the tectorial membrane, as indicated in Fig. 5.1 by the thin straight lines. Thus, the membrane's stiffness can be measured in both the transversal and radial directions by the method used earlier by Békésy (1953, 1960) and Flock and Strelioff (1984), who used bending of fine rods of known stiffness as a means for exploring the elastic properties of the tectorial membrane and the OHC stereocilia, respectively.

The experimental method was described earlier in several articles (Zwislocki, 1988; Zwislocki et al., 1988; Zwislocki & Cefaratti, 1989). It is

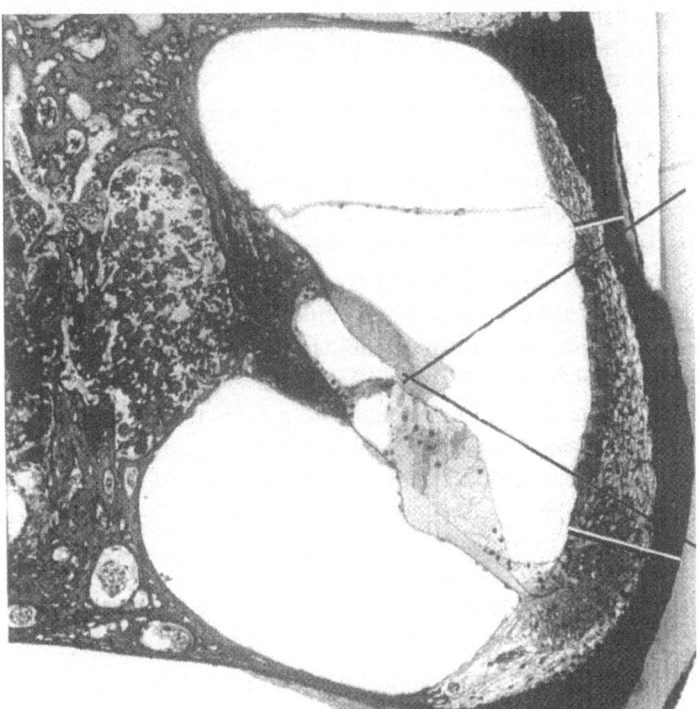

FIG. 5.1. Cross section of the Mongolian gerbil cochlear canal in the second turn. The double black-white lines indicate the size of the surgical opening; the black lines the locations of the micropipettes for the measurement of the tectorial membrane stiffness. (Modified from Zwislocki & Cefaratti, 1989). Reprinted from *Hearing Research, 41*, J. J. Zwislocki & L. K. Cefaratti, Tectorial membrane II: Stiffness measurements in vivo, 211–228, copyright © 1989, with permission from Elsevier Science.

described here in a somewhat abbreviated form but also with some additions not included before. In total, 50 Mongolian gerbils weighing between about 60 and 70 g were involved. They were anesthetized with sodium pentobarbital (35 mg/kg of body weight) and maintained at the body temperature of 37°C. They were tracheotomized, but no artificial respiration needed to be administered. The right pinna and the soft tissue over the right ventrolateral part of the bulla were removed and the bulla opened wide enough to visualize the round window niche. A silver wire introduced through an additional small opening was placed in the niche for recording cochlear microphonics (CM). The overall sensitivity of the cochlea was checked by means of sound produced by a miniature earphone coupled hermetically to the remaining ear canal or, simply, by whistling.

If the sensitivity was satisfactory, additional surgery was performed. For a sufficient access to the cochlea, it was necessary to remove the whole lateral part of the bulla, the ear canal, the tympanic ring with the tympanic membrane and the large middle ear ossicles. The stapes was removed next in an attempt at reducing the perilymphatic pressure (Tonndorf et al., 1962) and, in this way, minimizing seepage of the perilymph into scala media through the borders of the surgical opening made in its lateral wall. Its removal required great caution because, in the gerbil, the stapedial artery courses between the stapedial crura. First, the crura had to be broken, then, the spiral ligament holding the stapedial footplate in the oval window severed and the footplate removed with forceps or gently pressed into the vestibule of the inner ear labyrinth. Abrupt motion of the footplate had to be avoided in order not to send a traumatizing shock wave through the cochlea.

Opening the lateral wall of scala media presented substantial difficulties. No adequate commercially available surgical instruments could be found for the purpose, and it was necessary for me to make some of my own. They had to be appropriately small for surgery under 20× to 40× magnification. Some are shown in Fig. 5.2. They include from right to left: a microchisel, a microknife, a microspatula, a micropick, and a microhook. For size comparison, a human hair is included in the picture. In addition to the microinstruments, an adjustable small platform was constructed for hand support. In view of the necessary relatively high magnification under which the surgery had to be performed, the support was quite essential for steadying the hand and minimizing the finger tremor.

First, the cochlear bony capsule was scored with the microknife and microchisel, once along a landmark situated nearly opposite the attachment of Reissner's membrane but still over scala media, and once, nearly opposite the attachment of the basilar membrane, but somewhat on the scala media side. The score lines extended over almost a quarter of the second cochlear turn. The bone was also scored perpendicularly to the longi-

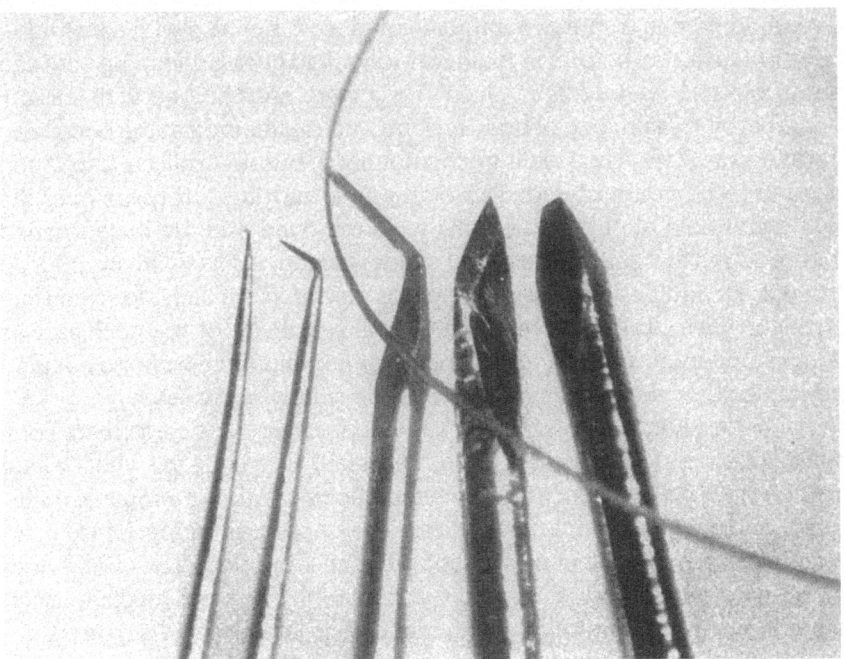

FIG. 5.2. Microinstruments for cochlear surgery. A curved human hair is placed over the instruments for size comparison.

tudinal score lines at their ends and gently broken near one end. Its section between the scorelines could then be lifted either with the microspatula or the micropick. A few remaining chips were removed with the micropick. The resulting opening was about 1 mm long and 300 um wide and provided a good view of the outer surface of the spiral ligament. Its margins are indicated in Fig. 5.1 by means of double black-white lines.

Removal of the spiral ligament together with stria vascularis attached to its inner surface constituted the hardest part of the operation. The ligament proved to be unexpectedly tough and only tenuously attached to the bony wall. These two features made it impossible to perforate it by pressing a sharp needle perpendicularly to its surface. Instead of becoming perforated, the ligament detached itself from the bone and sank toward the organ of Corti. I was only able to make a perforation in it by finding an imperfection in its rather smooth surface and driving the point of the micropick into it at a shallow angle, almost tangentially to it. A sideways scratching motion was also helpful. Once a small perforation was obtained, it could be enlarged enough to insert the microspatula. The latter was then moved along the bony window under the ligament, cutting it along the edges of the bone opening through a kind of scissor action. The removal was rarely clean, and remaining debris had to be lifted one piece at a time by the micro pick and the microhook. The worst enemy was bleeding into scala media. Invariably, one to three capillaries were opened during the ligament removal, and blood gushed from them. It was not possible to stop the flow chemically in fear of changing the mechanical properties of the tectorial membrane. Fortunately, it was possible to remove the semicoagulated blood as it flowed into the endolymph, and usually, the bleeding stopped within a few minutes. The remaining opacity created by residual diluted blood, which obscured the view of intracochlear structures was soon cleared by ongoing endolymph secretion in the cochlear parts adjacent to the opening. The process was sometimes helped along by injecting minute amounts of isotonic KCl solution with the help of a fine micropipette. This had to be done with extreme caution because any excessive streaming of fluid in the neighborhood of the tectorial membrane tended to dislodge it.

The cochlear structures are almost completely transparent and have to be visualized by staining. I used two traditional dyes—Janus green B and Alcian blue 8Gx. Both were dissolved in a 0.15 M KCl solution, isotonic with the endolymph, the first, at a concentration of 0.1%, the second, at a concentration of 3%. The concentrations were found to be optimum by trial and error. The stained solutions were placed in the endolymph in the same way as the unstained KCl solution. When the natural endolymph secretion cleared the fluid space above the tectorial membrane and the organ of Corti, these structures came into clear view. Sometimes, the process was accelerated by cautious administration of the clear KCl solution.

When stained with Janus green, the tectorial membrane appeared in various shades of purple, depending on the preparation. Its marginal zone was stained the darkest, and a narrow dark band often appeared at the spiral limbus. Hensen's cells became dirty greenish blue and, when the tectorial membrane was lifted partially, it revealed the pillars of Corti appearing as a dirty blue band and the rows of the OHCs appearing as feint dirty blue lines. These observations agree reasonably well with those described in the past by Tonndorf et al. (1962). Janus green also stained bits of the spiral ligament sometimes remaining in the opening. These were removed with the microhook, the micropick or an eyelash instrument consisting of a human eyelash attached to the end of a micropipette.

When stained with Alcian blue, the tectorial membrane appeared as a rather vividly blue band, the darkest in the marginal zone and the faintest near the spiral limbus, which remained unstained. The other cochlear structures were not stained either, and their visualization required counterstaining with Janus green.

The cochlear structures visible in the surgical window were viewed under fiber optic illumination through a Nikon SMZ-10 zoom stereo microscope with a magnification of 20× to 160× and photographed. Black and white versions of some of the photographs are shown in Fig. 5.3 (Zwislocki, Chamberlain, & Slepecky, 1988). In panel A of the figure, the tectorial membrane is stained with Alcian blue and appears as a broad, dark band going approximately diagonally from the upper left to the lower right of the photograph. Its marginal zone abutting Hensen's cells appears the darkest and the zone closest to the spiral limbus, which remained unstained, the lightest, probably because of its relative thinness. The lighter, somewhat mottled band, situated above the dark band, belongs to Hensen's cells counterstained with Janus green. No gap is visible between the two bands, so that the tectorial membrane appears to be in its normal place, close to the organ of Corti. The two black surfaces on both sides of the visible cochlear structures belong to the bony cochlear capsule. Their edges delineate the margins of the surgical opening. The same structures but belonging to another preparation and stained with Janus green alone are shown in Fig. 5.3, panel B. Here, the tectorial membrane appears as the lightest band and Hensen's cells as a darker one. Again, no gap between them is evident, suggesting that the tectorial membrane is in its normal place and is not lifted. Note, however, the difference in texture between panels A and B. Whereas in the former both the tectorial membrane and the Hensen's cell mass appear as smooth surfaces, in the latter, the tectorial membrane exhibits a fibrous structure, especially in its marginal zone, near Hensen's cells. The difference suggested to me and my coworkers that the two dyes stained different elements. This was subsequently confirmed in an auxiliary experiment. In the third part of Fig. 5.3, panel C, the tectorial membrane, stained with Janus green, is shown lifted up somewhat from the organ of Corti, with its marginal zone curled up, and revealing partially the row of the pillars of Corti appearing as a narrow darkish band. Somewhat above it, one row of the OHCs can be discerned as a faint dark line.

To check the suspicion that Alcian blue and Janus green stained different elements of the tectorial membrane anatomy, N. Slepecky performed a supplemental experiment involving 10 gerbil cochleas (Zwislocki, Chamberlain, & Slepecky, 1988). In 4, the tectorial membrane was stained with Alcian blue, in another 4, with Janus green, and in the remaining 2, with both dyes. The animals were killed and decapitated under pentobarbital anesthesia and the bullae removed for further processing. During the whole procedure the cochleae were bathed in 0. 15 M KCl solution. As each cochlear turn was opened and the spiral ligament and stria vascularis pulled away, a drop of 1% solution of either dye in 0.15 M KCl was applied to scala media. Finally, sections of the tectorial membrane were teased off

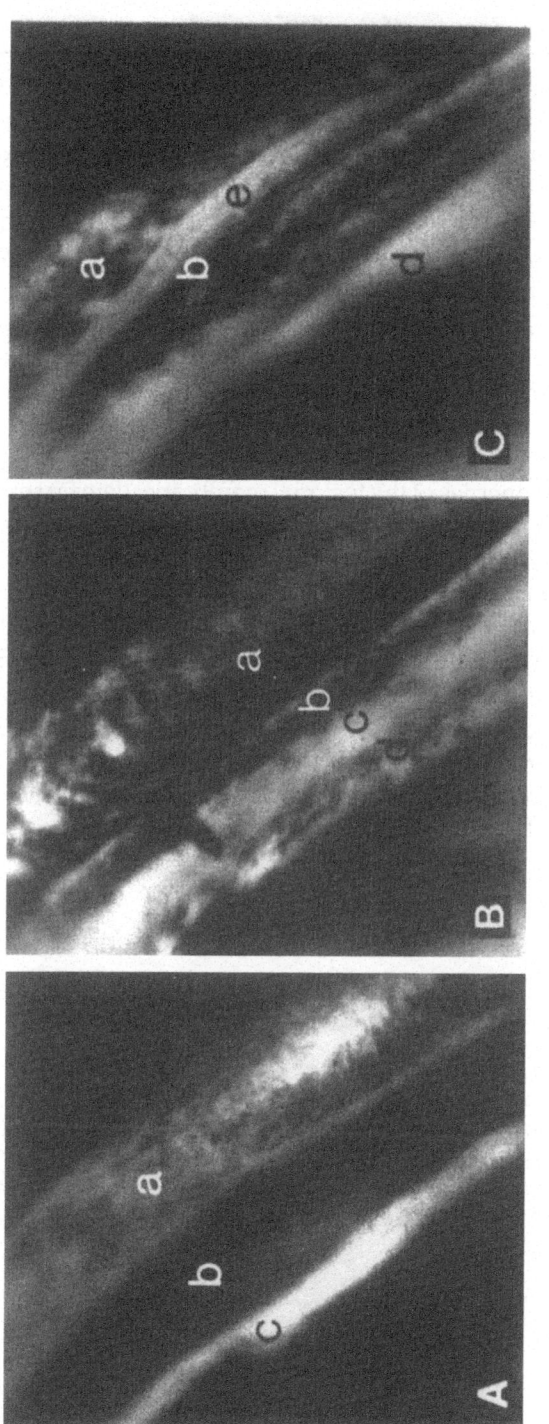

FIG. 5.3. The tectorial membrane and the mass of Hensen's cells, as seen through the surgical opening in the lateral wall of scala media. For panel A, both Alcian blue and Janus green stains were used, for the other two panels Janus green alone. In A, a indicates Hensen's cells, b, the tectorial membrane, and c, the spiral limbus. In B, a indicates Hensen's cells, b, c, and d, the marginal, middle, and limbal zones of the tectorial membrane. The small black protrusion is artifactual. In C, a indicates Hensen's cells, b, the marginal zone of the tectorial membrane, and d, the spiral limbus; e marks the organ of Corti visible through the gap between the lifted up tectorial membrane and Hensen's cells (From Zwislocki et al., 1988). Reprinted from Hearing Research, 33, J. J. Zwislocki, S. C. Chamberlain, & N. B. Slepecky, Tectorial Membrane I: Static mechanical properties in vivo. 207–222, copyright © 1988, with permission from Elsevier Science.

with the tip of a fine micropipette and placed in a drop of fresh 0.15 M KCl on a glass slide and coverslipped for microscopic observation under high magnification and photography. Three characteristic photomicrographs are shown in Fig. 5.4. In panel A, the tectorial membrane has been stained with Alcian blue, in panel B, with Janus green, and in panel C, with both. The rather amorphous appearance of the tectorial membrane in panel A indicates that Alcian blue binds to its ground substance. By contrast, its fibrous appearance in panel B leads to the conclusion that Janus green concentrates around the membrane's radial fibers in its middle zone and around whorled fibers in its marginal zone. The whorled fibers in the surface view correspond to the porous appearing part of the tectorial membrane in its radial section of Fig. 5.1.

The mechanical properties of the tectorial membrane were first explored qualitatively with the help of slender micropipettes having tip diameters of about 1 µm. The micropipettes were pulled on a Narishige vertical puller from Omega Dot 1-mm glass tubes and were quite flexible. When inserted into the tectorial membrane and moved laterally, they bent visibly, the size of the bend making it possible to gauge the force applied to the membrane. They were attached to a Narishige three axial hydraulic micromanipulator, which allowed accurate placement of the micropipettes within 1 µm and their precisely controlled movement. With proper maintenance, the micromanipulator produced little hysteresis. The experiments were performed on 28 animals, 22 in vivo, 2, postmortem, and 4 both in vivo and postmortem.

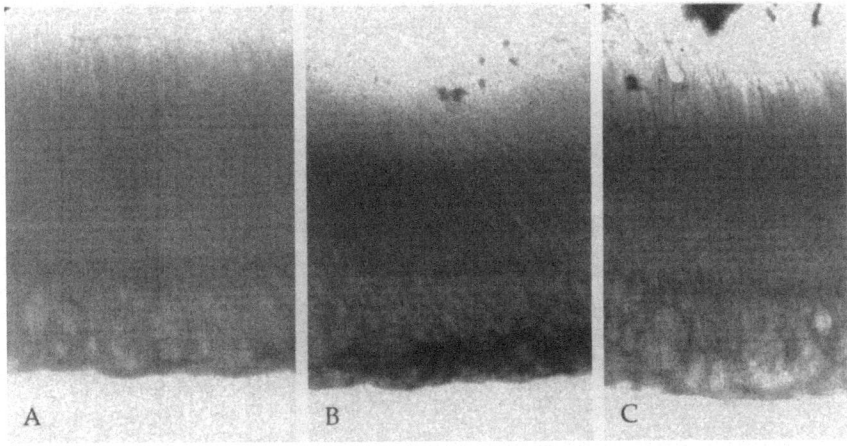

FIG. 5.4. Tectorial membrane gross structures made visible by Alcian blue in panel A, Janus green, in Panel B, and both dies in combination in Panel C.

The micropipettes were inserted into the tectorial membrane either frontally or dorsally, through its top surface, at the steepest possible angle, which was limited by the location of Reissner's membrane to about 55°. The frontal insertion was facilitated by the rather blunt margin of the tectorial membrane in the second turn of the Mongolian gerbil cochlea (Fig 5.1). When inserted dorsally, the tip of a micropipette was placed either in the marginal zone of the tectorial membrane, at the estimated location of the middle row of the OHCs, or near the spiral limbus. The micropipettes were moved transversally relative to the dorsal surface of the tectorial membrane, when inserted frontally, and nearly parallel to it, when inserted dorsally. In the latter situation, departure from a strictly parallel motion resulted from the insertion angle being smaller than 90°.

When a micropipette was inserted only in the superficial layers of the tectorial membrane, its sharp tip tore these layers rather easily when moved laterally. However, the micropipettes were inserted 30 µm deep, almost all the way through the approximately 35 to 40 µm thick membrane, in order to explore the mechanical properties of the membrane as a whole rather than of its individual layers.

Lateral motions of a deeply inserted micropipette were followed by adjacent portions of the tectorial membrane over distances almost as great as the membrane's width without tearing. This was especially true when the membrane was lifted above the organ of Corti only slightly and, presumably, in good condition. When the micropipette was withdrawn from it, a moderately deformed tectorial membrane sprang back to its resting position without any noticeable delay. Only after an excessive deformation, was it possible to detect a relatively slower recovery over the last few microns. In the presence of light to medium staining with Janus green, the membrane showed an astonishing resilience and toughness and was almost impossible to tear apart. This was less so in the presence of Alcian blue staining. It is possible that the difference was due to different binding sites of the two dyes within the membrane structure, as has been pointed out above. Because of the seemingly deleterious effect of Alcian blue, Janus green was used in most of the experiments on the mechanical properties of the tectorial membrane.

The membrane also tore more easily when it was strongly lifted up, perhaps due to an inadvertent effect of sodium ions, independent of the stain. In one preparation, a tear observed over several minutes was seen to heal within seconds. I did not study this perplexing phenomenon any further.

It has to be pointed out that the effect of sodium ions on the state of the tectorial membrane was not clear in our preparations. In one, for example, a moderately lifted tectorial membrane changed its position only slightly even after several injections of Ringer's solution into scala media. At the same time its mechanical properties appeared to remain normal by comparison to other preparations in which the tectorial membrane remained in closer proximity to the organ of Corti. This stability was consistent with the observations made by Frommer (1982).

No dramatic changes in the tectorial membrane position or elasticity were found within 45 minutes after death in any of our preparations studied postmortem. Only the viscosity seemed to increase gradually, as judged by the time it took a deformed tectorial membrane to return to its position of rest. Nevertheless, in one preparation in which the dye concentration was increased from the 0.1% used routinely in our experiments to 1%, the membrane became almost black, nearly opaque and stiff like a board, in agreement with Békésy's (1953, 1960) observations under similar conditions.

On several occasions, the tectorial membrane deformation by lateral displacement of the special purpose micropipettes was photographed by S. Chamberlain. The photographs are shown in Figs. 5.5 to 5.9 (Zwislocki, 1988; Zwislocki, Chamberlain, & Slepecky, 1988). In Fig. 5.5 (Zwislocki, 1988), the micropipette is inserted frontally into the marginal zone and displaced transversally, away from the organ of Corti. It can be seen in the middle top part of the photomicrograph, approaching vertically the co-

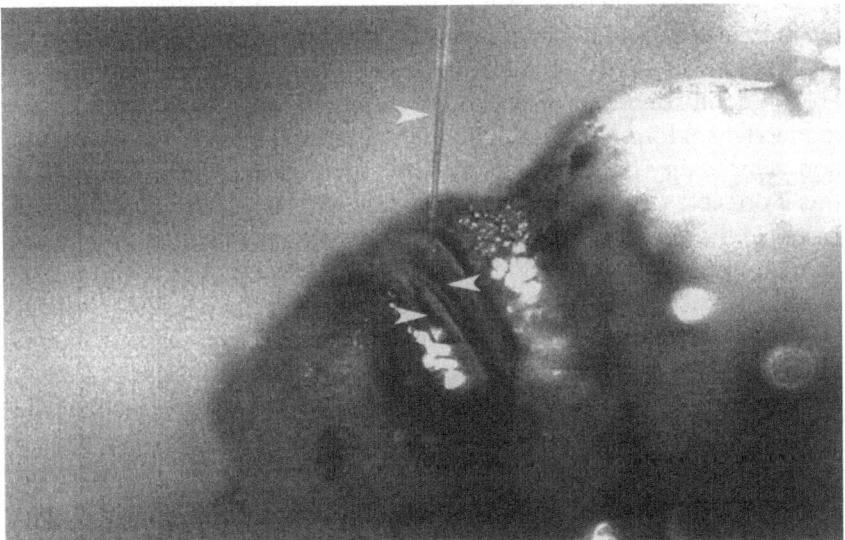

FIG. 5.5. Transversal indentation of the tectorial membrane (left-most arrow head) produced by the micropipette (uppermost arrow head) inserted frontally into the marginal zone of the membrane, as seen through the surgical opening in the lateral wall of scala media. The right-most arrow head points to Hensen's cells. Janus green stain (modified from Zwislocki & Cefaratti, 1989). Reprinted from *Hearing Research, 41*, J. J. Zwislocki & L. K. Cefaratti, Tectorial membrane II: Stiffness measurements in vivo, 211–228, copyright © 1989, with permission from Elsevier Science.

PHYSICAL CONSTANTS AND FUNDAMENTAL CHARACTERISTICS 189

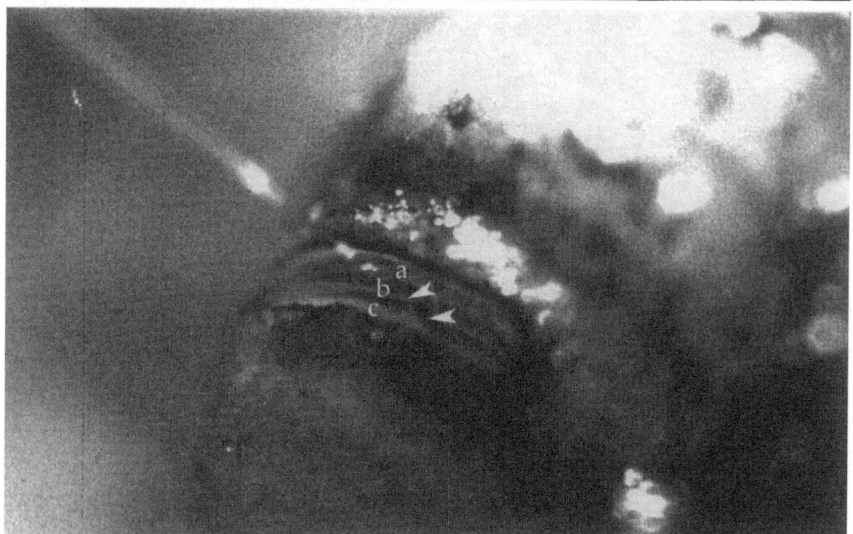

FIG. 5.6. Cochlear structures stained with Janus green, as seen through the surgical opening. The letters a, b, and c indicate Hensen's cells, marginal, and middle zones of the tectorial membrane, respectively. The tectorial membrane is indented radially toward Hensen's cells by the microelectrode inserted in its marginal zone, as indicated by the arrow heads (modified from Zwislocki, 1988). From "Mechanical Properties of the Tectorial Membrane in Situ," by J. J. Zwislocki, 1988, *Acta Oto-Laryngologica, 105,* pp. 450–456. Copyright © 1988 by Taylor & Francis. Reprinted with permission.

chlear opening and appearing bent to the left at the interface between the ambient air and the cochlear fluid. The marginal zone of the tectorial membrane appears as a dark band curved in its upper part away from the organ of Corti as a result of the micropipette pull. The organ of Corti is seen as a light band to the right of the tectorial membrane and to the left of the Hensen's cell mass. In Fig. 5.6 (Zwislocki, 1988), the micropipette approaches the cochlear opening from the upper left and penetrates the marginal zone dorsally, pulling the tectorial membrane toward Hensen's cells that appear as a mottled band in the upper part of the opening. The resulting indentation in the dark band of the marginal zone can be clearly seen. It is quite narrow and has exponential skirts on both of its sides. This shape of the skirts is significant. It indicates that bending moments are quite small in the longitudinal direction of the tectorial membrane, and that its elasticity is controlled mainly by shear forces in this direction.

Another example of dorsal insertion of the micropipette into the marginal zone is shown in Fig. 5.7A (Zwislocki, Chamberlain, & Slepecky, 1988). The micropipette, almost invisible, is marked by the highlighted ar-

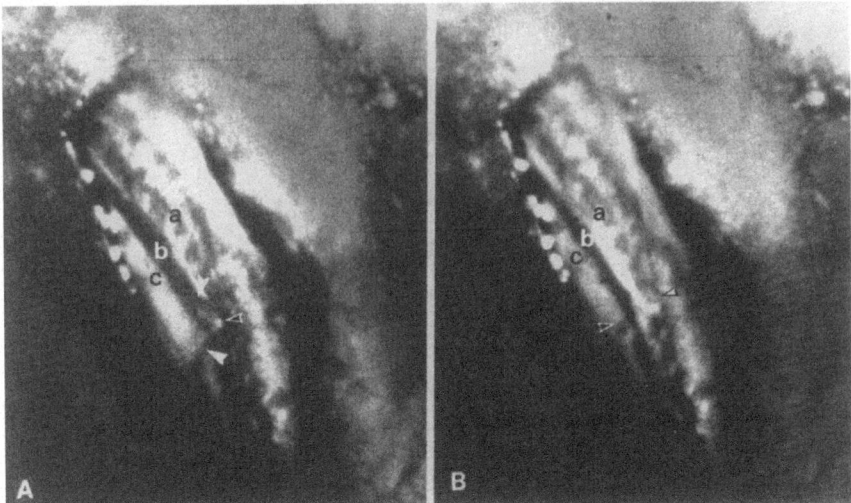

FIG. 5.7. Shape of radial indentation of the tectorial membrane produced by a micropipette inserted dorsally into its marginal zone. In both panels, the letters a, b, and c indicate Hensen's cells, and the marginal, and middle zones of the membrane, respectively. In panel A, the membrane is indented (white arrow heads) by the micropipette inserted at the place of the highlighted arrow head. In B, the membrane is in its position of rest after withdrawal of the micropipette. The lower highlighted arrow head shows the location of the micropipette insertion, the upper one, a lesion in Hensen's cell mass produced by the micropipette during its withdrawal (from Zwislocki et al., 1988). Reprinted from *Hearing Research, 33,* J. J. Zwislocki, S. C. Chamberlain, & N. B. Slepecky, Tectorial Membrane I: Static mechanical properties in vivo, 207–222, copyright © 1988, with permission from Elsevier Science.

rowhead at its point of insertion. It indents the tectorial membrane toward the mottled appearing band of Hensen's cells in the right part of the cochlear opening. In spite of only light Janus green staining, the lines of mechanical strain caused by the micropipette pull can be discerned, especially, below the indentation. Their approximately exponential pattern confirms the impression gained from Fig. 5.6 that bending moments are not important in the length direction of the tectorial membrane. Panel B of the figure shows the state of the tectorial membrane after the micropipette withdrawal. The tectorial membrane is back in its undisturbed position, as indicated by its marginal zone appearing as a smooth, dark, narrow band somewhat to the right of the left margin of the cochlear opening. The only trace of the past insertion of the micropipette is a faint dot marked by the highlighted arrowhead to the left. Because the micropipette was bent prior to its withdrawal from the tectorial membrane, it snapped back into Hensen's cells producing a lesion marked by the right-hand arrowhead.

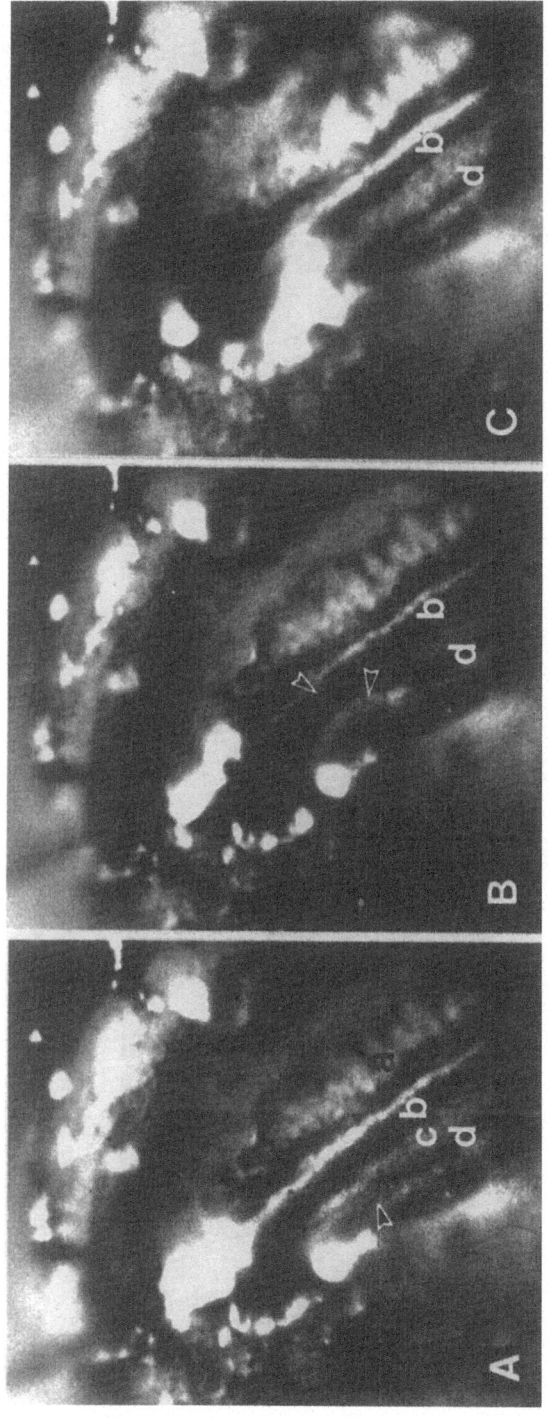

FIG. 5.8. Cochlear structures, as seen through the surgical opening in the lateral wall of scala media. In all panels, the letters a, b, c, and d indicate Hensen's cells, and the marginal, middle, and limbal zones of the tectorial membrane, respectively. In A, the micropipette is inserted into the limbal zone (arrow head) but not displaced; in B, it indents the limbal zone radially (lower arrow head) without indenting the marginal zone (upper arrow head); in C, the micropipette is withdrawn and the tectorial membrane is back in its rest position (from Zwislocki et al., 1988). Reprinted from *Hearing Research*, 33, J. J. Zwislocki, S. C. Chamberlain, & N. B. Slepecky, Tectorial Membrane I: Static mechanical properties in vivo, 207–222, copyright © 1988, with permission from Elsevier Science.

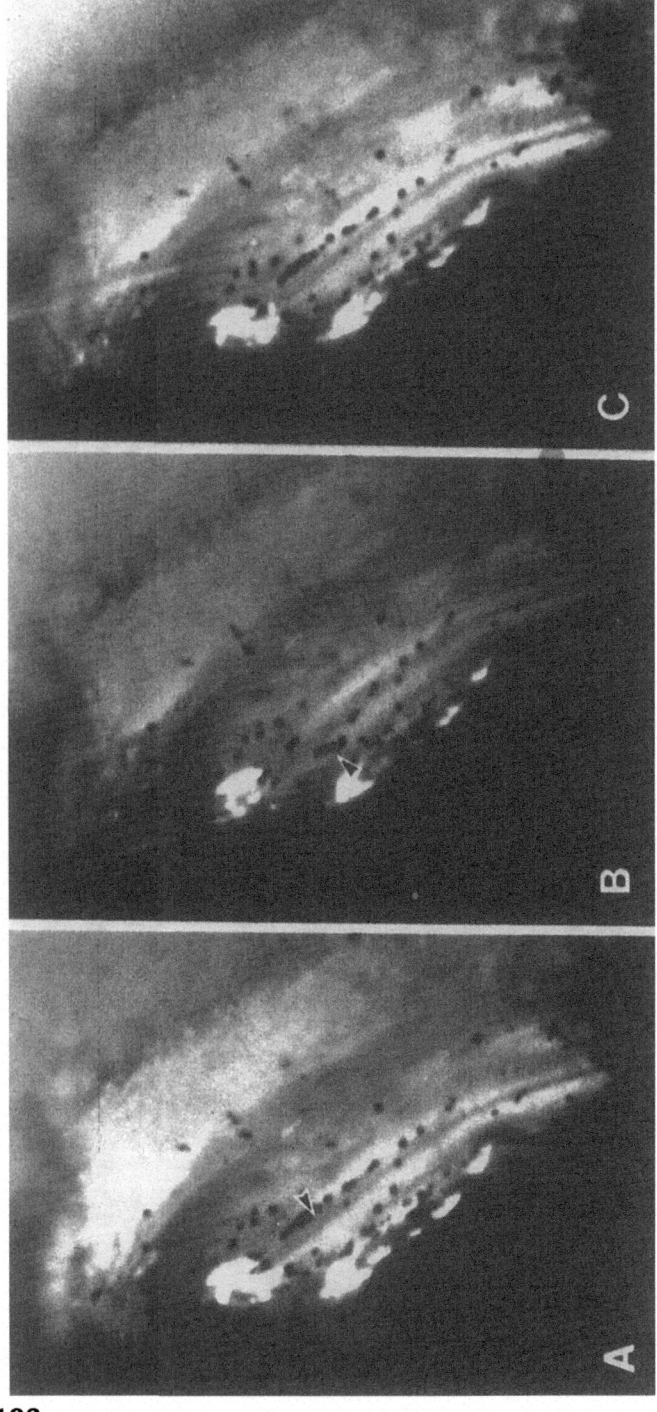

FIG. 5.9. Deformation of the tectorial membrane indicated by microspheres instead of a dye. In A, the arrow head points to microspheres aligned along the tectorial membrane marginal zone. In B, a micropipette inserted dorsally into the marginal zone pulls it toward the limbus (arrow head). In C, the micropipette is withdrawn, and the tectorial membrane is back at rest (from Zwislocki et al., 1988). Reprinted from *Hearing Research, 33,* J. J. Zwislocki, S. C. Chamberlain, & N. B. Slepecky, Tectorial Membrane I: Static mechanical properties in vivo, 207–222, copyright © 1988, with permission from Elsevier Science.

The photo sequence of Fig. 5.8 shows what happens when the micropipette is inserted into the tectorial membrane near the spiral limbus and moved toward Hensen's cells. The mottled appearing mass of Hensen's cells is on the right side of the opening, the dark band produced by the tectorial membrane's marginal zone is somewhat to the left of the middle of the opening. The point of insertion of the micropipette is indicated by the highlighted arrowhead to the left in panel A. In panel B, the micropipette has been displaced toward Hensen's cells, producing a bow shaped deformation pattern within the tectorial membrane, as indicated by the lower highlighted arrowhead. In panel C, the micropipette is withdrawn and the tectorial membrane appears completely undisturbed in its resting position. Note in panel B that the membrane is stretched in its limbal zone, as indicated by its lighter color, but compressed at the marginal zone, where it is dark. The marginal zone itself appears to be hardly affected at all. The pattern suggests that the marginal zone is considerably stiffer than the membrane's more limbal parts.

Because the mechanical properties of the tectorial membrane appeared to depend to some extent on the dye and its concentration, some experiments were undertaken without staining. Instead, latex microspheres 9 μm in diameter were used. The microspheres were introduced into the opened scala media with the help of the micropick. With gentle prodding, they sank onto the various visible structures of the cochlea, predominantly, the tectorial membrane. When there was no gap between the tectorial membrane and Hensen's cells, some of the microspheres gathered at the membrane's margin, others could be guided there by means of the eyelash instrument. To facilitate the procedure, the practically completely transparent cochlear structures were stained faintly with Janus green in a 0.03% solution. Subsequently, the dye was allowed to completely diffuse out of the tectorial membrane before its manipulation begun. Some faint staining remained at the locations of the pillars of Corti and Hensen's cells for the duration of the experiment and provided helpful landmarks. The experiments were successful in three preparations. An example is illustrated in Fig. 5.9 (Zwislocki, Chamberlain, & Slepecky, 1988). As is apparent in panel A of the figure, the microspheres are distributed over all the visible structures. Nevertheless, their greatest concentration, marked by the highlighted arrowhead, coincides with the margin of the tectorial membrane located slightly to the right of the pillars of Corti that appear as a narrow faint band to the left of the faintly stained mass of Hensen's cells. In panel B, a micropipette is inserted into the tectorial membrane and pulls it to the left, toward the spiral limbus. Its point of insertion is marked by the highlighted arrowhead. The deformation of the membrane is indicated best by the row of the microspheres aligned along the membrane's margin. Accordingly, the deformation is maximum at the arrowhead, and decreases exponentially on its sides. The pattern is similar to that obtained before with the Janus green dye in a 0.1% concentration used predominantly in our experiments and exemplified in Figs. 5.6 and 5.7. To verify the similarity more directly, Janus green in a 0.1% concentration was injected into the preparations containing the microspheres at the end of each experiment. When the micropipette was reinserted into the tectorial membrane and displaced toward the spiral limbus, the same deformation pattern resulted as without the dye. We were able to conclude, therefore, that the 0.1% concentration of Janus green did not noticeably affect the elastic properties of the tectorial membrane. Panel C shows the recovery of the tectorial membrane from its deformed state in panel B, after withdrawal of the micropipette. The recovery is complete since the microspheres are aligned in the same way as before the insertion of the micropipette (panel A).

Quantitative measurements of the tectorial membrane stiffness were performed with the same kind of micropipettes as the qualitative exploration of its elastic properties. For this purpose, the micropipettes' bend caused by their lateral pull after insertion into the tectorial membrane was calibrated. In the past, Flock and Strelioff (1984) used fine quartz rods of known modulus of elasticity, diameter and length for measurement of the stiffness of OHC stereocilia. They calculated the rods' stiffness from these parameters. However, we found the micropipettes to be easier to handle. It was possible to draw them on a standard micropipette puller and hold in a standard micropipette holder. Because their diameter, except near their tips, was relatively large, they could be seen and placed more easily. The one problem that had to be solved was their sufficiently accurate calibration.

The solution was found by inventing a new instrument. It was based on an almost paradoxical physical principle according to which a relatively small force is required to indent a string stretched under tension between two rigid supports. Such a small force was required for calibration of the very fine and flexible micropipettes. Accordingly, a thin nylon fiber was stretched between two pairs of crossed Teflon knife edges. Teflon was used to minimize friction. As shown in Fig. 5.10 (Zwislocki & Cefaratti, 1989), the fiber was fixed at one end to a support just beyond one pair of the knife edges and to a small weight over a pulley at the other. The weight stressed the fiber with a force

$$F = gM \qquad 5.1$$

where g means the gravity acceleration and M, the mass of the weight. According to known principles, the same force was present everywhere in the fiber. When a micropipette was made to press on the fiber in its middle between the knife edges, it produced a restoring force which, according to the geometric relationships drawn in the figure, amounted to

$$F_x = 2F \sin\Theta \qquad 5.2$$

For small Θ angles that occurred in the measurements, it is approximately true that $\sin\Theta = \tan\Theta = d/l$, where d is the deflection of the fiber from its resting position and 2l, the length of the fiber between the knife edges. Accordingly,

$$F_x = 2Fd/l \qquad 5.3$$

In the instrument used in the experiments, the nylon fiber had a diameter of 15 µm and a length between the knife edges of $2l = 6.34$ cm. The mass of the weight was $M = 0.277$ g. The weight consisted conveniently of a small screw and one to several nuts screwed onto it. In this way, the mass could be changed within useful limits by changing the number of nuts.

The string instrument, as we called the instrument schematized in Fig. 5.10, had several advantages. It was rather simple and stable, and its calibration required only two easy measures—the length of the fiber between the knife edges and the weighting mass, both within comfortable ranges of magnitude. Because it was possible to make the ratio d/l quite small, a large force

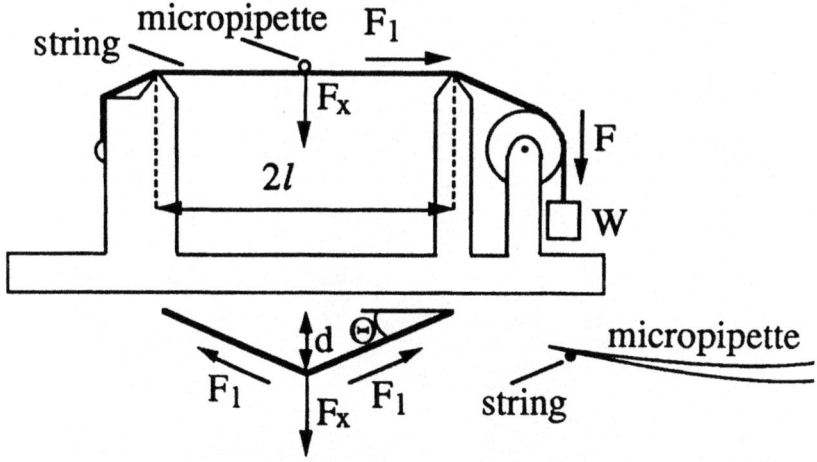

FIG. 5.10. Schematic drawing of the "String instrument" for calibration of the micropipette stiffness (explanation in text; from Zwislocki & Cefaratti, 1989). Reprinted from *Hearing Research, 41*, J. J. Zwislocki & L. K. Cefaratti, Tectorial membrane II: Stiffness measurements in vivo, 211–228, copyright © 1989, with permission from Elsevier Science.

reduction could be achieved. Fractions of a dyne could be measured. The lower limit was dictated by the optical resolution of the displacement, d, and by the mass, M, which had to be made large enough to overcome friction in the knife supports of the fiber. Although the string instrument did not require any direct calibration, we checked it by placing small paper wedges on the nylon fiber and measuring its resulting deflection. Subsequently, the wedges were weighed on a chemical balance. The difference between the indirect and direct calibration values was on the order of 0.2%.

Photographs of the calibration instrumentation are shown in Figs. 5.11 and 5.12. The first shows the string instrument, the Narishige micromanipulator with a micropipette behind it, and the microscope for measuring the fiber's deflection in front of it. The deflection was measured with the help of the microscope's graticule at a magnification of 80×. The second figure shows a magnified view of the string instrument with a micropipette held on the top of the stretched fiber. The micropipette was manipulated with the micromanipulator until it touched the fiber at a distance of 15 μm from its tip. The distance was chosen to match half the in-

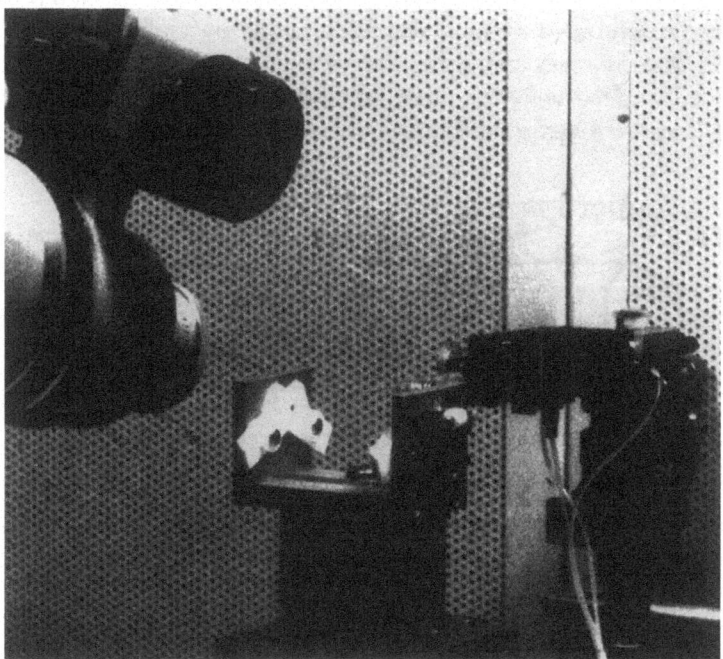

FIG. 5.11. Set up for calibration of the micropipette stiffness, with the microscope to the left, the string instrument in the middle, and the three axial micromanipulator to the right (from Zwislocki & Cefaratti, 1989). Reprinted from *Hearing Research, 41*, J. J. Zwislocki & L. K. Cefaratti, Tectorial membrane II: Stiffness measurements in vivo, 211–228, copyright © 1989, with permission from Elsevier Science.

PHYSICAL CONSTANTS AND FUNDAMENTAL CHARACTERISTICS

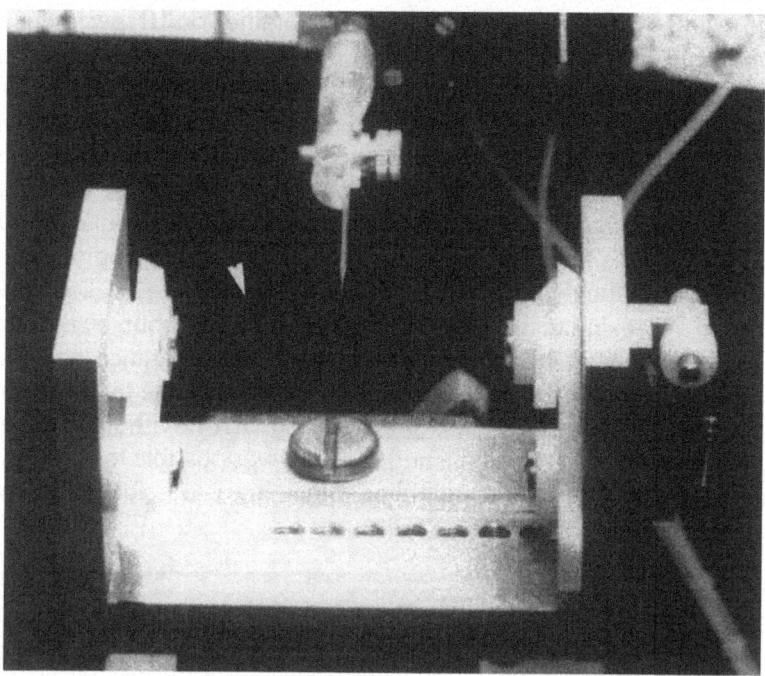

FIG. 5.12. Photograph of the string instrument with a micropipette placed on it from the top. The arrow head points to the string (from Zwislocki & Cefaratti, 1989). Reprinted from *Hearing Research, 41*, J. J. Zwislocki & L. K. Cefaratti, Tectorial membrane II: Stiffness measurements in vivo, 211–228, copyright © 1989, with permission from Elsevier Science.

sertion depth of the fiber in the tectorial membrane. As already mentioned, the micropipettes were inserted 30 μm deep, almost through the entire thickness of the tectorial membrane. An approximately homogeneous distribution of forces within the membrane was assumed, so that the equivalent lumped force should have acted at half the insertion depth. By changing the distance of the contact with the fiber from the micropipette tip by several μms, it was possible to demonstrate that the distance was not critical in the micropipette calibration. The position of the micromanipulator at which the micropipette first touched the fiber was noted within ± 1μm and the micropipette lowered in steps while measuring the fiber's indentation. The advance of the micromanipulator, x, minus the fiber's deflection gave the overall bend of the micropipette — $b = x - d$. It should be mentioned that best viewing of the micropipette tip and the fiber was achieved when the microscope axis was at an angle of 30° with respect to the micropipette axis. As a consequence, the fiber displacement seen through the microscope had to be divided by the cosine of this angle.

The results of the micropipette calibration obtained on 10 specimens are shown in Fig. 5.13. They are plotted in terms of the force exerted on a micropipette as a function of the micropipette's overall bend. The calibrated micropipettes were divided into two equal groups. One group (open symbols) consisted of fresh micropipettes pulled about one hour before their calibration, the other, of old ones (closed symbols), pulled 1 to 2 days before their calibration. The separation between the two populations was necessary because the impression was gained that the micropipettes became stiffer with the delay from the time at which they were pulled. It was possible to approximate the means of the data for each group by a straight line, its tangent giving the stiffness. There was a 20% difference in stiffness between the two groups. However, the difference was due mainly to one deviant fresh specimen (inverted triangles in Fig. 5.13). Without this specimen, the difference was insignificant, so that it was possible to average all the data together with one straight line. Since it was possible to average

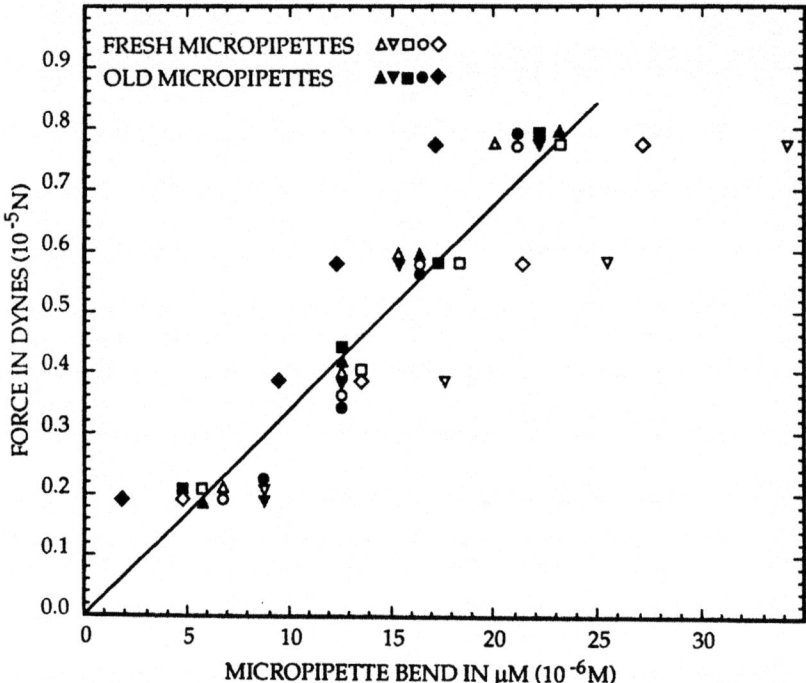

FIG. 5.13. Calibration curve for the measuring micropipettes. The points indicate individual measurements (from Zwislocki & Cefaratti, 1989). Reprinted from *Hearing Research, 41,* J. J. Zwislocki & L. K. Cefaratti, Tectorial membrane II: Stiffness measurements in vivo, 211–228, copyright © 1989, with permission from Elsevier Science.

them by a straight line, and a straight line fitted the individual data of each individual micropipette reasonably well, the stiffness of the micropipettes did not depend on the force acting on them within the range of our measurements. According to the straight line of Fig. 5.13, the average stiffness amounted to 340 dyne/cm (0.34 N/M). The stability of the micropipette stiffness was important in our experiments because, for technical reasons, the measurements of the tectorial membrane stiffness had to be made several hours after the micropipettes had been pulled.

The tectorial membrane stiffness was measured in two planes, transversally to the membrane and radially. As indicated in Fig. 5.1., it was possible to perform the measurement of the transversal stiffness directly, by inserting the measuring micropipette into the tectorial membrane frontally, parallel to its dorsal surface. The measurement of the radial stiffness, on the other hand, had to be obtained indirectly because it was not possible to insert the micropipette at a right angle to the dorsal surface. The best angle that could be achieved was on the order of 50°, so that the micropipette pulled the tectorial membrane in a direction that had both radial and transversal components. The radial stiffness had to be calculated from that measured in the direction of pull by subtracting vectorially the stiffness measured in the transversal direction. The situation is schematized in Fig. 5.14, where the micropipette is symbolized by the long arrowhead. In the uppermost drawing, the micropipette is inserted frontally and pulls the membrane downward with a force F_T, producing a displacement y_T. The corresponding point stiffness is calculated as a quotient of the two variables

$$S_{PT} = (F_T/y_T) \qquad 5.4$$

In the second and third drawings from the top, the micropipette is inserted into the tectorial membrane dorsally at an angle ϕ, producing the displacement y with the components y_T and y_R, as shown in the middle drawing, and the force F with the components F_T and F_R, as shown in the bottom drawing. The displacement components are related to the resultant displacement by the equations

$$y_T = y \cos \phi \qquad 5.5a$$

$$y_R = y \sin \phi \qquad 5.5b$$

The resultant force can be expressed in terms of the component forces as

$$F^2 = F_T^2 + F_R^2 \qquad 5.6$$

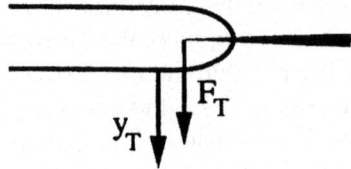

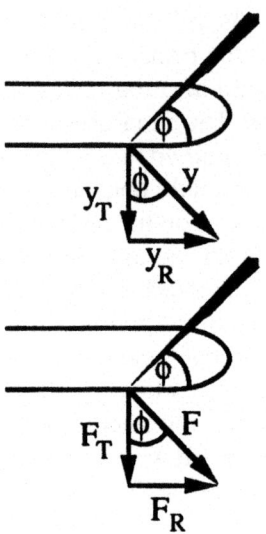

FIG. 5.14. Illustration of the vector analysis associated with the measurement of the tectorial membrane stiffness. At the top, frontal micropipette insertion (long arrow head). The two lower drawings are for dorsal insertion (from Zwislocki and Cefaratti, 1989). Reprinted from *Hearing Research, 41,* J. J. Zwislocki & L. K. Cefaratti, Tectorial membrane II: Stiffness measurements in vivo, 211–228, copyright © 1989, with permission from Elsevier Science.

The radial point stiffness is defined in analogy to the transversal point stiffness as

$$S_{PR} = (F_R/y_R) \qquad 5.7$$

It is possible to eliminate the force components, F_T and F_R, in equation 5.6 with the help of equations 5.4 and 5.7,

$$F^2 = S_{PT}^2 y_T^2 + S_{PR}^2 y_R^2 \qquad 5.8$$

Because of equations 5.5$_{a,b}$, equation 5.8 can be rewritten in the form

$$F^2 = S_{PT}^2 y^2 \cos^2\phi + S_{PR}^2 y^2 \sin^2\phi \qquad 5.9$$

Since the variables F, y, S_{PT} and ϕ can be measured directly, the radial stiffness can be obtained from the latter equation as follows

PHYSICAL CONSTANTS AND FUNDAMENTAL CHARACTERISTICS 201

$$S_{PR}^2 = (F^2/y^2\sin^2\phi) - (S_{PT}^2/\tan^2\phi) \qquad 5.10$$

or

$$S_{PR}^2 = [(F^2/y^2\sin^2\phi) - (S_{PT}^2/\tan^2\phi)]^{1/2} \qquad 5.11$$

Examples of tectorial membrane deformation by a micropipette are shown in Figs. 5.5 to 5.9. Raw numerical data for the transversal deformation expressed in terms of micropipette bend as a function of the deformation are shown in Fig. 5.15. The open circles indicate the individual values obtained on 4 specimens and the closed circles, their means. It should be noted that, although the individual data exhibit considerable scatter, every set of the individual values parallels the mean curve. Because the tectorial membrane stiffness is proportional to the slope of the curves and not to their vertical position, it appears that the individual variation of tectorial membrane stiffness was surprisingly small. To bring this better into evidence, the data of Fig. 5.15 have been normalized in Fig. 5.16 with the help of additive constants so that each set of individual points traces a curve going through the origin. The small scatter of the normalized data confirms the impression given by the raw data that the tectorial membrane stiffness in the transversal direction varies little among individual animals. The fact that the curve approximating the data is straight up to a tectorial membrane deformation of about 35 µm indicates that the stiffness is constant within this range. For greater deformations, the slope of the curve increases, indicating an increasing stiffness. As to the parallel deviations of the individual sets of the raw data from the mean curve in Fig. 5.15, it is likely that the deviations were caused in part by the hysteresis of the micromanipulator settings and in part by the difficulty in inserting a micropipette into the tectorial membrane without causing any change in its undisturbed position. It should be further noted that the data of Fig. 5.15 were obtained by pulling the tectorial membrane away from the organ of Corti. The reverse direction was blocked by this organ. It is not possible to decide on the basis of our data, therefore, if the transversal stiffness of the tectorial membrane is direction dependent.

Means of the raw data obtained for the radial stiffness of the tectorial membrane are shown in Fig. 5.17. The values indicated by the open circles have been obtained with the micropipette pulling the tectorial membrane toward the spiral limbus, those indicated by the closed circles, with the membrane being pulled toward Hensen's cells. The vertical bars show the standard deviations of the individual data obtained on 7 specimens. It is important to note that, up to a displacement of about 25 µm, the slope of

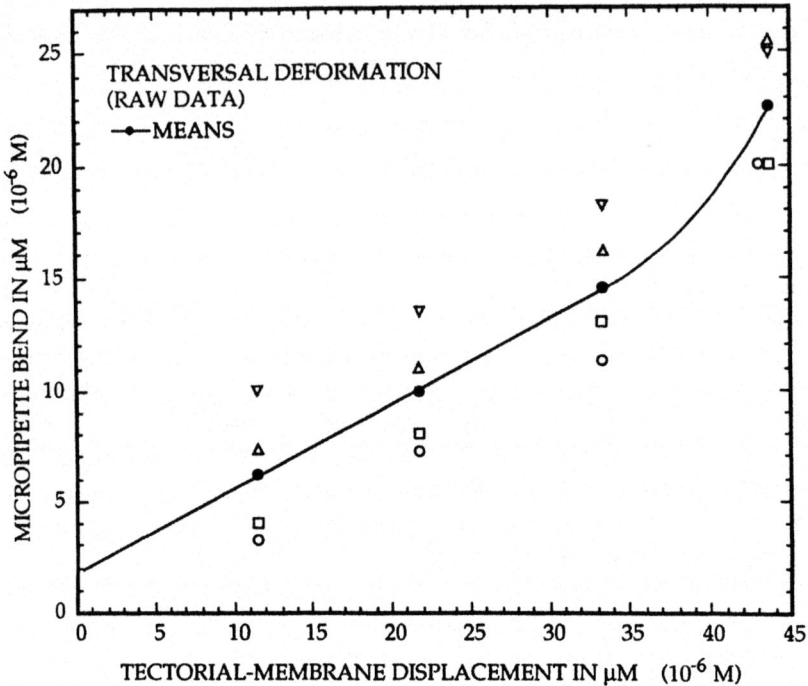

FIG. 5.15. Micropipette bend as a function of the tectorial membrane indentation for frontal micropipette insertion. Individual data and means (from Zwislocki and Cefaratti, 1989). Reprinted from *Hearing Research*, 41, J. J. Zwislocki & L. K. Cefaratti, Tectorial membrane II: Stiffness measurements in vivo, 211–228, copyright © 1989, with permission from Elsevier Science.

the curves fitted to the data is independent of the direction of pull, so that the stiffness of the tectorial membrane is the same, whether it is pulled toward the limbus or Hensen's cells. Note also that the mean curves go through the origin. Beyond a displacement of 25 μm, the curves for the limbus and Hensen's cell directions separate. In the former, the slope simply increases, indicating an increasing stiffness. In the latter, there appears a discontinuity producing a locally decreased slope. The discontinuity is very likely associated with damage to the membrane. In some specimens, a tear appeared in it.

It should be noted here that the numerical values of the tectorial membrane displacement plotted in Figs. 5.15 to 5.17 are those measured directly with the microscope graticule. Because they were viewed at an angle of about 30°, they appeared shorter by about 13% than they were in reality. However, since a correction would have applied almost equally to the

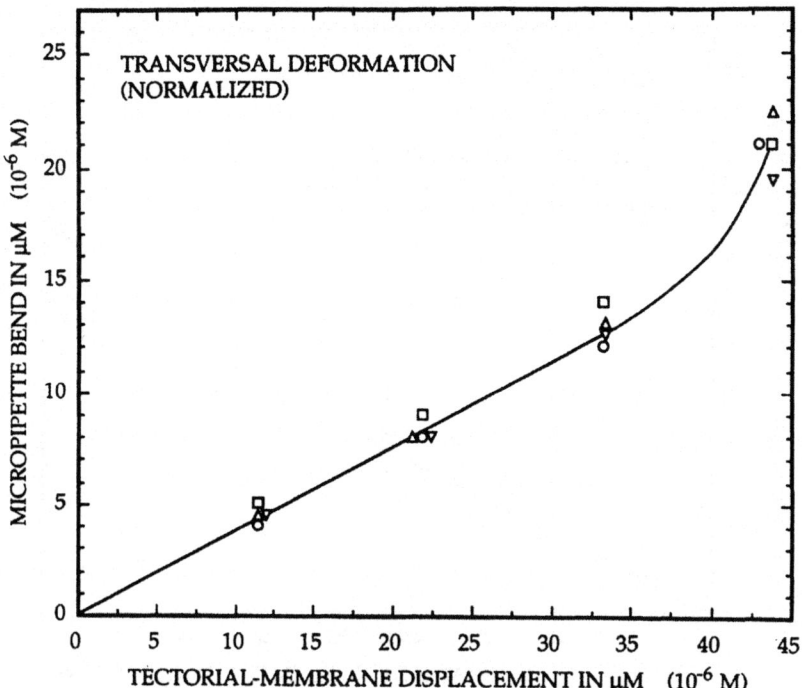

FIG. 5.16. Same points as in Fig. 5.15 normalized to the origin by means of subtractive constants (from Zwislocki and Cefaratti, 1989). Reprinted from *Hearing Research*, 41, J. J. Zwislocki & L. K. Cefaratti, Tectorial membrane II: Stiffness measurements in vivo, 211–228, copyright © 1989, with permission from Elsevier Science.

micropipette bend and would not have changed significantly the relationship between the bend and the displacement and, with it, the calculated stiffness values of the tectorial membrane, it was omitted.

The stiffness values of the tectorial membrane in both the transversal and radial directions were obtained from the curves of Figs. 5.16 and 5.17 with the help of the calibration curve of Fig. 5.13. For example, according to Fig. 5.16, a deformation of the tectorial membrane of 20 µm at the point of micropipette insertion was associated with a micropipette bend of 7.7 µm. According to the calibration curve, this required a force of 0.25 dyne. By associating this force with the membrane deformation, as required by equation 5.4, a point stiffness, S_{PT} = 0.25 dyne/20 µm or 125 dyne/cm (0.125 N/M) is obtained. With respect to the radial stiffness according to Fig. 5.17, a displacement of the tectorial membrane of 20 µm was associated with a micropipette bend of 7.1 µm. With the help of the micropipette calibration curve, we see that such a bend required a force of 0.24 dyne.

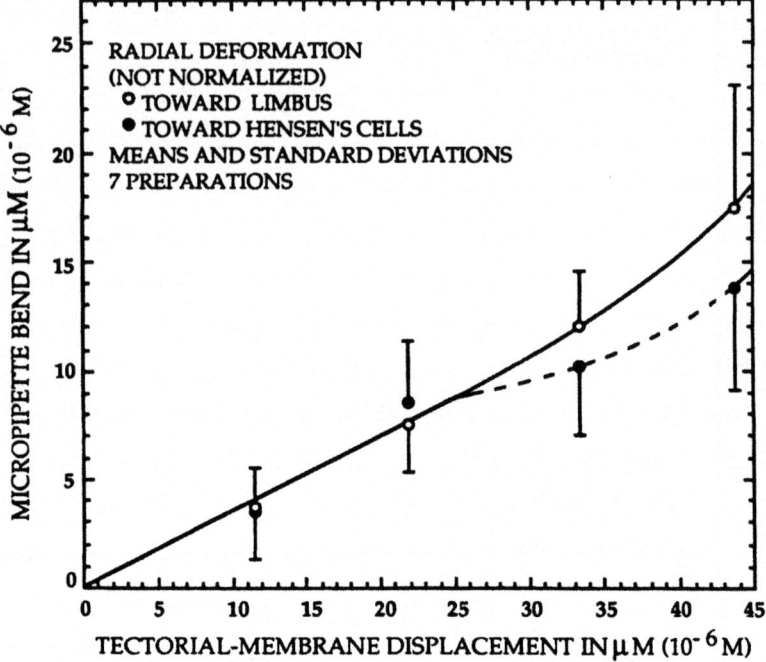

FIG. 5.17. Micropipette bend as a function of tectorial membrane indentation for dorsal micropipette insertion. Open circle are for the tectorial membrane pull toward the limbus, the closed circles, for the pull toward Hensen's cells (from Zwislocki and Cefaratti, 1989). Reprinted from Hearing Research, 41, J. J. Zwislocki & L. K. Cefaratti, Tectorial membrane II: Stiffness measurements in vivo, 211–228, copyright © 1989, with permission from Elsevier Science.

Because the micropipettes were not inserted perpendicularly into the tectorial membrane but at an angle of approximately 50°, the stiffness has to be determined from equation 5.11. With the values stated above, it becomes $S_{PR} = 116$ dyne/cm (0,116 N/M). According to equation 5.11, this value is quite insensitive to the angle of the micropipette insertion. Its variation from 45° to 60° produces a stiffness variation of only 8.6%. It is of some interest that the tectorial membrane stiffness was found to be practically the same in both the transversal and radial directions, in complete contradiction of the assumption implied in the classical model of the shear motion between the tectorial membrane and the reticular lamina. In that model the tacit assumption is made that the radial stiffness is much greater than the transversal one.

The measured point stiffnesses of the tectorial membrane cannot be applied directly to the calculation of cochlear events, such as the propagation of transversal waves on the cochlear partition. The fluid pressure involved acts on relatively large surfaces, and the cochlear parameters are specified accordingly per cm, or another convenient unit of length of the cochlear canals. It is necessary, therefore to convert the measured point stiffnesses to stiffnesses per unit length. This is possible, provided the mechanical coupling between the adjacent parts of the tectorial membrane can be specified. Of particular interest is the radial stiffness because the membrane is coupled to the organ of Corti and, through it, to the basilar membrane via the incompressible perilymph and the bundles of OHC stereocilia with high longitudinal (axial) stiffness and can hardly move transversally relative to the reticular lamina. As a consequence, the transversal stiffness becomes only a small component of the total partition stiffness.

The character of the longitudinal coupling within the tectorial membrane can be derived and its parameters determined from the shape of the membrane deformation when pulled by an inserted micropipette. As shown in Figs. 5.6 and 5.7, pulling the membrane toward Hensen's cells produces a displacement pattern with a sharp maximum at the micropipette and an exponentially decreasing magnitude on both sides of it. Such a deformation pattern signifies that bending moments in the membrane's longitudinal direction are negligible, and the elastic restoring forces are produced by radial normal and shear forces. The normal force may be due to bending in the radial plane.

The effective radial restoring force of the tectorial membrane per unit of its length may be derived with the help of the diagram of Fig. 5.18. In it the exponential deformation pattern of the membrane is schematized. For mathematical treatment, a Cartesian coordinate system is introduced with the horizontal, x, axis parallel to the margin of the membrane and the vertical, y, axis perpendicular to it and pointing in the radial direction. The origin of the x coordinate is located at the insertion point of the micropipette and that of the y coordinate at the undisturbed position of the membrane. Three dx elements bounded by a curve paralleling the margin of the deformed membrane are specified. In a static situation, the radial restoring force and the shear forces acting on the middle element from both sides must be in equilibrium. Therefore, if the shear force is defined as

$$F_{Sh}(x)dx = \sigma[d^2y(x)/dx^2]dx \qquad 5.12$$

with σ—a coupling constant and y—the magnitude of the deformation, and the radial restoring force as

$$F_{St}(x)dx = -S_R y(x)dx \qquad 5.13$$

with S_R—the radial stiffness per unit length, we must have

$$F_{Sh}(x)dx + F_{St}(x)dx = 0 \qquad 5.14$$

When the expressions of equations 5.12 and 5.13 are introduced for F_{Sh} and F_{St}, the latter equation becomes a differential equation of the form

$$\sigma\,[d^2y(x)/dx^2]dx - S_R y(x)dx = 0 \qquad 5.15$$

Its solution for $x>0$ is

$$y_a = y_a(0)\,\exp(-x/\lambda_a) \qquad 5.16$$

where $\lambda_a = (\sigma/S_R)^{1/2}$ is the longitudinal space constant of the tectorial membrane and defines the distance over which the deformation decays to 1/e—a little less than 1/3. For symmetry reasons, we must have

$$y_b = y_b(0)\,\exp(x/\lambda_b) \qquad 5.17$$

for $x<0$ with $\lambda_b = (\sigma/S_R)^{1/2}$. There is no reason to believe that the space constants should be different on the two sides of the micropipette insertion, and no asymmetry of tectorial membrane deformation was observed after

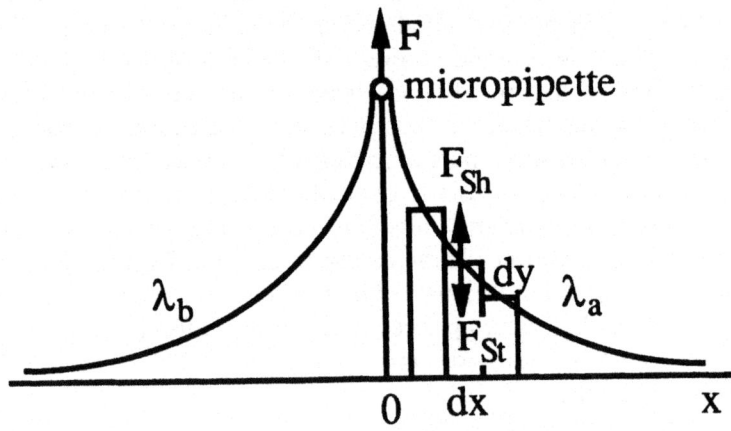

FIG. 5.18. Graphic illustration of the deformation of the tectorial membrane for the derivation of its radial stiffness per unit of cochlear length (from Zwislocki & Cefaratti, 1989). Reprinted from *Hearing Research, 41*, J. J. Zwislocki & L. K. Cefaratti, Tectorial membrane II: Stiffness measurements in vivo, 211–228, copyright © 1989, with permission from Elsevier Science.

correction for the angle of viewing, so that $\lambda_a = \lambda_b = \lambda$. The total reaction force on the micropipette must be equal to the sum of the elemental forces, $F_{St}(x)dx$, as defined by equation 5.13. It can be obtained by multiplying both sides of equations 5.16 and 5.17 by the stiffness constant, S_R, and integrating over x. The resultant equation for both sides of the micropipette location taken together is

$$F = S_R y(0)\, 2\lambda \qquad 5.18$$

Because F, $y(0)$, $S_{PR} = F/y(0)$ and λ are known through measurement, the radial stiffness of the tectorial membrane per unit length can be calculated from the latter equation. After its appropriate transformation, we obtain

$$S_R = S_{PR}/2\lambda \qquad 5.19$$

Measurement of the space constant of the tectorial membrane with the microscope graticule on both sides of the micropipette insertion, when corrected for the angle of viewing, gave an approximate measure of the space constant of $\lambda = 100\ \mu m$ which varied somewhat from preparation to preparation. As an example, the space constant can be estimated from Fig. V. 19 on the basis of the microsphere diameter that averaged 9 µm, and the angle of viewing represented by the angle between the tectorial membrane margin delineated by the microspheres and a horizontal. Accordingly, the radial stiffness per cm length of the tectorial membrane in the second turn of the Mongolian gerbil cochlea becomes roughly $S_R = 6 \times 10^3$ dyne/cm² (6×10^2 N/M²). The analogous transversal stiffness is almost the same.

As is shown in the subsequent section of this chapter and the next, it is possible to verify the value of the radial stiffness through measurement of dynamic processes in the cochlea. The first such verification is introduced in the next section. Unfortunately, I was able to determine it at only one cochlear location and do not know how it changes with the distance from the cochlear base.

Effective Mass of the Tectorial Membrane

The mass of the tectorial membrane can be estimated from its cross sectional area and its density (e.g., Zwislocki & Cefaratti, 1989). As judged from our histological preparations, the width of the tectorial membrane approximated 130 µm at our experimental location and its thickness, 35 µm. Because of shrinkage present in such preparations (e.g., Kronester-Frei, 1979), these dimensions have to be increased by about 20%. With this correction, the numerical value of the cross sectional area comes out to be

6.4×10^{-5} cm^2. Because the tectorial membrane consists almost entirely of hydrated proteins (Iurato, 1962; Kronester-Frei, 1978) its density should be around 1.2 g/cm^3, which, when multiplied by the cross sectional area, produces a mass per cm length of the membrane of 7.7×10^{-5} g/cm.

To verify the result based on histological preparations, we measured the cross sectional dimensions of the tectorial membrane in one of our in vivo preparations. The tectorial membrane was cut approximately radially and bent with the help of the micropick so that its cross section faced the microscope, as illustrated in Fig. 5.19. Under these conditions, the width of the membrane was estimated at 150 µm and its thickness at 40 µm, giving a cross sectional area of 6×10^{-5} cm^2, almost exactly the same as derived from histological preparations.

Having determined the tectorial membrane stiffness that should be representative of its elastic attachment to the spiral limbus, and the tectorial membrane mass, we can calculate the membrane's series resonance in the radial plane on the assumption that the stiffness and the mass can be treated as lumped elements. In view of the observation exemplified in Fig. 5.8 that the membrane is softer near the limbus than toward its marginal zone, this appears permissible. The resonance frequency is determined by the simple equation

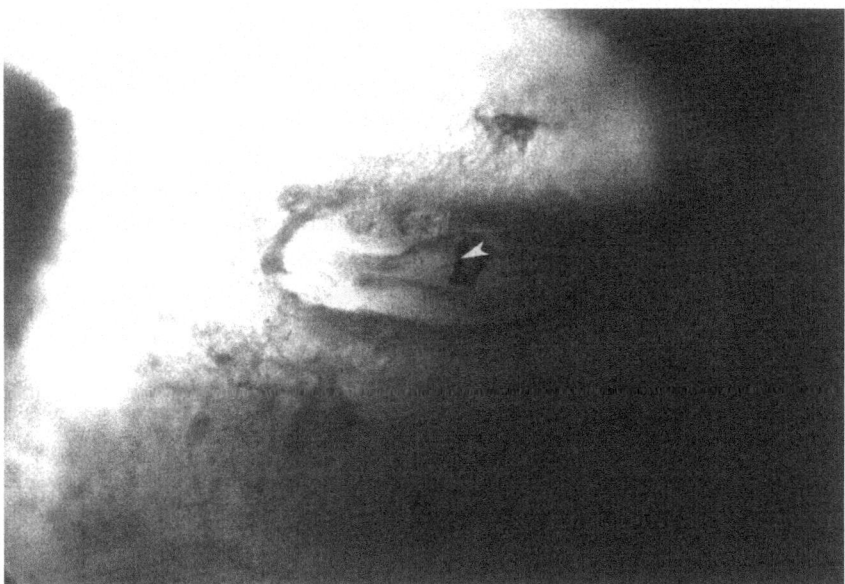

FIG. 5.19. Tectorial membrane seen in the surgical opening in the lateral wall of scala media, bent so as to make its cross section (arrow head) visible.

PHYSICAL CONSTANTS AND FUNDAMENTAL CHARACTERISTICS

$$f_S = \frac{1}{2\pi} * \left(\frac{S_R}{M_T}\right)^{1/2} \qquad 5.20$$

When the numerical values specified above are introduced, a resonance frequency f_S = 1.4 kHz results. As was first shown theoretically by Allen (1980) and confirmed experimentally by Gummer, Hemmert, & Zenner (1996) and is discussed extensively in the next chapter, the resonance should produce a minimum in the shear motion between the reticular lamina and the tectorial membrane and an associated minimum in the hair cell transfer functions, such as shown in Figs. 5.45, 5.46, 5.47, and especially, in Fig. 5.54. Depending on the exact location of recording in the second turn of the Mongolian gerbil cochlea, the frequency of the minimum varied in our experiments from about 0.4 to 0.9 kHz. In association with the experiments on the stiffness of the tectorial membrane, the minimum was found at a frequency of approximately 0.8 kHz (Zwislocki & Cefaratti, 1989), smaller by a factor of 1.75 than the resonance frequency derived from the stiffness measurements of the tectorial membrane. This implies, because of the square root relationship of equation 5.20, that the measured stiffness was too large by a factor of about 3 and should be close to $2.0*10^3$ dyne/ cm². Because, during the experiments, it was observed that the dyes used had the tendency of increasing the stiffness, they could be blamed. A somewhat unnatural way in which the tectorial membrane was deformed by the measuring micropipette could have been another contributing factor. However, it is shown in chapter 6 that the active feedback, as modeled in that chapter, increases the radial vibration amplitude of the tectorial membrane more than it does the corresponding vibration amplitude of the reticular lamina. This produces a shift of the effective shear displacement minimum to lower sound frequencies relative to the resonance frequency of the tectorial membrane. As a consequence, the minimum response at 0.8 kHz or an even lower frequency could be compatible with the tectorial membrane resonance frequency of 1.4 kHz, and the static stiffness measurement correct. It should be added that the low frequency minima found in the hair cell and related transfer functions were always quite flat and hardly discernible in the presence of recording noise. It is shown in the next chapter that such patterns are consistent with the theoretical ones in the presence of the active feedback.

Because, in a good part of the analysis of the cochlear dynamics performed in chapter 6, the feedback is not taken into account, the three times smaller stiffness value of the tectorial membrane is used to better mimic the conditions existing when it is present. However, its interaction with the tectorial membrane stiffness is taken into account, when the feedback is introduced.

PHYSICAL PARAMETER VALUES ASSOCIATED WITH OHC STEREOCILIA

Effective Stiffness of the OHC Stereocilia in Aggregate

There are no systematic measurements of the stiffness of the OHC stereocilia in Mongolian gerbil, our experimental model. Therefore, it was necessary to borrow them from guinea pigs. Appropriate measurements were made by Strelioff and Flock (1984) on individual stereocilia. It is unlikely that, for a corresponding characteristic frequency region, the stiffness is substantially different in the gerbils. Because their tectorial membranes appear to be somewhat larger, however, it is possible that the stiffness is somewhat larger also. Strelioff and Flock made their measurements in vitro on excised organs of Corti. The current body of knowledge suggests that, if any deterioration of the stereocilia occurred under these conditions, it would have led to a decreased stiffness. Furthermore, the stereocilia were detached from the tectorial membrane. The effective stiffness of elastic beams, which the stereocilia resemble, depends on the way their ends are held. A beam held at both ends appears stiffer than a beam that is free at one end. All these considerations suggest that Strelioff and Flock may have measured the lower bound of the stereocilia stiffness.

On the basis of the stiffness measurements of individual stereocilia of Strelioff and Flock (1984), Strelioff, Flock, & Minser (1985) estimated the stiffness of stereocilia bundles. For the cochlear CF region between 1.5 and 2 kHz, corresponding roughly to our experimental region in the gerbils, they obtained a combined stiffness of 40 dyne/cm for a set of 3 OHCs distributed among the three OHC rows. According to their measurements, the contribution of the IHC stereocilia was negligible. They also estimated the space occupied by one OHC, arriving at a cylindrical space of 10 µm diameter. Accordingly, there must be approximately 3×10^3 OHCs per cm of reticular lamina length, approximately the same as the tectorial membrane length, which brings the total aggregate stiffness of the stereocilia bundles to $S_{Hm} = 4 \times 10^4$ dyne/cm^2. If this value is close to being correct, the corresponding radial stiffness of the tectorial membrane appears to be about 7 times smaller according to the static measurements and 20 times smaller according to the dynamic ones. In any event it appears to be smaller by at least one order of magnitude, in complete contradiction of the classical model of the shear motion between the reticular lamina and the tectorial membrane.

It was possible to confirm this conclusion for Mongolian gerbils during our stiffness measurements of the tectorial membrane. In a small number of preparations, the tectorial membrane was found in its normal position, not lifted up from the organ of Corti. This was signaled by the absence of an unstained gap between the stained tectorial membrane and the stained Hensen's cells and by unstained pillars of Corti, when the Janus green dye was used. Already Tonndorf et al. (1962) had found that the tectorial membrane prevented penetration of the dye to the organ of Corti, when in its normal position and sealed to Hensen's cells.

Successful measurement of tectorial membrane stiffness in a specimen, in which it appeared to be in its normal place and, presumably, attached to the OHC stereocilia, revealed a significant phenomenon. When the stiffness measuring micropipette was inserted in it through its dorsal surface, as already described, and moved slowly laterally toward the spiral limbus, no perceptible deformation of the membrane took place until the micropipette bend reached 6 µm. Then, a sudden jump occurred, and the membrane became visibly deformed. Subsequent measurement revealed that, after the jump, a micropipette bend of 6 µm corresponded to a tectorial membrane displacement of approximately 12 µm. This relationship was consistent with preceding measurements on tectorial membranes that were lifted up. It is most likely that the discontinuity in the membrane deformation at the micropipette bend of 6 µm occurred because the membrane was torn from the OHC stereocilia, and that the apparent immobility of the membrane associated with smaller bends was due to a relatively large stiffness of the stereocilia.

The existence of the discontinuity in the tectorial membrane deformation was confirmed in another preparation in which the tectorial membrane seemed to be lifted up near the basal end of the surgical opening in scala media but in its normal position near the apical end of the opening. A measuring micropipette inserted in the tectorial membrane at the former location and moved toward the limbus produced a smoothly increasing deformation. However, when it was inserted at the latter location and moved in the same way, the deformation jump occurred. A perceptible deformation did not occur here until the micropipette bend reached 12 µm.

It may be pointed out that, in these experiments, the micropipette was not inserted in the tectorial membrane perpendicularly but at an angle of approximately 50°, as already described, and its motion toward the spiral limbus pulled the membrane not only toward the limbus but also upward. Under these conditions, the membrane's initial deformation could not have been hampered significantly by any other structure than the OHC stereocilia. Because no perceptible deformation occurred until the micropipette bend reached 6 or 12 µm, a bend associated with a deformation of at least 12 µm in a detached tectorial membrane, and a deformation change in the neighborhood of 1 µm was discernible, it can be concluded that the stiffness of the aggregate stereocilia bundles holding the tectorial membrane was greater by at least one order of magnitude than the stiffness of the membrane.

Mechanical Resistance Associated with the Stereocilia

The internal resistance of the stereocilia is probably negligible, and the main resistance component associated with their motion most likely stems from the fluid friction (Zwislocki, 1980) due to the shear motion between the reticular lamina and the tectorial membrane. Because of the narrow gap between the two structures, a laminar flow can be assumed. Perturbation of this flow by the stereocilia protruding into the gap should be negligible because their motion must be practically identical to the shear motion.

For a laminar flow, the fluid resistance in the subtectorial gap can be calculated according to the following well established formula (Zwislocki, 1980)

$$R_G = d_G \eta_E / h_G \qquad 5.21$$

where R_G means the resistance, d_G, the gap width, η_E, the coefficient of viscosity of the endolymph, and h_G, the height of the gap. From photomicrographs, such as in Fig. 5.1, d_G can be estimated for the cochlear midturn to be on the order of 60 µm, and h_G, on the order of 3 µm (Lim, 1980, for the chinchilla). The coefficient of viscosity of the endolymph, η_G, is nearly the same as of water, which is close to 1×10^{-2} cgs units (Rausch, 1964). With these numerical values, the resistance becomes $R_G = 0.2$ dyne sec/cm^2 per cm length of the cochlea.

PHYSICAL PARAMETER VALUES OF THE BASILAR MEMBRANE AND COCHLEAR WAVE VELOCITY

Stiffness of the Basilar Membrane and Wave Velocity in Mongolian Gerbil Cochlea

To be able to analyze the dynamics of the Mongolian gerbil cochlea, it is necessary to know the stiffness and effective mass of the basilar membrane as well as the geometric dimensions of the cochlea in addition to the parameters already specified. To the best of my knowledge, the stiffness of the Mongolian gerbil basilar membrane was never measured. However, it can be derived from the wave travel time, which was measured (Schmiedt & Zwislocki, 1977), assuming that the waves are sufficiently long relative to the canal depth. For this purpose, it is necessary to first convert the delay time to wave velocity. The fundamental equation specifying the relationship between the two variables is given in chapter 4 as equation 4.45. For a

PHYSICAL CONSTANTS AND FUNDAMENTAL CHARACTERISTICS

velocity that decays exponentially with the distance from the cochlear base, as specified in equation 4.44, the equation can be written as

$$\Delta t = (1/c_{Lo}v)(e^{vx}-1) \qquad 5.22$$

where

$$c_L(x) = c_{Lo}e^{-vx} \qquad 5.23$$

The constants c_{Lo} and v can be determined by fitting the curve described by equation 5.22 to the experimental data. Both are displayed in Fig 5.20. Note that the exponential fit is quite good. It yields a correlation coefficient of 0.999. According to the least squares fit, the following numerical values are associated with it: $c_{Lo} = 94*10^2$ cm/sec or 94 m/sec and $v = 3.1$ cm^{-1}, respectively. The velocity curve according to equation 5.23 and the numerical constants derived above is plotted in Fig. 5.21 as a function of the distance from cochlear base expressed in % of the total cochlear length of about 1.2 cm. It reaches almost 100 m/sec in the base and decays to about 2.5 m/sec at the apex.

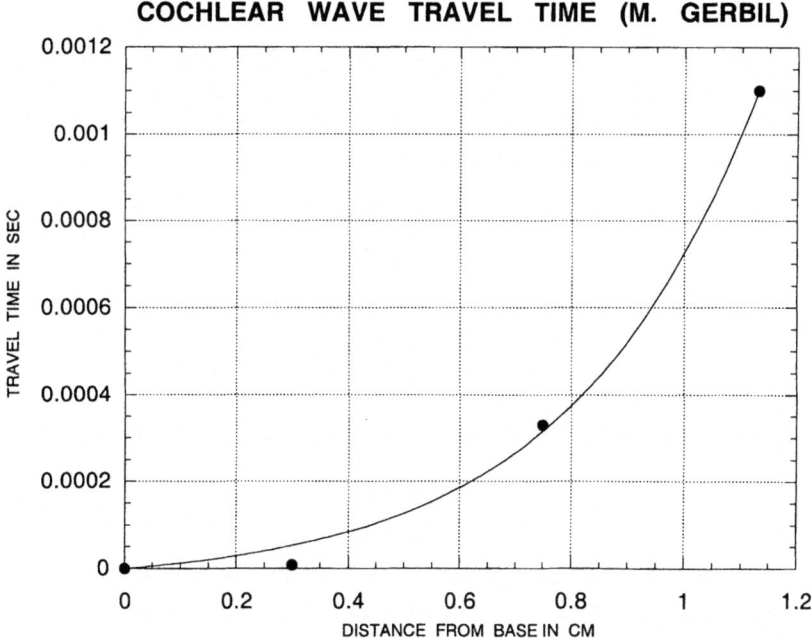

FIG. 5.20. Travel time of waves in the Mongolian gerbil cochlea, as measured by means of CM (circles) and approximated by an exponential function.

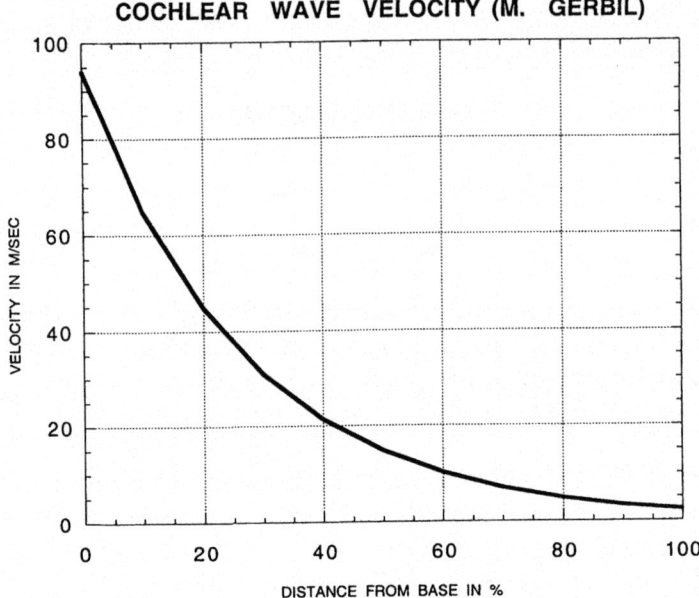

FIG. 5.21. Cochlear wave velocity in Mongolian gerbil, as derived from the wave travel time of Fig. 5.20.

The relationship between the wave velocity and the stiffness, or rather its inverse, the compliance of the basilar membrane for long waves is given in the preceding chapter as equation 4.42. For the purpose of deriving the compliance, the equation can be rewritten as

$$C_{Ba} = \frac{Q}{\rho c_L^2} \qquad 5.24$$

To determine the compliance, the effective cross section of the cochlea, Q, as well as the specific density, ρ, of the cochlear fluids have to be specified in addition to the long wave velocity. As already mentioned, the density is close to that of water and is approximately 1 g/cm^3. The effective cross section, as defined in the preceding chapter by equation 4.34, rewritten here as

$$Q = \frac{Q_V Q_T}{Q_V + Q_T} \qquad 5.25$$

can be determined by measuring the cross sections of scalae vestibuli, including that of scala media, Q_V, and tympani, Q_T, respectively. This can be done with the help of photomicrographs, such as that shown in Fig. 5.1. For the cochlear location of our experiments, which was estimated as being at a distance of about 0.77 cm from the cochlear base, or equivalently, at 63% of the total length, we obtain: $Q_V = 1.42*10^{-3}$ cm^2 and $Q_T = 0.7*10^{-3}$

cm², giving $Q = 0.47*10^{-3}$ cm². At the same location, according to Fig. 5.21, the wave velocity is $c_L = 9*10^2$ cm/sec, so that the acoustic compliance of the basilar membrane becomes $C_{Ba} = 0.58*10^{-9}$ cm⁴/dyne and its inverse, the acoustic stiffness, $S_{Ba} = 1.724*10^9$ dyne/cm⁴.

To compare the stiffness of the basilar membrane to that of the stereocilia aggregate defined in a preceding section, both have to be expressed in the same units. For this reason the acoustic stiffness of the basilar membrane has to be converted to a mechanical one. To achieve the conversion, the acoustic stiffness has to be multiplied by the effective width of the basilar membrane squared. According to Fig. 5.1, the geometric width at the location of our experiments was $w_B = 2.37*10^{-2}$ cm. It is shown in the preceding chapter that the effective width of the basilar membrane is about one half of its geometric width (Fig. 4.9), so that, for the gerbil, $w_{Be} = 1.2*10^{-2}$ cm and $w_{Be}^2 = 1.4*10^{-4}$ cm². With this numerical value, the mechanical stiffness of the basilar membrane becomes $S_{Bm} = 2.41*10^5$ dyne/cm² per cm of its length. The corresponding mechanical compliance is $C_{Bm} = 0.41*10^{-5}$ cm²/dyne. The mechanical stiffness of a corresponding stereocilia aggregate has been derived above as being $S_{Hm} = 4*10^4$ dyne/cm², so that it is smaller than that of the basilar membrane by roughly a factor of 6—about one order of magnitude. The statically measured stiffness of the tectorial membrane, $S_R = 6*10^3$ dyne/cm², is smaller by another order of magnitude, and the ratio is further increased when the latter stiffness is corrected according to cochlear transfer functions at the level of the OHCs. These relationships indicate that the stereocilia-tectorial membrane complex should have little effect on the vibration of the basilar membrane, except at frequency locations of resonance, as discussed in detail in chapter 6.

Cochlear Wave Velocity: A Generalization

One of the purposes of this book is to determine the characteristics of sound transmission in the human ear. Accordingly, it is necessary to determine the relationships between the key cochlear constants and variables of Mongolian gerbils, our animal models, and of humans. The best opportunity for a comparison is offered by the velocity of the cochlear waves. This velocity as a function of cochlear location has been calculated for the human cochlea in the preceding chapter, based on Békésy's (1960) postmortem measurements of basilar membrane compliance and of Wever's (1949) and my (Zwislocki-Moscicki, 1948) measurements of cochlear geometry. It is compared in Fig. 5.22 to that determined for live Mongolian gerbil and plotted in Fig. 5.21. Clearly, the velocity magnitude obtained for the human cochlea is less than half that obtained for the gerbil cochlea. Interestingly, the functional relationship between the velocity and the cochlear distance in % is roughly the same. Both curves are governed by an exponential function having an exponent of –0.037 for the gerbil cochlea and almost the same, –0.039, for the human. It has not been affected either by species difference or possible postmortem changes.

The difference in the absolute velocity magnitude between the live gerbil and dead human cochleae may not appear intuitively surprising, but what is it due to? Is it due to species differences or to the fact that one velocity was derived for postmortem conditions and the other for in vivo conditions? The latter possibility cannot be easily rejected in view of Kohllöffel's (1972a, 1972b, 1972c) demonstration that, in guinea pigs, the stiffness of the basilar membrane decreases gradually after death, and a related demonstration of Rhode's (1973) in squirrel monkeys that, at a given cochlear location, the phase of the basilar membrane vibration changes more rapidly with sound frequency after death than in vivo. This means an increased wave travel time, therefore, a decreased velocity of wave propagation.

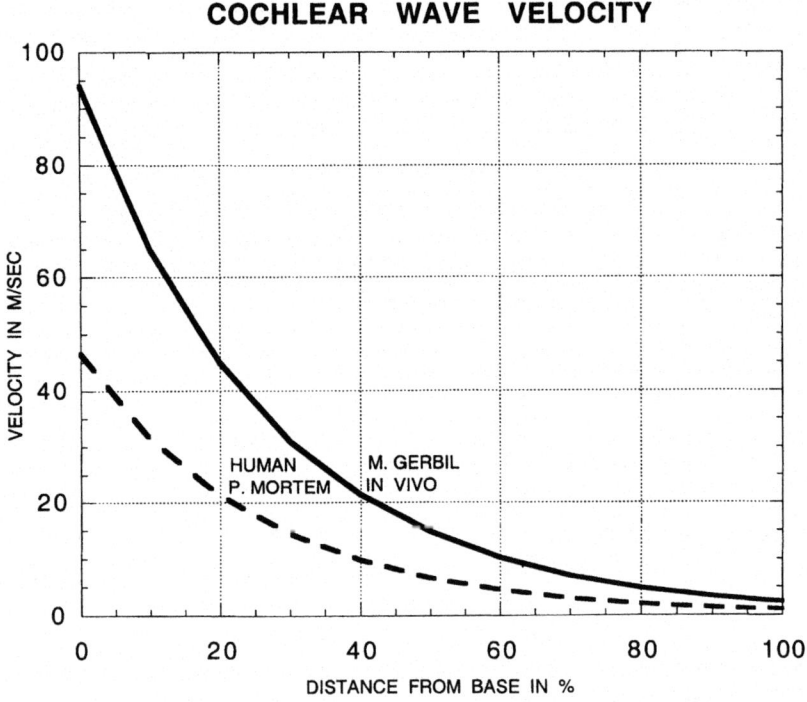

FIG. 5.22. Cochlear wave velocity for live Mongolian gerbil reproduced from Fig. 5.21 in comparison to the cochlear wave velocity in humans postmortem according to Békésy's measurements.

An indirect test of the reason for the velocity difference in Fig. 5.22 can be derived from measurements on guinea pigs for which extensive data on wave travel time as well as Békésy's (1960) postmortem measurements of basilar membrane compliance are available. Most of the travel time data were obtained with the help of CM measurements but two sets were generated by direct measurements of basilar membrane vibration (Johnstone & Taylor, 1970; Wilson & Johnstone, 1975). The travel time data were reviewed by Schmiedt and Zwislocki (1977), who also contributed their own. After small corrections for inaccuracies in data processing, all the CM data were found to be in good agreement. However, the basilar membrane measurements yielded significantly shorter travel times. It was possible to account for this discrepancy with the help of the theory of cochlear waves described in the preceding chapter on the realization that both teams, Johnstone and Taylor and Wilson and Johnstone, had to empty scala tympani of perilymph in the region of their measurements. According to equation 5.25 defining the effective cross section of the cochlear canals, emptying the perilymph from scala tympani is equivalent to eliminating the term Q_T and increasing the effective cross sectional area. If $Q_T = Q_V$, the area is increased by a factor of two. Because, according to equation 4.42 the wave velocity is directly proportional to the square root of the effective cross sectional area, the velocity is increased by a factor of $\sqrt{2}$, and the travel time decreased correspondingly according to equation 5.22. Corrected accordingly, the basilar membrane data became entirely consistent with the CM data. The means of these data have been used here to calculate a best fit function according to the least squares method. The resulting curve is shown in Fig. 5.23. According to this function, the travel time to the end of the cochlea is about twice as long in guinea pig as in Mongolian gerbil, which is not surprising, since the cochlea is about 0.6 cm longer in the former. The least squares fit of equation 5.22 produces the following parameter values: $c_{L_o} = 110*10^2$ cm/sec for the wave velocity at the cochlear base and $v = -0.042$ for the exponent of the exponential function describing the velocity decay toward the apex when the cochlear length is expressed in percent of the total length. In agreement with Fernandez's (1952) measurements, a total length of 1.8 cm has been assumed. Note that the velocity parameter values are nearly equal to the corresponding parameter values of Mongolian gerbil. The velocity function calculated on their basis is plotted in Fig. 5.24 as a solid curve. For comparison, the dashed curve shows the guinea pig velocity function calculated on the basis of Békésy's postmortem measurements of basilar membrane compliance with the help of Fernandez's anatomical measurements (Zwislocki, 1974). Its parameters are: $c_{L_o} = 42*10^2$ cm/sec and $v = -0.043$. They are quite similar to those obtained on the basis of Békésy's (1960) postmortem

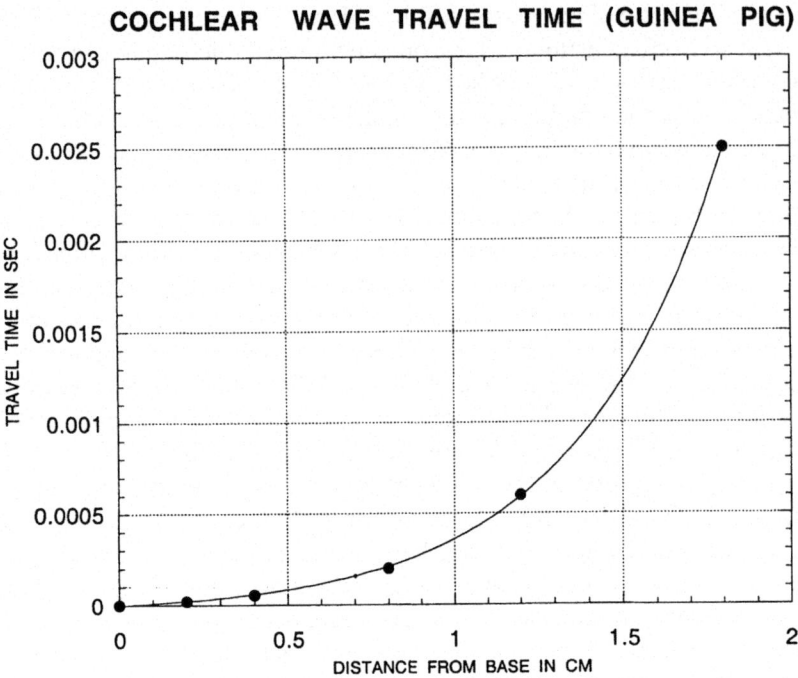

FIG. 5.23. Guinea pig cochlear wave travel time, as measured empirically and shown by the circles, approximated by an exponential function.

data for humans. Comparison of Figs. 5.22 and 5.24 should make it clear that the difference between the in vivo and postmortem velocity curves within guinea pig is roughly the same as the difference between the Mongolian gerbil curve in vivo and the human postmortem curve. Accordingly, the likelihood that the gerbil and human velocity curves of Fig. 5.22 differ because of postmortem changes in the human cochlea rather than because of species differences is increased. The experiments of Kohllöffel (1972c) and Rhode (1973) suggest that the changes occur in the compliance measured on the basilar membrane. Apparently, they are distributed uniformly throughout the cochlear length because the exponent of the velocity decay function remains unchanged. An informal observation to this effect was already made by Kohllöffel. The invariance of the velocity decay function pre- and postmortem strongly suggests that the decay function in live humans is the same as postmortem and the same as in Mongolian gerbils and guinea pigs in vivo.

Russell and Nilsen (1997) made direct phase measurements of basilar membrane vibration in the basal turn of the live guinea pig cochlea. From

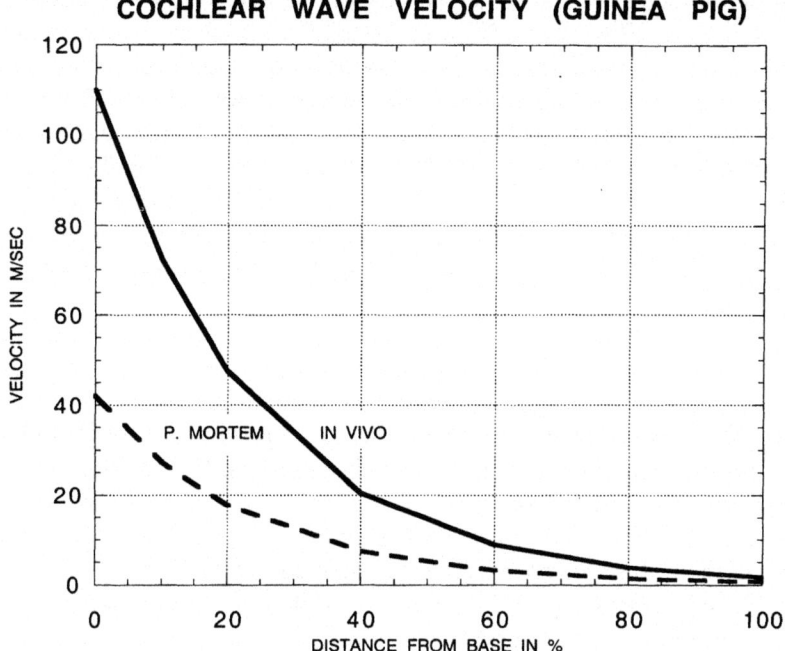

FIG. 5.24. Cochlear wave velocity for live guinea pigs, as derived from the wave travel time function of Fig. 5.23 and the wave velocity obtained from Békésy's measurements postmortem.

their data, it is possible to compute the wave velocity. According to the theory of wave propagation discussed in chapter 4, the phase is related to the wave velocity as follows:

$$c_L = 2\pi f \frac{\Delta x}{\Delta \phi} \qquad 5.26$$

For a location of 1.47 cm from the cochlear apex, which corresponds to a distance of 0.33 cm from the cochlear base, given a 1.8 cm long cochlea, their measurements, shown in Fig. 5.36, indicate a phase increment, $\Delta\phi$, of approximately $\pi/2$ for a distance increment, Δx, of approximately 0.022 cm. Introduced into equation 5.26, these numerical values yield a wave velocity, $c_L = 13.2$ m/sec. A cochlear distance of 0.33 cm corresponds to approximately 18% of cochlear length. At this location, the wave velocity derived from the CM and the Johnstone and Taylor's and Wilson and Johnstone's basilar membrane measurements amounts to approximately

53 m/sec, which is 4 times larger. At least one reason for the discrepancy is easily apparent. Russell and Nilsen used relatively high sound frequencies that produced wavelengths of the same order of magnitude as the canal depth, so that the requirement of long waves was not satisfied. From their phase and corresponding distance increments an approximate wavelength of only 0.9 mm can be computed. In addition, the opening they had to make in scala tympani to visualize the basilar membrane must have produced a shunt controlled by surface tension in the opening. Thus, an additional compliance was introduced paralleling the compliance of the basilar membrane.

Some confirmation of the proposition that opening scala tympani increases the effective cochlear compliance per unit length can be found in the measurements of basilar membrane vibration in chinchillas by Ruggero et al. (1997). For these measurements, scala tympani was opened, and the opening was sometimes covered with a glass sliver and sometimes not. Unfortunately, they did not specify which measurements were performed with a covered opening and which were performed without the cover. Still, some of their phase measurements exemplified in one of their figures and reproduced here as Fig. 5.25 show clear differences in the slope of the curves of phase versus sound frequency. Of 6 curves, 5 agree with each other at low sound frequencies, in the region of long waves. The 6th is steeper. It agrees with phase curves collected on a preparation that was the most extensively studied and which are reproduced in Fig. 5.37. The best agreement among the 5 curves occurs in the frequency region between 4 and 5 kHz. When they are approximated there by an averaging line, a phase increment of 0.16π, corresponding to a frequency increment of 1 kHz, can be computed. According to the wave theory, the corresponding wave travel time can be obtained from the equation

$$\Delta t = \frac{\Delta \phi}{2\pi \Delta f} \qquad \textbf{5.27}$$

For the five curves, it amounts roughly to $0.07*10^{-3}$ sec. The location of the measurement was given as being 0.35 cm away from the basal end of the cochlea. Since the total length of the chinchilla cochlea measures approximately 1.84 cm (Eldredge, Miller, & Bohne, 1981), and is only slightly greater than that of the guinea pig cochlea, the travel time should agree roughly with that found at a corresponding location in the latter, if the wave velocities are the same. Comparison with the curve of Fig. 5.23 reveals that this is true, indeed. Accordingly, the velocities appear to be the same. On the other hand, the deviant curve, when subjected to the same computation, yields a travel time that is twice as long. It would be of some interest to

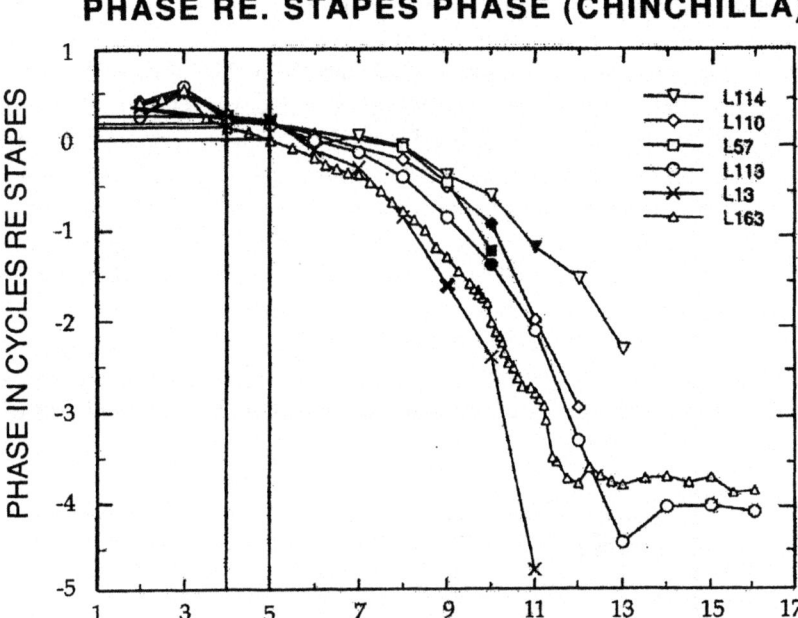

FIG. 5.25. Response phase of the basilar membrane in the basal cochlear turn, as measured by Ruggero et al. in several chinchilla specimens in vivo. The frequency interval marked by the vertical lines has been used for calculation of the wave travel time (modified from Ruggero et al., 1997). Reprinted with permission from M. A. Ruggero, N. E. Rich, A. Recio. S. Narayan, & L. Robles (1997). Basilar-membrane responses to tones at the base of the chinchilla cochlea. *Journal of the Acoustical Society of America, 101,* 2151–2163. Copyright © 1997, Acoustical Society of America.

verify if the shorter time corresponds to a coverslipped surgical opening in scala tympani, and the longer time to one that is not. However, other factors may have played a role. For example, incidental loosening of the spiral ligament could have increased the effective compliance of the basilar membrane. In general, injuries to the basilar membrane and tissues attached to it would tend to increase its effective compliance and reduce the wave velocity. Accordingly, it is likely that the greater measured velocities represent more closely the undisturbed state of the cochlea.

The interspecies invariance of the cochlear wave velocity documented in preceding paragraphs is striking, but its extrapolation to live humans would be strengthened if it could be shown to hold for still other species belonging to different mammalian orders. All three, Mongolian gerbils, guinea pigs, and chinchillas are rodents. It has been possible to find sufficient data for this purpose on domestic cats, representing carnivores, and squirrel monkeys, representing primates.

With respect to domestic cats, Pfeiffer and Kim (1975) recorded firing rates in populations of auditory nerve fibers with various CFs for constant sound frequencies. They converted the CF scale to a cochlear distance scale according to Schuknecht's (1960) tonotopic map and were able in this way to construct spatial magnitude and phase transfer functions. Their phase results obtained on one cat, apparently their best preparation, at a low SPL of 20 dB are shown in Fig. 5.26. Their data for the sound frequencies of 0.6 and 1.2 kHz are approximated by smooth curves drawn by hand. With the help of these curves, phase differences associated with small distance differences have been graphically constructed. However, the distance scale of Schuknecht was modified ac-

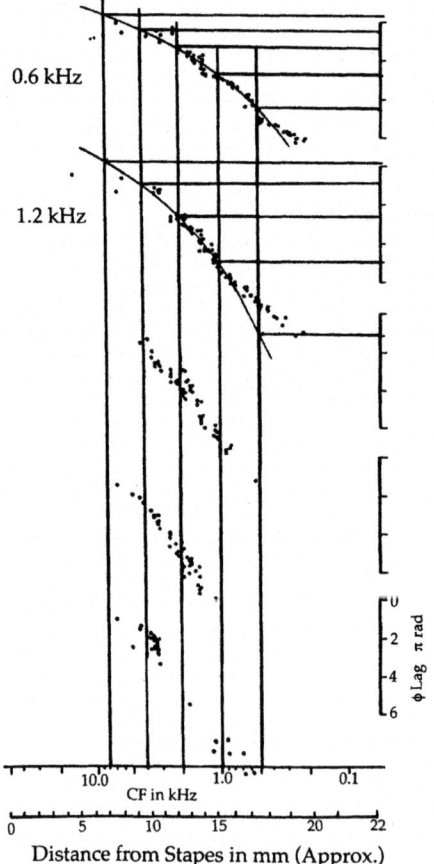

FIG. 5.26. Response phases of auditory nerve fibers of domestic cats as functions of the fiber CFs for several sound frequencies, as recorded by Pfeiffer and Kim. The solid curves approximate the response medians at 0.6 and 1.2 kHz (modified from Pfeiffer and Kim, 1975). Reprinted with permission from R. R. Pfeiffer and D. O. Kim (1975). Cochlear nerve fiber responses: Distribution along the cochlear partition. *Journal of the Acoustical Society of America, 58,* 867–869. Copyright © 1975, Acoustical Society of America.

cording to more recent anatomical results obtained by Liberman (1982) and replotted by Greenwood (1990). The relationship between phase increments related to distance increments and wave velocity is specified in equation 5.26. Velocity values extracted with the help of this equation from the Pfeiffer and Kim data for the frequencies of 0.6 and 1.2 kHz are plotted in Fig. 5.27 by means of circles and triangles, respectively, over cochlear distances in % of the total cochlear length. Note that the computed values for both sound frequencies are in agreement. The solid curve has been fitted to them by the method of least squares. Its parameter values are: $c_{Lo} = 115$ m/sec; $v = -0.037$, quite similar to those for the rodents.

With respect to squirrel monkeys, Rhode (1971) measured the vibration phase of the basilar membrane at a location corresponding to a CF of between 7 and 8 kHz as a function of sound frequency. According to Greenwood (1977), this CF range corresponds to a distance of about 0.7 cm from cochlear base. The opening in scala tympani through which the basilar membrane was approached was covered, so that wave travel effects that may arise from the opening were kept at a minimum. Because the distance

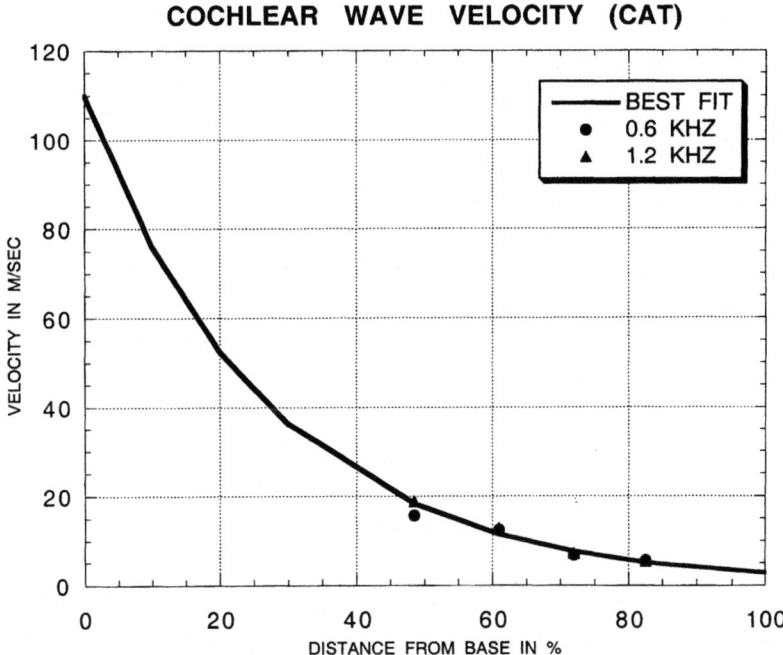

FIG. 5.27. Cochlear wave velocity calculated from the averaged data of Fig. 5.26. Circles and triangles show the calculated values, the curve provides their least squares exponential fit.

from the measuring point to the base was greater than was possible to achieve in guinea pigs or chinchillas, Rhode was able to measure the phase over an extended range of low sound frequencies. In this range, he found direct proportionality between the phase and sound frequency, making an accurate determination of their relationship possible. As shown in Fig. 5.28, a frequency increment of 2 kHz produced a phase increment of approximately 0.67π. In agreement with equation 5.27, their ratio yields a wave travel time of $0.16*10^{-3}$ sec. Because of cochlear length similarity, this travel time should be similar to that found in guinea pig at a similar location, if the wave velocities are approximately the same. Indeed, from Fig. 5.23, we find a travel time of $0.16*10^{-3}$ at a distance of 0.7 cm from the cochlear base. Accordingly, the wave velocities appear to be the same.

According to the evidence presented here, the cochlear wave velocity in vivo in Mongolian gerbils, guinea pigs, and chinchillas, which are rodents, domestic cats, which are carnivores, and squirrel monkeys, which are primates, is approximately the same. By extrapolation, it appears likely that it is roughly the same in live humans. Generalizing, it seems possible that it is roughly the same in all mammals who vocalize in the same frequency

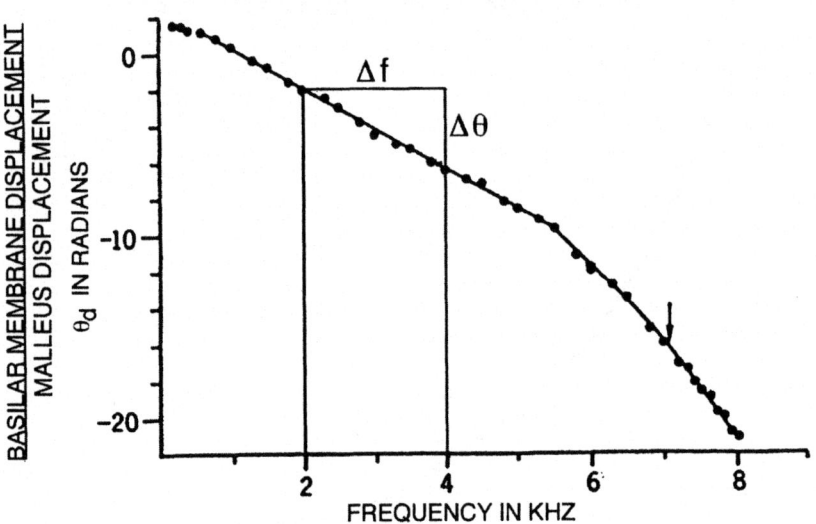

FIG. 5.28. Basilar membrane displacement phase in the cochlear base of a live squirrel monkey relative to the stapes displacement. The phase and frequency increments, $\Delta\Theta$ and Δf, have been determined for calculation of wave travel time (modified from Rhode, 1971). Reprinted with permission from W. S. Rhode (1971). Observations of the vibration of the basilar membrane in squirrel monkey using the Mössbauer technique. *Journal of the Acoustical Society of America, 49,* 1218–1231. Copyright © 1971, Acoustical Society of America.

Cochlear Wave Length

If the generalization of a constant cochlear wave velocity is added to that proposed by Greenwood (1961, 1974, 1990) for the cochlear tonotopic map, a further generalization can be derived, which concerns the wave length. The wavelength is simply equal to the wave velocity divided by the frequency of oscillation,

$$\lambda = \frac{c}{f} \qquad 5.28$$

For this purpose, generalizing the data presented here, the wave velocity can be assumed to be on the order of $c_{Lo} = 100$ m/sec at the cochlear base, and the exponential constant of the velocity decay function to be $\nu = 0.04$ when the cochlear distances are expressed in percent of the total length. Furthermore, reorganizing Greenwood's formula (e.g., Greenwood, 1990) for the tonotopic map and neglecting a relatively small constant that seems to make itself felt only near the cochlear apex, it is possible to write

$$f_c = f_{co} e^{-\kappa x} \qquad 5.29$$

where f_c is the characteristic frequency, f_{co} the characteristic frequency at the cochlear base, and κ, the exponential constant associated with cochlear distance from the base expressed in percent. From Greenwood's equations and graphs, it appears reasonable to assume $f_{co} = 30$ kHz and $\kappa = 0.05$. The long-wave velocity cannot be used to calculate the wave length at the CF location because basilar membrane impedance is no longer controlled predominantly by compliance there, and the effect of the mass associated with the basilar membrane makes itself felt. As a result, the wave velocity becomes dependent on sound frequency. However, at the location corresponding to $(2/3) f_c$, the mass effect is still negligible. For this location, the tonotopic map can be expressed as

$$\frac{2}{3} f_c = \frac{2}{3} f_{co} e^{-\kappa x} \qquad 5.30$$

and the equation for the wave length as

$$\lambda_{2/3} = \frac{c_L(x)}{\frac{2}{3} f_c(x)} \qquad 5.31$$

When $c_L(x)$ and $(2/3)f_c$ are replaced by their defining formulas, equations 5.23 and 5.30, respectively, equation 5.31 becomes

$$\lambda_{2/3} = \frac{c_{Lo}(x)}{\frac{2}{3}f_{co}} e^{-(\nu-\kappa)x} \qquad 5.32$$

and, after introduction of the numerical values already specified,

$$\lambda_{2/3} = 0.5 e^{0.01x} \quad (cm) \qquad 5.33$$

This equation means that, at the cochlear location corresponding to $(2/3)f_c$ $((2/3)CF)$, the wave length is on the order of 0.5 cm in the cochlear base and increases slowly with distance from the base to reach 1.36 cm near the apex in all the mammals mentioned in this chapter and, possibly, in others as well. The $\lambda_{2/3}$ function is plotted in Fig. 5.29. as well as the λ_c function that would be true if the conditions for c_L were satisfied there.

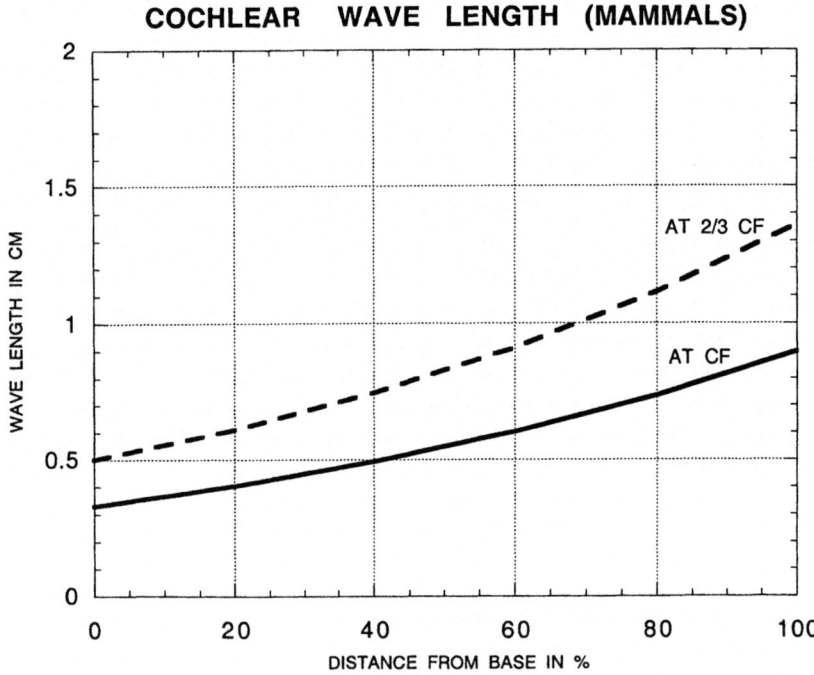

FIG. 5.29. Cochlear wave length calculated from the generalized wave velocity function as a function of cochlear CF location and a location of 2/3 CF.

Cochlear Wave Velocity and Basilar Membrane Parameter Values in Live Humans

If the absolute magnitude of wave velocity in live human cochleas is roughly the same as in the mammals included in preceding sections, it should amount to about

$$c_L = 100e^{-0.039x} \ (m/sec) \qquad 5.34$$

where the exponential constant of 0.039 is the same as for the dead cochlea but the absolute magnitude is 2.15 times larger. The inferred velocity curve for the human cochlea in vivo is plotted in Fig. 5.30 and the wave travel time obtained through its integration, in Fig. 5.31.

According to equation 5.24, the compliance of the basilar membrane is inversely proportional to the wave velocity squared. As a consequence, it should be 4.62 times smaller than postmortem. For the latter condition, it has been derived in the preceding chapter from Békésy's measurements as $C_{Ba}(x) = 5.4*10^{-10} \, e^{0.0622x}$ (cm^4/dyne) for x expressed as percent

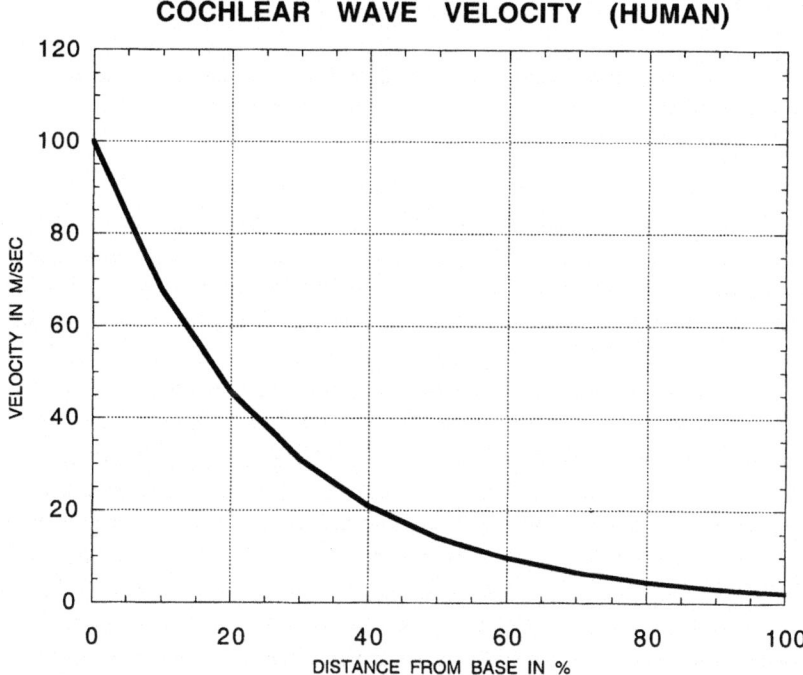

FIG. 5.30. Cochlear wave velocity for live humans based on mammalian interspecies similarity of the velocity.

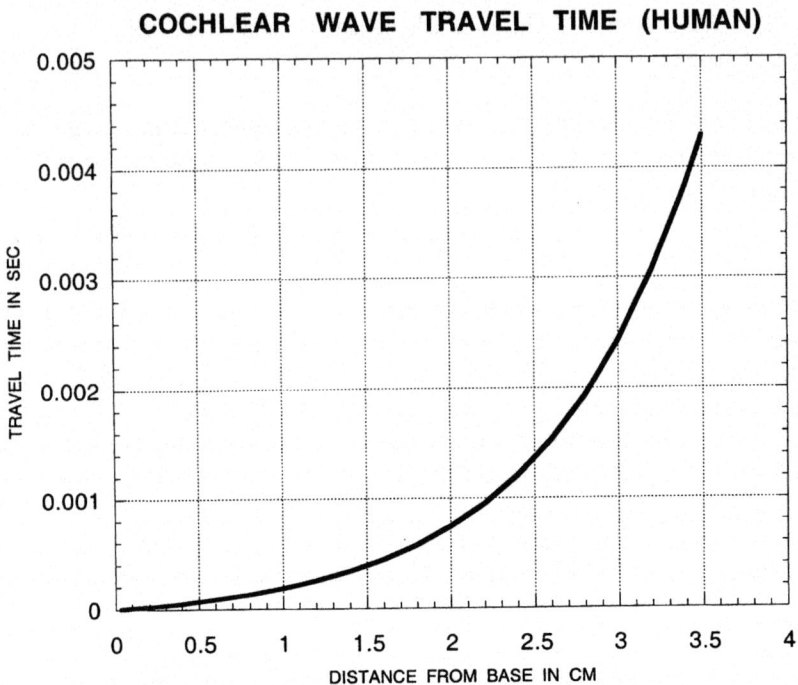

FIG. 5.31. Cochlear wave travel time calculated for live humans from the wave velocity of Fig. 5.30.

of the total cochlear length. Therefore, in a live human cochlea, it should amount to $C_{Ba}(x) = 1.17*10^{-10} \, e^{0.0622x}$ (cm^4/dyne). To compare it with the available data for the compliance of the Mongolian gerbil basilar membrane, it has to be calculated for the 63% location. We obtain: $C_{Ba}(63\%) = 5.89*10^{-9}$ (cm^4/dyne). For the gerbil, a value $C_{Ba}(63\%) = 0.58*10^{-9}$ (cm4/dyne), 10 times smaller, has been derived above. Because, according to equation 4.42, the wave velocity is directly proportional to the square root of the ratio of the effective cross-sectional area and the compliance, the 10 times smaller compliance compensates for the 10 times smaller effective cross-sectional area to produce practically the same wave velocity. It is not surprising that, in relatively small Mongolian gerbils, the cochlear dimensions are smaller than in humans. However, the nature's substantial effort to compensate for the smallness by decreasing the compliance of the basilar membrane by an order of magnitude, so as to maintain a constant wave velocity is remarkable and may have an evolutionary significance.

The conclusion is weakened somewhat by the fact that the wave velocity in the human cochlea was adjusted from the postmortem value so as to equal the velocity found in live experimental mammals. However, similar relationships are found in guinea pigs, based on both postmortem and in vivo measurements. According to a past article (Zwislocki, 1974) and measurements of Fernandez (1952), a cochlear effective cross section function can be derived as follows: $Q = 0.6*10^{-2} e^{-0.03x}$ (cm^2), and a basilar membrane compliance function: $C_{Ba} = 3.4*10^{-10} e^{0.055x}$ (cm^4/dyne). When combined according to equation 4.42, they produce the wave velocity function: $c_L = 42 e^{-0.0426x}$ (m/sec). As already introduced in the preceding chapter, for humans postmortem, we have: $Q = 1.16*10^{-2} e^{-0.015x}$ (cm^2) and $C_{Ba} = 5.38*10^{-10} e^{0.0622x}$ (cm^4/dyne). Their combination produces the wave velocity function: $c_L = 46 e^{-0.039x}$ (m/sec). The two velocity functions are almost identical in spite of the fact that the cross section and the compliance functions are clearly different. This means that, at every cochlear location, differences in the cross sectional area are compensated for by appropriate differences in the basilar membrane compliance.

The effect of the mass associated with the basilar membrane is negligible in the region of long waves but, as shown in particular in the following chapter, plays an important role in the vicinity of the CF. As discussed in chapter 4, the effect is produced mainly by the cell mass of the organ of Corti. The effective acoustic mass has been derived for the human cochlea in that chapter and shown to obey the equation $M_{ea} = 4A\rho/3w^2$, where A—the cross-sectional area of the organ of Corti, ρ—its specific density, w—the width of the basilar membrane, and A and w are empirical functions of the cochlear distance from the base.

An effective resistance has also been derived in chapter 4 for the human basilar membrane postmortem. It has been shown there that the resistance calculated from Békésy's damping measurements is inappropriate and that the resistance has to be derived from plots of basilar membrane vibration magnitude versus cochlear location. The acoustic resistance obtained in this way amounts to $R_{Ba} = 1.2*10^3$ dyne sec/cm^4 and is independent of cochlear location.

Cochlear Input Impedance In Vivo—Comparative Considerations

In chapter 4, a simple relationship between the wave velocity and the characteristic impedance was introduced for the cochlea, which is valid in the broad mid frequency range of audible sound. According to equation 4.52, the relationship can be expressed as $Z_{Ca}(x) = \rho c_L(x)/Q(x)$. It becomes immediately apparent that, if the wave velocity, c_L is invariant across animal species but Q varies, the cochlear input impedance $Z_{Ca}(0)$ must vary inversely with it. For the human cochlea postmortem, a numerical value of $Z_{Ca}(0) = 0.4*10^6$ dyne sec/cm^5 has been derived on the basis of Békésy's (1960) measurements. Since, according to equation 4.52, the impedance covaries with the wave velocity, and the velocity is expected to be 2.15 larger in vivo than postmortem, the impedance should become $Z_{Ca} = 0.86*10^6$ dyne sec/cm^5 for in vivo conditions. This value is based on a rather rough, exponential approximation of the cochlear effective cross section, Q. When the directly measured cross sectional areas of scalae vestibuli and tympani are used, a slightly smaller impedance of $Z_{Ca} = 0.762*10^6$ dyne sec/cm^5 results. Somewhat surprisingly, this values is close to the value of about $0.6*10^6$ dyne sec/cm^5 obtained by Merchant, Ravicz, and Rosowski (1996) postmortem.

Lynch, Nedzelnitsky, and Peake (1982) obtained a mid frequency input impedance of $1.2*10^6$ dyne sec/cm^5 for the domestic cat cochlea in vivo, and the question may arise whether the higher numerical value of this impedance, as compared to the corresponding human impedance, is due to theoretical and experimental errors or to species differences. Because the cat cochlea is smaller than the human cochlea and may be expected to have a smaller Q, the inverse relationship between Q and the impedance of equation 4.52 suggests that the difference may be due to species variation. However, no sufficient data are available for an unequivocal decision. Such data are available for guinea pigs. According to an exponential approximation of the effective cross sectional area as a function of distance from the cochlear base, the area at the cochlear input is $0.6*10^{-2}$ cm^2, about half the area found in the human cochlea (Zwislocki, 1974). With the wave velocity of 110 m/sec already determined independently, this area produces an input impedance of $1.83*10^6$ dyne sec/cm^5, more than twice the value obtained for humans. When the actual cross sectional area is used (Zwislocki, 1974), the impedance decreases to $1.4*10^6$ dyne sec/cm^5 nearly equaling the impedance obtained by Lynch et al. (1982) for domestic cats.

Dancer and Franke (1980) measured the cochlear input impedance in live guinea pigs and obtained an average value of about $0.45*10^6$ dyne sec/cm5, more than three times smaller. However, because this value is also almost three times smaller than that obtained by Lynch et al. (1982) for domestic cats, it seems to be unrealistic. In their measurements, they connected scala vestibuli and perhaps also scala tympani to rather large tubes for pressure measurement. It is possible that these tubes artifactually increased the effective cross sectional area of the cochlea. It is also possible that the tubes were not located directly at the cochlear input. The sound pressure should decrease as the distance from this input increases. Interestingly, Nedzelnitsky, who measured the cochlear sound pressure in a similar way to that of Dancer and Franke, obtained an input impedance for the cat cochlea, whose magnitude is very similar to that Dancer and Franke obtained for the guinea pig cochlea and is about half that obtained for the cat cochlea by Lynch et al. The latter measured the intracochlear sound pressure in the vestibule rather than in scala vestibuli.

Key Cochlear Parameter Values for Humans and Mongolian Gerbils

It may be useful to have a list of key cochlear parameter values arrived at above for live humans and Mongolian gerbils. The cochlea of the latter is used in the next chapter as a model for the cochlea of the former, and all the model experiments have been performed on it.

The constants have not been derived in a uniform way. Most of the human cochlea constants have been derived in the preceding chapter concerning the cochlea postmortem and adjusted to the higher wave velocity found in experimental mammals. Some of the Mongolian gerbil constants have been obtained in direct measurements described in preceding sections; others have been inferred from the human constants. Reciprocally, some of the human constants are inferred from the gerbil constants, some of the latter, on the basis of the analysis described in chapter 6. As already discussed, the numerical values of th C_{Ta} constants have been increased by a factor of approximately 3 relative to their static measurements.

SOME FUNDAMENTAL CHARACTERISTICS OF COCHLEAR DYNAMICS

To analyze the cochlear dynamics in vivo, it is first necessary to know some of its phenomenology. Perhaps the most fundamental mechanical event is the vibration of the basilar membrane associated with the propagation of transversal cochlear waves. The nature of these waves in a live cochlea appears to be the same as postmortem. However, there are differences in amplitude and phase patterns, as has been revealed in numerous measurements of basilar membrane vibration. Perhaps most important, whereas the dead cochlea acts as a linear system in the sense that the vibration amplitude of the basilar membrane increases in direct proportion to sound pressure, and the wave velocity is independent of it, the live cochlea does not, unless it is severely damaged (e.g., Sellick, Patuzzi, & Johnstone, 1982). In a healthy cochlea, neither does the vibration amplitude of the basilar membrane increase in direct proportion to sound pressure nor is the wave velocity entirely independent of it, except at basal locations sufficiently removed from the location of maximum response (e.g., Rhode, 1971; Ruggero et al., 1997).

HUMAN		MONGOLIAN GERBIL	
$Q = 1.16*10^{-2} * \exp(-0.015x)$	(cm^2)		
$Q(63\%) = 0.45*10^{-2}$	(cm^2)	$Q(63\%) = 0.47*10^{-3}$	(cm^2)
$w_{Be} = 0.66*10^{-2}*\exp(0.0135x)$	(cm)		
$w_{Be}(63\%) = 1.54*10^{-2}$	(cm)	$w_{Be}(63\%) = 1.2*10^{-2}$	(cm)
$C_{Ba} = 1.17*10^{-10}*\exp(0.0622x)$	$(cm^4/dyne)$		
$C_{Ba}(63\%) = 5.89*10^{-9}$	$(cm^4/dyne)$	$C_{Ba}(63\%) = 5.8*10^{-10}$	$(cm^4/dyne)$
$c_L = 100*10^2*\exp(-0.039x)$	(cm/sec)		
$c_L(63\%) = 8.6*10^2$	(cm/sec)	$c_L(63\%) = 9.0*10^2$	(cm/sec)
$M_{Ba} = 0.54*\exp(-0.0122x)$	(g/cm^3)		
$M_{Ba}(63\%) = 0.25$	(g/cm^3)	$M_{Ba}(63\%) = 3.0$	(g/cm^3)
$R_B = 5.5*10^3$	$(dyne\ sec/cm^4)$	$R_{Ba} = 5.0*10^4$	$(dyne\ sec/cm^4)$
$C_{Ha} = 7.0*10^{-10}*\exp(0.0622x)$	$(cm^4/dyne)$		
$C_{Ha}(63\%) = 3.5*10^{-8}$	$(cm^4/dyne)$	$C_{Ha}(63\%) = 0.35*10^{-8}$	$(cm^4/dyne)$
$R_{Ha} = 140$	$(dyne\ sec/cm^4)$	$R_{Ha}(63\%) = 1.4*10^3$	$(dyne\ sec/cm^4)$
$C_{Ta} = 1.4*10^{-8}*\exp(0.0622x)$	$(cm^4/dyne)$		
$C_{Ta}(63\%) = 7*10^{-7}$	$(cm^4/dyne)$	$C_{Ta}(63\%) = 0.7*10^{-7}$	$(cm^4/dyne)$
$M_{Ta} = 0.18*\exp(-0.0122x)$	(g/cm^3)		
$M_{Ta}(63\%) = 0.083$	(g/cm^3)	$M_{Ta}(63\%) = 0.55$	(g/cm^3)
$R_{Ta} = 800$	$(dyne\ sec/cm^4)$	$R_{Ta}(63\%) = 9*10^3$	$(dyne\ sec/cm^4)$

In addition to basilar membrane vibration, in a live cochlea, the dynamic characteristics of the shear motion between the reticular lamina and the tectorial membrane become relevant. After all, the cochlear hair cells are not stimulated directly by the basilar membrane vibration but by the shear motion, which controls the deflection of their stereocilia. Often the assumption is made that the difference is trivial, and that the shear motion is directly proportional and in constant phase relationship to basilar membrane vibration. I show below that this is not true in general.

Because no measurements of basilar membrane vibration were performed in my laboratory, but its vibration characteristics are essential for the analysis of cochlear function attempted in the next chapter, I have to borrow here from the work of others. This is not necessary for the shear motion. Although the shear motion does not seem to have ever been measured directly, it is possible to show that it can be inferred safely from the responses of the OHCs and the related responses of Hensen's cells recorded in several laboratories, including mine. This is so in particular because the stereocilia of the OHCs are held at both ends in the apical parts of the cells firmly lodged in the reticular lamina and in the tectorial membrane, respectively, and their deflection follows directly the shear displacement. In addition, the alternating receptor potentials of these cells are directly proportional to the stereocilia deflection over a useful range of SPLs. This follows directly from the generic characteristics of vibration sensing hair cells of the inner ear, as determined by Hudspeth and Corey (1977) and, indirectly, from recordings of OHC and related Hensen's cell alternating potentials (Zwislocki, 1991; Zwislocki, Szymko, & Hertig, 1996; Zhang & Zwislocki, 1995, 1996). Up to a SPL of 70 dB, these potentials grow in direct proportion to SPL in the same frequency region below the frequency of maximum response in which basilar membrane vibration also grows in direct proportion to it. This means by deduction that the potentials are directly proportional to basilar membrane vibration. Furthermore, sinusoidal sound pressure wave forms produce sinusoidal potentials with very little distortion. Above 70 dB SPL, the proportionally gives way to a gradually increasing compressive nonlinearity, and the wave form distortion grows.

Basilar Membrane Vibration

Measurement of basilar membrane vibration in live animals is fraught with difficulties. First of all, there is the problem of accessibility. A direct access to the basilar membrane can be gained only through scala tympani, and scala tympani is accessible sufficiently for this purpose only in the basal turn of the cochlea. In higher turns, it is hidden almost entirely by the lower turns.

Access through scala vestibuli is practically possible only in the apical turn, access to lower turns being blocked by the higher ones. In addition, the access usually requires puncturing Reissner's membrane and allowing the perilymph to invade scala media and possibly change the mechanical properties of the tectorial membrane. As is shown further on, the mechanical properties of the tectorial membrane affect the vibration of the basilar membrane. Although Khanna, Flock, & Ulfendahl (1989) developed an elegant method combining an optical sectioning microscope with a laser interferometer, which allows vibration measurements of organ of Corti structures through an intact Reissner's membrane, its application has been limited thus far almost exclusively to cochleas postmortem. Consequently, attempts have been made at determining basilar membrane vibration indirectly through measurement of the vibration of Reissner's membrane (e.g., Cooper & Rhode, 1995). It follows from the mathematical analysis of the preceding chapter that, indeed, its vibration should be directly proportional to that of the basilar membrane in the cochlear region where the wavelength exceeds the canal depth. This should be true in the apical turn for all sound frequencies that produce a measurable vibration amplitude.

In interpreting the results of the measurements of basilar membrane vibration, it is important to realize that the vibration is sensitive to the biological state of the cochlea, which can deteriorate during surgery of an experimental animal and subsequent unnatural conditions of its survival. Several studies have shown that a decreasing compression in cochlear input-output functions is an early sign of trouble (e.g., Sellick et al., 1982; Robles, Ruggero, & Rich, 1986; Ruggero, Rich, Shivapuja, & Temchin, 1996). It is shown in this chapter that decreasing dependence of the frequency of maximum response on sound intensity constitutes another sign (Zhang & Zwislocki, 1996).

The amplitude of basilar membrane vibration as a function of the distance from the membrane's basal or apical end, respectively, measured at a single sound frequency may be regarded as the most fundamental dynamic characteristic of the cochlea. It is independent of sound transmission characteristics of other parts of the ear. Békésy (1960) measured it on several occasions in postmortem preparations. Unfortunately, I am aware of only one set of measurements performed in vivo. It has been obtained by Russell and Nilsen (1997) on guinea pigs in the very basal end of the cochlea. They accessed the basilar membrane through an opening in scala tympani long enough to make measurements over a basilar membrane length of about 3 mm. A sensitive method based on laser interferometry allowed them to make the measurements without placing reflecting objects on the basilar membrane. Light reflection provided by natural structures was sufficient. This allowed them to measure basilar membrane vibration at numerous points along the cochlea—up to 11 in one specimen. The results obtained on this specimen with a 15 kHz tone are shown in Fig. 5.32, reproduced from their Fig. 1D where they were plotted to the right of the vertical dotted line. The data points to the left of the line were obtained on 4 animals, each vertical set on a different one. The data in terms of displacement amplitude were plotted over a double axis referring to both the distance from the cochlear apex and the characteristic frequency determined for each measurement location. On the whole, the lines joining the data points trace characteristics resembling qualitatively those observed by Békésy (1960) on postmortem preparations. There is a vibration maximum whose location is determined by the frequency of the stimulating tone. The vibration amplitude decays more rapidly on the apical side of the

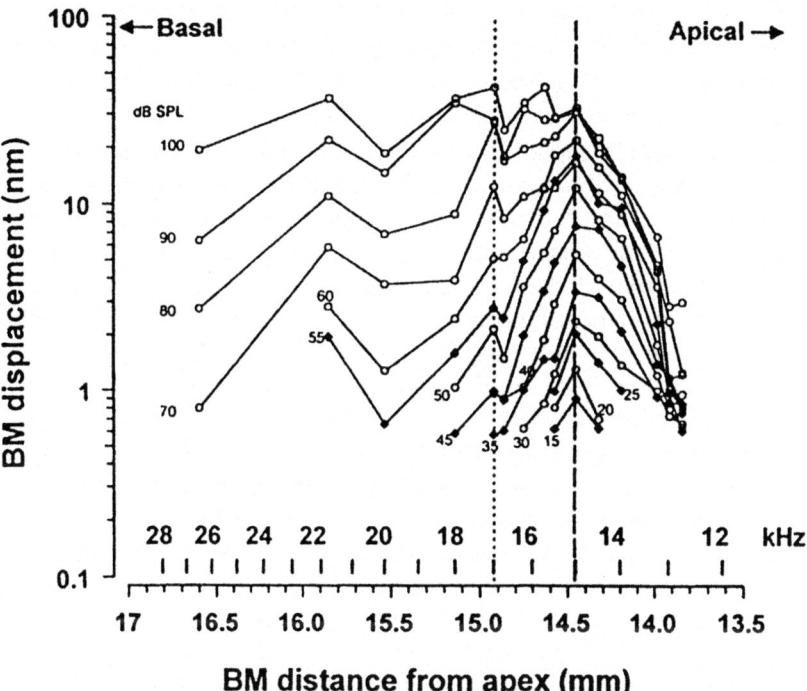

FIG. 5.32. Basilar membrane displacement amplitudes as functions of cochlear location in the guinea pig basal turn for several SPLs. The horizontal scale is expressed in terms of distance from the apex and in terms of corresponding CF. The data to the right of the dotted vertical line are from one specimen, those to the left are from several specimens (from Russell & Nilsen, 1997). From "The Location of the Cochlear Amplifier: Spatial Representations of a Single Tone on the Guinea Pig Basilar Membrane," by I. J. Russell and K. E. Nilsen, 1997, *Proceedings of the National Academy of Sciences, 94*, pp. 2660–2664. Copyright © by National Academy of Sciences, U.S.A. Reprinted with permission.

maximum than on the basal side. The asymmetry is particularly evident at high SPLs corresponding more closely to Békésy's experimental conditions. There are some clear differences as well. Perhaps the most important resides in the nonlinearity of the amplitude growth with SPL. Whereas, at the more basal locations, the amplitude appears to grow in direct proportion to SPL, the growth follows a saturating nonlinearity in the broad vicinity of the maximum, the compression growing toward the apex. As a result of uneven compression, the vibration maximum is much sharper at low SPLs than at the higher ones.

There is a question concerning independence of the location of the maximum vibration of SPL. The authors seemed to think that the independence is true. However, the maximum was not at a constant sound frequency when they measured the frequency dependence of the vibration amplitude at a single location and obtained in this way transfer functions. As SPL increased, the maximum moved toward lower frequencies. Because of the reciprocal relationship between the characteristics determined as functions of cochlear location and sound frequency, respectively, such a displacement would suggest a corresponding displacement of the maximum in the space domain toward the cochlear base, as is explained in chapter 4. Russell and Nilsen attempted to account for the apparent inconsistency between the space and frequency domains in a somewhat speculative way, I fail to understand. It is true nevertheless, that the shape of the transfer function in the frequency domain depends on the transfer function of the middle ear, which has a low-pass characteristic in the frequency region of the experiments. In this region, the characteristic has a negative slope of slightly less than 12 dB per octave for a constant stapes displacement (Schmiedt & Zwislocki, 1977). Whereas, for a constant stapes displacement, the sound pressure at the cochlear input increases in direct proportion to sound frequency, the effective slope is decreased to 6 dB per octave. The results of Sellick and Nilsen shown in their Fig. 4 for a cochlear location with a characteristic frequency of 13.25 kHz are replotted in Fig. 5.33 without and with the slope correction. Clearly, the correction has a substantial effect. Nevertheless, the peak shift is not abolished. This is especially evident in the 50 dB curve. The 60 dB curve is very flat, and it is difficult to decide where the peak lies. It appears shifted from 13.25 kHz to about 11.0 kHz but a small perturbation at 13.0 kHz obscures the situation.

The existence of a SPL dependent peak shift in the frequency domain is confirmed in another part of Russell and Nilsen's Fig. 1, in which they plot the input–output functions for three sound frequencies—the characteristic frequency at 13.25 kHz, a higher frequency of 14 kHz, and a lower one of 11 kHz obtained at one location. The 11 kHz curve crosses the characteristic-frequency curve at about 60 dB and reaches higher amplitudes. This in-

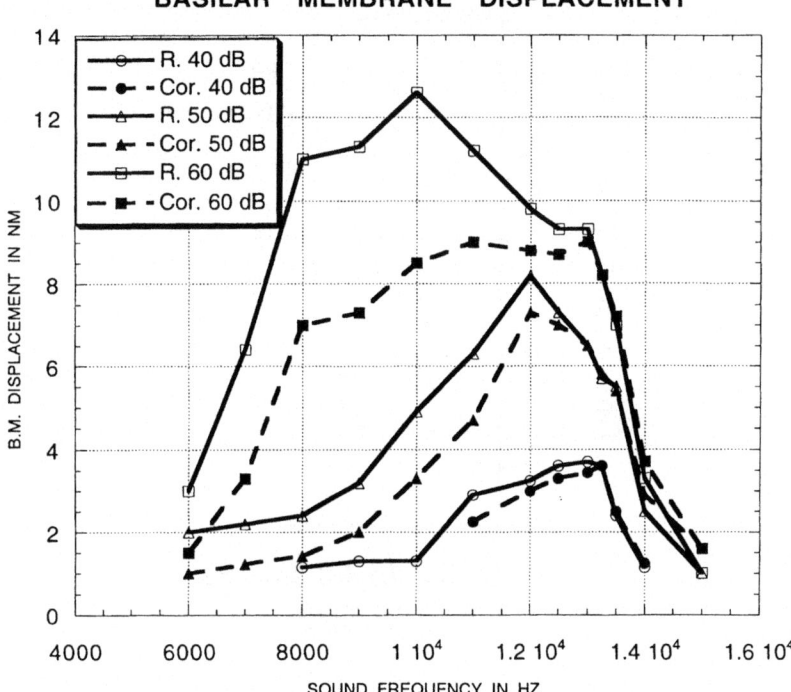

FIG. 5.33. Data from Russell and Nilsen (1997) reported by them as functions of sound frequency are replotted with (intermittent lines) and without (solid lines) correction for the middle ear transfer function. From "The Location of the Cochlear Amplifier: Spatial Representations of a Single Tone on the Guinea Pig Basilar Membrane," by I. J. Russell and K. E. Nilsen, 1997, *Proceedings of the National Academy of Sciences, 94*, pp. 2660–2664. Copyright © by National Academy of Sciences, U.S.A. Reprinted with permission.

dicates a shift of the maximum response to lower frequencies. The cross-over is preserved even after correction for the middle ear transfer function, as shown in Fig. 5.34 reproducing the Russell and Nilsen plot. The correction is indicated by the dot under the 11 kHz curve. The dot is still above the CF curve.

The discrepancy found by Russell and Nilsen between the frequency and space domains with respect to the SPL dependent peak shift may be more apparent than real. The spatial transfer functions of Fig. 5.32 appear somewhat jugged, revealing considerable variability in measurement of basilar membrane vibration at different locations. This is understandable in view of the difficulties in setting precisely the laser beam and calibrating the

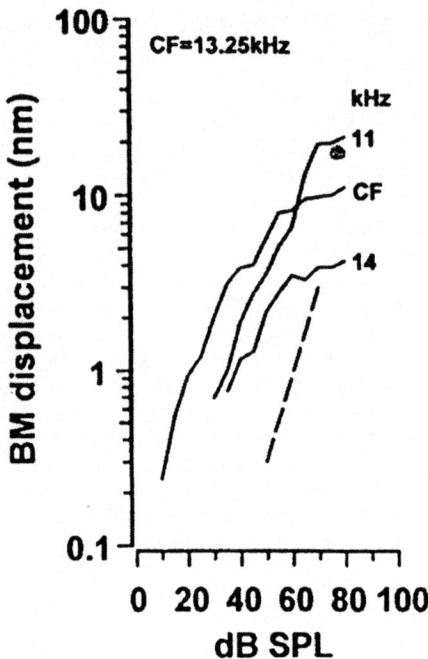

FIG. 5.34. Input–output functions of basilar membrane displacement in the base of a guinea pig cochlea for several sound frequencies surrounding the CF. The intermittent line shows direct proportionality. The dot shows correction for the middle-ear transfer function (modified from Russell & Nilsen, 1997). From "The Location of the Cochlear Amplifier: Spatial Representations of a Single Tone on the Guinea Pig Basilar Membrane," by I. J. Russell and K. E. Nilsen, 1997, *Proceedings of the National Academy of Sciences, 94,* pp. 2660–2664. Copyright © by National Academy of Sciences, U.S.A. Reprinted with permission.

setup anew at every location. For example, relative peaks at locations corresponding to characteristic frequencies of 15 and 17 kHz suggest relatively greater sensitivity than was achieved in the frequency region between the two peaks, especially at the frequency of about 16.5 kHz, where a trough is apparent. Also, at the highest SPLs of 90 and 100 dB, the greatest vibration amplitudes appear to have been encountered at locations basal to the characteristic frequency location. As a consequence, the spatial curves of Fig. 5.32 can hardly be used to demonstrate that the location of maximum basilar membrane vibration was independent of SPL.

While this monograph was being written, there appeared an article of Cooper (2000) in which he reports measurements of basilar membrane vibration at several radial locations of the cochlear base in guinea pigs. According to these measurements, the dynamic characteristics of basilar

PHYSICAL CONSTANTS AND FUNDAMENTAL CHARACTERISTICS 239

membrane vibration in this region do not differ fundamentally from those in more apical regions. Especially, the location of the maximum response is shifted toward the cochlear base, as SPL is increased.

A clear evidence of the SPL dependent location of maximum basilar membrane vibration follows from the experiments of Ruggero et al. (1997) who measured the vibration velocity, rather than displacement, at single cochlear locations as functions of sound frequency. They used chinchillas and laser interferometry based on the Doppler shift effect for the purpose. Their set up required placement of light reflecting microspheres on the basilar membrane but it is unlikely that they affected appreciably its vibration. One set of their measurements plotted in the frequency domain in their Fig. 8 is reproduced in Fig. 5.35. Because the recorded amplitudes are reported in terms of vibration velocities rather than displacements, the resulting curves have an upward tilt relative to the displacement curves, such as obtained by Russell and Nilsen and reproduced in Fig. 5.33. This means that the peak shift toward lower sound frequencies at high SPLs should have

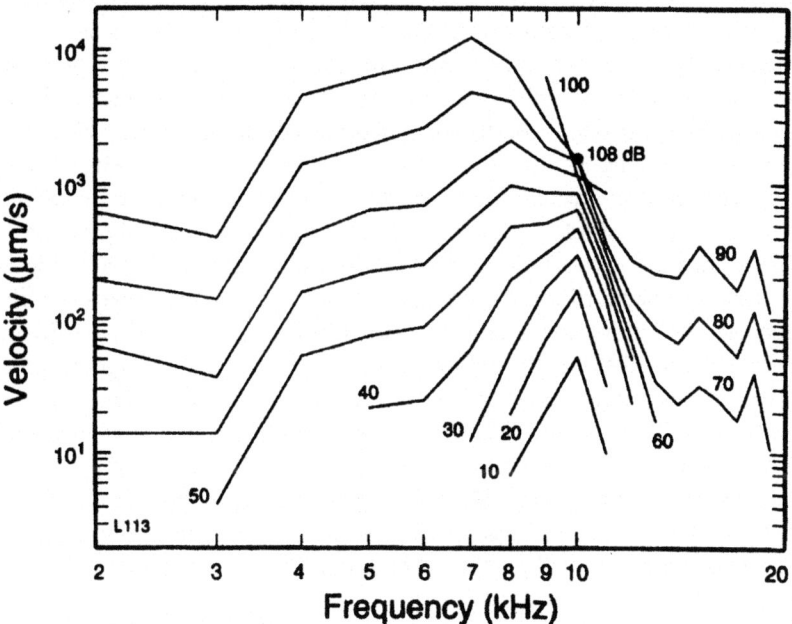

FIG. 5.35. Basilar membrane velocity in the cochlear base of a chinchilla at one location as a function of sound frequency for several SPLs (from Ruggero et al., 1997). Reprinted with permission from M. A. Ruggero, N. E. Rich, A. Recio. S. Narayan, and L. Robles (1997). Basilar-membrane responses to tones at the base of the chinchilla cochlea. *Journal of the Acoustical Society of America, 101,* 2151–2163. Copyright © 1997, Acoustical Society of America.

been deemphasized. Even so, it is quite obvious. Recalculation in terms of displacement would make it even more apparent. Because the stapes transfer function Ruggero et al. show in their Fig. 9 in terms of vibration velocity is quite flat, no further correction is required, and a corresponding peak shift in the cochlear space domain must have taken place in their experiments.

Two more features of the family of curves in Fig. 5.35 are worth mentioning. One is the uneven distribution of compressive nonlinearity, which can be judged from the spacing of the curves. Near the low SPL peak and at higher sound frequencies, the spacing is much smaller than at low sound frequencies, where it indicates hardly any compression at all. A similar pattern is evident in Russell and Nilsen's amplitude distributions in the space domain. The other is the appearance of secondary peaks, absent in Russell and Nilsen's curves. The locations of these peaks appear to be independent of SPL. Although small, the peaks play an important role in the analysis of cochlear dynamics discussed in the next chapter.

Both groups, Russell and Nilsen (1997) and Ruggero et al. (1997), measured the phase of basilar membrane vibration, the former in guinea pigs as a function of the distance from the cochlear apex for a constant sound frequency, the latter, in chinchillas as a function of the frequency for a constant cochlear location. The results of Russell and Nilsen are reproduced from their Fig. 2 in Fig 5.36, those of Ruggero et al., from their Fig. 13 in Fig. 5.37. In both figures, there is a phase lag that increases as a function of the corresponding independent variable at an accelerating rate and reaches a plateau at relatively high values of this variable. Just past the location where the vibration amplitude reaches its maximum, the CF location, there is a slight dip in the curves and the phase lag decreases as SPL increases. The latter phenomenon is particularly pronounced in the Russell and Nilsen's results. On the other hand, according to the Ruggero et al. measurements but not those of Russell and Nilsen, the phase lag increases ahead of the CF location. The SPL dependent phase lag ahead of the CF and a corresponding relative phase lead past the CF was found by several other investigators and appears to be a prevailing phenomenon, but the size of the effect found by Ruggero et al. was larger than found by others in the most basal and apical portions of the cochlea (e.g., Cooper & Rhode, 1992, 1995).

The phase lag that increases with the cochlear distance at an accelerating rate was already encountered by Békésy in dead cochleas and is accounted for theoretically in the preceding chapter. However, the phase dependence on SPL is found uniquely in live cochleas. It is shown in the next chapter to result mainly from dynamic interactions between the OHC stereocilia and the tectorial membrane and to a lesser extent from the mechanical changes

PHYSICAL CONSTANTS AND FUNDAMENTAL CHARACTERISTICS 241

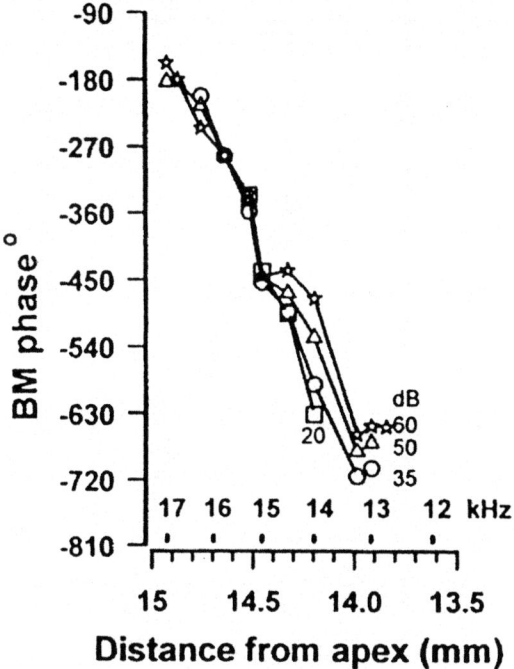

FIG. 5.36. Phases of basilar membrane displacement in the cochlear base of a guinea pig as functions of the distance from the apex and CF for several SPLs (from Russell & Nilsen, 1997). From "The Location of the Cochlear Amplifier: Spatial Representations of a Single Tone on the Guinea Pig Basilar Membrane," by I. J. Russell and K. E. Nilsen, 1997, *Proceedings of the National Academy of Sciences, 94*, pp. 2660–2664. Copyright © by National Academy of Sciences, U.S.A. Reprinted with permission.

in the organ of Corti, that affect the effective compliance of the basilar membrane. The phase dependence appears to be correlated with the health of the cochlea—the healthier the cochlea, the bigger the change.

The basilar membrane characteristics discussed thus far in this section have been found in the cochlear base, and it is important to know if they apply to the rest of the cochlea. The only experimental results extensive enough to shed some light on this question, as far as live cochleas are concerned, seem to be those of Cooper and Rhode (1995). Unfortunately, the extensive surgery and the use of a potentially toxic cyanoacrylate glue required by their experiments may have affected the health of their cochlear specimens. Almost complete absence of amplitude compression at the location of the peak response and of an intensity dependent shift of this location suggest, as explained further on, that the health of the specimens was

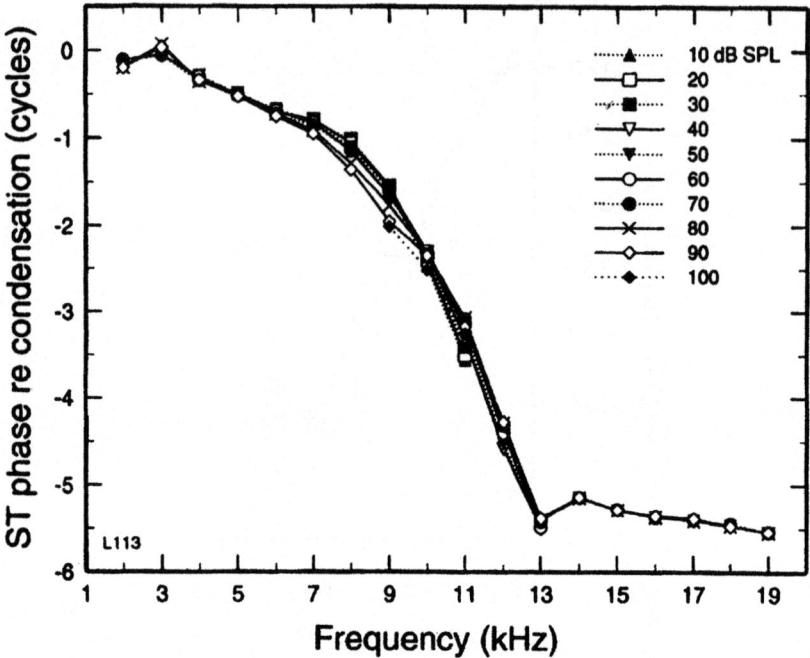

FIG. 5.37. Phases of basilar membrane displacement velocity in the cochlear base of a chinchilla at one location as functions of sound frequency for several SPLs (from Ruggero et al., 1997). Reprinted with permission from M. A. Ruggero, N. E. Rich, A. Recio. S. Narayan, and L. Robles (1997). Basilar-membrane responses to tones at the base of the chinchilla cochlea. *Journal of the Acoustical Society of America, 101*, 2151–2163. Copyright © 1997, Acoustical Society of America.

not good. It is true that several indices obtained in the presence of intact cochleae and discussed by the authors imply that these nonlinearities are reduced near the cochlear apex, and reduced compression is even evident in low frequency loudness functions (e.g., Hellman & Zwislocki, 1968). But it is unlikely that this reduction provides a sufficient explanation for their practically complete absence.

There are further difficulties with the experiments of Cooper and Rhode (1995). Their laser interferometric technique required light reflecting microbeads, so that their measurements had to be performed on the Reissner's membrane rather than on the basilar membrane that was not directly accessible. In the few instances in which they were performed on the basilar membrane, Reissner's membrane had to be torn for the purpose of introducing the microbeads, allowing perilymph into scala media. The fact that this massive chemical change in the tectorial membrane environment appeared to have little effect on the basilar membrane vibration suggests that the cochlear sensitivity was reduced prior to the change.

For all these reasons and some additional ones, the results of Cooper and Rhode (1995) have to be accepted with caution. Nevertheless, at least to the extent they have features in common with the characteristics obtained in the cochlear base, they suggest that these features prevail throughout the cochlea.

Some of Cooper and Rhode's (1995) results are reproduced in Figs. 5.38 to 5.40. In Fig. 5.38 are displayed Reissner's membrane vibration amplitudes as functions of sound frequency obtained at several sound pressure levels. They show rather broad tuning and little compression at any sound frequency up to 90 dB, the highest SPL used. They also show a secondary peak above the frequency of the main peak. These features, except the lack of compression, are qualitatively similar to those found in the cochlear base. The results of Fig. 5.38 recalculated as gain functions relative to the incus vibration are plotted in Fig. 5.39. Their superposition confirms

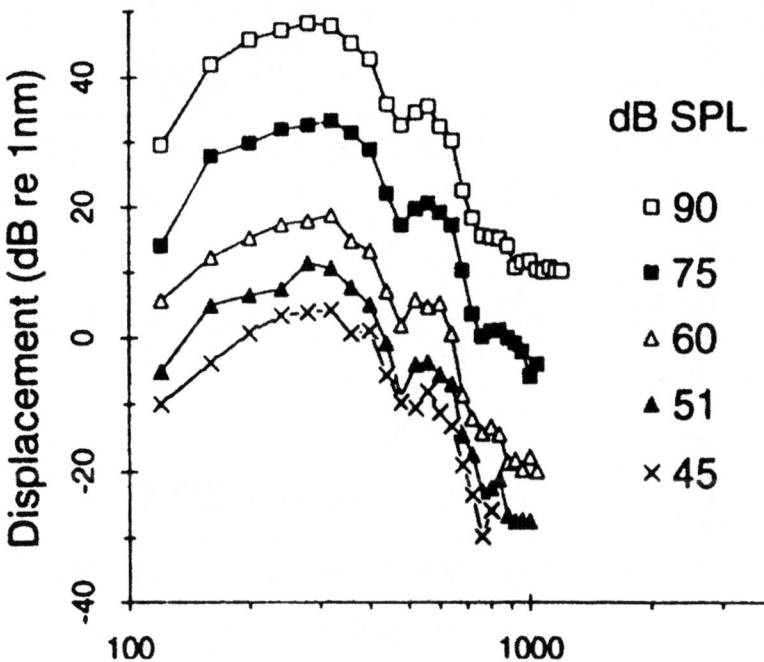

FIG. 5.38. Displacement amplitudes of Reissner's membrane as functions of sound frequency in the apical portion of a guinea pig cochlea at one location for several SPLs (from Cooper & Rhode, 1995). Reprinted from *Hearing Research, 82,* N. P. Cooper and W. S. Rhode, Nonlinear mechanics at the apex of the guinea-pig cochlea, 225–243. Copyright © 1995, with permission from Elsevier Science.

the conclusion that there is little compression and makes the secondary peak, separated from the first by a shallow notch, more apparent. The bottom part of the figure shows the corresponding response phases that seem completely independent of SPL, unlike what was found in the cochlear base. The phase plateau above the frequency band of the secondary peak indicates that there is no wave propagation beyond it. This phenomenon is explained in the preceding chapter. It has also been found in the cochlear base (e.g., Fig. 5.37). Finally, in Fig. 5.40 vibration measurements made in

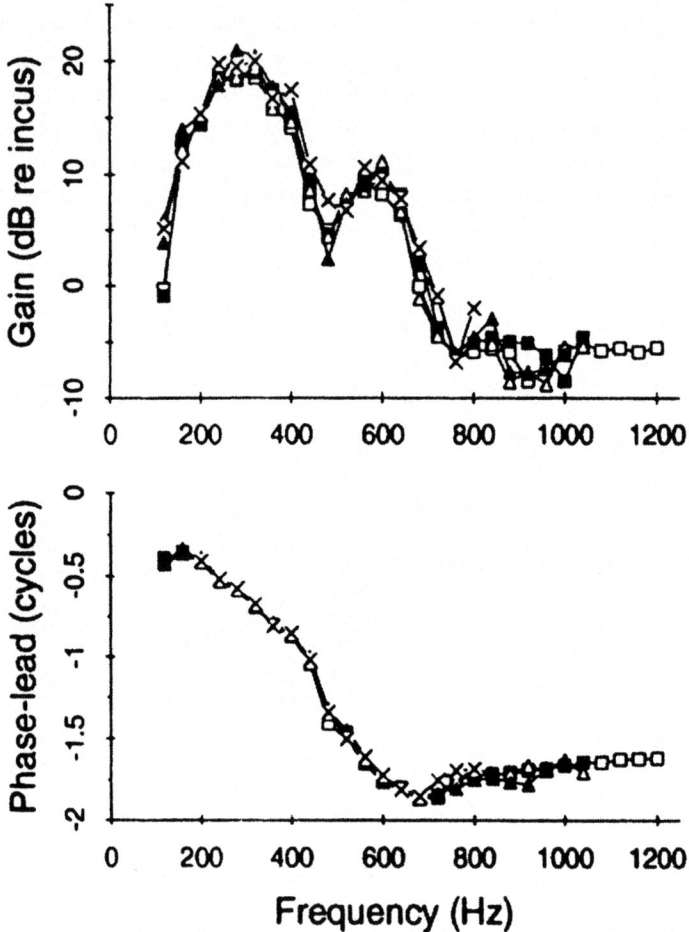

FIG. 5.39. Magnitude and phase gain functions of Reissner's membrane displacement in the apical portion of the guinea pig cochlea calculated from the results of Fig. 5.38. The functions are referred to the incus vibration (from Cooper & Rhode, 1995). Reprinted from *Hearing Research, 82*, N. P. Cooper and W. S. Rhode, Nonlinear mechanics at the apex of the guinea-pig cochlea, 225–243. Copyright © 1995, with permission from Elsevier Science.

the same preparation at three locations—Reissner's membrane, tectorial membrane, and basilar membrane, are compared. Except for one set of the data (RM 4) referring to a location on Reissner's membrane very close to the spiral limbus, the similarity among the rest of the amplitude and phase characteristics is close. During all the measurements, a tear was present in Reissner's membrane, which allowed for the placement of microbeads on the basilar membrane and the tectorial membrane. The first measurement (BM 2) was made on the basilar membrane between the

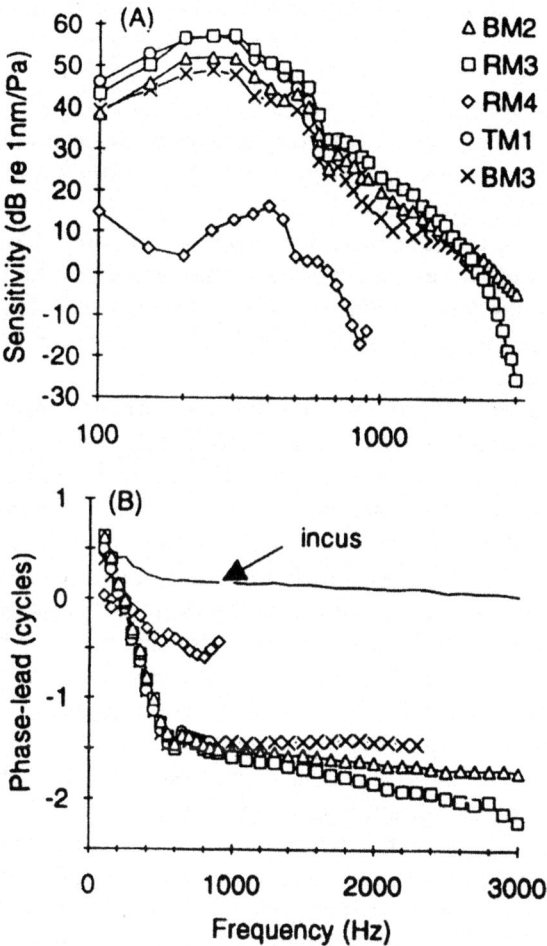

FIG. 5.40. Displacement amplitudes and phases of several parts of a guinea pig cochlear partition as functions of sound frequency at one apical location (further explanation in text; from Cooper & Rhode, 1995). Reprinted from *Hearing Research, 82,* N. P. Cooper and W. S. Rhode, Nonlinear mechanics at the apex of the guinea-pig cochlea, 225–243, copyright © 1995, with permission from Elsevier Science.

margin of Hensen's cells and the spiral ligament. At this location, the vibration of the basilar membrane is not expected to be maximum. The second (RM 3) was made on Reissner's membrane above the outer hair cells, where the vibration amplitude can be expected to be relatively large. For the third (RM 4), the reflecting bead was located on Reissner's membrane near its insertion into the spiral limbus, and the measured vibration probably does not reflect properly Reissner's membrane vibration. For the fourth (TM 1), the bead was pressed onto the tectorial membrane hard enough to indent visibly adjacent Hensen's cells. It is not clear what damage this produced in the organ of Corti. In any event, the associated changes in the recorded characteristics were small. The final measurement (BM 3) was made after complete removal of the organ of Corti. According to the authors, even this radical intervention did not produce clear changes in the characteristics. Nevertheless, it is possible to note a questionable presence of the secondary maximum, a decreased response amplitude in its vicinity and a flatter phase plateau than in the presence of a less damaged organ of Corti. The finding that complete removal of the organ of Corti had an almost negligible effect on the vibration characteristics of the basilar membrane supports the theory of the preceding chapter according to which the vibration maximum of the basilar membrane in a passive cochlea is not due to a resonance effect.

A note should be added concerning the deep notch in basilar membrane vibration amplitude found by Cooper and Rhode between the main and the secondary maxima. Subsequent experiments of the authors (Cooper & Rhode, 1996), confirmed by Hemmert et al. (2000) on in vitro preparations, revealed that the notch was due at least in part to an artifact produced by the rather large opening in the outer wall of scala vestibuli. This opening appears to have generated an interfering wave originating in the compressional wave that accompanies the transversal wave on the basilar membrane when sound is introduced into the cochlea through the stapes. Physical or mathematical removal of the interfering wave was found to substantially reduce the notch.

The legitimacy of inferring basilar membrane vibration characteristics from vibration measurements on Reissner's membrane has been questioned. Some experiments performed on excised cochlear specimens and, in one instance, even in vivo, suggested that the vibration characteristics of Reissner's membrane are quite different from those of the basilar membrane (e.g., Ulfendahl, Khanna, & Decraemer, 1996). However, in the original excised preparations, Reissner's membrane was coupled hydraulically to the tympanic membrane through the fluid-filled space of the bulla, so that the membrane was driven not only by the fluid motion generated by the basilar membrane but also by that generated directly by the tympanic membrane. Curiously, isolating the cochlear space from the bulla space by placing a small cylinder over the surgical opening in the cochlea and closing it with a glass window while the bulla was drained and filled with air failed to make Reissner's membrane mode of vibration entirely similar to that of the basilar membrane. According to Békésy's experiments (1960) and the most fundamental theory of surface waves, as applied to the cochlea in the preceding chapter, Reissner's membrane must be expected to vibrate in phase with the basilar membrane although with a somewhat smaller amplitude. Cooper and Rhode's experiments are consistent with the theory and suggest that the effect of Reissner's membrane stiffness, which is much smaller than that of the basilar membrane, as was shown already by Békésy (1960), does not affect noticeably the fluid flow in scalae media and vestibuli. Hemmert et al. (2000) came to the same conclusion.

If Cooper and Rhode's experiments are approximately correct, the dynamics in the cochlear apex differs from that in the base in quantitative relationships but not in the fundamental processes. At a given location, when sound frequency is increased, the basilar membrane vibration amplitude increases to a maximum, then decreases to a shallow notch and increases somewhat again to produce a secondary maximum and finally decreases for good. At the same time, the phase lag of the vibration increases monotonically until, past the secondary maximum, it reaches a plateau. The secondary maximum had been first discovered by Rhode (1971) during his measurements of basilar membrane vibration in the base of the squirrel monkey cochlea and was absent in postmortem preparations of Békésy's. It was confirmed since then on several occasions (e.g., Robles et. al., 1986; Ruggero et al., 1997) and appears to be a property of the living cochlea. Because of its smallness, it is often missed. It appears to be more pronounced in the cochlear apex than in the cochlear base. The apical maximum is even evident in neural tuning curves recorded in the auditory nerve (e.g., Liberman, 1976). As already mentioned, although small, it is of substantial theoretical interest.

Shear Motion Between the Tectorial Membrane and the Reticular Lamina

As mentioned in the introduction to this section, my coworkers and I used OHC as little microphones detecting the shear motion. We did so directly by recording their alternating receptor potentials and, indirectly, by measuring the reflection of these potentials in Hensen's cells. The legitimacy of both procedures is explained briefly in the introduction. Access to the organ of Corti was gained through the lateral wall of scala media following the pioneering work of Dallos, Santos-Sacchi, and Flock (1982) who were the first to record intracellularly the OHC potentials. We modified slightly their method and adapted it to Mongolian gerbils. The measurements of the cellular potentials we made in the organ of Corti overlapped in some instances with those of the Dallos group (e.g., Dallos, 1986). Our interest was focussed mainly on the mechanics of the cochlea, however, whereas theirs seemed to focus increasingly on the electrophysiology of the hair cells themselves (e.g., Dallos & Evans, 1995).

Our experiments were performed in a large Industrial Acoustics Company double wall, double floor soundproofed booth. The experimental animal was held in home modified guinea pig holder placed on an Ealing pneumatic table. The surgery was performed on the same table while the pneumatic suspension was disconnected. The surgical microscope was also placed on the table. All the required electronics and electroacoustics equipment was located in the booth, near the table, so that the experimenter was able to control it from the same seat he/she occupied while advancing the electrode. Electrical artifacts in recording cochlear potentials were prevented by careful grounding.

Our method was first described in 1988 (Zwislocki & Smith, 1988) but in the greatest detail, in 1992 (Zwislocki, Slepecky, Cefaratti, & Smith). The preliminary surgery was the same as used in connection with the measurement of tectorial membrane stiffness already described. However, the opening made in the ventrolateral part of the right bulla was smaller and the middle ear apparatus was left intact. Nevertheless, the opening allowed convenient access to all three turns of the gerbil cochlea. A silver wire electrode was inserted through a separate small opening and placed in the niche of the round window for monitoring cochlear microphonics (CM). Its purpose was to check on sound transmission from the ear canal to the cochlea and on the overall sensitivity of the cochlea. Békésy (1951, 1960) has demonstrated that, within a useful SPL range, CM is directly proportional to basilar membrane displacement, and his observation was confirmed indirectly on several occasions (e.g., Dallos, Billone, Durrant, Wang, & Raynor, 1972; Dallos & Durant, 1972).

Sound was delivered to the ear canal through a miniature earphone (Knowles, ED-1932), to which was added a home made corrective acoustic filter. Up to 100 dB SPL, its wave distortion did not exceed 5%. The sound was monitored with a miniature microphone (Knowles, EA-1934) with a practically flat transfer characteristic within the useful frequency range of the experiments, ranging approximately from 0.25 to 5 kHz. Both transducers were coupled to the ear canal by means of short tubes which did not affect appreciably the sound transmission. They were secured in the ear canal by means of a special conical adapter made of teflon and sealed at the entrance to the bony part of the ear canal by means of a malleable plastic compound used otherwise for ear protection against excessive noise. The sinusoidal electrical signals energizing the earphone were swept logarithmically in frequency. This allowed a rapid determination of transfer functions referred to an approximately constant sound pressure at the tympanic membrane.

A typical example of sound pressure generated in the ear canal of a gerbil as a function of sound frequency for a constant input voltage to the earphone is shown in Fig. 5.41 by the middle trace. Its width is proportional to the sound pressure amplitude on a linear scale. Over the frequency range shown from about 1 to almost 15 kHz, the trace varies by no more than 50%, except for the sharp peak near 15 kHz. On the decibel scale 50% amounts to 6 dB, so that the SPL remains constant within a ± 3 dB tolerance. The lowest trace shows the constant voltage, and the uppermost one, the CM at the round window. The latter indicates a constant CM amplitude up to about 4 kHz and an amplitude descent at a rate of roughly 6 dB per octave at higher frequencies. Because the CM near the round window remains constant for a constant amplitude of basilar membrane vibration up to about 4 kHz and decays at a rate of 6 dB per octave above this frequency according to our determinations (Schmiedt & Zwislocki, 1977), it is possible to conclude that, in our ex-

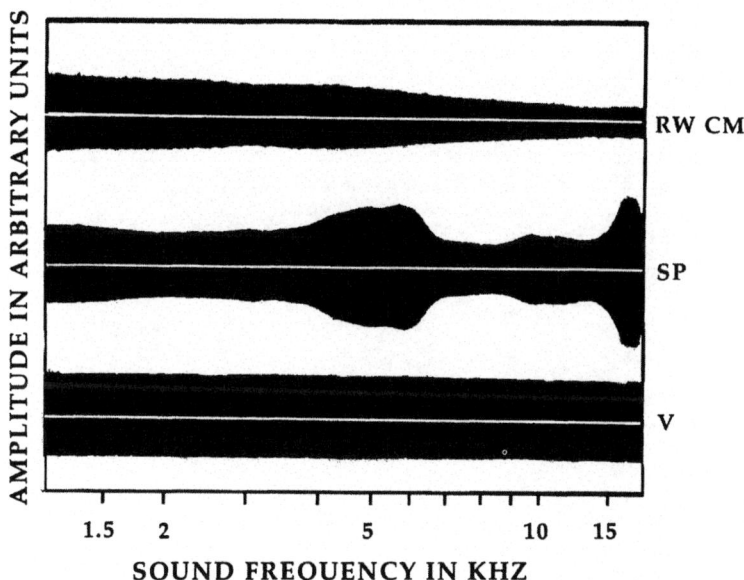

FIG. 5.41. Typical input signal magnitude for the Mongolian gerbil cochlea as a function of sound frequency. The frequency was swept logarithmically. The widths of the black bands indicate the double amplitudes—at the bottom, of the electrical signal to the earphone; in the middle, of the sound pressure in the ear canal; at the top, of the round window CM.

periments, the displacement amplitude of the basilar membrane at the cochlear input remained approximately constant for a constant sound pressure in the ear canal. Thus, the transfer functions of cochlear responses shown below refer indirectly to a constant displacement amplitude of the basilar membrane at the cochlear base. Perhaps it may be useful to point out here that Békésy (1960) measured his transfer functions of basilar membrane displacement keeping the amplitude of stapedial displacement constant. This is very likely to have produced in the cochlear base a basilar membrane displacement amplitude that increased in direct proportion to sound frequency. As a consequence our transfer functions are tilted downward toward higher sound frequencies relative to his.

To place microelectrodes in the organ of Corti a hole of about 100 μm diameter was drilled in the cochlear capsule over the lateral wall of scala media. Depending on need, up to 3 holes were drilled on occasion, as shown in Fig. 5.42. To improve visibility of the landmarks on the capsule, according to which the holes were drilled, the cochlear surface was stained

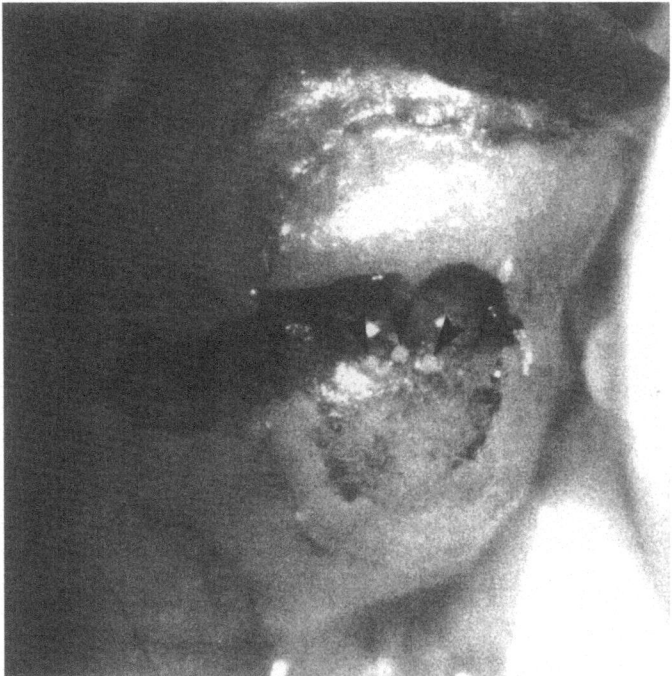

FIG. 5.42. The cochlear capsule of a Mongolian gerbil, stained with Alcian blue and perforated at three places along the lateral wall of scala media (modified from Zwislocki et al., 1992). Reprinted from *Hearing Research, 57*, J. J. Zwislocki, N. B. Slepecky, L. K. Cefaratti, and R. L. Smith, Ionic coupling among cells of the organ of Corti, 175–194, copyright © 1992, with permission from Elsevier Science.

PHYSICAL CONSTANTS AND FUNDAMENTAL CHARACTERISTICS   251

with Alcian blue, as can be seen in the figure. We made sure that the dye stained only the superficial bone layers and did not penetrate to the cochlear fluids. When the holes were placed correctly with sufficient precision, the electrodes inserted through them at a right angle to the cochlear surface usually penetrated the organ of Corti, as illustrated in Fig. 5.43. The angle could be varied, nevertheless, over a range of about 30°. This was helpful in finding the hair cells.

Precise placement of the electrodes was made possible by a precision drill developed especially for the purpose. It is shown in Fig. 5.44, mounted on a three-axial micromanipulator, the same that was used for microelectrode placement. It was equipped with a hydraulic Narishige microdrive featuring only a small hysteresis. The drill was energized by a DC motor with a variable speed and a high gear ratio to achieve a slow drill bit rotation. To minimize wobble at the bit tip, the bit was coupled to the motor assembly by a flexible joint and was held in a narrow tube that extended almost to its tip. The remaining wobble was on the order of 5 µm and was hardly visible under a

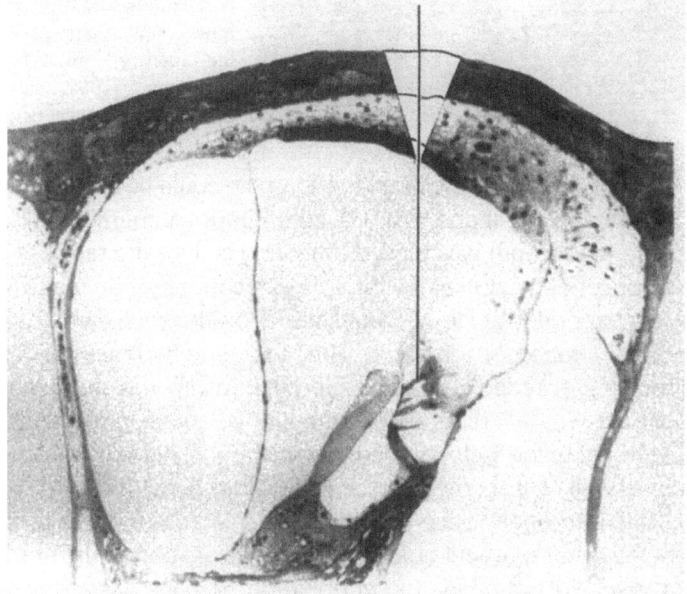

FIG. 5.43. Cross section of the cochlear canal at the location of the experiments. The access opening to scala media and the electrode track to the organ of Corti are shown schematically. Note the nearly vertical electrode track, reflecting the actual experimental situation (modified from Zwislocki et al., 1992). Reprinted from *Hearing Research, 57*, J. J. Zwislocki, N. B. Slepecky, L. K. Cefaratti, and R. L. Smith, Ionic coupling among cells of the organ of Corti, 175–194, copyright © 1992, with permission from Elsevier Science.

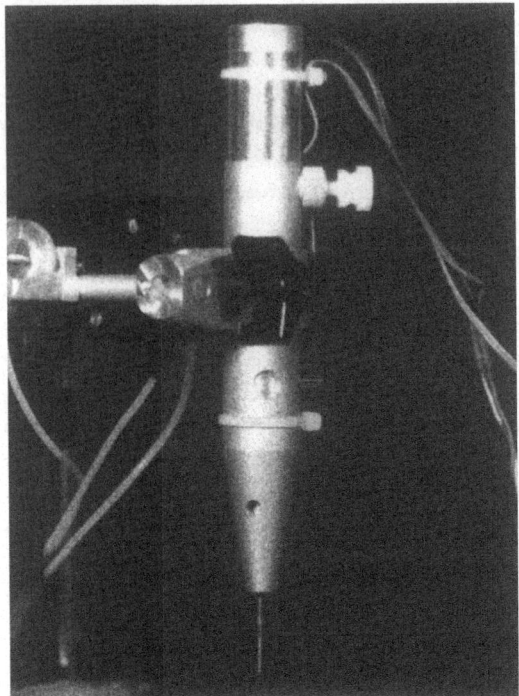

FIG. 5.44. The precision drill mounted in the three-axial micromanipulator for making openings in cochlear capsule at precise locations. Note the fine tube at the bottom of the drill for guiding the drill bit (modified from Zwislocki et al., 1992). Reprinted from *Hearing Research, 57,* J. J. Zwislocki, N. B. Slepecky, L. K. Cefaratti, and R. L. Smith, Ionic coupling among cells of the organ of Corti, 175–194, copyright © 1992, with permission from Elsevier Science.

magnification of 40x. For protection, the sharp tip could be retracted into the tube. It had a triangular shape that proved optimum in manual drills. In the middle of the tube length was placed a heating coil for the purpose of preventing vapor condensation at the bit tip. Such condensation makes drilling precise holes very difficult. To prevent stresses on the cochlear capsule and too deep a penetration of soft tissue, once the bone had been pierced, the drilling process proceeded very slowly and the drill tip was sharpened each time a new hole was drilled. As already mentioned above in connection with the measurement of the tectorial membrane stiffness, the spiral ligament lining the lateral wall of scala media proved tough to penetrate, and I found it helpful to drill it through, once the drill bit pierced the bony capsule. The opening in the spiral ligament quickly sealed itself as soon as the drill bit was withdrawn and did not allow any endolymph seepage. Nevertheless, the weakened tissue at its location facilitated electrode penetration. On several occasions, the noise produced in the cochlea by the drill was checked by measuring the round window CM and comparing it to the CM produced by an external sound in the mid frequency range, whose SPL was measured in the ear canal. The cochlear drill noise was found to be equal to that produced by a 60 dB sinusoid. Such sound levels are known not to be injurious to the cochlea.

The electrodes we used routinely consisted of micropipettes filled with 1.5- to 3-M KCl and, on occasion, with a millipore filtered 6% solution of Lucifer yellow in 1-M LiCl for marking cells impelled by a microelectrode. The micropipettes were pulled on a Brown-Flaming horizontal puller set to produce short tapers so as to keep their stiffness as high as possible. Stiff microelectrodes penetrated the spiral ligament more easily and could be placed more accurately in the organ of Corti. When filled with the electrolyte, the microelectrodes had resistances between 20 and 60 MOhms, which increased to between 50 and 110 MOhms when the dye solution was added. The electrodes were held in a WPI holder with a silver chloride interface, and the same interface was provided for the reference ground electrode embedded in a cheek muscle to cancel interface potentials. The holder replaced the microdrill on the three axial micromanipulator, which was provided with a device for changing the electrode angle, once the drilling was finished.

The WPI holder containing a shielded preamplifier was connected to a WPI KS-700 amplifier that allowed measurement of both DC and AC potentials and also of resistances at the end of a microelectrode. The recorded signal was further led through an EG&G Park amplifier band limited between 0.3 and 10 kHz and fed to an Ithaco 393 Lock-in-Amplifier set to a narrow filter bandwidth of 3 Hz for effective noise rejection. As a consequence, only the first harmonic of the signal was further processed. Because the recorded signal showed little distortion up to at least 70 to 80 dB SPL, as already mentioned, this was not a severe limitation. The signal wave form was checked in separate experiments in which the Lock-in-Amplifier was eliminated and a pass band of .3 to 10 kHz used. Some results of these experiments are shown in the next chapter. The lock-in-Amplifier was locked to the signal generator energizing the miniature earphone and provided either the amplitude or the phase of the signal as a function of its frequency. Because the signal frequency was swept logarithmically, the obtained functions followed the frequency on a logarithmic scale. This was done in the recognition that the tonotopic map of the cochlea was approximately logarithmic, except in the vicinity of the cochlear apex. The output of the Lock-in-Amplifier was fed to a Tectronix storage oscilloscope and displayed on its screen over a horizontal coordinate proportional to log frequency. The log scale was obtained with the help of a home-built frequency discriminator whose output voltage was proportional to log frequency. To obtain families of curves on the oscilloscope screen the frequency sweeps were repeated at a rate of one per 6 sec, each lasting 5 sec with a 1 sec interval between them. This rate was compatible with the narrow bandwidth of the Lock-in-Amplifier.

The sweep frequency system proved very efficient as it allowed us to record 10 frequency characteristics in one minute. This was essential since it is difficult to record intracellularly hair cell potentials over time periods exceeding a minute. When the curves were recorded with SPL as a discrete parameter, an intensity series of the curves was generated, which completely defined a cell's signal processing characteristics. By making vertical cuts through the series, it was possible to construct input–output functions for any desired sound frequency. Because the curves were recorded in effect for constant vibration amplitudes of the basilar membrane in the cochlear base, as already mentioned, they paralleled what is defined as transfer functions, and I refer to them as such to simplify their descriptions.

A microelectrode was advanced through the opening drilled in the cochlear capsule with the help of the Narishige hydraulic drive. The advancement was monitored on the scale of the drive with the help of the encountered potentials, as had been described by Dallos, Santos-Sacchi, & Flock (1982). As soon as the spiral ligament was traversed, in which the electrodes encountered a variety of negative and positive potentials, a stable positive potential on the order of 80 mV was encountered signaling that the electrode entered scala media. At the same time the amplitude of CM potentials, which were recorded even within the spiral ligament, increased by an order of magnitude. In the gerbil preparation, the organ of Corti was encountered about 100 µm further. Its margin was signaled by a slightly increased, then decreased endolymphatic potential (EP). Negative capacitance ringing of the electrode produced by negative capacitance overcompensation of the electrode capacitance usually was sufficient to make the electrode penetrate a border Hensen's cell. The penetration was signaled by a negative potential on the order of –80 mV, an increment in the amplitude of the alternating potential and a change in its phase by 180°. Once the electrode entered the first Hensen's cell, and the electrode angle was correct, it was usually possible to advance it further through the whole population of Hensen's cells without much change in the potentials until it reached the approximately zero potential of the outer tunnel of Corti. This happened at a distance of 50 to 80 µm and was associated with a somewhat decreased amplitude of the alternating potential. Sometimes the zero potential was not found at this distance but continued to be negative. This meant that the electrode was too low and was going through the population of Deiter's cells. If the electrode was too high, it popped out of Hensen's cells back into the endolymph above the organ of Corti. It then happened sometimes that the electrode entered the tectorial membrane. This was indicated by decreased CM recorded at the same time without any appreciable change in the positive endolymphatic potential. In some experiments, the tectorial membrane was impaled on purpose and manipulated with the microelectrode in an attempt at weakening its attachment to OHC stereocilia. Some results of such experiments are described in chapter 6. If the electrode found the approximately zero potential of the outer tunnel, it was expected to be in the vicinity of the OHCs about 10 µm further. Nearness of an OHC was often signaled by a slight potential negativity. An effort was then made to penetrate it with the help of negative capacitance ringing. If the effort succeeded, the decision that the electrode tip was lodged in an OHC, indeed, was made on the basis of the following criteria: (1) distance from the last Hensen's cell in excess of 10 µm; (2) an approximately zero potential surrounding the cell; (3) an alternating potential amplitude distinctly larger than in the surrounding intercellular space and in Hensen's cells; (4) intracellular potential of around −70 mV. By contrast to Hensen's cells, which could be held for many minutes, sometimes beyond experimental needs, rarely was it possible to hold an OHC for more than 1 minute.

The search for OHCs extended over a distance of about 40 µm. Beyond it laid the tunnel of Corti with the pillars of Corti, impaled on occasion, and the IHCs surrounded by inner supporting cells. This group of cells was encountered at a distance of about 200 µm from border Hensen's cells and about 50 µm from the place where the zero potential of the outer tunnel was first encountered. The electrical characteristics of the inner supporting cells were similar to those of Hensen's cells, but their potentials seemed to be somewhat larger. On the other hand, IHCs were characterized by markedly higher alternating potentials and rather modest resting potentials, rarely exceeding 40 mV. We were able to hold them for periods of time similar to those for the OHCs.

Several examples of recorded cochlear transfer functions are shown below. Typical transfer functions of an OHC recorded intracellularly (upper curve) and extracellularly (lower curve) are shown in the upper panel of Fig. 5.45 (Zwislocki et al., 1992). The lower panel shows a Hensen's cell transfer function. They were all recorded on the same electrode penetration at an SPL of 40 dB and have the same best frequency (BF) of 1.8 kHz. Note that they have practically identical shapes and are almost symmetrical relative to the BF. Their Q(10 dB) factor, defined as a ratio between the BF and the bandwidth 10 dB below the peak amplitude, is only slightly larger than unity and is similar to the corresponding basilar membrane Q factor evident in Cooper and Rhode's apical transfer functions. For basal regions, Ruggero's et al. curves reproduced in Fig. 5.35 suggest a somewhat higher Q of about 2. Another set of transfer functions obtained at 30 dB SPL on the same electrode penetration is shown in Fig. 5.46. The curve with the highest amplitudes was obtained intracellularly in a Hensen's cell, the lowest, extracellularly in the outer tunnel of Corti. Note that both indicate the same BF but that the Hensen's cell curve follows substantially higher amplitudes than the extracellular curve. The middle curve shows the CM recorded in the endolymph outside the organ of Corti. Note that it points to a slightly lower BF and broader bandwidth, indicating a somewhat smaller Q factor, than for the other two curves.

The somewhat unexpectedly low Q factor aroused the suspicion that the microelectrode affected negatively the cochlear tuning. To check on this possibility, two-electrode experiments were performed on a small number of gerbils. Either two openings were drilled in the lateral wall of scala media or both electrodes were inserted through the same opening. They were manipulated by means of individual micromanipulators. One electrode was held on the top surface of the spiral ligament, the other was inserted

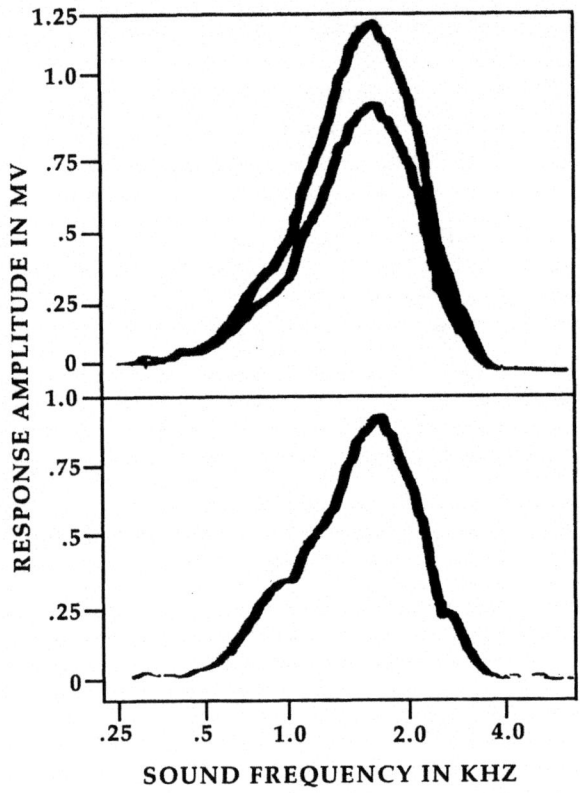

FIG. 5.45. Cochlear alternating potentials as functions of log frequency. Top panel: intra- and extracellular (smaller magnitude) recordings from an OHC; bottom panel: from a Hensens's cell (the same electrode track; modified from Zwislocki et al., 1992). Reprinted from *Hearing Research, 57*, J. J. Zwislocki, N. B. Slepecky, L. K. Cefaratti, and R. L. Smith, Ionic coupling among cells of the organ of Corti, 175–194, copyright © 1992, with permission from Elsevier Science.

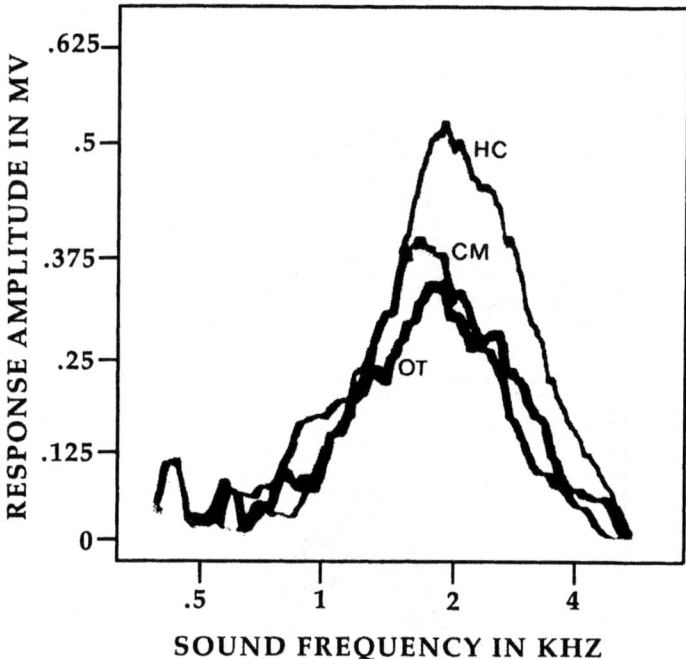

FIG. 5.46. Cochlear alternating potentials as functions of log frequency recorded from: HC—Hensen's cell, OT—outer tunnel space, CM—endolymph outside the organ of Corti (modified from Zwislocki et al., 1992). Reprinted from *Hearing Research, 57*, J. J. Zwislocki, N. B. Slepecky, L. K. Cefaratti, and R. L. Smith, Ionic coupling among cells of the organ of Corti, 175–194, copyright © 1992, with permission from Elsevier Science.

into scala media, as for organ of Corti recordings. The first electrode was being held on the surface of the spiral ligament rather than being inserted into scala media because the effective space constant is smaller there than in the endolymph and a better resolution of the details of cochlear transfer functions is achieved, albeit at the price of a smaller response magnitude. The second electrode was advanced toward the organ of Corti and, ultimately, inserted into it. Transfer functions recorded by means of the first electrode with the second electrode in and out of the organ of Corti were compared to each other. An example of the obtained results is illustrated in Figs 5.47 and 5.48. In Fig. 5.47A, the upper, thin curve shows a spiral ligament transfer function with a BF of about 1.8 kHz obtained with the second electrode in the endolymph, the lower, thick curve, with the electrode advanced intracellularly through the population of Hensen's cells up to the outer tunnel of Corti. To check if the clearly decreased response amplitudes

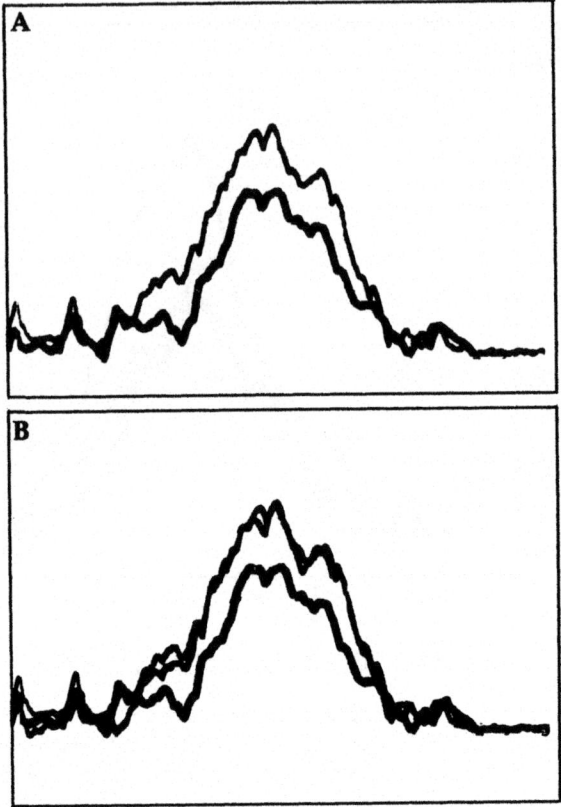

FIG. 5.47. CM recordings from the top of the spiral ligament. Top panel: thin line, before insertion of the second electrode into Hensen's cells; thick line, after. Bottom panel: thin lines, before insertion of the electrode into Hensen's cells and after its withdrawal; thick line, with the electrode in a Hensen's cell.

associated with the electrode presence in the organ of Corti were due to a mechanical interference produced by the electrode or to a damage of the organ of Corti, the following sequence of recordings was executed. The first recording on the spiral ligament was made before the second electrode reached the organ of Corti, the second, while the electrode was in the organ of Corti, the third, after it had been withdrawn from it. The results obtained with the second electrode in the organ of Corti are shown by the lower, thick curve of Fig. 5.47B, those obtained with the electrode in the endolymph, by the upper thin curves. Their coincidence with the curves of panel A indicates that no appreciable damage of the organ of Corti occurred, and that the decreased response magnitude was probably due to a

PHYSICAL CONSTANTS AND FUNDAMENTAL CHARACTERISTICS 259

mechanical electrode interference. It reduced the response magnitude by about 30%. The Q(10 dB) factor was reduced only slightly. When the electrode was reinserted into the Organ of Corti, this time intercellularly, almost no CM reduction took place. Only when the electrode was withdrawn and reinserted again intracellularly did the CM magnitude decrease by about the same amount as during the first intracellular insertion. The relationships are displayed graphically in Fig. 5.48, in panel A with respect to the intercellular electrode insertion, in panel B, with respect to the intracellular insertion. The latter results are almost identical to those of Fig. 5.47B. The difference between the effects of the intracellular and intercellular electrode insertions brings up the possibility that the interference effect may have been electrical rather than mechanical. We did not pursue this problem any

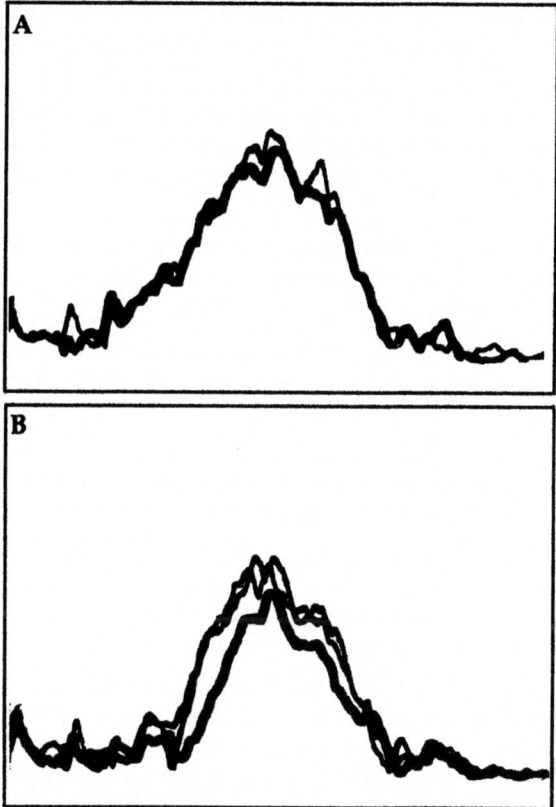

FIG. 5.48. CM recorded from the top of the spiral ligament. Top panel: thin line, before second electrode insertion into the organ of Corti; thick line, after its intercellular insertion. Bottom panel: thin lines, before second electrode intracellular insertion into the organ of Corti and after its withdrawal; thick line, during its intracellular insertion.

further. Important is the demonstration that electrode insertion into the organ of Corti can have a small but noticeable effect on CM and, probably, related potentials, so that the measured potentials can represent the potentials of an intact organ of Corti only to the first order of approximation.

The decreased sensitivity due to electrode insertion into the organ of Corti seemed to be almost evenly distributed over all the sound frequencies surrounding the BF and did not affect the BF. If the effect was electrical, such a pattern could be readily understood but, if it was mechanical, it was somewhat surprising. It was expected, perhaps naively, that the electrode would increase the stiffness of the basilar membrane–organ of Corti aggregate. To check out the situation, I decided to explore the mechanical properties of the electrode. For this purpose typical electrodes, held in their usual holder containing the electrolytic interface, were mounted on a vibrator and vibrated laterally. Their pattern of vibration was observed under stroboscopic illumination. To my surprise, the electrodes had their lowest mechanical resonance at frequencies around 125 Hz. This meant that, at higher frequencies used in cochlear recordings, they did not act through their stiffness but through a complex impedance with a relatively high resistive component. This was consistent with an approximately evenly distributed damping around the BF.

The findings that the alternating response potential was higher in a Hensen's cells than outside of it and that the CM transfer function had only a slightly broader transfer bandwidth than did Hensen's cell's seemed intriguing but they were confirmed in our laboratory on many occasions and were considered routine. They were in conflict with the belief that the alternating potentials originating in OHCs were induced in Hensen's cells via the intercellular fluid and with the accepted rather large electrical space constant measured in the fluids. They suggested a direct ionic communication between the OHCs and the supporting cells. None had been found until then (e.g., Santos-Sacchi & Dallos, 1983; Santos-Sacchi, 1986), and we decided to make an attempt at resolving the conflict (Zwislocki et al., 1992).

The electrodes were filled with a Lucifer yellow solution in 1-M LiCl instead of the KCl solution, as already mentioned. The dye was injected into the impaled cells electrophoretically by means of a square wave with an amplitude of 5 nA and a 16 Hz fundamental frequency. The injection always stopped before the resting potential sunk to one half. In most experiments, sound was delivered to the cochlea at a SPL of 40 dB during the injection. Usually, only one cell was injected per hole in the cochlear capsule. However, on some occasions, two electrode tracks were made at two different angles in a plane parallel to the basilar membrane so as to achieve nonoverlapping staining regions. The dye injection did not usually affect the resting and alternating potentials of the cells being injected, in agreement with the experience of Oesterle and Dallos (1989) concerning cell marking with horseradish peroxidase. The staining results were mostly evaluated in surface preparations of the organ of Corti, and some examples are shown in the photomicrographs of Figs. 5.49 to 5.52. Details of the histological procedure are described in Zwislocki, et al. (1992). The photomicrographs shown were obtained by superposition of two pictures, one obtained with phase-contrast, the other with fluorescence microscopy with a filter set for fluoresceine. A 35 mm camera with a 2× adapter and a 10× glycerin microscope objective were used. The first figure (Fig. 5.49) shows a rather typical pattern occurring in the absence of the acoustic stimulus, although there was substantial variability. It consists of a spherical spread of Lucifer yellow to a small population of Hensen's cells resulting from its injection into a Hensen's cell for 1 minute. For orientation, the cylindrical appearing structures near the top of the picture are the pillars of Corti. Another pattern, produced by a 2 minute injection of the dye into a Hensen's cell in the presence of the acoustic stimulus is shown in Fig. 5.50. The lowermost arrowhead points to the injected cell; the middle one to

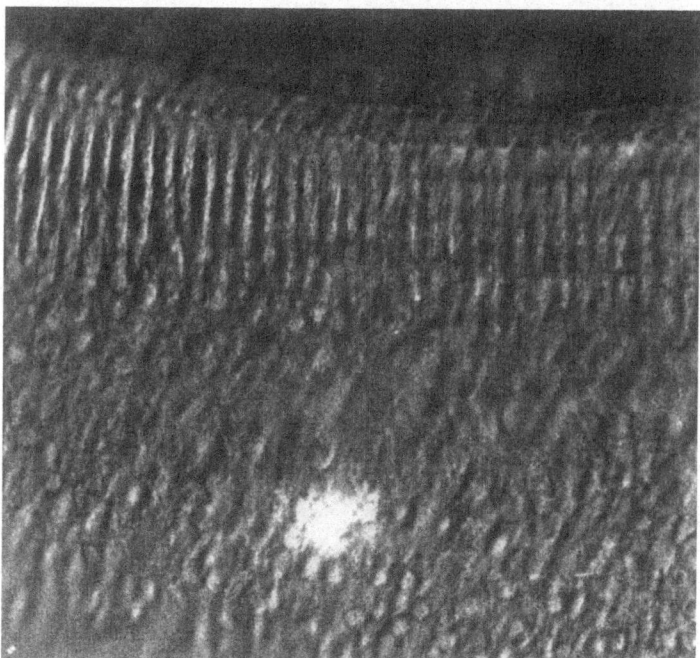

FIG. 5.49. Lucifer yellow diffusion from an injected Hensen's cell to neighboring Hensen's cells in the absence of acoustic stimulation (modified from Zwislocki et al., 1992). Reprinted from *Hearing Research, 57*, J. J. Zwislocki, N. B. Slepecky, L. K. Cefaratti, and R. L. Smith, Ionic coupling among cells of the organ of Corti, 175–194, copyright © 1992, with permission from Elsevier Science.

stained Deiter's cells and the top one to a nerve fiber crossing the tunnel of Corti. Evidently, the dye must have crossed from Hensen's cells to Deiter's cells and pillar cells. This result is in agreement with some previous investigations indicating gap junctions among the supporting cells of the organ of Corti (e.g., Iurato et al., 1976) . Also in agreement with previous investigations (e.g., Santos-Sacchi, 1986), the dye did not invade the Hair cells. However, the photomicrograph reproduced in Fig. 5.51 shows clear evidence of simultaneously stained OHCs of the second and third rows and of Hensen's cells. The result was obtained after a 1.5-minute injection of Lucifer yellow into an OHC of the third row, as indicated by the distance of the micropipette tip beyond Hensen's cell population and by the encountered direct and alternating potentials (0 potential surrounding the cell, resting potential of –71 mV, alternating potential clearly larger than in Hensen's cells). Presence of the dye in the OHCs, provides indisputable confirmation of this evidence. In no experiment of ours were we able to clearly demonstrate staining of OHCs after dye injection into any of the supporting cells

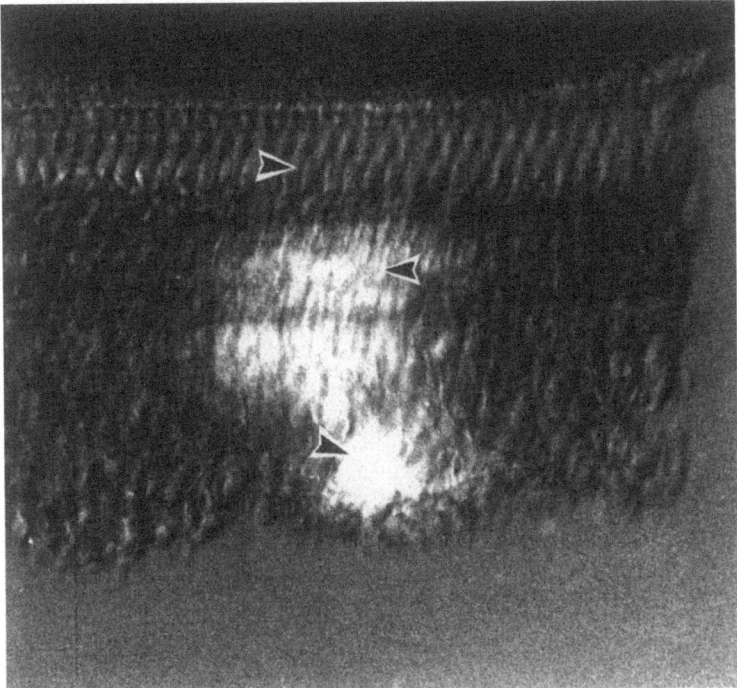

FIG. 5.50. Lucifer yellow diffusion from an injected Hensen's cell to neighboring Hensen's cells and Deiter's cells in the presence of 40 dB SPL acoustic stimulation (modified from Zwislocki et al., 1992). Reprinted from *Hearing Research, 57,* J. J. Zwislocki, N. B. Slepecky, L. K. Cefaratti, and R. L. Smith, Ionic coupling among cells of the organ of Corti, 175–194, copyright © 1992, with permission from Elsevier Science.

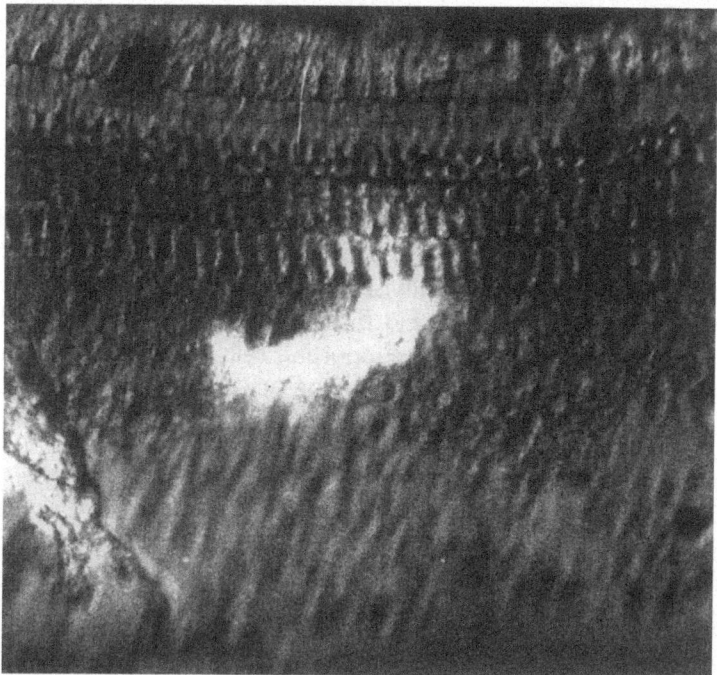

FIG. 5.51. Lucifer yellow diffusion from an injected OHC to neighboring Deiter's cells and OHCs and also Hensen's cells during acoustic stimulation at 40 dB SPL (modified from Zwislocki et al., 1992). Reprinted from *Hearing Research, 57,* J. J. Zwislocki, N. B. Slepecky, L. K. Cefaratti, and R. L. Smith, Ionic coupling among cells of the organ of Corti, 175–194, copyright © 1992, with permission from Elsevier Science.

of the organ of Corti. Nevertheless, some evidence of structures resembling gap junction structures had been found in the past between the OHCs and the supporting cells not only in submammalian vertebrates (e.g., Nadol, Mulroy, Goodenough, & Weiss, 1976) but even in mammals (Gulley & Reese, 1976; Nadol, 1978). Was it possible for Lucifer yellow to penetrate Hensen's cells when injected into an OHC but not the other way around? Is a rectifying gap junction possible? Indeed, such gap junctions had been found before (e.g., Jaslove & Brink, 1987). However, alternating potentials recorded in Hensen's cells are not rectified. Perhaps, direct current flowing from OHCs to supporting cells is modulated producing the alternating potentials seen in Hensen's cells.

To make sure that the dye could not have diffused into OHCs from intercellular space, we injected it first intercellularly into the outer tunnel of Corti, then, in the same preparation, into a Hensen's cell. The results are shown in Fig. 5.52. In the left panel, the dye was injected intercellularly, and no diffusion into any cells situated in the tunnel or adjacent to it is evident. The small spot in the Hensen's cell population was produced almost certainly through negative capacitance ringing while the electrode was traversing intracellularly the population before it reached the outer tunnel. In the right panel, intracellular injection into a Hensen's cell produced staining of several Hensen's cells.

On the basis of the existing knowledge of the electrophysiology of the organ of Corti (e.g., Dallos, 1983) and our finding of ionic coupling between OHCs and supporting cells we devised a partial network model of an organ of Corti cross section, shown in Fig. 5.53. The model includes the pillars of Corti but not the IHCs and the supporting cells surrounding them. The omission of the latter was dictated by our lack of sufficient knowledge of ionic connections in this complex of cells. It is assumed in the model that

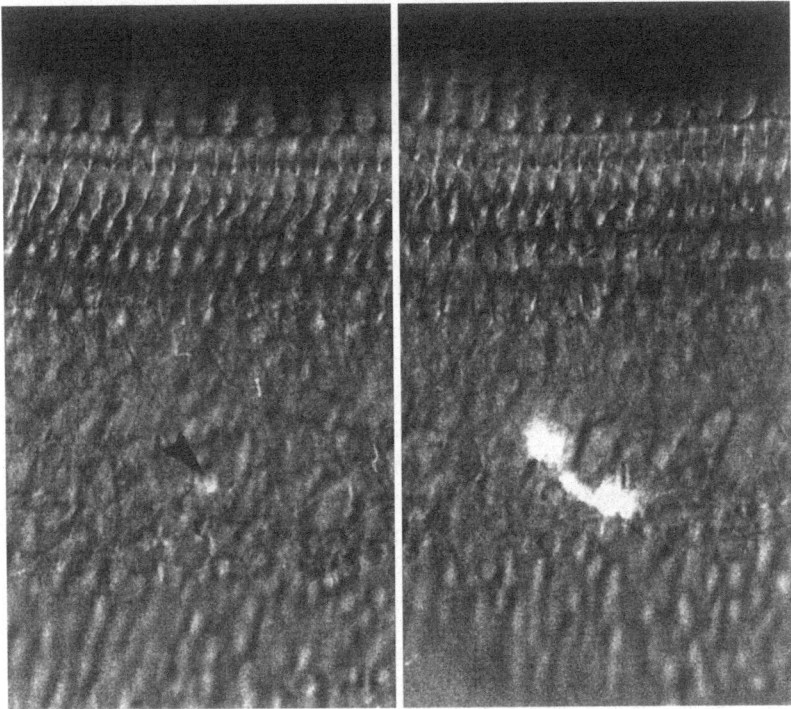

FIG. 5.52. Left panel: lack of the dye diffusion to Hensen's cells as a result of intercellular dye injection (the small dye spot is artifactual and due to a short burst of negative capacitance ringing of the electrode). Right panel: dye diffusion to neighboring Hensen's cells from an injected Hensen's cell in the same preparation but a different electrode track (modified from Zwislocki et al., 1992). Reprinted from *Hearing Research, 57*, J. J. Zwislocki, N. B. Slepecky, L. K. Cefaratti, and R. L. Smith, Ionic coupling among cells of the organ of Corti, 175–194, copyright © 1992, with permission from Elsevier Science.

PHYSICAL CONSTANTS AND FUNDAMENTAL CHARACTERISTICS 265

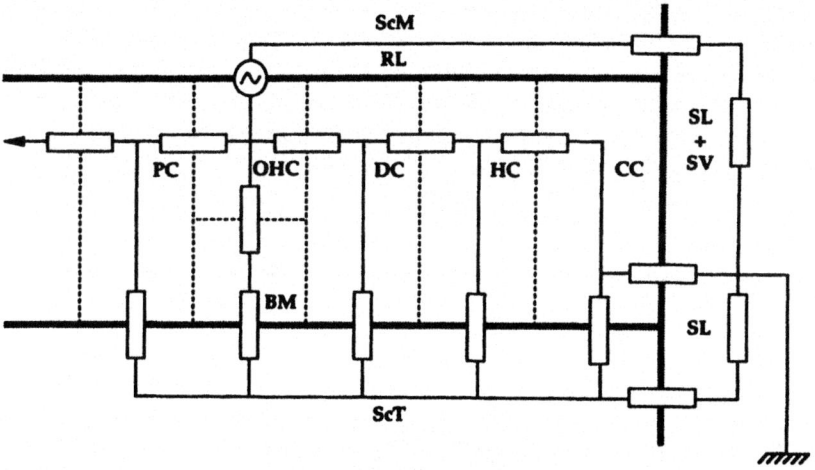

FIG. 5.53. Hypothetical electrical connectivity network of a part of the organ of Corti, based in part on the experiments described in the text above (from Zwislocki et al., 1992). Reprinted from *Hearing Research, 57*, J. J. Zwislocki, N. B. Slepecky, L. K. Cefaratti, and R. L. Smith, Ionic coupling among cells of the organ of Corti, 175–194, copyright © 1992, with permission from Elsevier Science.

OHCs are connected to Deiter's cells and the pillars of Corti through functional gap junction, and that gap junctions are present among all the supporting cells between the pillars of Corti and stria vascularis. The model may explain qualitatively the shape of the Hensen's cell and CM transfer functions in Figs. 5.45 and 5.46, which suggest narrower bandwidth than implied by the static cochlear space constants (e.g., Békésy, 1960). In particular, it may account for the almost exact similarity between the Hensen's cell and OHC transfer functions. As mentioned above, this similarity makes it possible to study the responses of OHCs indirectly by recording alternating potentials in Hensen's cells.

Because of the known cochlear nonlinearity it is necessary to determine cochlear transfer functions at a sufficient number of SPLs. Resulting intensity series for the basilar membrane are shown in Figs. 5.35 and 5.38. A corresponding series for an OHC in the middle turn of a Mongolian gerbil cochlea is reproduced in Fig 5.54 (Zwislocki, 1991). It covers a SPL range extending from 20 to 50 dB in the lower panel and from 50 to 80 dB in the upper. The curves are plotted on log frequency and linear magnitude coordinates. As for the basilar membrane in the basal turn (Fig. 5.35), the peak response (BF) moves toward lower sound frequencies as SPL is increased, but its shift is even more pronounced—in the SPL range from 20 to 80 dB, it extends over more than one octave, as indicated by the vertical lines, one going through the 20 dB peak, the other, through the 80 dB one. The functional relationship between the BF and SPL is plotted explicitly in Fig. 5.55. The curve approximating the data indicates that the BF dependence on SPL is small at low SPL and increases as SPL increases. There are other features of the intensity series of Fig. 5.54 worth mentioning. One is the

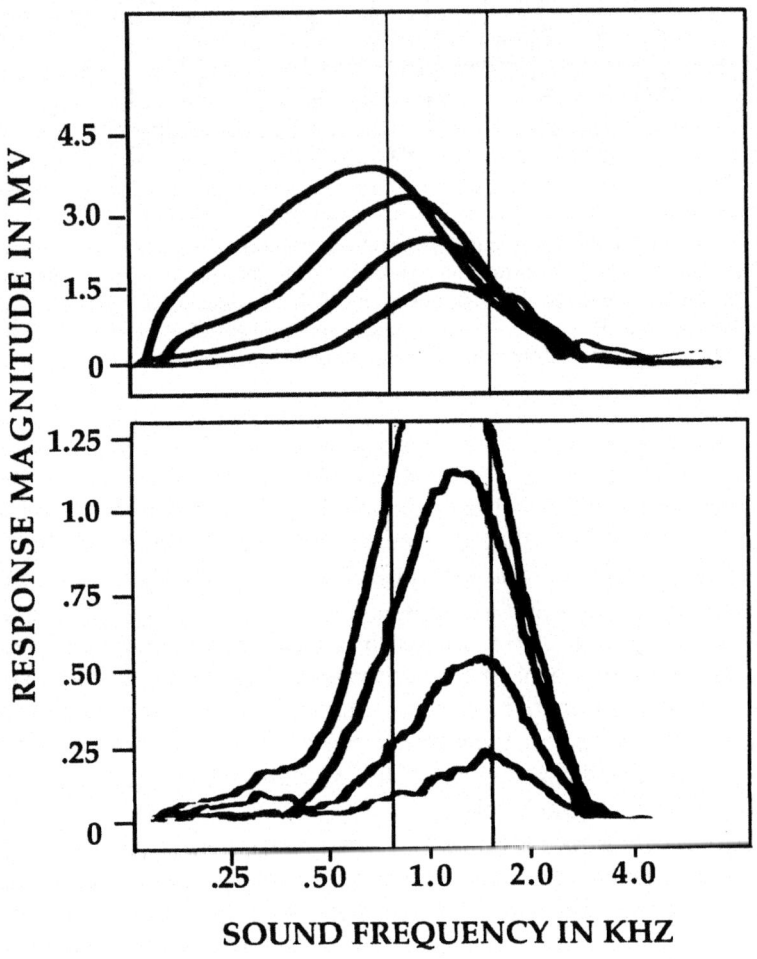

FIG. 5.54. Intensity series of OHC intracellular transfer functions for alternating receptor potentials. Lower panel: SPLs of 20 to 50 dB; upper panel, from 50 to 80 dB in 10 dB steps. The vertical lines mark the frequencies of the 20 and 80 dB response maxima (modified from Zwislocki, 1991). From "What is the Cochlear Place Code for Pitch?" by J. J. Zwislocki, 1991, *Acta Oto-Laryngologica, 111*, pp. 256–262. Copyright © 1991 by Taylor and Francis. Reprinted with permission.

PHYSICAL CONSTANTS AND FUNDAMENTAL CHARACTERISTICS 267

compression that can be judged from the vertical distances among the curves. It increases from the low to high sound frequencies, being almost nonexistent at the very low frequencies and increasing to almost hard saturation at the high ones. The pattern is similar to that found in basilar membrane vibration in the basal turn, but rather more pronounced. In part, the variable saturation is due to the peak shift toward lower frequencies. Some authors ascribed the peak shift to the variable compression, but it is shown in the next chapter that the relationships work the other way around—the variable compression is due to the peak shift. The latter makes the curves nearly coincide at the high frequency skirts of the transfer functions, so that they converge to a common foot. Furthermore, it should be noted that the 80 dB curve undershoots the other curves in their high frequency portions. This is consistent with the negative slope found in CM at high SPLs (Dallos, 1973). The curves belonging to the highest SPLs also go through a secondary maximum near the high frequency cutoff. Sometimes even a third maximum can be detected. Similar maxima were found in basilar membrane vibration, as is evident in Fig. 5.35.

For comparison purposes intensity series of Hensen's cell transfer functions are shown in Figs. 5.56 and 5.57, recorded in the middle and apical turns of two Mongolian gerbil cochleas, respectively. The pattern of the

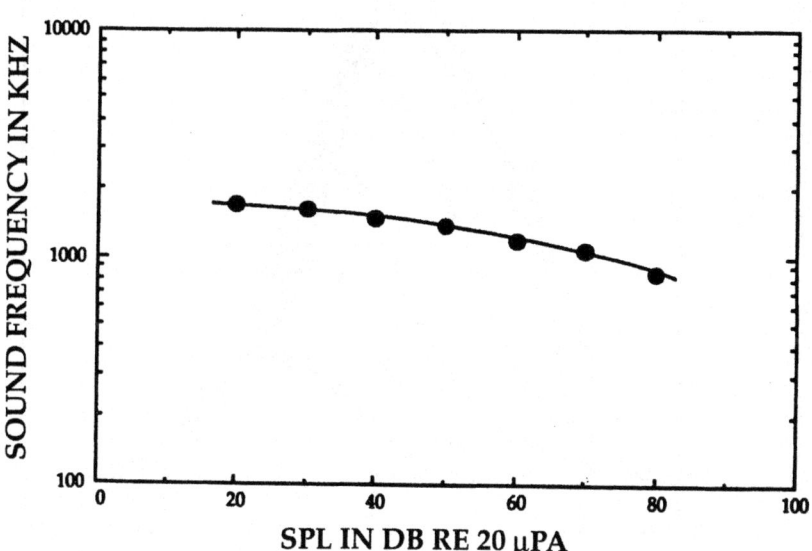

FIG. 5.55. Locus of the response maxima in Fig. 5.54 (modified from Zwislocki, 1991). From "What is the Cochlear Place Code for Pitch?" by J. J. Zwislocki, 1991, *Acta Oto-Laryngologica, 111,* pp. 256–262. Copyright © 1991 by Taylor and Francis. Reprinted with permission.

268 CHAPTER 5

middle turn curves is quite similar to the pattern of the OHC curves reproduced in Fig. 5.54. The principal response maximum moves toward lower sound frequencies as SPL is increased, the apparent compression increases with the frequency, and the curves tend to converge at the high frequency cutoff. The highest SPL curves undershoot the lower SPL curves at their high frequency skirts. Note that, at the lowest frequencies, the 90 dB curve descends abruptly and crosses the lower SPL curves. This is artifactual and results from the instability of the fixed filter in the Lock-in-Amplifier. For the same reason, the shapes of the curves below 0.25 kHz are unreliable. The apical intensity series of Fig. 5.57 displays the same fundamental characteristics as the middle turn curves, but its quantitative relationships are somewhat different. The intensity dependent peak shift is somewhat reduced, and the secondary and tertiary maxima are more prominent. Because of its size, the growth of the secondary maximum with SPL becomes apparent. It begins as a flattening of the high frequency skirt at 40dB that becomes increasingly more pronounced, becomes a plateau and, finally, a relative maximum. Its frequency location seems to remain approximately

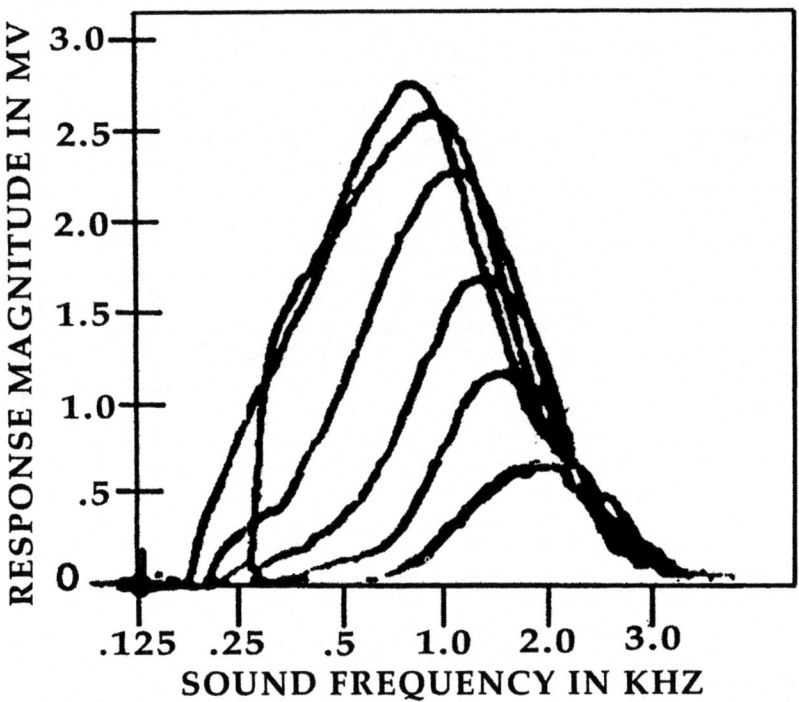

FIG. 5.56. Intensity series of a Hensen's cell intracellular transfer functions in the second cochlear turn of a Mongolian gerbil. SPL ranges from 40 to 90 dB in 10 dB steps.

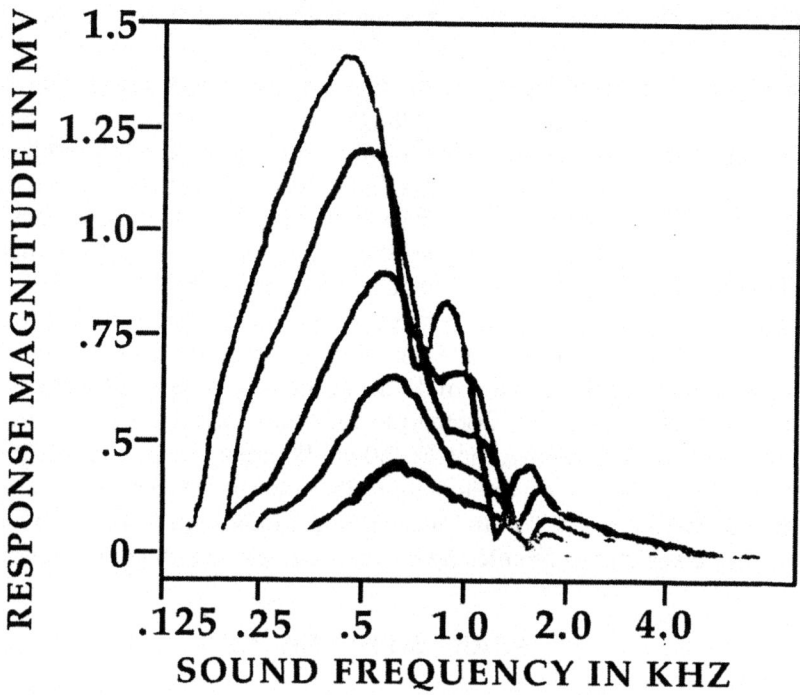

FIG. 5.57. Intensity series of a Hensen's cell intracellular transfer functions in the apical cochlear turn of a Mongolian gerbil. SPL ranges from 40 to 80 dB in 10 dB steps.

constant up to 70 dB, but it moves toward lower frequencies at higher SPLs. The tertiary maximum appears to move toward lower frequencies from the beginning of its appearance at 50 dB. The relatively pronounced secondary maximum is in agreement with the pattern of the apical basilar membrane transfer functions shown in Figs. 5.38 and 5.39.

Because the intensity dependent peak shift aroused some controversy in the past, it may deserve a special consideration. For basilar membrane vibration, Russell and Nilsen (1997) found only a small peak shift in the basal end of the guinea pig cochlea, and Cooper and Rhode found none in the apical turn. Ruggero et al. (1997) found a larger one in the basal turn of the chinchilla cochlea. Recording hair cell and Hensen's cell responses, we found a very large one in the mid-turn of the Mongolian gerbil cochlea and a smaller one in the lower part of the apical turn (Chatterjee and Zwislocki, 1997). For comparison, the various results, as derived from the published data, are plotted in Fig. 5.58. The solid lines belong to basilar membrane vibration, the one with the circles having been derived from the data of Ruggero and the one with the squares from those of Cooper and Rhode. The dashed lines refer to cell responses, the one with the upright triangles, in the mid turn, the one with the inverted triangles, in the apical turn. An orderly pattern seems to emerge—the peak shift appears to be large in the middle of the cochlear canal and small at its ends. It should be mentioned here that the various amounts of peak shift do not seem to affect the convergence of the transfer functions at their high frequency skirts, as defined in this book, making the response magnitude there almost independent of SPL.

To make certain that the peak shift found in the frequency domain has its counterpart in the cochlear space domain, we recorded Hensen's cell responses at two locations of the mid turn of the Mongolian gerbil cochlea (Zwislocki & Nguyen, 1999). For this purpose, two openings were drilled in the cochlear capsule, approximately 250 or 500 µm apart along the lateral wall of scala media. An example of intensity series of transfer functions recorded at two locations 226 µm apart in one preparation is shown in Fig. 5.59. The black curves refer to the more basal opening, the gray ones, to

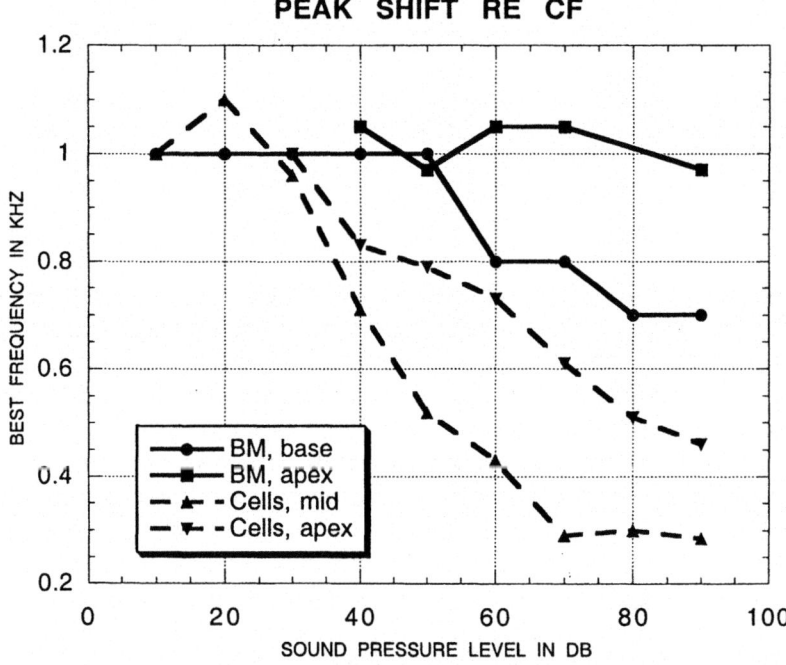

FIG. 5.58. The locus of maximum response for the basilar membrane and the organ of Corti cells at various cochlear locations, derived from several studies.

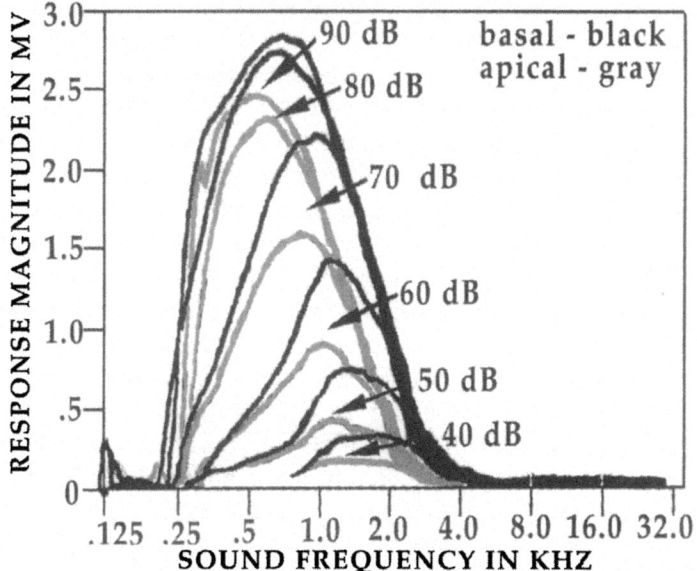

FIG. 5.59. Intensity series of two Hensen's cells' intracellular transfer functions in the second cochlear turn of a Mongolian gerbil. The Hensen's cells were separated by a distance of approximately 226 µm along the cochlear canal. Note the clear separation of the high frequency skirts between the two locations, and the overlap of the maximum responses (from Zwislocki & Nguyen, 1999). From "Place Code for Pitch: A Necessary Revision," by J. J. Zwislocki and M. Nguyen, 1999, *Acta Oto-Laryngologica, 119*, pp. 140–145. Copyright © 1999 by Taylor & Francis. Reprinted with permission.

the more apical opening. The high frequency skirts of the functions corresponding to the two locations are clearly separated along the frequency axis, reflecting the tonotopic organization of the cochlea. However the peaks overlap. For instance, near 1 kHz, the peak that, at 50 dB, coincides with the more apical location is shifted to the more basal location at 60 dB. Similarly the peak that, at 60 dB, coincides with the former location is moved to the more basal location at 70 dB. From these relationships we can see that the peak moved basalward by about 226 µm when the SPL was increased by 10 dB. This finding should provide indisputable evidence for an intensity dependent spatial peak shift along the cochlea.

The gross similarity between the magnitude transfer functions determined for basilar membrane vibration and, indirectly, for the shear motion between the reticular lamina and the tectorial membrane suggests that the phase transfer functions should be similar also. However, we found a clear, perhaps fundamental, dissimilarity. In the same way as for the shear motion magnitudes, we derived the response phases from the alternating potentials of the OHCs and Hensen's cells recorded in nearly intact cochleas in vivo. The Lock-in-Amplifier, through which these potentials were led, provided either their magnitudes or sin phases, depending on the selection made by a flip of a switch. A typical example of sin phase responses of a Hensen's cell as functions of log sound frequency is shown in Fig. 5.60. One trace has been obtained at a stimulus level of 40 dB SPL and is plotted by means of the thin line, the other, at a level of 90 dB and is plotted by means of the heavy line. The corresponding amplifications were adjusted to partially compensate for the difference in the resulting magnitudes. The sound frequency of the best response at low SPLs, the CF, was in the region of 2 kHz. The BF decreased dramatically at 90 dB, as can be judged from the amplitude pattern in the figure. At the same time, a large phase lag showed up. It increased with sound frequency and reached 180° around the CF. This can be clearly concluded from the phase opposition between the two curves. The pattern is in agreement with that found originally on the occasion of the discovery of this SPL dependent phase reversal (Zwislocki & Smith, 1988). The origin of the phase difference seems to reside in the low frequency plateau in the 40 dB curve, where the alternating

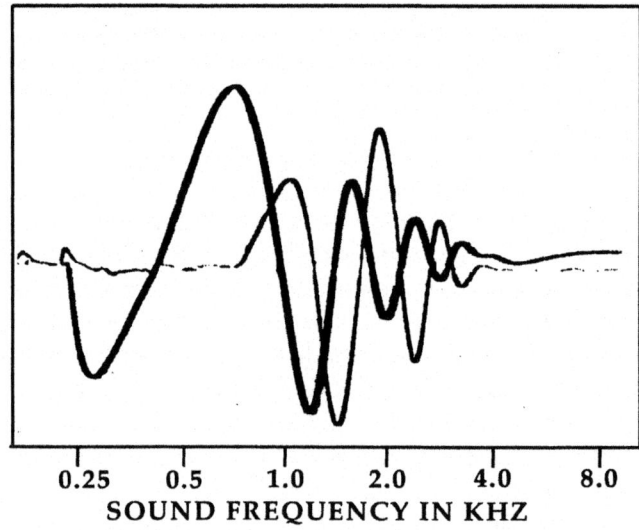

FIG. 5.60. Intracellular alternating potentials of a Hensen's cell in a Mongolian gerbil's cochlear mid-turn as functions of sound frequency, obtained by multiplying the amplitude envelopes with the sins of the response phases. The thin curve was obtained at 40 dB, the thick one, at 90 dB SPL. The amplifications were adjusted to obtain comparable amplitudes. Note the phase shift between the two SPLs and the suppressed low frequency response in the 40 dB curve.

potential appears to have been cancelled out almost entirely. The plateau has been found by us routinely and is not artifactual. Its likely mechanism is described in chapter 6.

Phase functions can be derived from sin, or cos functions, such as those of Fig. 5.60, by plotting the sound frequencies of zero crossings, at which the phases must be equal to multiples of a half cycle. Normalized phase curves so obtained on an individual preparation are plotted in Fig. 5.61 for 4 SPLs—40, 60, 80, and 90 dB. They are referred to a common origin, the first ascending zero crossing, as seen in Fig. 5.60. For a constant time delay associated with a constant wave velocity, the phase lag should increase linearly with sound frequency, and the curves should coincide along a straight line, independent of SPL. This is clearly not true, although a straight line course is grossly approximated. The 40 dB curve shows a slight perturbation around 2 kHz, however, which decreases slightly the phase lag at the higher frequencies. In the same frequency region, the 60 dB curve crosses the 40 dB curve, further decreasing the phase lag. It does the opposite at the lower frequencies. This pattern resembles the detailed basilar membrane phase patterns determined by Ruggero et al. (1997) and others. However, the similarity ends there. Already at 80 dB, no phase lead relative to the 40 dB curve takes place. Instead, a phase lag of about 90° extends over at least 3 octaves, including the CF region. At 90 dB, the phase lag is further increased, especially at high sound frequencies, reaching about 180° in the vicinity of the CF. Similar patterns were obtained on 6 additional animals and also on other occasions (Zhang & Zwislocki, 1996; Szymko, Zwislocki, & Hertig, 1997).

It should be pointed out here that a dependence of the cochlear response phase on SPL was first described by Anderson, Rose, Hind, and Brugge (1970) on the occasion of their recordings of firing rates of auditory nerve fibers. They had found that very little phase shift occurred at the CF and that it grew as the frequency decreased or increased. This pattern is consistent with that found by Ruggero et al. and with the low SPL curves in Fig. 5.61. Both groups, Anderson et al. and Ruggero et al. (1997), reported their results relative to the phases at a high SPL, 90 and 80 dB, respectively. As a consequence, their plots show phase differences that increase as SPL decreases. It should be pointed out that the results of Anderson et al. must depend on the shear motion between the reticular lamina and the tectorial membrane and should, therefore, be more closely related to our results exemplified in Fig 5.61 than to Ruggero et al.'s results. However, they should also depend on the timing of neural spike generation during the vibration cycle, which must be expected to depend on sound intensity. The relationships appear to be quite complicated. They are discussed more extensively in chapter 6.

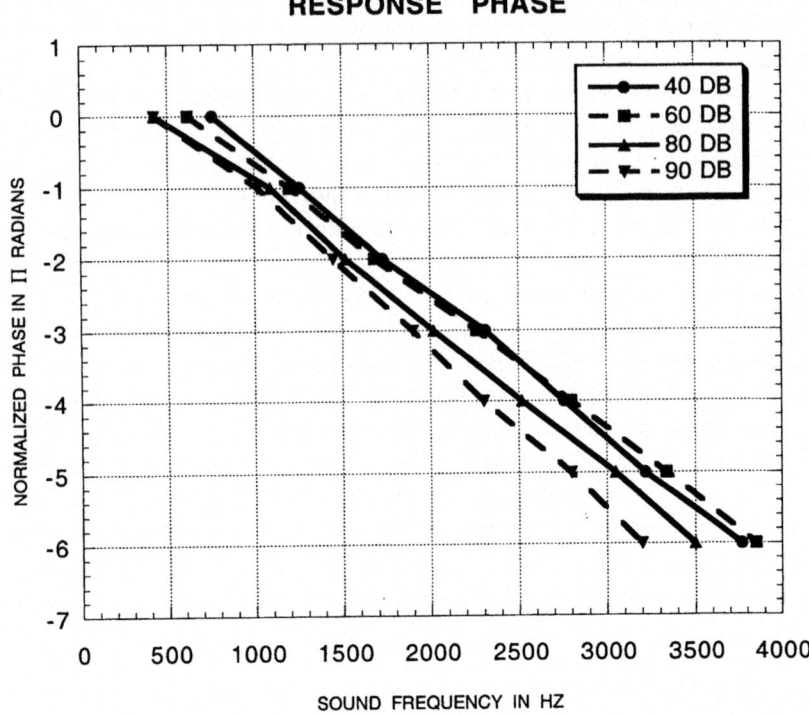

FIG. 5.61. A Hensen's cell response phases as functions of sound frequency at several SPLs. The phases were obtained from zero crossings of curve patterns exemplified in Fig. 5.60. Note the SPL dependent shifts among the phase curves.

REFERENCES

Allen, J. B. (1980). Cochlear micromechanics—A physical model of transduction. *Journal of the Acoustical Society of America, 68,* 1660–1670.
Anderson, D. J., Rose, J. E., Hind, J. E., & Brugge, J. F. (1970). Temporal position of discharges in single auditory nerve fibers within the cycle of sine-wave stimulus: Frequency and intensity effects. *Journal of the Acoustical Society of America, 49,* 1131–1139.
Ashmore, J. F. (1987). A fast motile response in guinea-pig outer hair cells: The molecular basis of the cochlear amplifier. *Journal of Physiology (London), 388,* 323–347.
Békésy, G. v. (1951). Microphonics produced by touching the cochlear partition with vibrating electrode. *Journal of the Acoustical Society of America, 23,* 29–35.
Békésy, G. v. (1953). Shearing microphonics produced by vibrations near the inner and outer hair cells. *Journal of the Acoustical Society of America, 25,* 786–790.
Békésy, G. v. (1960). *Experiments in hearing.* New York: McGraw-Hill.
Brownell, W. E., Bader, C. R., Bertrand, D., & Ribaupiere, Y. (1985). Evoked mechanical responses of isolated hair cells. *Science, 227,* 194–196.

Chatterjee, M., & Zwislocki, J. J. (1997). Cochlear mechanisms of frequency and intensity coding. I. The place code for pitch. *Hearing Research, 111*, 65–75.
Cooper, N. P. (2000). Radial variations in the vibrations of the cochlear partition. In H. Wada, T. Takasaka, K. Ikeda, K. Ohyama, & T. Koike (Eds.), *Recent Developments in Auditory Mechanics* (pp. 109–115). Singapore: World Scientific.
Cooper, N. P., & Rhode, W. S. (1992). Basilar membrane mechanics in the hook region of cat and guinea-pig cochleae: Sharp tuning and nonlinearity in the absence of baseline position shifts. *Hearing Research, 63*, 163–190.
Cooper, N. P., & Rhode, W. S. (1995) Nonlinear mechanics at the apex of the guinea-pig cochlea. *Hearing Research, 82*, 225–243.
Dallos, P. (1973). *The Auditory Periphery*. New York: Academic Press.
Dallos, P. (1983). Some electrical circuit properties of the organ of Corti I: Analysis without reactive elements. *Hearing Research, 12*, 89–119.
Dallos, P. (1986). Neurobiology of cochlear inner and outer hair cells: Intracellular recordings. *Hearing Research, 22*, 185–198.
Dallos, P., & Durrant, J. D. (1972). On the derivative relationship between stapes movement and cochlear microphonic. *Journal of the Acoustical Society of America, 52*, 1263–1265.
Dallos, P., & Evans, B. N. (1995). High-frequency motility of outer hair cells and the cochlear amplifier. *Science, 267*, 2006–2009.
Dallos, P., Billone, M. C., Durrant, J. D., Wang, C-Y., & Raynor, S. (1972). Cochlear inner and outer hair cells, functional differences. *Science, 177*, 356–358.
Dallos, P., Santos-Sacchi, J., & Flock, A. (1982). Intracellular recordings from cochlear outer hair cells. *Science, 218*, 582–584.
Dancer, A., & Franke, R. (1980). Intracochlear sound pressure measurements in guinea pigs. *Hearing Research, 2*, 191–205.
Eldredge, D. H., Miller, J. D., & Bohne, B. A. (1981). A frequency-position map for the chinchilla cochlea. *Journal of the Acoustical Society of America, 69*, 1091–1095.
Fernandez, C. (1952). Dimensions of the cochlea (guinea pig). *Journal of the Acoustical Society of America, 24*, 519–523.
Flock, Å. (1977). Physiological properties of sensory hairs in the ear. In E. F. Evans & J. P. Wilson (Eds.), *Psychophysics and physiology of hearing*. London: Academic Press.
Flock, Å, & Strelioff, D. (1984). Studies on hair cells in isolated coils from guinea pig cochlea. *Hearing Research, 15*, 11–18.
Frommer, G. H. (1982). Observations of the organ of Corti under in vivo-like conditions. *Acta Oto-Laryngologica, 94*, 451–460.
Gold, T. (1948). The physical basis of the action of the cochlea. *Proceedings of the Royal Society, B 145*, 492–498.
Gold, T., & Pumphrey, R. J. (1948). Hearing I: The cochlea as a frequency analyzer. *Proceedings of the Royal Society, B 135*, 462–491.
Greenwood, D. D. (1961). Critical bandwidth and the frequency coordinates of the basilar membrane. *Journal of the Acoustical Society of America, 33*, 1344–1356.
Greenwood, D. D. (1974). Critical bandwidth in man and some other species. In A. R. Moskowitz, B. Scharf, & S. S. Stevens (Eds.), *Sensation and Measurement: Papers in Honor of S.S. Stevens* (pp. 231–239). The Netherlands: Reidel, Dordrecht.
Greenwood, D. D. (1977). Empirical travel time functions on the basilar membrane. In E. F. Evans & J. P. Wilson (Eds.), *Psychophysics and physiology of hearing* (pp. 43–53). New York: Academic Press.
Greenwood, D. D. (1990). A cochlear frequency-position function for several species—29 years later. *Journal of the Acoustical Society of America, 87*, 2592–2605.
Gully, R. L., & Reese, T. S. (1976). Intercellular junctions in the reticular lamina of the organ of Corti. *Journal of Neurophysiology, 5*, 479–507.

Gummer, A. W., Hemmert, W., & Zenner, H. P. (1996). Resonant tectorial membrane motion in the inner ear: Its crucial role in frequency tuning. *Proceedings of the National Academy of Sciences U.S.A. (Neurobiology), 93,* 8727–8732.

He, D. Z. Z., & Dallos, P. (1999). Somatic stiffness of cochlear outer hair cells is voltage dependent. *Proceedings of the National Academy of Sciences, U.S.A., 96,* 8223–8228.

Hellman, R. P., & Zwislocki, J. J. (1968). Loudness determination at low sound frequencies. *Journal of the Acoustical Society of America, 43,* 60–64.

Hemmert, W., Zenner, H. P., & Gummer, A. W. (2000). Characteristics of the travelling wave in the low-frequency region of a temporal-bone preparation of the guinea-pig cochlea. *Hearing Research, 142,* 184–202.

Hudspeth, A. J., & Corey, D. P. (1977). Sensitivity, polarity, and conductance change in the response of vertebrate hair cells to controlled mechanical stimuli. *Proceedings of the National Academy of Sciences (Biophysics), 74,* 2407–2411.

Iurato, S. (1962). Functional implications of the nature and submicroscopic structure of the tectorial and basilar membranes. *Journal of the Acoustical Society of America, 34,* 1386–1395.

Iurato, S., Franke, K., Luciano, L., Wermbter, G., Pannese, E., & Reale, E. (1976). Intercellular junctions in the organ of Corti as revealed by freeze fracturing. *Acta Oto-Laryngologica, 82,* 57–69.

Jaslove, S. W., & Brink, P. R. (1987). Electronic coupling in the nervous system. In W. C. De Mello (Ed.), *Cell-to-Cell Communication* (pp. 103–147). New York: Plenum Publishing Corp.

Johnstone, B. M., & Boyle, A. J. T. (1967). Basilar membrane vibration examined with the Mössbauer technique. *Science, 158,* 389–390.

Johnstone, B. M., & Taylor K. (1970). Mechanical aspects of cochlear function. In R. Plomp & G. F. Smoorenburg (Eds.), *Frequency Analysis and Periodicity Detection in Hearing* (pp. 81–93). Leiden: Sijthoff

Kemp, D. T. (1978). Stimulated acoustic emissions from within the human auditory system. *Journal of the Acoustical Society of America, 64,* 1386–1391.

Kemp, D. T. (1979). The evoked cochlear mechanical response and the auditory microstructure—Evidence for a new element in cochlear mechanics. Scandinavian Audiology Suppl. *Proceedings of the workshop on Models of the Auditory System* (pp. 35–47). Münster, Germany.

Khanna, S. M., & Leonard, D. G. B. (1982). Basilar membrane tuning in the cat cochlea. *Science, 215,* 305–306.

Khanna, S. M., Flock, Å., & Ulfendahl, M. (1989). Comparison of the tuning of outer hair cells and the basilar membrane in the isolated cochlea. *Acta-Oto-Laryngologica. Suppl., 467,* 151–156.

Kohllöffel, L. U. E. (1972a). A study of basilar membrane vibrations I. Fuzziness-detection: a new method of analysis of micro-vibrations with laser light. *Acustica, 27,* 49.

Kohllöffel, L. U. E. (1972b). A study of basilar membrane vibrations II: The vibratory amplitude and phase pattern along the basilar membrane (post-mortem). *Acustica, 27,* 66–89.

Kohllöffel, L. U. E. (1972c). A study of basilar membrane vibrations III: The basilar membrane frequency response curve in the living guinea pig. *Acustica, 27,* 82–89.

Kronester-Frei, A. (1978). Ultrastructure of the different zones of the tectorial membrane. *Cell Tissue Research, 193,* 11–23.

Kronester-Frei, A. (1979). The effect of changes in the endolymphatic ion concentrations on the tectorial membrane. *Hearing Research, 1,* 81–94.

Liberman, M. C. (1976). *Abnormal discharge patterns of the auditory-nerve fibers in acoustically-traumatized cats.* Unpublished doctoral dissertation, Harvard University, Cambridge, Massachusetts.

Liberman, M. C. (1982). The cochlear frequency map of the cat. *Journal of the Acoustical Society of America, 72,* 1441–1449.
Lim, D. J. (1980). Cochlear anatomy related to cochlear micromechanics. A review. *Journal of the Acoustical Society of America, 67,* 1686–1695.
Lynch, T. J., Nedzelnitsky, V., & Peake, W. T. (1982). Input impedance of the cochlea in cat. *Journal of the Acoustical Society of America, 72,* 108–130.
Merchant, S. N., Ravicz, M. E., & Rosowski, J. J. (1996). Acoustic input impedance of the stapes and cochlea in human temporal bones. *Hearing Research, 97,* 30–45.
Nadol, J. B., Jr. (1978). Intercellular junctions in the organ of Corti. *Annals Otology, Rhinology, and Laryngology, 87,* 70–80.
Nadol, J. B., Jr., Mulroy, J. J., Goodenough, D. H., & Weiss, T. F. (1976). Tight and gap junctions in a vertebrate inner ear. *American Journal of Anatomy, 147,* 281–302.
Oesterle, E. C., & Dallos, P. (1989). Intracellular recordings from supporting cells in the guinea-pig cochlea: AC potentials. *Journal of the Acoustical Society of America, 86,* 1013–1030.
Pfeiffer, R. R., & Kim, D. O. (1975). Cochlear nerve fiber responses: Distribution along the cochlear partition. *Journal of the Acoustical Society of America, 58,* 867–869.
Rausch, S. (1964). *Biochemie des Hörorgans* [Biochemistry of the Hearing Organ]. Stuttgart, Germany: Thieme.
Rhode, W. S. (1971). Observations of the vibration of the basilar membrane in squirrel monkey using the Mössbauer technique. *Journal of the Acoustical Society of America, 49,* 1218–1231.
Rhode, W. S. (1973). An investigation of postmortem cochlear mechanics using the Mössbauer effect. In A. R. Møller (Ed.), *Basic Mechanics in Hearing* (pp. 49–67). New York: Academic Press.
Robles, L., Ruggero, M. A., & Rich, N. C. (1986). Basilar membrane mechanics at the base of the chinchilla cochlea. I: Input–output functions, tuning curves, and response phases. *Journal of the Acoustical Society of America, 80,* 1364–1374.
Ruggero, M. A., Rich, N. C., Recio, A., Narayan, S., & Robles, L. (1997). Basilar-membrane responses to tones at the base of the chinchilla cochlea. *Journal of the Acoustical Society of America, 101,* 2151–2163.
Ruggero, M. A., Rich, N. C., Shivapuja, B. G., & Temchin, A. N. (1996). Auditory-nerve responses to low-frequency tones: Intensity dependence. *Auditory Neuroscience, 2,* 159–185.
Russell, I. J., & Nilsen, K. E. (1997). The location of the cochlear amplifier: Spatial representation of a single tone on the guinea pig basilar membrane. *Proceedings of the National Academy of Sciences, 94,* 2660–2664.
Russell, I. J., & Sellick, P. M. (1977). The tuning properties of cochlear hair cells. In E. F. Evans & J. P. Wilson (Eds.), *Psychophysics and Physiology of Hearing.* London: Academic Press.
Russell, I. J., Kössl, M., & Murugasu, E. (1995). A comparison between tone-evoked voltage responses of hair cells and basilar membrane displacements recorded in the basal turn of the guinea pig cochlea. In G. A. Manley, G. M. Klump, C. Köppl, H. Fastle, & H. Oeckinghaus (Eds.), *Advances in Hearing Research.* Singapore: World Scientific.
Santos-Sacchi, J. (1986). Dye coupling in the organ of Corti. *Cell Tissue Research, 245,* 525–529.
Santos-Sacchi, J. (1990). Fast outer hair cell motility: How fast is fast? In P. Dallos, C. D. Geisler, J. W. Matthews, M. A. Ruggero, & C. R. Steele (Eds.), *The Mechanics and Biophysics of Hearing* (pp. 69–75). Berlin: Springer.
Santos-Sacchi, J., & Dallos, P. (1983). Intercellular communication in the supporting cells of the organ of Corti. *Hearing Research, 9,* 317–326.

Schmeidt, R. A., & Zwislocki, J. J. (1977). Comparison of sound-transmission and cochlear-microphonic characteristics in Mongolian gerbil and guinea pig. *Journal of the Acoustical Society of America, 61,* 133–149.

Schuknecht, H. F. (1960). Neuroanatomical correlates of auditory sensitivity and pitch discrimination in the cat. In G. L. Rasmussen & W. F. Windle (Eds.), *Neural Mechanisms of the Auditory and Vestibular Systems.* Symposia in Neuroanatomical Sciences, Conference on the Neural Mechanisms of the Auditory and Vestibular Systems. NIH, Springfield, IL.

Sellick, P. M., Patuzzi, R., & Johnstone, B. M. (1982). Measurement of the basilar membrane motion in the guinea pig using the Mössbauer technique. *The Journal of the Acoustical Society of America, 72,* 131–141.

Spoendlin, H. (1966). The organization of the cochlear receptor. *Advances in Oto-Rhino-Laryngology, 13,* 1–227.

Spoendlin, H. (1970). Structural basis of peripheral frequency analysis. In R. Plomp & G. F. Smoorenburg (Eds.), *Frequency Analysis and Periodicity Detection in Hearing* (pp. 2–36). Leiden, Netherlands: A.W. Sitjthoff.

Strelioff, D., & Flock, Å. (1984). Stiffness of sensory-cell hair bundles in the isolated guinea pig cochlea. *Hearing Research, 15,* 19–28.

Strelioff, D., Flock, Å., & Minser, K. E. (1985). Role of inner and outer hair cells in mechanical frequency selectivity of the cochlea. *Hearing Research, 18,* 169–175.

Szymko, Y. M., Zwislocki, J. J., & Hertig, L. (1997). Enhanced cochlear responses after sound exposure. *Hearing Research, 110,* 164–178.

Tonndorf, J., Duvall, A. J., III, & Reneau, J. P. (1962). Permeability of intracochlear membranes to various vital stains. *Annals of Otology, Rhinology, and Laryngology, 71,* 801–841.

Ulfendahl, M., Khanna, S. M., & Decraemer, W. F. (1996). Acoustically induced vibrations of the Reissner's membrane in the guinea-pig inner ear. *Acta Physiologica Scandinavica, 158,* 275–285.

Wever, E. G. (1949). *Theory of Hearing.* New York: Wiley.

Wilson, J. P. (1980). Evidence for a cochlear origin for acoustic emissions, threshold fine-structure and tonal tinnitus. *Hearing Research, 2,* 233–252.

Wilson, J. P., & Johnstone, J. R. (1975). Basilar membrane and middle ear vibration in guinea pig measured by capacitance probe. *Journal of the Acoustical Society of America, 57,* 705–723.

Zhang, M., & Zwislocki, J. J. (1995). OHC response recruitment and its correlation with loudness recruitment. *Hearing Research, 85,* 1–10.

Zhang, M., & Zwislocki, J. J. (1996). Intensity-dependent peak shift in cochlear transfer functions at the cellular level, its elimination by sound exposure, and its possible underlying mechanisms. *Hearing Research, 96,* 46–58.

Zurek, P. M. (1981). Spontaneous narrowband acoustic signals emitted by human ears. *Journal of the Acoustical Society of America, 69,* 514–523.

Zwislocki, J. J. (1974). Cochlear waves: Interaction between theory and experiments. *Journal of the Acoustical Society of America, 55,* 578–583.

Zwislocki, J. J. (1980). Five decades of research on cochlear mechanics. *Journal of the Acoustical Society of America, 67,* 1679–1685.

Zwislocki, J. J. (1988). Mechanical properties of the tectorial membrane in situ. *Acta Oto-Laryngologica, 105,* 450–456.

Zwislocki, J. J. (1990). Active cochlear feedback: Required structure and response phase. In P. Dallos, C. D. Geisler, J. W. Matthews, M. A. Ruggero, & C. R. Steele (Eds.), *The Mechanics and Biophysics of Hearing* (pp. 114–120). Berlin: Springer.

Zwislocki, J. J. (1991). What is the cochlear place code for pitch? *Acta Oto-Laryngologica (Stockholm), 111,* 256–262.

Zwislocki, J. J., & Cefaratti, L. K. (1989). Tectorial membrane II: Stiffness measurements in vivo. *Hearing Research, 41,* 211–228.

Zwislocki, J. J., & Nguyen, M. (1999). Place Code for Pitch: A Necessary Revision. *Acta Oto-Laryngologica (Stockholm), 119,* 140–145

Zwislocki, J. J., & Smith, R. L. (1988). Phase reversal in OHC response at high sound intensities. In J. P. Wilson & D. T. Kemp (Eds.), *Mechanics of Hearing—A NATO Advanced Research Workshop.* University of Keele, England.

Zwislocki, J. J., Chamberlain, S. C., & Slepecky, N. B. (1988). Tectorial membrane I: Static mechanical properties in vivo. *Hearing Research, 33,* 207–222.

Zwislocki, J. J., Slepecky, N. B., Cefaratti, L. K., & Smith, R. L. (1992). Ionic coupling among cells of the organ of Corti. *Hearing Research, 57,* 175–194.

Zwislocki, J. J., Slepecky, N. B., Chamberlain, S. C., & Cefaratti, L. K. (1988). Elastic properties of the tectorial membrane in vivo. In J. Syka & R. B. Masterton (Eds.), *Auditory Pathway* (pp. 17–21). New York: Plenum Publications Corp.

Zwislocki, J .J., Szymko, Y. M., & Hertig, L. Y. (1996). The cochlea is an automatic-gain control system after all. In E. R. Lewis, G. R. Long, R. F. Lyon, P. M. Narins, C. R. Steele, & E. Hecht-Poinar (Eds.), *Diversity in Auditory Mechanics.* Singapore: World Scientific.

Zwislocki-Moscicki, J. J. (1948). Theorie der Schneckenmechanik: Qualitative und Quantitative Analyse. [Theory of cochlear mechanics: Qualitative and quantitative analysis] *Acta Oto-Laryngologica,* Suppl. 72, 1–76.

chapter

Live Cochlea: Analysis

From my perspective, based on mechanical measurements, electrophysiological experiments, physical models, and mathematical analysis, the main difference in the mechanics between a mammalian cochlea postmortem and in vivo resides in the dynamics of the OHC stereocilia-tectorial membrane complex. Whereas, in the former, the tectorial membrane is nearly immobilized by viscous forces consistent with the high damping found by Békésy and brought forth analytically in chapter 4, in the latter, it must be able to oscillate radially with a sizable amplitude, as dictated by its great flexibility and appreciable mass both introduced in the preceding chapter. Such oscillation is consistent throughout with experimental results obtained by myself and my coworkers, as well as by several other scientists, and has been demonstrated by direct optical measurements of Gummer, Hemmert, and Zenner (1996). It accounts for some observed phenomena that otherwise appear paradoxical. The oscillation is discussed extensively in this chapter.

The first signal that something may be wrong with the classical concept of the shear motion between the reticular lamina and the tectorial membrane, which demands that the hair cells be always depolarized and the auditory nerve fibers excited during basilar membrane displacement or motion toward scala vestibuli, was given by the experiments of Konishi and Nielsen in 1973. They closed the helicotrema in a guinea pig cochlea, gain-ing access to it through a surgical opening near the cochlear apex. This allowed them to produce low-frequency, almost static, trapezoidal displacements of the basilar membrane. To their astonishment, they found that the firing rate of the auditory nerve fibers they recorded from did not always occur during basilar membrane displacement toward scala vestibuli. More often than not, the firing increased during basilar membrane displacement in the opposite direction, toward scala tympani. These inconvenient results were roundly discarded by colleagues in the field on the grounds that they were artifactual. The way the cochlea was tampered with was blamed for them.

I became curious, however, and persuaded my then assistant, W. G. Sokolich, to verify them with a more refined method. The method did not require any tampering with the cochlea and relied on our knowledge that the coupling between the middle ear and the cochlea produced approximate signal differentiation. Accordingly, it was possible to produce trapezoidal displacements of the basilar membrane by introducing into the ear canal their integrated wave form, which is triangular. As was confirmed indirectly by recording round window CM, it was possible to obtain trapezoidal displacements of the basilar membrane at repetition rates as low as 40 per sec. The firing rates of the auditory nerve fibers Sokolich recorded from in our model animal, Mongolian gerbil, confirmed the results of Konishi and Nielsen obtained on guinea pigs, extending them and systematizing somewhat (Zwislocki & Sokolich, 1973). The fibers were excited predominantly during basilar membrane displacement toward scala tympani but they could be excited even more strongly during its motion. In fibers with low CFs, maximum excitation occurred during basilar membrane motion toward scala vestibuli, in higher CF fibers, toward scala tympani (Sokolich, Hamernik, Zwislocki, & Schmiedt, 1976). The complex patterns were found subsequently to be consistent with phase relationships resulting from sinusoidal stimulation (e.g., Ruggero & Rich, 1983).

Explanation of the patterns was found to require two components that counteracted each other. It was rather plausible to assume that one component was associated with the IHCs and the other with the OHC. To test this possibility, the OHCs were destroyed by noise in the basal turns of several cochleae, leaving the populations of the IHCs reasonably well preserved (Sokolich et al., 1976). The nerve fibers with CFs corresponding to these cochlear parts were excited during basilar membrane displacements toward scala vestibuli. These results were subsequently confirmed by extensive recordings of R. A. Schmiedt (Schmiedt, Zwislocki, & Hamernik, 1980). Still accepting the classical concept of hair cell stimulation, I concluded that the OHCs provided inhibitory inputs that counteracted excitatory inputs originating in the IHCs, and that both interacted at the level of habenula perforata of the spiral osseous lamina. If the two components where somewhat offset in time, the coexistence of excitatory responses occurring during basilar membrane motion toward scala vestibuli with those occurring during basilar membrane displacement toward scala tympani could be explained. The scheme was also able to account for the sharper frequency resolution at the level of the auditory nerve than was seen in the cochlea (Zwislocki, 1974), in agreement with experimental evidence to this effect (Evans & Wilson, 1973; Geisler, Rhode, & Kennedy, 1974).

The idea of a neural interaction between the OHCs and the IHCs had a short life. Russell and Sellick (1977) showed that frequency resolution in the IHCs was as sharp as in the nerve fibers and Spoendlin (1973) was unable to find any interneuronal synapses in the habenula perforata. These experimental results indicated that the two interacting components required for an explanation of the complex response patterns produced by trapezoidal stimulation had to reside in the cochlea itself and be mechanical in nature. No such components could be identified if the classical concept of the shear motion between the reticular lamina and the tectorial membrane were to be conserved. The tectorial membrane had to be allowed to oscillate radially (Zwislocki & Kletsky, 1979). Such an oscillation was made more probable by Flock's (1977) finding that the OHC stereocilia were stiff and, as a consequence, capable of driving the tectorial membrane radially against its viscoelastic attachment to the spiral limbus. As discussed extensively in the preceding chapter, this assumption was later confirmed by direct measurements. The resulting amplitude of the radial oscillation of the tectorial membrane had to depend on the relationship between the wavelength of the transversal waves on the basilar membrane and the mechanical longitudinal space constant of the tectorial membrane. In the presence of relatively long waves, the tectorial membrane would be driven in the same direction over distances exceeding the space constant and its motion would be opposed only by its viscoelastic attachment to the spiral limbus. As a consequence, it would be relatively large. On the other hand, in the presence of relatively short waves, the direction of the force exerted on the tectorial membrane by the stereocilia would rapidly change direction over distances shorter than the space constant. Under such conditions, the effect of the force would be nullified and the radial motion of the tectorial membrane minimized. Because the wavelength is relatively large in the cochlear base and decreases gradually toward the maximum of basilar membrane vibration, the radial oscillation would gradually decrease toward this maximum. Because the shear motion between the reticular lamina and the tectorial membrane would increase as the radial motion of the tectorial membrane decreased, the maximum of the basilar membrane vibration would be enhanced (Zwislocki & Kletsky, 1979).

The shear motion was analyzed mathematically and modeled on an electrical transmission line on the basis of reasonable numerical values of the physical constants involved (Zwislocki & Kletsky, 1979). An available, classical transmission line model of the cochlea was used for this purpose, in which the basilar membrane impedance consisted of electrical analogs of three acoustic elements—compliance, mass, and resistance. The tectorial membrane and the stereocilia were represented by a cascaded network consisting of three capacitances, one representing the aggregate stereocilia compliance, one, the tectorial membrane attachment to the spiral limbus, and one, the longitudinal coupling within it. The mass of the tectorial membrane was neglected, as this was customary at the time. In an effort at representing the relationships at the IHCs innervated by a high percentage of afferent fibers of the auditory nerve (Spoendlin, 1967) rather than at the OHCs, reasonable hypothetical decrements of the radial motions between the OHC and the IHC locations were assumed, one for the hair cells and one for the tectorial membrane. By changing the ratio between the decrements it was possible to make the model tectorial membrane motion equal to, smaller than, or larger than that of the IHCs at low sound frequencies. In the first instance, the shear motion disappeared, the second instance corresponded to IHC depolarization during basilar membrane displacement toward scala vestibuli, the third, to their depolarization during basilar membrane displacement toward scala tympani. Thus, it became possible not only to increase the sharpness of tuning relative to the basilar membrane tuning but also to mimic the experimentally found variable phase of the responses of auditory nerve fibers at low sound frequencies. The model results are illustrated in Fig. 6.1 as functions of time. The curve with the broader amplitude envelope represents the basilar membrane oscillation, the one with the narrower envelope, the shear motion. In panel A, the shear motion at the low frequencies was nullified, in panel B, the tectorial membrane oscillation amplitude was made larger than that of the IHCs. The latter relationship is consistent with the IHC depolarization during basilar membrane displacement toward scala tympani. Note that, in the second instance, the response phase was reversed in the broad vicinity of maximum response (CF), and the amplitude reached a minimum at the location of phase reversal.

Russell and Sellick (1983) resuscitated the idea of a destructive interaction between the IHCs and the OHCs by suggesting that the interaction takes place at the basal part of the IHC membrane where the synapses of the afferent nerve fibers are located. They found that extracellular potentials produced there by the OHCs as cochlear microphonics had greater magnitudes at low sound frequencies than did the intracellular receptor potentials. Because both had the same phase, the potential gradient across

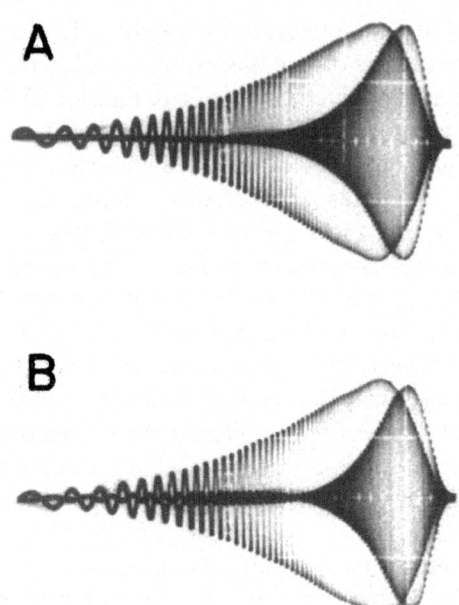

FIG. 6.1. Model basilar membrane (broader envelope) and shear displacement amplitude envelopes as functions of sound frequency at one cochlear location. For the parameter values in panel A, the shear displacement goes to zero at low sound frequencies, for those in panel B, an overshoot occurs.

the cell membrane was reversed. As a consequence, the nerve endings were excited when the cells were hyperpolarized. This was assumed to occur during basilar membrane displacement toward scala tympani. At higher frequencies, the potential gradient was reversed, and nerve excitation occurred during basilar membrane displacement toward scala vestibuli, in agreement with the classical model. The model of Russell and Sellick had the potential of accounting for the variable response phase found at the low sound frequencies in the auditory nerve fibers but it is not clear if it could account for the sharpening of frequency selectivity. It also made an incorrect phase prediction for the frequency region of CF, as is shown in the following sections. Finally, it predicted a difference between the receptor and neural tuning curves at low sound frequencies. Such a difference has not been demonstrated.

The model based on the wave-length relationship to the tectorial membrane space constant seemed to satisfy the experimental evidence existing at its inception—it accounted for the sharpening of frequency analysis provided by the basilar membrane ahead of the IHCs and for the variable response phase at low sound frequencies as well. It soon became clear, however, that it was not applicable to the lizard ear, in particular to that of the alligator lizard whose papilla began to be used at MIT as a simplified model of the mammalian cochlea. In this papilla, the hair cells are sharply tuned (Weiss, Mulroy, Turner, & Pike, 1976) in the absence of waves on the basilar membrane that vibrates uniformly over its whole length (Peake & Ling, 1978). The papilla is divided along the short basilar membrane in two sections. In one section, the hair cell stereocilia are loaded by a tectorial membrane, as in a mammalian cochlea. But, in the remaining section, the tectorial membrane is missing. In spite of the crucial anatomical difference, the hair cells exhibit sharp frequency tuning in both sections. Significantly, the CFs are lower in the section of short stereocilia endowed with the tectorial membrane than in the remaining section, although the stereocilia are considerably longer there. These relationships invited the speculation that the stereocilia bundles themselves were mechanically tuned, especially since the CFs were found to be inversely related to the stereocilia lengths in the section with free standing stereocilia. The speculation was made plausible by the finding that the stereocilia were stiff (Flock, 1977; Hudspeth & Corey, 1977), so that they could resonate with sound frequencies within the audible frequency range. It was further strengthened by the finding that the presence of the tectorial membrane lowered the CF presumably by lowering the resonance frequency. Such an effect is well known in mechanics of elastic beams, which the stereocilia resemble. There were further considerations. Extensive studies of E. G. Wever (1967) revealed a great variety of stereocilia-tectorial membrane arrangements among the various lizard species. In some, there was no tectorial membrane, in others it was partially present and solidly attached to bone, in still others the attachment was quite flimsy or entirely absent. In the latter, the effect on the stereocilia had to be that of mass loading.

In view of all these considerations, I began exploring the possibility that, the main effect of the tectorial membrane on the stereocilia bundles was that of mass loading with the effect of lowering their natural vibration frequency. The possibility was rendered more likely by the finding of Kronester-Frei (1979) that, in chemically well-preserved histological preparations, the tectorial membrane was much larger than was seen previously, when no special attention was paid to the chemical environment. The shrinkage artifact was subsequently confirmed on many occasions. In addition, Zwicker (1971) found in domestic pigs tectorial membranes that, in the apical cochlear turn, were larger than the organs of Corti. Similar observations had been made already at the beginning of the 20th century (e.g., Shambough, 1907)

In an attempt at finding out if stereocilia bundles loaded with the tectorial membrane mass had a chance to resonate, I calculated the damping resulting from the viscosity of the fluid enclosed between the tectorial membrane and the reticular lamina (Zwislocki, 1980). The answer was affirmative for sound frequencies above some 140 Hz. However, the validity of this conclusion must be questioned because the internal friction within the tectorial membrane was not taken into account. On the basis of our current knowledge, it seems that the damping may be too large, and resonance with less than a critical damping may only be possible in the presence of the now known active feedback that compensates for it.

Unfortunately, the stiffness of the stereocilia bundles had not yet been measured, so that it was not possible to calculate the resonance frequency for any given cochlear location. This was done later by Strelioff, Flock, and Minser (1985) on the basis of their measurements of the stereocilia stiffness and my estimates of the tectorial membrane mass. Their result agreed with the known tonotopic map of their experimental animal, guinea pig.

A SIMPLE MECHANICAL MODEL

To have a preliminary insight into the effects of a possible stereocilia resonance, we constructed a very simple mechanical model of an incremental section of the basilar membrane with the organ of Corti and OHC stereocilia mass loaded by a corresponding section of the tectorial membrane (Zwislocki, 1980). The main question was whether the stereocilia could be excited to oscillate transversally in a direction almost perpendicular to the basilar membrane vibration. The model consisted of the elastically suspended armature of an electrodynamic vibrator, an elastic reed attached to it almost parallel to the direction of its vibration and a small weight attached to the tip of the reed. Its photograph is shown in Fig. 6.2. The vertical vibration of the armature was measured with the help of a Fotonic Sensor whose probe can be seen to the right of the vibrator. The horizontal vibration of the reed could be seen with the naked eye and was studied under stroboscopic illumination. It is easily apparent in the photograph. This large an amplitude occurred only in the vicinity of the reed's resonance frequency, however. At frequencies sufficiently removed from it, no horizontal vibration of the reed could be discerned. This phenomenon suggested a very sharp frequency resolution.

When measured as a function of oscillation frequency in the absence of the reed, the vertical vibration of the armature exhibited a lowpass characteristic, as can be seen in Fig. 6.3. Addition of the read with an appropriate stiffness and loaded with an appropriate mass perturbed the lowpass characteristic, introducing a sharp peak just below the corner frequency, followed by a narrow dip. The dip, not the peak, coincided with the resonance frequency of the read. Why this had to be so should become evi-

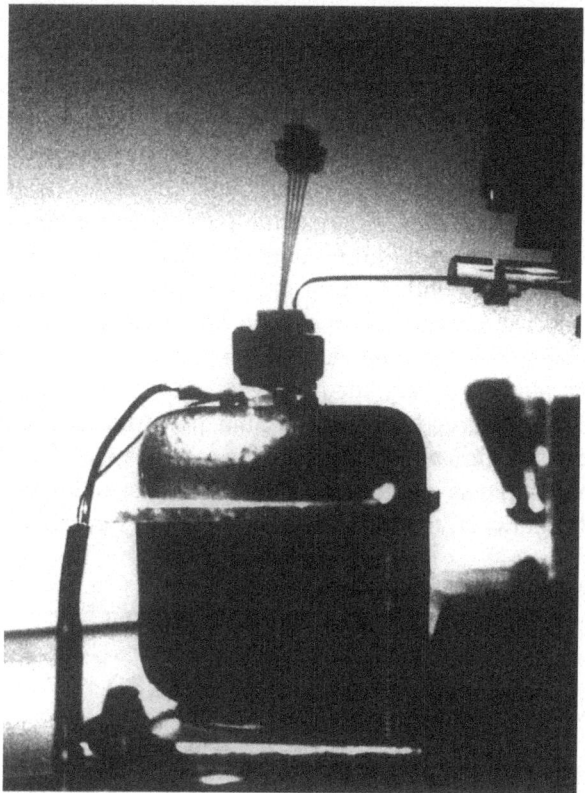

FIG. 6.2. Vibrator model of a short portion of the cochlear partition. The oscillating reed represents a stereocilia aggregate, the small mass at its tip, the tectorial membrane. The narrow tube to the right is a light guide of a Photonic sensor measuring the vertical vibration of the vibrator armature that represents the mass associated with the basilar membrane.

dent from the mechanical network equivalent of the vibrator assembly and its electrical analog shown in Fig. 6.4. In the mechanical network on the left, F indicates the mechanical force driving the vibrator armature, M_V, the mass of the armature, C_V, its elastic suspension and R_V the associated mechanical resistance. Furthermore, C_R indicates the compliance of the reed, R_R, the associated resistance, and M_R, the mass at the reed's tip. In the electrical analog on the right, V_V indicates the driving voltage, L_V, C_V, and R_V, the analogs of the armature's mass and compliance and the resistance of its suspension, respectively. The elements C_R, R_R, and L_R, are the electrical analogs of the mechanical elements C_R, R_R, and M_R. In the mechanical vibrator network, the elements are in parallel, in its electrical analog, in series. Both have a minimum input impedance at the resonance frequency.

By contrast, in the mechanical network representing the reed, the compliance and mass are in series, and the capacitance and inductance in its electrical analog, in parallel. At its resonance frequency, the resistance goes through a maximum whose sharpness is inversely related to the network resistance. The maximum resistance loads the armature system of the vibrator and produces a relative vibration minimum clearly apparent in Fig. 6.3. Below the resonance frequency, the reed input reactance is positive and tends toward a maximum near the resonance frequency. This reactance interacts with the negative reactance of the armature assembly to produce an overall impedance minimum and a vibration maximum at a frequency where their absolute values are equal. The maximum precedes the minimum along the frequency scale, as is evident in Fig. 6.3.

It should be observed that the sizable perturbation of the vibrator lowpass characteristic was produced by the reed in spite of the fact that the mass load at its tip was much smaller than the mass of the vibrator armature, the ratio between the two being much larger than that between the

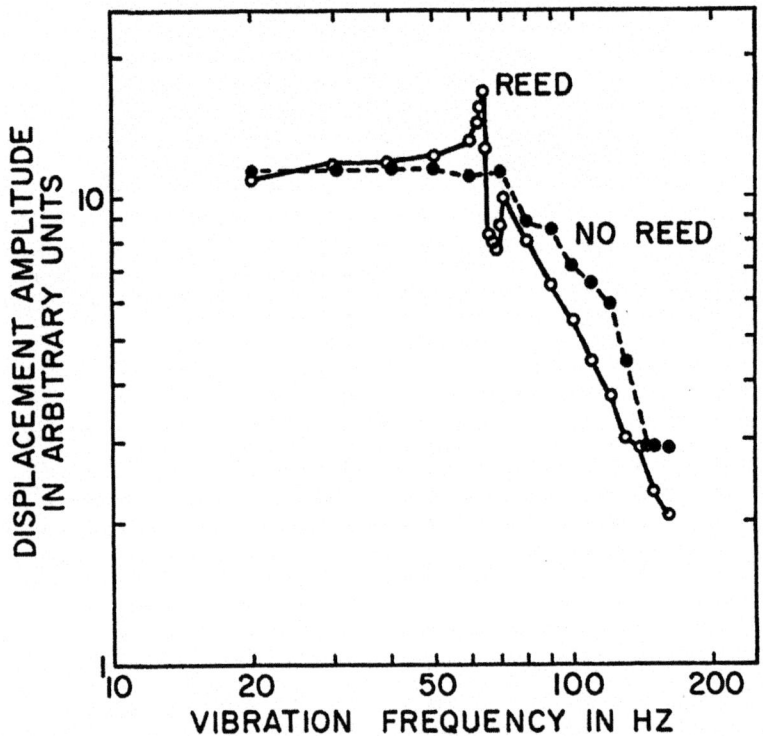

FIG. 6.3. Vertical vibration amplitude of the vibrator armature as a function of vibration frequency. For the intermittent curve, the reed was removed.

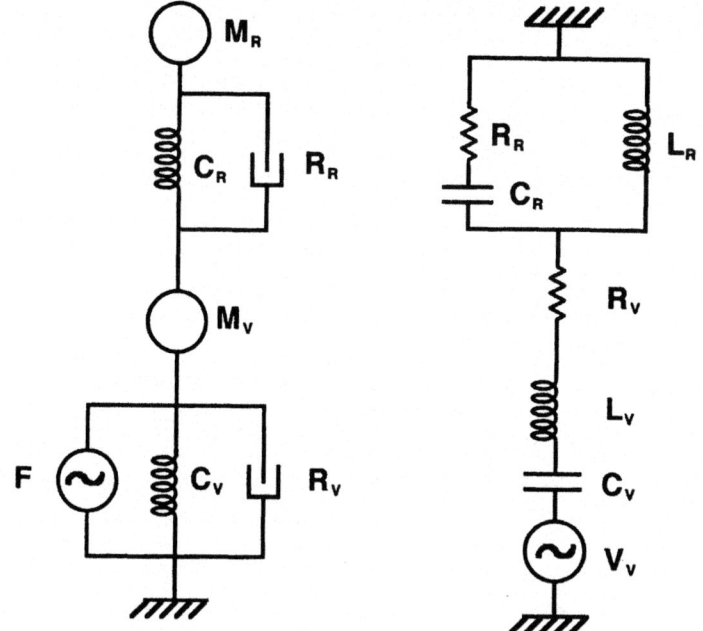

FIG. 6.4. Mechanical network of the vibrator model (on the left) and its electrical analog. Explanation of the symbols in the text.

masses of the tectorial membrane and the organ of Corti. It should be observed also that the pattern of a maximum followed by a minimum at a short frequency interval is similar to that found in basilar membrane vibration, as demonstrated for the first time by Rhode (1971).

Measurement of the vertical vibration of the vibrator armature revealed still another phenomenon, this one, relevant to the understanding of cochlear nonlinear distortions. Such distortions are inherent in the nearly perpendicular orientation of the reed's vibration relative to the vibration of the armature. For simplicity, imagine that the reed stands exactly vertically on the armature. When it oscillates, it is deflected from this orientation, and its motion vector is no longer entirely horizontal but slightly inclined and acquires a vertical component. Acceleration of the loading mass along this component produces an inertial force acting on the armature. Because of the symmetry of the configuration, the force is exerted in the same direction on both sides of the reed, doubling the frequency of the armature oscillation and introducing a second harmonic. The effect must be the strongest when the reed oscillates maximally. This occurs at its resonance frequency that coincides with the dip in the armature vibration characteristic. When the reed does not stand entirely vertically on the armature, an asymmetry arises that introduces odd harmonics in addition to the even ones, as is evident in the bottom trace of Fig. 6.5. The top and third traces of the figure show sinusoidal electrical wave forms at the vibrator input, the second and fourth, corresponding wave forms of the vertical armature vibration—the second one, at a frequency below the reed resonance, the fourth one, near the resonance frequency. Clearly, the nonlinear distortions are much greater in the latter. LePage and Johnstone (1980a, 1980b) found that the nonlinear distortions in the basilar membrane vibration were the strongest just above the frequency of maximum vibration, at the location of the dip.

The idea of a stereocilia-tectorial membrane resonance and its effect on basilar membrane vibration, together with its physical model, was first presented at the 50th anniversary meeting of the Acoustical Society of America in the spring of 1979. Unfortunately, the symposium that included the paper was not published until a year later because of technical difficulties with another symposium paper. This produced a chronological confusion.

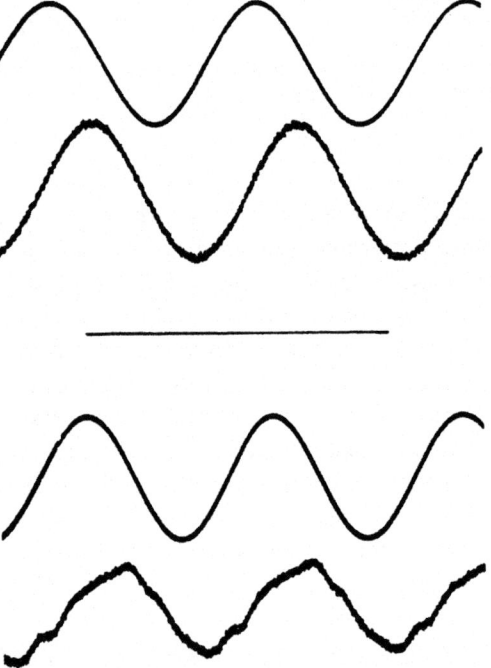

FIG. 6.5. Wave forms of the vertical armature oscillation in the vibrator model. The upper and third traces belong to the electrical input to the vibrator; the second and bottom traces, to the armature oscillation—the former at a frequency below the maximum reed oscillation, the latter, at the frequency of the notch.

My idea of a stereocilia-tectorial membrane resonance was immediately criticized by a group of prominent auditory scientists present in the audience. They did not think that the cochlear damping would allow such a resonance. Possibly, the notion of a high cochlear damping may have lingered since the time of Békésy's postmortem experiments, and the significance of just-discovered cochlear acoustic emissions, consistent with a low damping, did not yet take hold. Even so, one of the scientists published two theoretical articles within about 1 year of the symposium, in which he tacitly accepted my resonance idea (Allen, 1980a, 1980b). In spite of these theoretical articles supporting the resonance idea and several following ones presenting experimental evidence for it, the idea has remained controversial. One of the reasons for it may have been that the radial vibration of the tectorial membrane is difficult to detect, as is explained in the next section. Nevertheless, in one experiment in which conditions approximating the normal state of the cochlea were sufficiently preserved and appropriate techniques applied, both the radial vibration and the resonance were clearly demonstrated by optical means (Gummer et al., 1996).

Phase relationships required by the resonance were extensively demonstrated in recordings of OHC alternating potentials and potentials induced by them in Hensen's cells. They are discussed in the following section.

DYNAMICS OF THE STEREOCILIA-TECTORIAL MEMBRANE COMPLEX

The dynamic characteristics of a system can be calculated mathematically once its structure and physical constants are known. On the other hand, our knowledge of these parameters, when uncertain or incomplete, can be verified by comparing the calculated characteristics to the measured ones. The calculation is done here for the mammalian stereocilia-tectorial membrane system with the help of parameter values obtained for Mongolian gerbil, our animal model, in the preceding chapter. The stereocilia-tectorial membrane system is defined as consisting of the OHC stereocilia attached at their tips to the tectorial membrane, and of the tectorial membrane attached viscoelastically to the spiral bony limbus. In agreement with the measurements of Strelioff and Flock (1984), the effect of the IHC stereocilia is neglected. It is assumed that the OHC stereocilia provide a viscoelastic coupling between the reticular lamina that holds the apical parts of the OHCs practically rigidly and the tectorial membrane.

As is generally accepted, because of the geometry of the organ of Corti, the transversal motion of the basilar membrane generates a radial motion component of the reticular lamina. Only this component in the stereocilia-tectorial membrane complex is considered. As already stated in the preceding chapter, practically no relative transversal motion between the tectorial membrane and the reticular lamina can be expected, because the two structures are separated only by a narrow gap filled with nearly incompressible fluid and containing bundles of stereocilia oriented nearly vertically to their surfaces. The axial stiffness of the stereocilia is accepted as being very high. The system is assumed to be driven radially by the reticular lamina, acting as a high impedance (current) source. This is justified on the basis of stiffness determinations of cells of the organ of Corti, in particular, the pillars of Corti (e.g., Tolomeo & Holley, 1997). The reticular lamina may be regarded as an extension of the outer pillars of Corti. In addition, the large stiffness of the pillars of Corti and also of Deiter's cells must provide a tight elastic coupling between the reticular lamina and the basilar membrane whose stiffness has been determined in the preceding chapter to be much greater than that of the corresponding stereocilia aggregate, the latter in the radial direction. As a consequence, the vibration of the basilar membrane should be transferred practically without loss and phase change to the reticular lamina.

The mechanical network of a unit section of the stereocilia-tectorial membrane complex, as well as its electrical analog, are illustrated in Fig. 6.6 in its upper and lower parts, respectively. Analysis of the performance of such a section is meaningful because the cochlear parameters vary slowly by comparison to the cochlear wave length, except at the very base of the cochlea, and the longitudinal space constant of the tectorial membrane has been found to be small by comparison to the wavelength. In the upper part of the figure, the aggregate stereocilia compliance is represented by the spring, C_{Ha}, which is driven by the reticular lamina with the radial velocity V_R. The resistance element, R_{Ha}, represents the resistance associated with the stereocilia movement and probably stemming for the most part from fluid friction in the subtectorial space. The stereocilia drive the mass of the tectorial membrane represented as the rigid body, M_{Ta}, which is suspended visco-elastically at the spiral limbus, the compliance of the attachment being represented by the spring, C_{Ta}, and the associated resistance by the dash pot R_{Ta}.

In the electrical analog, the radial velocity of the reticular lamina is represented by the current Is1, the stereocilia compliance, by the capacitance C_{Ha}, the associated resistance by the resistance R_{Ha}, and the velocity of the stereocilia bending motion, by the current I1. In the third branch, the inductance, L_{Ta}, represents the tectorial membrane mass, and the capacitance, C_{Ta}, and resistance, R_{Ta}, its attachment to the limbus. The radial motion of the tectorial membrane is represented as the current I2.

The base numerical values of all the elements, except the resistance, R_{Ta}, have already been given in the preceding chapter, as derived from independent measurements. They should be valid for a healthy but passive cochlea, where the active feedback is absent. Such a situation appears to

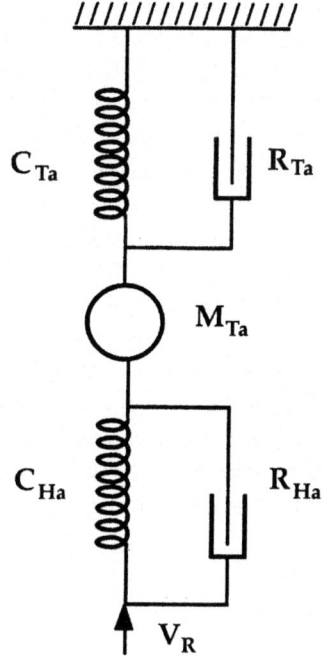

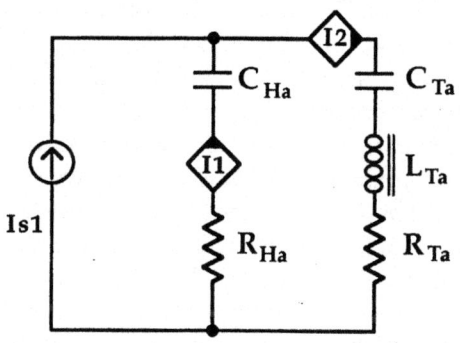

FIG. 6.6. Mechanical network of the stereocilia-tectorial membrane complex (top) and its electrical network analog (explanation in the text).

arise in the gerbil cochlea at SPLs around 80 dB SPL (e.g., Dallos, 1992; Zhang & Zwislocki, 1995). The values are listed here for convenience in acoustic units referred to the effective width of the basilar membrane, as defined in chapter 4. They apply to our experimental location, which was at approximately 63% of the total cochlear length from the cochlear base. The width amounted there to $w_{Be} = 1.2*10^{-2}$ cm.

$C_{Ha} = 0.35*10^{-8}$ cm^4/dyne; $C_{Ta} = 0.7*10^{-7}$ cm^4/dyne; $M_{Ta} = 0.55$ g/cm^3; $R_{Ha} = 1.4*10^3$ dyne sec/cm^4; $R_{Ta} = 3.5*10^3$ dyne sec/cm^4.

Note that the numerical value of C_{Ta} is increased by a factor of about 3 relative to that measured directly, as explained in chapter 5. It must also be pointed out that the resistance, R_{Ha} was originally calculated from the geometry of the subtectorial gap and the viscosity coefficient of the endolymph as a mechanical resistance amounting to 0.2 dyne sec/cm^2 (Zwislocki, 1980b). To obtain its acoustic value, it has been divided by the width of the basilar membrane, $w_{Be} = 1.2*10^{-2}$ cm, squared.

According to the networks of Fig. 6.6, the stereocilia-tectorial membrane complex must have two resonances—one arising from interaction between the tectorial membrane mass and the compliance of its attachment to the limbus, and one, from interaction of this mass with both the compliance of its attachment to the limbus and of the compliance of the stereocilia, acting together. The first resonance frequency is much lower than the second, and it is shown further that it is situated well below the CF, whereas the other is situated somewhat above it. The lower resonance frequency is calculated with the help of the known formula

$$f_{Tl} = \frac{1}{\sqrt{M_{Ta} C_{Ta}}} \qquad 6.1$$

that, with the constants given above, produces the value of $f_{Tl} = 800$ Hz. To calculate the higher resonance frequency, the compliance C_{Ta} has to be replaced in Eq. VI.1 by the compliance C_{THa} that results from the sum of inverse values of this compliance and the compliance C_{Ha}— $C_{THa} = C_{Ha} * C_{Ta}/(C_{Ha}+C_{Ta})$. With the numerican values given above, the numerical value of the combined compliance becomes $C_{THa} = 0.33*10^{-8}$ cm^4/dyne, and the high frequency resonance, $f_{Th} = 3,700$ Hz. The ratio between the two resonance frequencies is approximately 4.6 and exceeds two octaves. A similar frequency ratio is found between the low frequency notch sometimes encountered in transfer characteristics of cochlear hair cells and supporting cells, often called tuning curves, and the high frequency notch often found in basilar membrane transfer characteristics (e.g., Zwislocki & Cefaratti, 1989). These relationships are explained more fully in the following paragraphs.

To understand the interaction of the dynamics of the stereocilia-tectorial membrane complex with those of the basilar membrane and the organ of Corti, it is necessary to know the input impedance of the former. Such an impedance per unit length of the complex, as calculated on the Macintosh computer with the help of the MacAC II program on the basis of the constants specified above, is shown in Fig. 6.7 in terms of its resistance and reactance components. Every panel corresponds to a different damping—the upper panel, to the values of R_{Ha} and R_{Ta} derived for the passive

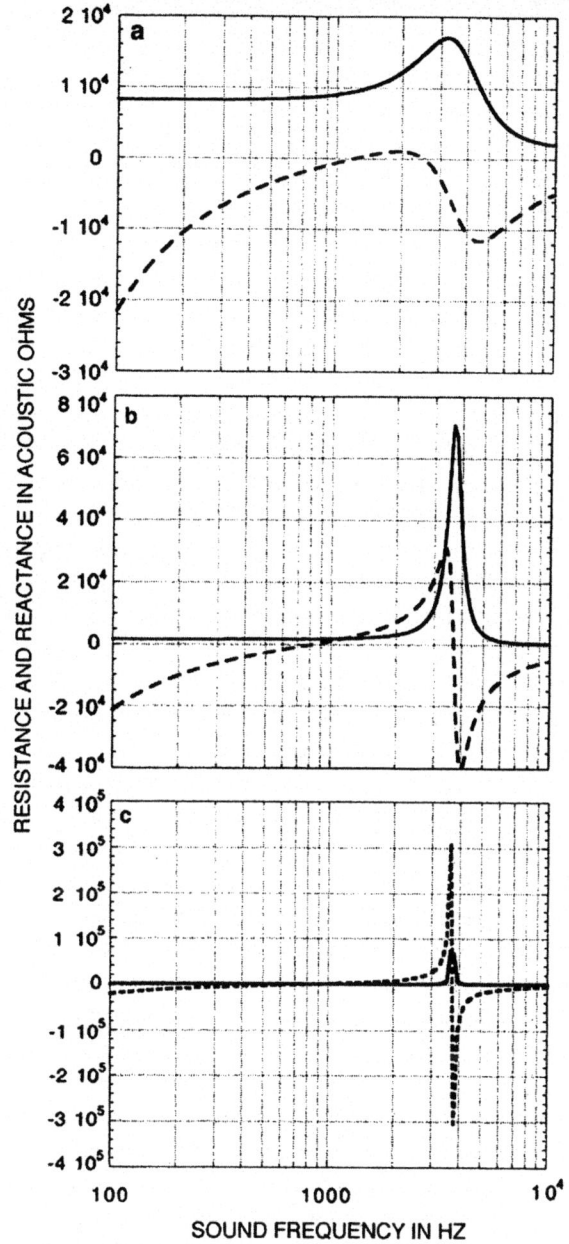

FIG. 6.7. Resistance (solid line) and reactance (dashed) components of the input impedance of the stereocilia-tectorial membrane complex as functions of sound frequency for three sets of network resistance values, decreasing from the top. Model results.

295

cochlea and specified above; the middle panel, to 5 times smaller values, and the lowest panel, to 100 times smaller values. Note that both resistances have been changed by the same factor, a procedure maintained in the following two figures. The solid lines indicate the values of the resistance component, the intermittent ones, those of the reactance component. In every panel, the resistance component goes through a maximum near the high frequency resonance, being shifted somewhat to a lower frequency in panel a, in the presence of the highest damping. The maximum becomes sharper as the damping decreases. At the maximum, the resistance component must exert the greatest load on the basilar membrane. It is shown in the next section that this load produces a notch in the basilar membrane transfer functions just above the CF. Such a notch was found in many empirical transfer functions, as exemplified in the preceding chapter in Figs. 5.35, 5.38, and 5.39. The notch is also evident in the vibrator model transfer function of Fig. 6.3. According to the intermittent lines, the reactance component is negative at low sound frequencies and increases toward the lower resonance frequency. Slightly above this frequency, it becomes positive, reaching a maximum just below the high frequency resonance location. Near this location, it sharply decreases and goes through a negative extremum. At the higher frequencies, it increases asymptotically toward zero. As may be expected, the maxima and the transition points become sharper, as the damping decreases. They also become more accurately aligned with the resonance frequencies.

The resistance and reactance patterns of Fig. 6.7 must appear well familiar to engineers. They are characteristic of series and parallel resonances, the former taking place in the low frequency region the latter, in the high frequency one. It should be pointed out that the reactance component must interact with the reactance component of the basilar membrane organ of Corti complex to produce new maxima, as is shown in the next section. Such an interaction was already demonstrated in the vibrator model of Fig. 6.2. It produced the principal maximum in the vertical oscillation of the vibrator armature and also, the smaller secondary maximum, both evident in Fig. 6.3.

The next parameter, essential in producing the shear motion between the reticular lamina and the tectorial membrane, is the radial motion of the tectorial membrane. It is shown in Fig. 6.8 relative to the motion of the reticular lamina for two levels of damping—the damping associated with the passive cochlea in panel a, and a five times smaller damping in panel b. At lower damping values, the characteristic features seen in panel b become exaggerated but no new features appear, so that these damping values have been omitted. In both panels, a and b, the magnitude of the tectorial membrane motion is almost equal to that of the reticular lamina up to about 1000

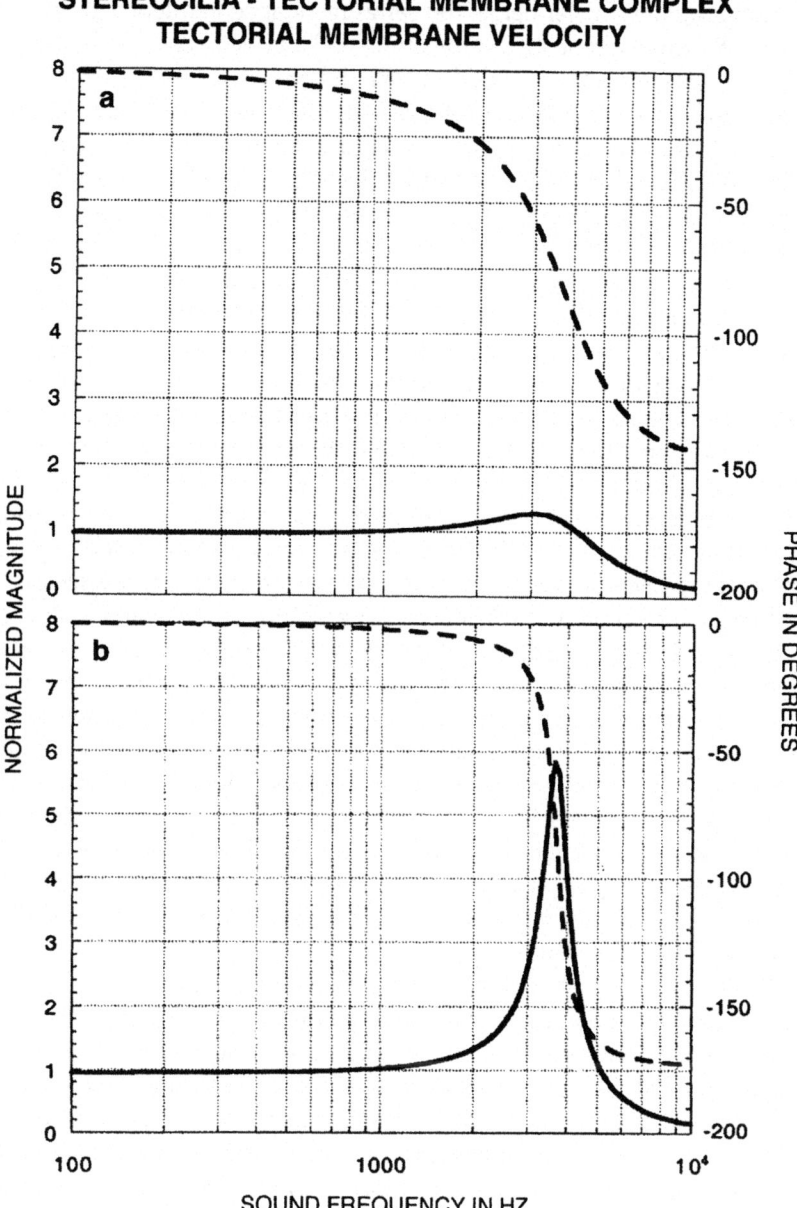

FIG. 6.8. Tectorial membrane radial oscillation velocity as a function of sound frequency for high (a) and low (b) damping of the stereocilia-tectorial membrane complex (magnitude—solid line; phase—dashed line). Model results.

Hz, being slightly smaller below the series resonance frequency at 800 Hz, and slightly larger above it. The maximum velocity is reached in the vicinity of the parallel resonance at 3700 Hz. With the passive damping of panel a, it exceeds there the magnitude of the reticular lamina motion by only 25%, but, with the smaller damping, it is greater by a factor of almost 6. It would be even larger if the damping were further decreased. The large magnitude at the parallel resonance frequency is a characteristic of the resonance and it is clearly present in the motion of the reed in the vibrator model of Fig. 6.2. It should be noted, however, that the velocity maximum coincides with the resistance maximum of the stereocilia-tectorial membrane complex, which produces a notch in the basilar membrane transfer function, as already mentioned and is discussed more fully in the next section. When the entire cochlea is considered, this notch occurs in the basilar membrane region where the vibration amplitude is quite small, so that the velocity of the tectorial membrane is reduced.

It may be surprising to some readers that the amplitude of the tectorial membrane radial velocity should be nearly equal to that of the reticular lamina at the lower frequencies. This is due to the large compliance of the viscoelastic attachment of the tectorial membrane to the spiral limbus by comparison to the compliance of the stereocilia coupling, as is discussed extensively in chapter 5. The near equality of the two velocities in the presence of the already mentioned equality of the transversal velocities of the two structures may create the optical illusion that the tectorial membrane does not vibrate radially. It certainly hardly does so relative to the reticular lamina. Because most observations of the reticular lamina vibration are performed under conditions where the active feedback is weakened or absent and the damping relatively high, the tectorial membrane velocity should not be expected to exceed that of the reticular lamina by more than 25% even at its maximum.

The illusion that the tectorial membrane does not move radially relative to the reticular lamina may be strengthened further by the fact that both vibrate nearly in phase at the lower frequencies, as indicated by the intermittent curve of Fig. 6.8. The relative phase increases in the lag direction with sound frequency and reaches 90° at the frequency location of the parallel resonance. At still higher frequencies, it approaches asymptotically 180°. The phase opposition increases the relative motion between the reticular lamina and the tectorial membrane. However, the increment is inconsequential, since the basilar membrane vibration amplitude approaches zero in this frequency region.

From the point of view of hair cell stimulation, the relative motion between the reticular lamina and the tectorial membrane, the shear motion, is the most directly relevant. It is shown for three damping levels in Fig. 6.9, for the damping corresponding to the passive state of the cochlea in panel a, for a 5 times smaller damping, in panel b, and for a 25 times smaller one, in panel c. Again, the radial motion amplitude of the reticular lamina serves as a reference unit. In every panel, the magnitude of the shear motion is shown by the solid line, the phase, by the intermittent one. At all three damping levels, the shear motion magnitude is extremely small at low sound frequencies compared to that of the reticular lamina. This is in agreement, of course, with the near equality of the radial reticular lamina and tectorial membrane motions shown in Fig. 6.8 and indicates a substantial reduction in hair cell stimulation. The reduction is consistent with the finding of Narayan, Temchin, Recio, and Ruggero (1998), according to which the sensitivity of the auditory nerve fibers is decreased at low sound frequencies by about 20 dB relative to that of the basilar membrane, when both sensitivities are equated at the CF. The numerical data underlying the magnitude curves of Fig. 6.9 indicate approximately the same numerical relationship between the magnitudes of the radial reticular lamina motion and the shear motion. It is shown in the next section that the amplitude of the radial motion of the reticular lamina is about equal to that of the transversal basilar membrane motion at the location of the OHCs. To take a specific example, the decibel difference at 200 Hz amounts to 27 dB in panel c, assumed to correspond to a healthy cochlea. In the two examples of Narayan et al., the corresponding differences amount to approximately 11 and 28 dB, respectively. At the higher frequencies, the difference decreases both theoretically and empirically and disappears completely near the CF. According to Fig. 6.9, the shear motion becomes large compared to the reticular lamina motion near the parallel resonance of the stereocilia-tectorial membrane complex, but this is of little consequence for the stimulation of the OHCs because the basilar membrane vibrates with a very small amplitude there. These relationships are discussed more fully further on.

The phase relationships shown in Fig 6.9 are of substantial theoretical interest, in part, because they test directly the phase relationship predicted by the classical model of the shear motion. The phase ordinates in the figure are placed so that zero phase coincides with the deflection of the stereocilia toward the spiral ligament, the excitatory direction, during basilar membrane displacement toward scala tympani and, therefore, the radial displacement of the reticular lamina toward the spiral ligament. As has been already mentioned on many occasions in this monograph and elsewhere, the classical model predicts that the hair cell stereocilia have to be deflected toward the spiral ligament and the cells depolarized during basilar membrane displacement or motion toward scala vestibuli. During this displacement, the reticular lamina is supposed to move radially toward the

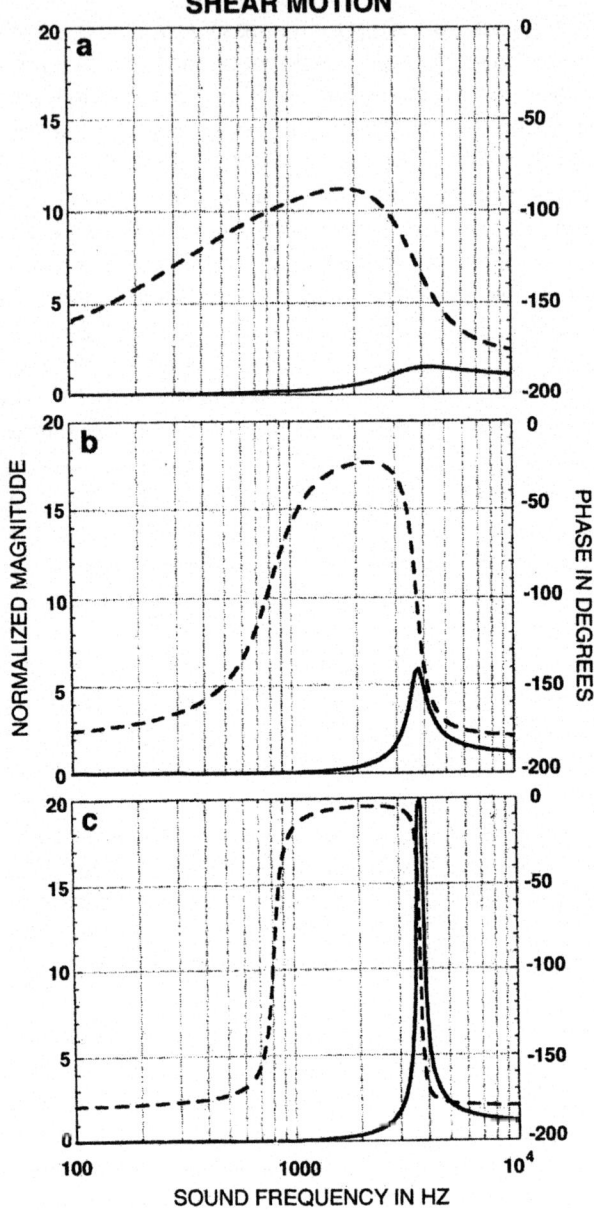

FIG. 6.9. Magnitude (solid line) and phase (dashed) of the shear motion between the tectorial membrane and reticular lamina as functions of sound frequency for three damping values of the stereocilia-tectorial membrane complex decreasing from the top. Model results.

cochlear modiolus, with a phase corresponding to 180° in Fig. 6.9. The phase curves of the figure indicate that this prediction is incorrect. First of all, the shear motion phase is not at all constant but varies with sound frequency, no matter what the damping. The variation is increased as the damping is decreased. In panel a, representing the conditions predicted for the passive cochlea, the phase is around −90° in the vicinity of the CF, signifying that the stereocilia are deflected toward the spiral ligament, the excitatory direction, when the reticular lamina moves radially toward the modiolus, and the basilar membrane, toward scala vestibuli. At lower and higher sound frequencies, the phase approaches −180°, signifying that the stereocilia are deflected in the excitatory direction when the basilar membrane is near its scala vestibuli position. This is in vague agreement with the classical model. The departure from it is the most drastic in the CF region when the cochlear damping is small, as is expected in a healthy cochlea. This is evident in panel c. Here the phase approaches 0°, signifying a depolarizing deflection of the stereocilia during basilar membrane displacement toward scala tympani, not vestibuli. Agreement with the classical model is achieved only at the very low and the very high frequencies, where the shear motion phase approaches −180°. For an intermediate damping, the phase relationships are intermediate. Still, the phase varies between almost 0° near the CF and −180° at the very low and high frequencies.

In Fig. 6.9 only the values of the resistances R_{Ha} and R_{Ta} have been varied, keeping the compliance values constant, on the assumption that the active feedback does just that. However, there is experimental evidence indicating that the stereocilia compliance, C_{Ha}, is decreased at very high SPLs by other mechanisms (e.g., Saunders & Flock, 1986; Szymko, Nelson-Adesokan, & Saunders, 1995). At these levels, the active feedback seems to be ineffective, and the resistances may remain constant. A similar softening of the stereocilia coupling with the resulting increased C_{Ha} can be expected after acoustic overstimulation that may not only damage the stereocilia but even reduce the population of the OHCs (e.g., Liberman & Kiang, 1978). Accordingly, Fig 6.10 shows magnitude and phase curves of tectorial membrane motion and the resulting shear motion for an increased stereocilia compliance, so that $C_{Ha} = C_{Ta}$. Clearly, the amplitude of radial tectorial membrane motion is now severely reduced, amounting to only about one half the radial reticular lamina motion at low sound frequencies and decreases further toward the higher frequencies, whereas the associated phase decreases from almost zero to almost −90°. These characteristics lead to a drastically increased shear motion between the reticular lamina and the tectorial membrane, which grows with sound frequency from about one half the radial motion of the reticular lamina to almost equality with it. The associated phase changes little with the frequency and hovers slightly above −180°.

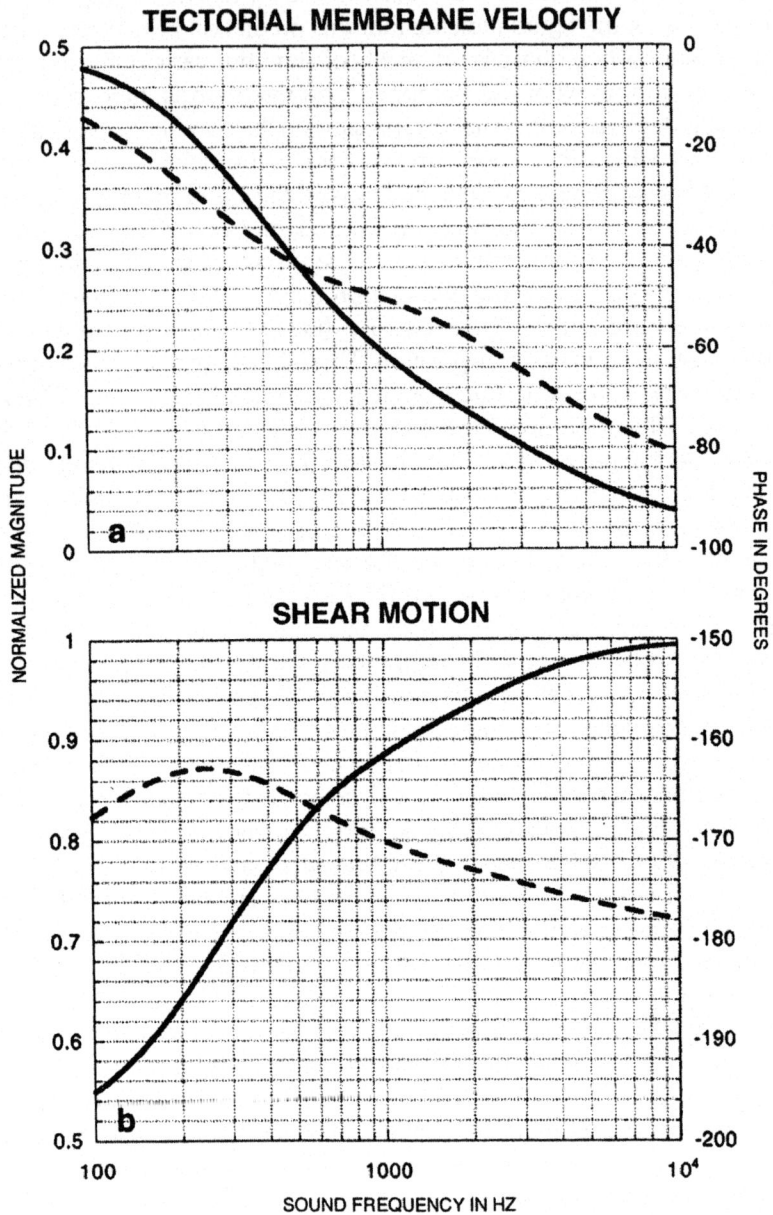

FIG. 6.10. Tectorial membrane radial oscillation velocity and the shear motion as functions of sound frequency for an increased compliance of the stereocilia. Magnitudes indicated by solid curves, phases, by the intermittent ones. Model results.

This means, according to the phase coordinates of Fig. 6.10, that the stereocilia are deflected toward the spiral ligament, the excitatory direction, during basilar membrane displacement toward scala vestibuli, in agreement with the classical model. The agreement is produced by the relatively small radial motion of the tectorial membrane. The increased shear motion signifies a paradoxically increased hair cell and neuronal response at low sound frequencies in damaged cochleas whose CF response may be substantially reduced, as was suggested earlier on related grounds (Zwislocki, 1984). A corresponding sensitization was found experimentally on several occasions (e.g., Dallos, Ryan, Harris, McGee, & Özdamar, 1977; Liberman & Kiang, 1978). In particular, Liberman and Kiang found a low frequency neuronal sensitization of about 25 dB in slightly damaged cochleas of domestic cats. The computed data underlying the graphs of Fig. 6.10 suggest a similar sensitization (23 dB at 200 Hz).

It may be of interest to point out that, according to the derivations already discussed, the reduction of the neuronal low frequency response relative to the basilar membrane response in healthy cochleas appears to be closely related to the relatively enhanced response in the damaged ones. The small response in healthy cochleas results from almost equal radial reticular lamina and tectorial membrane motions leading to a small shear motion. In pathological cochleas in which the tectorial membrane is driven less strongly by damaged or depleted stereocilia, its radial motion becomes substantially smaller than that of the reticular lamina, and the shear motion is increased. In other words, the sensitivity suppression taking place at sound frequencies well below the CF is abolished. It is likely that the suppression effectively increases the cochlear frequency selectivity.

The theoretical phase characteristics derived in preceding paragraphs for the stereocilia-tectorial membrane complex can be validated to some extent by comparing the obtained shear motion phases to the measured response phases of the OHC or Hensen's cells (Zwislocki, 2000). The assumption has to be made that the latter are in a constant relationship to the shear motion phases, independent of SPL or sound frequency within a reasonable range. This seems to be justified on the strength of the finding that hair cells are always depolarized when their stereocilia are deflected toward their kinocilium or basal body (Wersäll, Flock, & Lindquist, 1965; Flock, 1971). The measured phases include the basilar membrane and organ of Corti structures, and a further assumption has to be made that, within a useful sound frequency range, these structures are practically rigidly coupled to the reticular lamina. This assumption is justified in chapter 5. It is necessary because the calculated phases are referred to the reticular lamina motion. Finally, physiological and equipment time delays as well as the cochlear wave travel time included in the measurements have to be taken into account. This can be done by dividing the measured phases by a variable proportional to the product of the time delay and sound frequency. The product determines almost entirely the slope of the phase curves in Fig. 5.61 of the preceding chapter. For want of better ways, the slope correction has been executed by making the average slope of the 40 dB phase curve of this figure horizontal, and referring the data obtained at other SPLs to the corrected 40 dB data. This procedure, although not exact because of the phase perturbations in the basilar membrane vibration discussed in chapter 5, has produced useful results.

The next problem encountered in comparing the theoretical and empirical data has been the lack of an absolute reference phase in the empirical data. It has been solved by varying the parameters of the theoretical computations until the shape of the theoretical curve fitted the empirical data corresponding to the 40 dB SPL. Once this was achieved, the empirical data have been adjusted to the absolute ordinate position of the theoretical curve. For best results the base values of the capacitances, C_{Ha} and C_{Ta} in the analog network of the stereocilia-tectorial membrane complex had to be adjusted somewhat to shift the low frequency resonance from the originally estimated sound frequency of 800 Hz to one of 600 Hz and the high frequency resonance from 3.7 kHz to 3.8 kHz. These adjustments must be considered as small in view of individual differences among animals and the only approximate determinations of the original values specified above and used in Figs. 6.7 to 6.10.

The theoretical fits to the modified empirical phase data of Fig. 5.61 are illustrated in Fig. 6.11. The data are indicated by filled circles, the fitted curves, by solid lines. For the 40 dB data, resistance values, $R_{Ha} = 56$ and $R_{Ta} = 360$ dyne sec/cm4 have been used—the same as in Fig. 6.9c. They are 25 times smaller than derived for the passive cochlea, presumably, reflecting the effect of the active feedback. Other parameter values, in particular the compliances, have been left unchanged in the belief that they are not affected by the active process. For a good match with the empirical data obtained at 80 dB, the resistance values have been increased to $R_{Ha} = 1.3*10^3$ and $R_{Ta} = 7.7*10^3$ dyne sec/cm^4. They are only slightly smaller than estimated independently for the passive cochlea ($R_{Ha} = 1.4*10^3$ and $R_{Ta} = 9.0*10^3$ dyne sec/cm^4). As already mentioned, there is strong empirical evidence indicating that the active feedback is no longer active at the SPL of 80 dB. To demonstrate that this was true for the animal on which the data of Fig. 5.61 were determined, the data points obtained on the same animal at 40 dB in the presence of sodium salicylate perfusion have been included (Zwislocki, 1990). They are marked by unfilled squares that show excellent agreement with both the 80 dB empirical data and the fitting theoretical curve. This confirms the results of others indicating that sodium salicylate in-

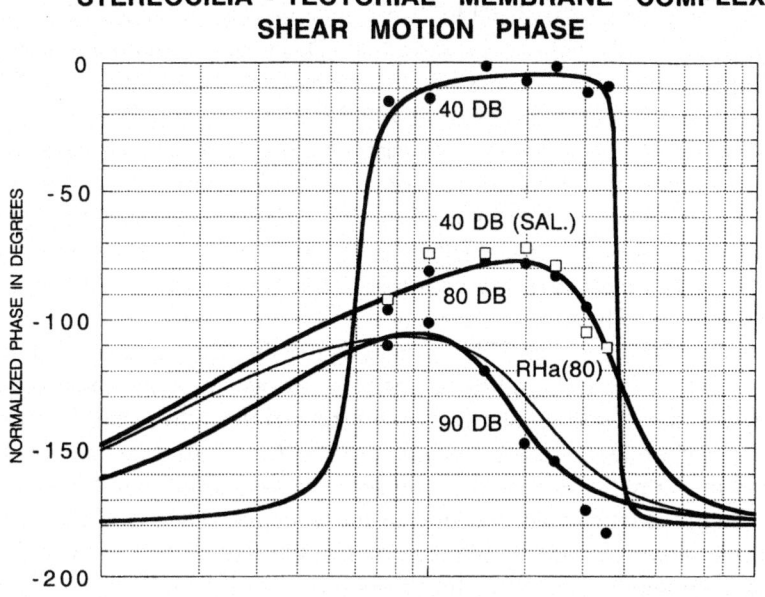

FIG. 6.11. Theoretical (lines) and empirical phases of the shear motion as functions of sound frequency for three damping levels associated with different SPLs. The empirical results indicated by squares were obtained in the presence of salicylate perfusion. Further explanation in text.

hibits the OHC motility necessary for the active process (e.g., Long & Tubis, 1988; Shehata, Brownell, Cousillas, & Imredy, 1990; Stypulkowski, 1990; Dieler, Shehata-Dieler, & Brownell, 1991; Zhang & Zwislocki, 1995).

The sodium salicylate experiments were performed successfully on 6 Mongolian gerbils with the help of surgical and recording techniques described in the preceding chapter. In addition, a small hole was drilled in scala tympani of the basal turn with the precision drill described in that chapter. It was possible to seal the perfusing micropipette almost hermetically in that hole. Another hole was drilled in scala vestibuli of the apical turn for fluid release. In some animals the perfusate was injected manually and in some, by an automatic micropump. The perfusate consisted of 5 mM solution of sodium salicylate in artificial perilymph (in mM, 150 NaCl, 5KCl, 1.5 CaCl2, 1.5 MgCl2, 10 glucose, 10 Hepes, 1 Na Peruvate, brought to pH 7.35 with 2mM NaOH, osmolarity 315 mOsm). The perfusion was maintained until the monitored CM settled at a low level—about

50% of the original value at 40 dB SPL. When this happened, the phase changed toward lag by about 90° and matched almost exactly the phase obtained at 80 dB without the salicylate. This was in agreement with the finding that the active process becomes ineffective at about 80 dB SPL.

To fit the data points obtained at 90 dB both the stereocilia compliance, C_{Ha}, and the tectorial membrane resistance, R_{Ta}, had to be changed from passive values, the compliance from $3.3*10^{-9}$ cm^4/dyne to as much as $17*10^{-9}$ cm^4/dyne, the resistance, from $7.7*10^3$ dyne sec/cm^4 to $4.8*10^3$ dyne/cm^4. As shown by the thick curve in Fig. 6.11, the achieved fit is reasonably good. However, the changes in the C_{Ha} and R_{Ta} values appear questionable. There is empirical evidence for decreased effective stereocilia stiffness at high SPLs, as already mentioned, but a 5-fold change from 3.3 to $17*10^{-9}$ cm^4/dyne specified above appears unexpectedly large. The decreased resistance is even more difficult to justify. It is likely that these changes are artifactual and are affected by another factor not included in the network model of the stereocilia-tectorial membrane complex. Such a factor is easy to find in the effective stiffness of the basilar membrane, which must be affected by the stiffness of the organ of Corti. There is convincing evidence that the stiffness of the OHCs is reduced at high SPLs (Flock, Flock, Fridberger, Scarfone, & Ulfendahl, 1999). The reduced effective stiffness of the basilar membrane would reduce the cochlear wave velocity and increase the slope of the 90 dB curve in Fig. 5.61, from which the 90 dB data points of Fig. 6.11 have been derived. An increased slope of the 90 dB curve in Fig. 5.61 is clearly evident. When an attempt is made at fitting the data points exclusively by changing the compliance of the stereocilia coupling and keeping the tectorial membrane resistance constant, as suggested by an ineffective active feedback at the high SPLs, a reasonable result is obtained with a $C_{Ha} = 10*10^{-9}$ cm^4/dyne, as indicated by the thin curve in Fig. 6.11. The curve approximates the empirical phase maximum of about $-105°$ but falls off somewhat too slowly above the frequency of the maximum. This is understandable in view of the relatively steep 90 dB phase curve in Fig. 5.61. A correction for the increased steepness would place the empirical data points near the thin curve. Accordingly, the hypothesis that the difference between the 80 and 90 dB phases within the stereocilia-tectorial membrane complex is due practically exclusively to a changed compliance of the stereocilia coupling does not have to be rejected.

To better visualize the dynamic phase relationships within the stereocilia-tectorial membrane complex, they are represented by vector diagrams in Fig. 6.12, as this was done in an earlier article (Zwislocki, 1990). There are three rows of the diagrams, each drawn for a different state of the cochlea. The top row is assumed to hold for a healthy cochlea stimulated at a low SPL and having accordingly small resistance values, R_{Ha} and R_{Ta}; the middle row, for a SPL of about 80 dB at which the active feedback is no longer effective, and the resistance values are increased to passive values; the bottom row, for a cochlea having the same resistance values as the passive cochlea but in which the compliance of the stereocilia coupling is increased either due to a high SPL or to stereocilia damage. At the top of the figure, two arrows with a common origin but opposite orientations give the reference directions. The direction to the right is assumed to coincide with reticular lamina motion or displacement toward the spiral ligament. Because of the geometry of the organ of Corti, this direction is associated with basilar membrane motion or displacement toward scala tympani. The opposite direction is assumed to coincide with reticular lamina motion or dis-

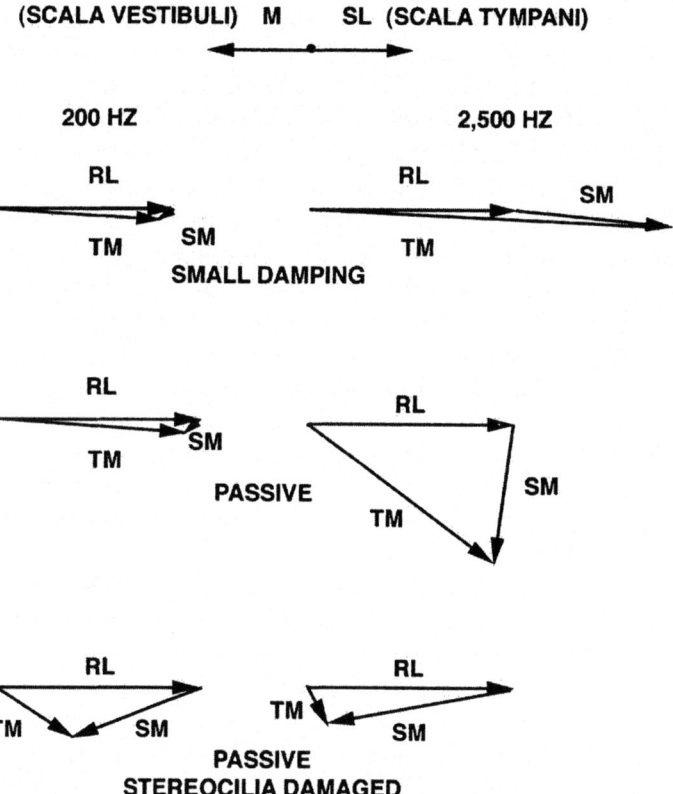

FIG. 6.12. Vector diagrams showing the relative radial tectorial membrane and reticular lamina vibration amplitudes and phases for two sound frequencies and two levels of damping, in addition, for damaged stereocilia. Further explanation in text. Model results.

placement toward the modiolus, associated with basilar membrane motion or displacement toward scala vestibuli. The diagrams in the left column correspond to the low frequency region, with 200 Hz being a typical example. The diagrams in the right column correspond to the CF region around 2500 Hz. In every diagram, the horizontal vector, labeled RL, indicates the motion or displacement of the reticular lamina toward the spiral ligament. The remaining two vectors are referred to it. One of them, labeled TM, represents the motion/displacement of the tectorial membrane driven by the OHC stereocilia; the other, labeled SM, the difference between them, which is the shear motion/displacement. The diagrams hold for both motion and velocity, provided all vectors are interpreted either as motion or displacement vectors.

Several features of the diagrams are notable. At low sound frequencies (200 Hz), in a healthy cochlea, whether active or passive, the tectorial membrane motion/displacement amplitude is almost the same as that of the reticular lamina, and the resulting shear motion extremely small. Because the tectorial membrane vector is slightly smaller, the shear motion vector is oriented toward the limbus, the inhibitory direction for the OHCs. This means that the OHCs are hyperpolarized during basilar membrane displacement toward scala tympani and depolarized during basilar membrane displacement toward scala vestibuli, as prescribed by the classical shear motion model. It should be noted that the tectorial membrane vector in the upper two diagrams on the left has been drawn as somewhat longer and at a somewhat greater angle relative to the reticular lamina vector than is true. This was necessitated by drawing limitations.

The 2,500 Hz diagram corresponding to a small damping (active cochlea) shows very different relationships than the corresponding 200 Hz diagram. Here, the tectorial membrane vector is longer than the reticular lamina vector, and the shear motion vector is oriented toward the spiral ligament, a direction associated with basilar membrane displacement toward scala tympani. Because the direction is excitatory for the OHCs, it means that their depolarization occurs during basilar membrane displacement toward scala tympani, contrary to the prediction of the classical model. The turnaround of this vector occurs at the location of the series resonance of the tectorial membrane, as is evident in Fig. 6.8. The relationships are changed at 2,500 Hz when the cochlea becomes passive, and the damping is increased. The tectorial membrane vector becomes shorter and its lag angle with respect to the reticular lamina vector increased. Under these conditions, the shear motion vector is rotated by about 90° toward lag, in agreement with the phase functions of Figs. 5.61 and 6.11.

When the stereocilia are damaged, the tectorial membrane becomes driven less strongly, and correspondingly, its vectors in both bottom diagrams of Fig. 6.12 are relatively short. As a result, the shear motion vectors become oriented toward the modiolus. This means hyperpolarization of the OHCs during basilar membrane displacement toward scala tympani, and depolarization, during its displacement toward scala vestibuli at all sound frequencies, in approximate agreement with the classical model.

It must be pointed out that the diagrams of Fig. 6.12 hold for the OHC location of the reticular lamina. At the location of the IHCs, the relationships may be expected to be somewhat different because of radial motion gradients in both the reticular lamina and the tectorial membrane (Zwislocki, 1984). Because these hair cells are located near the spiral osseous lamina, where the motion of the basilar membrane is very small, the corresponding radial motion of the reticular lamina must be decreased also. A similar decrement must occur in the tectorial membrane near its anchor location at the spiral limbus. Because this location is recessed somewhat relative to the margin of the spiral lamina, the decrement may be somewhat smaller, however. As a result, the radial motion amplitude of the tectorial membrane can become larger there than that of the reticular lamina. It should be easy to see from the upper 200 Hz diagram that such an occurrence would reverse the direction of the shear motion vector, so that it would point toward the spiral ligament rather than toward the modiolus. This would mean depolarization of the IHCs during basilar membrane displacement or motion toward scala tympani, while the OHCs are depolarized during its displacement toward scala vestibuli. Such a direction reversal would be unlikely in the presence of stereocilia damage, however, because of the reduced motion amplitude of the tectorial membrane, which becomes substantially smaller than that of the reticular lamina. These conclusions are consistent with past empirical findings (e.g., Sokolich et al., 1976, Schmiedt et al., 1980) in which the low frequency response phase at the level of the auditory nerve fibers was found to be variable in healthy cochleas, but excitation always occurred during basilar membrane motion or displacement toward scala vestibuli in cochleas with organ of Corti damage.

At the end of this section, several crucial insights should be highlighted. First of all, it has been possible to derive dynamical characteristics of the stereocilia-tectorial membrane complex solely on the basis of the known structure and independently measured physical constants of its elements, with one exception—the resistance of the tectorial membrane, which had to be fitted ad hoc. The derived characteristics have been found to agree with the available empirical characteristics. This is especially true for the response phases of the OHCs and Hensen's cells, where the comparison has been possible over wide sound frequency and SPL ranges. In particular, it has been possible to show that the difference between the response phases measured at 40 and 80 dB, respectively, can be fully duplicated by changing the resistance values associated with the stereocilia and the tectorial membrane by equal amounts. The change has been ascribed to the active feedback, which has been assumed not to affect appreciably the associated compliance values. On the other hand, the difference between the phases measured at 80 and 90 dB, respectively, has been accounted for almost entirely by changing the compliance of the OHC stereocilia, in agreement with the empirical findings. A small, experimentally justified change in the basilar membrane compliance has been added. The derived characteristics indicate that, at low sound frequencies relative to the CF, the OHCs are depolarized during basilar membrane displacement toward scala vestibuli, in agreement with the evidence derived from cochlear microphonics (e.g., Dallos, 1973; Zwislocki & Sokolich, 1973; Sokolich et al., 1976) but that this is not true in the broad vicinity of the CF, where, in normal cochleas, the phase is reversed. Accordingly, the often made assumption that the OHCs are depolarized during basilar membrane displacement toward scala vestibuli at all sound frequencies, including the CF, appears to be incorrect. The derived characteristics are also quantitatively consistent with the empirically found difference between neural and basilar membrane sensitivities at low sound frequencies. Furthermore, they account for the paradoxical phenomenon of an increased neural sensitivity at low sound frequencies in cochleas with damaged OHCs but preserved IHCs, in which the sensitivity at the CF is decreased.

DYNAMICS OF THE COCHLEAR PARTITION

In this section, the dynamics of a unit section of the cochlear partition is analyzed with the purpose of determining the characteristics of interaction between the stereocilia-tectorial membrane complex and the organ of Corti-basilar membrane complex. The elements involved and their geometric relationships are shown in the upper panel of Fig. 6.13 (Zwislocki, 1990). The simplified contours of the organ of Corti-basilar membrane complex, as well as those of the tectorial membrane, should be easily recognizable. The four short lines drawn between the top surface of the organ of Corti and the bottom surface of the tectorial membrane indicate the four rows of stereocilia bundles, the one on the left belonging to the IHCs, the three on the right, to the OHCs. The IHC stereocilia are omitted in the analysis because, according to Strelioff et al. (1985), their mechanical effect is negligible. Reissner's membrane is also omitted for the same reason. For reasons stated in chapter 4, the motion of the basilar membrane under the organ of Corti is approximated in the cross section by that of a stiff beam. The resulting motion, V_B, of the reticular lamina at the location of the sec-

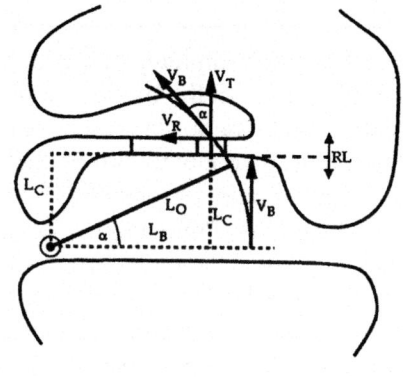
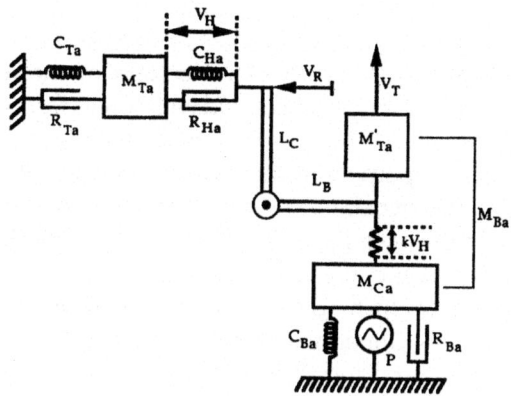
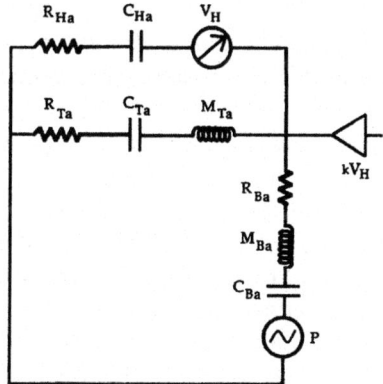

FIG. 6.13. Structural network analysis of the cochlear partition including the basilar membrane, the organ of Corti and the tectorial membrane. Schematic representation (top), mechanical network (middle), its electrical analog (bottom). Explanation in text.

311

ond row of the OHC must be approximately perpendicular to the radius, L_O, connecting the point of rotation of the basilar membrane, assumed to be located at the tip of the spiral osseous lamina, and the location of the second row. Its magnitude must be equal to that of the basilar membrane motion at a distance L_O from the point of rotation. The motion vector, V_B, at the reticular lamina can be decomposed into two components, one, V_T, perpendicular to the surface of the reticular lamina, a direction referred to as *transversal*, and one, V_R, parallel to the surface, a direction called *radial*. It should be clear from the drawing that the reticular lamina moves toward the modiolus, parallel to vector V_R, when the basilar membrane moves upward, in the direction of scala vestibuli. It moves toward the spiral ligament when the basilar membrane is being displaced downward, toward scala tympani. Note that the origin of the vectors has been moved from the reticular lamina to the undersurface of the tectorial membrane for clarity of the drawing. The velocities are associated with forces through appropriate mechanical impedances. The forces interact with each other through torques determined by perpendicular distances between the points of their application and the point of basilar membrane rotation, the velocity V_T being associated with the lever L_B, and the velocity V_R, with the lever L_C. The levers are indicated by the intermittent lines.

The middle panel of Fig. 6.13 shows the mechanical network representing the structures of the uppermost panel (Zwislocki, 1990). From the bottom, the spring, C_{Ba}, and the dash pot, R_{Ba}, symbolize the viscoelastic properties of the basilar membrane, and M_{Ca} stands for the mass of the organ of Corti, including that of the basilar membrane. The source, P, symbolizes the sound pressure difference across the cochlear partition. The short spring above the mass M_{Ca} stands for the motility of the OHCs. It is assumed that the velocity of the alternating contractions and dilations of the OHCs is directly proportional to the velocity of the stereocilia deflections, V_H, the proportionality constant being denoted by k. It can be complex and change with the magnitude of the deflections. It is also assumed that the OHC vibration constitutes a high impedance source justified by empirical measurements (Tolomeo & Holley, 1997) indicating that the longitudinal stiffness of the OHCs and other organ of Corti structures is substantially greater than the lateral stiffness of the stereocilia and the tectorial membrane attachment to the spiral limbus. The mass M'_{Ta} stands for the mass of the tectorial membrane associated with the transversal motion of the membrane. Because of the relatively high impedance of the source, kV_H, and of the longitudinal rigidity of the stereocilia, this mass, together with the mass M_{Ca} can be considered as one mass, M_{Ba}, loading the basilar membrane.

In the network, the transversal motion of the reticular lamina generates the radial motion of the stereocilia via the lever pair, L_B and L_C. The associated viscoelastic properties of the stereocilia bundles and surrounding fluid are symbolized by the spring C_{Ha} and the dash pot, R_{Ha}. The stereocilia move with the velocity V_H. At low sound frequencies, where the mass effects are negligible, they are deflected toward the spiral ligament, the excitatory direction, when the basilar membrane moves toward scala vestibuli and the reticular lamina, toward the modiolus. This deflection is represented by compression of spring, C_{Ha}. During the motion of the basilar membrane toward scala tympani, and of the reticular lamina toward the spiral ligament, they are deflected toward the modiolus, the inhibitory direction. This is reflected in the network model by a decompression of spring, C_{Ha}. As the spring is compressed, the tectorial membrane mass, M_{Ta}, is moved toward the spiral limbus, and the spring, C_{Ta}, standing together with the dash pot R_{Ta} for the viscoelastic attachment of the tectorial membrane to the spiral limbus, is compressed also.

The bottom panel of Fig. 6.13 shows the electrical analog of the first kind, the impedance analog, of the mechanical network of the middle panel. The network should be self-explanatory, except for the lack of a transformer that would correspond to the levers L_B and L_C and the feedback arrangement. The omission of the transformer has been made possible by the fact that, in a real cochlea, the levers have about equal lengths, and their ratio is equal to unity. The top drawing of the figure is not quite accurate in this respect. Concerning the feedback, the meter designated, V_H, measures the current analog of the velocity of the stereocilia deflection, and the measured quantity is fed back to the network as a current multiplied by the constant, k, that has the same properties as in the mechanical network above. The derived current is fed back at a nod corresponding to the reticular lamina. Of course, the change of the direction of motion so prominent in the mechanical network is meaningless in its electrical analog. As a result, the analog cannot simulate the nonlinear distortions discussed in connection with the vibrator model and illustrated in Fig. 6.5.

The dynamics of the partition have been analyzed for the same cochlear location as for the stereocilia-tectorial membrane complex (63% of the cochlear length from the cochlear base), and the same numerical constants have been used, except for the additional constants referring to the basilar membrane and the organ of Corti. According to the determinations of chapter 5, a basilar membrane acoustic compliance of $C_{Ba} = 5.8*10^{-10}$ cm^4/dyne has been accepted. The organ of Corti mass has been estimated at $M_{Ca} = 1.6$ g/cm^3 and that of the tectorial membrane, at $M_{Ta} = 0.55$ g/cm^3, producing a total mass load of the basilar membrane of $M_{Ba} = 2.15$ g/cm^3 in the transversal direction. Because of the added mass of the basilar membrane itself and of the adhering fluid, the numerical value of the total mass associated with the basilar membrane has been rounded off to 3.0 g/cm^3. To obtain the passive resistance value of the Mongolian gerbil basilar membrane, the resistance value obtained for the human cochlea in chapter 4 has been extrapolated according to the relationships listed in chapter 5. The numerical value of $R_{Ba} = 5.0*10^4$ dyne sec/cm^4 has resulted.

The characteristic pattern of the transversal waves in the cochlea is controlled to a large extent by the acoustic impedance of the cochlear partition. This impedance for the 63% cochlear location, as obtained with the help of the electrical analog of Fig. 6.13, is shown in Fig. 6.14 in terms of its resistance and reactance components. A small damping value, corresponding to 1/25 of the passive resistance values of the stereocilia and the tectorial membrane, R_{Ha} and R_{Ta}, respectively, has been used. The impedance has been calculated for the complete partition section (solid lines) and in the absence of the stereocilia-tectorial membrane complex (intermittent lines) to highlight the effect of the complex. Both the resistance component appearing at the bottom of the graph and the reactance component have been plotted on scales with equal units to show their quantitative relationships. Without the stereocilia-tectorial membrane complex, the partition

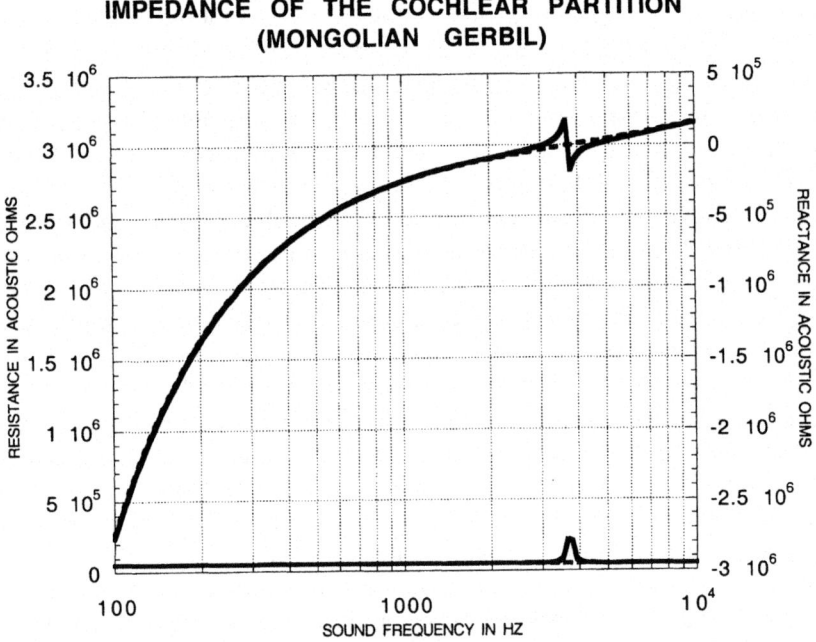

FIG. 6.14. Resistance (bottom) and reactance (top) components of the impedance of the Mongolian gerbil cochlear partition as functions of sound frequency, when the stereocilia-tectorial membrane complex is included (solid lines) and eliminated (intermittent lines).

impedance is that of a series circuit consisting of a capacitance, inductance and resistance. The resistance component is independent of sound frequency, and the reactance component increases from strongly negative values at the low frequencies to positive ones at the high ones, crossing the zero coordinate around 4 kHz, the approximate resonance frequency. Addition of the stereocilia-tectorial membrane complex has a negligible effect at the low and high frequencies but a substantial one in the neighborhood of the resonance frequency of the basilar membrane–organ of Corti complex, where its reactance is small. The relatively large magnitude of the effect is due to an approximate coincidence of two resonances—the series resonance of the basilar membrane–organ of Corti complex, which minimizes its impedance value, and the parallel resonance of the stereocilia-tectorial membrane complex, which maximizes its respective impedance value. The effect consists of a resistance peak and two reactance peaks, one positive and one negative. The latter, in combination with the reactance of the basilar membrane–organ of Corti complex, produce three zero crossings, two associated with response maxima and one with a relative response minimum, as shown in Fig. 6.15. It should be emphasized that the approximate coincidence of the two resonances in the network model of the partition has been obtained with independently determined numerical parameter values of the two complexes and is very likely to represent the actual empirical conditions.

The response magnitudes of the cochlear partition plotted in Fig. 6.15 as functions of sound frequency correspond to a sound pressure difference of 1 dyne/cm^2 across the partition and the same small damping as used for the impedance characteristics of Fig. 6.14. The solid line refers to the complete partition section, the intermittent one, to one without the stereocilia-tectorial membrane complex. Note the powerful effect of the latter. A simple maximum is replaced by a pair of maxima separated from each other by a deep minimum and a frequency difference of almost 2 kHz. The peaks are considerably sharper than the simple peak. This effect is preserved in a modified form when a full cochlea is considered rather than a unit element of its length. As is shown in the next section, the modification consists of a strong reduction of the second, high frequency peak.

The location of the dip and the peaks in the solid curve of Fig. 6.15 depends critically on the stereocilia compliance, C_{Ha}, as is shown in Fig. 6.16. Here, the solid curve corresponds to a compliance, $C_{Ha} = 0.8*10^{-9}$ cm^4/dyne, the intermittent one, to a compliance twice as large, and the dotted one to a compliance four times as large. The damping has been maintained at one fifth the passive value. Clearly, the peaks and the dip are shifted toward the lower frequencies as the compliance is increased. This is particularly true for the higher frequency peak which also becomes sub-

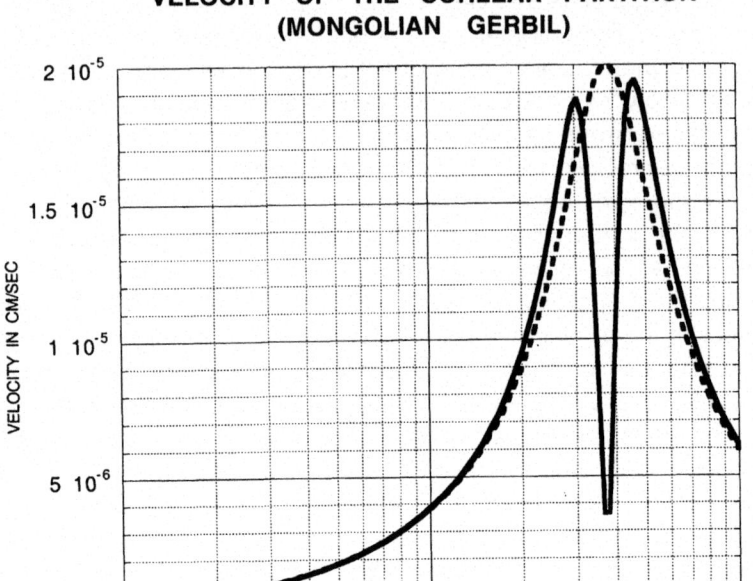

FIG. 6.15. Oscillation velocity magnitude of the cochlear partition (basilar membrane) when the stereocilia-tectorial membrane complex is included (solid line) and when it is not (intermittent line).

stantially larger. The effect of damping is shown in Fig. 6.17, in which the curve for the $C_{Ha} = 0.8*10^{-9}$ cm^4/dyne compliance and one fifth passive damping has been reproduced from Fig. 6.16. The intermittent curve in the same figure shows what happens when the damping is increased to the passive value. All the extrema are reduced, but their location remains unchanged. It is shown in the next section that these characteristics are maintained when the cochlea is treated as a whole.

Phase characteristics corresponding in part to the magnitude characteristics of Figs. 6.16 and 6.17 are shown in Fig. 6.18. More specifically, both the solid and intermittent curves hold for the compliance, $C_{Ha} = 0.8*10^{-9}$ cm^4/dyne but, whereas the former is associated with one fifth of the passive damping value, the latter is associated with passive damping. For the dotted curve, the compliance has been increased by a factor of three to $C_{Ha} = 2.4*10^{-9}$ cm^4/dyne in the presence of passive damping. According to the preceding section, especially Fig. 6.11, the parameter values chosen corre-

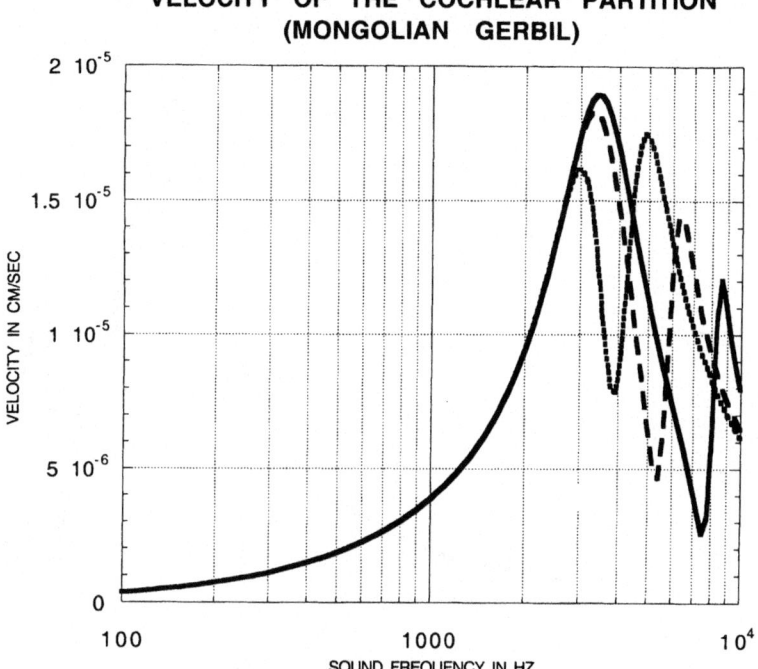

FIG. 6.16. Oscillation velocity of the cochlear partition for three values of the stereocilia compliance. The greater the compliance the larger is the secondary maximum.

spond to changes occurring in the stereocilia-tectorial membrane complex, as SPL is increased from about 40 to 90 dB. The pattern of the phase curves of Fig. 6.18 is quite significant in that it is reminiscent of some phase patterns determined empirically in basilar membrane vibration. In particular, it was found that the phase lag, which grows with sound frequency, is enhanced below the frequency of maximum vibration, the best frequency (BF), as SPL is increased, but is decreased above it. The balance between the two phenomena seems to vary. As shown in the preceding chapter, sometimes, they are of equal size (Ruggero, Rich, Recio, Narayan, & Robles, 1997), sometimes, the enhancement is much smaller than the decrement (Russell & Nilsen, 1997).

A comment is due with respect to the complicated phase pattern of Fig. 6.18 at high sound frequencies. This pattern corresponds to the double peaked magnitude patterns of Figs. 6.16 and 6.17. Empirically, it is difficult to study because the amplitude of basilar membrane vibration becomes

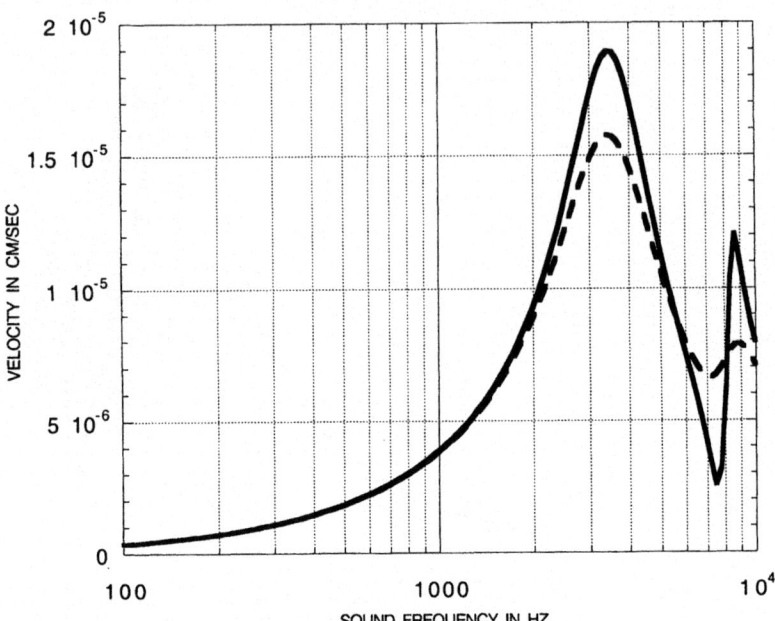

FIG. 6.17. Oscillation velocity of the cochlear partition for two damping values. Solid curve, the same as in Fig. 6.16; intermittent curve—increased damping.

very small in the vicinity of the second peak as a result of energy losses in the process of wave propagation. The relationships are especially evident in Fig. 5.35 of the preceding chapter.

This book targets the human ear. However, because of the unavailability of various key physiological data for the human cochlea, animal models have to be used to determine some of its properties indirectly. For this reason, much of the preceding chapter and the first part of this one are concerned with animal cochleas. Here, the transition is made from the Mongolian gerbil ear to the human. This transition is possible because of the availability of key anatomical data for the human cochlea, as described in chapter 4, and the inter-animal cochlear similarities discussed in chapter 5. Especially, it is shown there that cochleas of all mammals used as animal models for the human cochlea have practically the same wave velocity that is described by the same function of cochlear location when the location is expressed in percent of total cochlear length. By inference, it has been con-

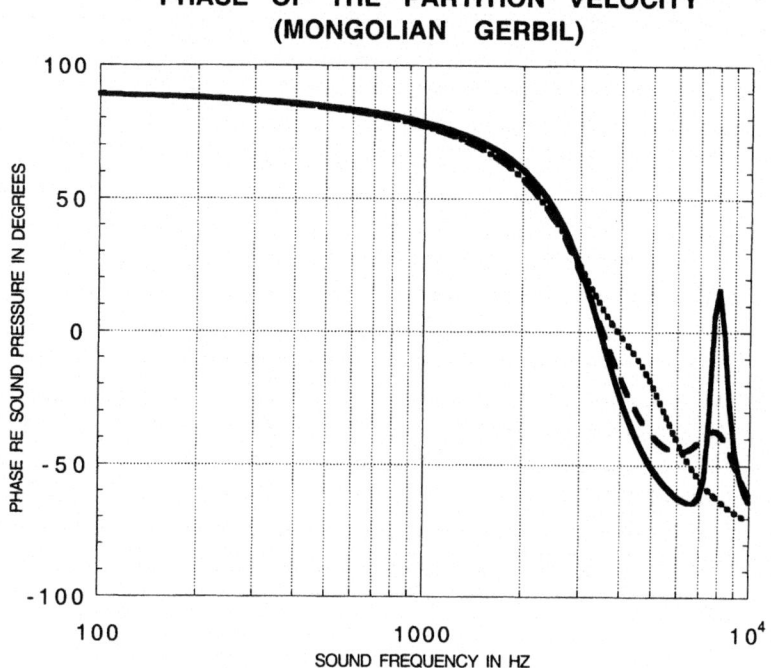

FIG. 6.18. Phases of the oscillation velocity of the cochlear partition for two damping values, as in Fig. 6.17 (solid and intermittent curves), and for an increased stereocilia compliance (dotted curve).

cluded that the wave velocity in the human cochlea is also approximately the same. This velocity is about twice as large as inferred from Békésy's measurements (1960, for review) performed on postmortem preparations. About the same ratio between the postmortem and in vivo cochlear wave velocities has been obtained empirically for guinea pigs. Relationships between numerical parameter values of the human and Mongolian gerbil cochleas established on the basis of the above considerations are listed in chapter 5. Here, some of the human ones, necessary for the analysis of cochlear partition dynamics are repeated for convenience of the reader with the use of the same nomenclature as for the gerbil. They hold for the same proportional cochlear location of 63% of the cochlear length. $C_{Ba} = 5.89*10^{-9}$ cm^4/dyne; $C_{Ha} = 3.5*10^{-8}$ cm^4/dyne; $C_{Ta} = 7*10^{-7}$ cm^4/dyne; $M_{Ba} = 0.185$ g/cm^3; $M_{Ta} = 0.06$ g/cm^3; $R_{Ba} = 5.5*10^3$ dyne sec/cm^4; $R_{Ha} = 700$ dyne sec/cm^4; $R_{Ta} = 350$ dyne sec/cm^4. For the same reasons as for the gerbil, the effective acoustic mass of the basilar membrane has been increased by about 35% to 0.25 g/cm³ for further calculations.

With these numerical values, and the electrical network model of Fig. 6.13, structurally the same as for the gerbil cochlea, it has been possible to calculate the input impedance of the unit section of the human cochlear partition at the 63% location. To emphasize the extrema, the calculation has been performed for a small damping obtain by reducing the resistances, R_{Ha} and R_{Ta} to 1/25 of their passive values, just as for the gerbil cochlea. The calculated impedance is shown in Fig. 6.19 in terms of its resistance and reactance components. The frequency pattern of the components is very similar to that of the Mongolian gerbil cochlea, shown in Fig. 6.14. The resistance is flat across the frequency range from 0.1 to 10 kHz, except for a narrow peak between 3 and 4 kHz. In the same region, the reactance exhibits two extrema, one positive at around 2.9 kHz and one negative at a somewhat higher frequency. The relative sizes of all the extrema are about the same as for the gerbil cochlea. However, the absolute values of the impedance components are approximately 10 times smaller than in Mongolian gerbil. This is noteworthy because the velocity of wave propagation, which depends critically on the reactance of the co-

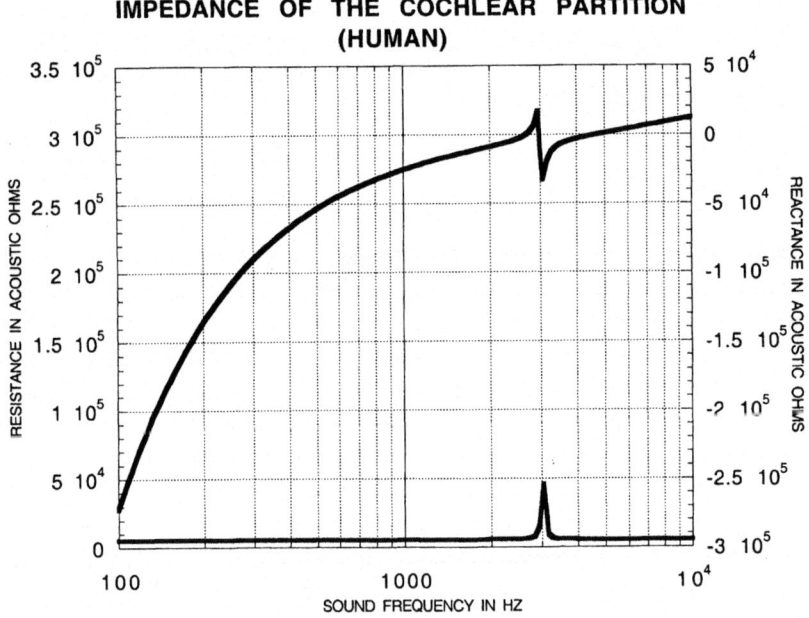

FIG. 6.19. Resistance (bottom) and reactance (top) components of the cochlear partition with human parameter values and a resonance of the stereocilia-tectorial membrane complex at about 3,000 Hz.

chlear partition, appears to be practically the same in humans and gerbils. The explanation is that the velocity also depends on the cross sectional area of the cochlea, which is about 10 times smaller in Mongolian gerbil. The smaller partition reactance in humans compensates for this difference.

The calculated magnitude and phase characteristics of the human cochlear partition are shown in Fig. 6.20 for R_{Ha} and R_{Ta} values 1/5 of the passive ones. The similarity of the magnitude pattern to that for the gerbil cochlea obtained under similar conditions and shown in Fig. 6.16 by the dotted curve should be evident. Note, however, that the magnitudes are approximately 10 times larger, corresponding to smaller impedance values. This may invite the speculation that the human auditory sensitivity is greater than that of Mongolian gerbil. However, psychophysical experiments belie such a conclusion. The auditory system of the latter must have some compensatory mechanisms.

The similarity of the magnitude response patterns of the human and Mongolian gerbil cochlear partitions must also hold for the corresponding phase patterns. As a consequence, a separate illustration of the phase pattern obtained for the gerbil cochlea under corresponding conditions has

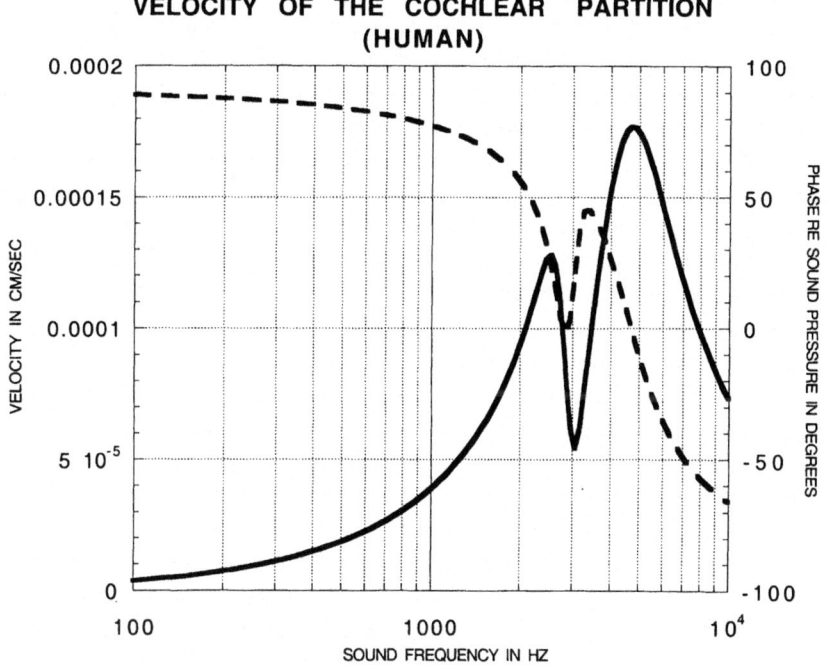

FIG. 6.20. Magnitude and phase of the cochlear partition oscillation velocity with the same reactance parameters as in Fig. 6.19.

been omitted. The one of Fig. 6.20 can be regarded as representing both the human and gerbil patterns.

At low sound frequencies, the phase of the partition velocity nears a 90° lead that decreases as the frequency is increased and reaches zero near the location of the magnitude dip. As the frequency is further increased, it rebounds to produce a relative maximum and decreases past it to asymptote on −90°. As shown in Fig. 6.18, the relative maximum becomes flatter as the damping is increased and nears a plateau. A corresponding phase plateau was first observed by Rhode (1971) in live squirrel monkey cochleas above the frequency of the maximum of basilar membrane vibration. The phenomenon is considered more directly in the next section.

PASSIVE COCHLEAR DYNAMICS

In this section, the dynamics of the live passive cochlea as a whole is analyzed. The effect of the active feedback is introduced separately in the following section. The two step procedure is chosen to highlight the role of the feedback and to show explicitly the effects of the variable parameter values. It is also true, as emphasized in the two preceding sections, that the cochlea does appear to act as a passive system at SPLs exceeding 70 or 80 dB.

Because of cochlear interspecies similarity, it is fundamentally not important for which animal species the whole-cochlea analysis is performed. For several reasons stated in the preceding section, it is applied here to humans. Significantly, it has been possible to adapt Békésy's empirical data obtained on postmortem human preparations to in vivo conditions , as shown in chapter 5, and to complement them with the help of analogies from the Mongolian gerbil cochlea. As a result, it has become possible for the first time, I believe, to develop an approximate, structurally and quantitatively explicit model of human cochlear dynamics in vivo.

The model is a transmission line model, similar to that developed in chapter 4 for the cochlea postmortem. Although a model of this type lacks precision in the vicinity of the CF because the assumption of long waves compared to the canal depth is not sufficiently well satisfied, it does preserve the fundamental relationships and is the easiest to interpret in terms of physical events. It is shown in chapter 4 that the numerical calculation errors are not great and manifest themselves mainly in a somewhat reduced vibration magnitude and increased wave length. Models that do not make the assumption of long waves show that the propagation velocity of the cochlear waves depends on sound frequency but such a nonlinearity is already introduced by the higher-order impedance of the cochlear partition. It is suggested in chapter 4 how the inaccuracy of the long wave assumption can be corrected by making the effective cross-sectional areas of the cochlear canals on both sides of the cochlear partition dependent on sound frequency. In view of increased mathematical complexity, the correction has not appeared important enough to be incorporated here. It would not have affected materially the fundamental characteristics included in this monograph.

Only the electrical analog of the transmission line model is treated in this section. The electrical transmission line is obtained by coupling the analogs of the incremental sections of the cochlear partition exemplified in Fig. 6.13 by means of inductances representing the incremental volumes of the fluids occupying the cochlear canals. This is illustrated in Fig. 6.21. As for the dead cochlea, only the fluid coupling is considered, the elastic coupling within the basilar membrane being negligible. In Fig. 6.21, the analog transmission line for the live cochlea is compared to that for the dead cochlea treated in chapter 4. It should be clearly apparent that, in the latter, shown on the left side of the figure, the shunt networks are relatively simple and contain only two energy storing elements, the capacitance, C_{Ba}, standing for the compliance of the basilar membrane, and the inductance, M_{Ba}, standing for its mass plus the mass of the cells associated with it. The two storage elements make the network a second order system. The resistance, R_{Ba} accounts for the damping of the partition. The network analog on the right, belonging to the live cochlea, is substantially more complicated and represents a fifth order system because three energy storing elements are added to those contained in the shunt network of the dead cochlea. The added storage elements are: the inductance, M_{Ta}, standing for the mass of the tectorial membrane, the capacitance, C_{Ta}, standing for the compliance of the tectorial membrane attachment to the limbus, and the capacitance, C_{Ha}, standing for the effective compliance of the OHC stereocilia. In addition, the resistances R_{Ta} and R_{Ha} are the analogs of the acoustic resistances associated with the attachment of the tectorial membrane to the limbus and the stereocilia motion, respectively.

Because of the complexity of the transmission line model of the live cochlea, no closed mathematical solution of its performance has been attempted. Instead, the network has been programmed on a computer in terms of finite network sections. The currents and voltages in every section have been obtained by solving the Kirchhof node and loop equations. The process begins at the helicotrema and proceeds backward toward the cochlear base. For this purpose, a starting voltage at the helicotrema and the acoustic impedance of the helicotrema has to be assumed. Because the cochlear waves reach the helicotrema only at very low sound frequencies only a low frequency impedance has to be considered. It has been estimated to be mainly resistive and have a numerical value of about 100 acoustic Ohms. This value has a significant effect only on the most apical

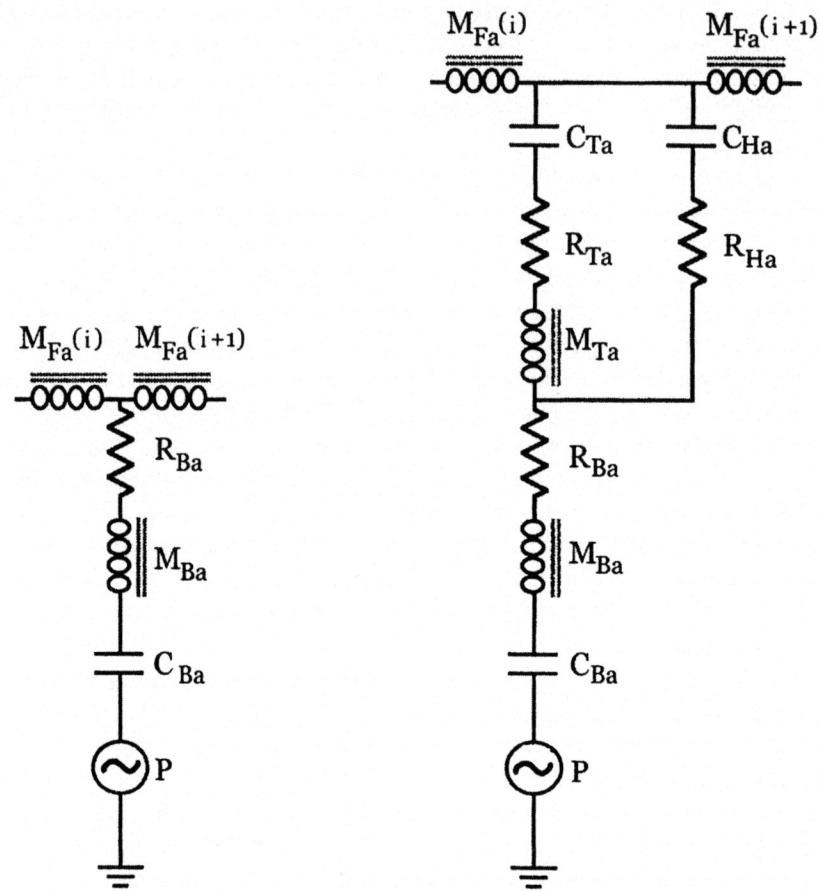

FIG. 6.21. Network analogs of cochlear incremental sections. Left, for the cochlea simplified by death; right, for the cochlea in vivo. Further explanation in text.

portion of the transmission line and decreases rapidly with the distance from the helicotrema. When the computation process reaches the cochlear input, the voltage there is normalized and the corresponding fluid input calculated, which is directly proportional to the displacement of the stapes footplate. In this way, it is possible to calculate all the cochlear characteristics relative to a constant stapes displacement. As has been explained in chapter 4, this stimulus definition has a fundamental effect on the appearance of the characteristics.

The electrical transmission line model of the cochlea used in this monograph consists of N final length sections short enough to avoid significant wave reflections at their junctions. Various numbers were tried successfully, and N = 960 has been chosen on the basis of gained experience. Assuming a cochlear length of 3.5 cm, this number produces n = N/3.5 = 274.3 sections/cm or about 27 sections per millimeter, so that a section amounts to a small fraction of the shortest wave length. The recoding of the cochlear length in terms of the lumped sections means that the cochlear length coordinate is expressed as $x = k/n$, or $x = 3.65*10^{-3}k$ for N = 960. According to the transformation, all the cochlear functions have been expressed in terms of the proportional number of sections. For example, the compliance of the basilar membrane as a function of x has been rewritten as $C_{Ba} = 1.17*10^{-10}*(3.5/N)*exp(1.79*3.5*k/N)$.

The computerized model of the cochlea has had a long history. Its physical principles and theoretical results obtained with it were published in a number of articles (e.g., Zwislocki & Kletsky, 1982; Zwislocki, 1983, 1984, 1986a, 1986b). Its first computer program was written by E. J. Kletsky (Zwislocki & Kletsky, 1981) in Fortran on a PDP 11-34 computer. Subsequently, it was rewritten in C on several Macintosh computers by several programmers, in particular M. Schachter and D. Arpajian. For this monograph, a Power Macintosh G3 has been used, and the program has been revised by Arpajian with helpful suggestions and calculations of Kletsky's. The program provides the following user outputs: magnitude and phase of basilar membrane displacement, magnitude and phase of the radial tectorial membrane displacement, and magnitude and phase of the shear displacement between the tectorial membrane and the reticular lamina, all as functions of cochlear location and sound frequency. It should be noted that, in the electrical analog, voltage is an analog of sound pressure, but current is an analog of mechanical velocity rather than displacement that is mimicked by electrical charge. As a consequence, an additional transformation has had to be included to obtain displacements. In most figures, the stapes displacement is said to serve as magnitude and phase reference. This nomenclature has been introduced for convenience but is not entirely accurate with respect to the magnitude. In fact, the computer program refers to the longitudinal volume displacement of cochlear fluid at the cochlear input, as measured either in scala vestibuli or tympani—both are expected to be nearly the same. This volume displacement is about the same as that of the stapes. However, the point displacement is inversely proportional to the cross sectional area of one or the other scala. Because this area is much larger than the width of the basilar membrane, an effective amplification of the displacement magnitude of the basilar membrane takes place. Since this amplification is difficult to determine in absolute terms, the magnitude unit used in the graphs must be considered as arbitrary. Nevertheless, it is constant in all the graphs, so that the numerical relationships among them should be roughly the same as the true physical relationships.

The first set of the model cochlear transfer functions is shown in Fig 6.22. It concerns the basilar membrane at several cochlear locations expressed in numbers of sections. The right most magnitude curve and the top phase curve belong to a basal most location. The second magnitude curve from the left and the second phase curve from the bottom correspond approximately to the 63% location. Because this location was used in gerbil experiments and offers the best opportunity for comparisons between theoretical and experimental results, it is used predominantly in the analysis. The characteristics of Fig. 6.22 have been obtained with parameter values corresponding to a passive cochlea believed to prevail at SPLs between 70 and 80 dB, which have been specified in chapter 5. This is true in particular for the 63% curves. The magnitude functions acquire larger amplitudes and become narrower toward the model cochlear base. The effect results from the constancy of the resistance values along the cochlea in the presence of an increasing reactance of the cochlear partition. The constancy has been assumed for lack of more specific empirical information. It should be pointed out, nevertheless, that the resulting narrowing of the transfer functions and increasing sensitivity toward the cochlear base are evident in empirical neural tuning curves (e.g., Schmiedt et al., 1980). Except for the artifactual upturn at high sound frequencies, the phase curves appear to be consistent with empirical data obtained on guinea pigs and Mongolian gerbils in terms of CM (e.g., Schmiedt & Zwislocki, 1977). In the region of the upturn, the cochlear wave magnitudes are too small for empirical measurements.

The next set of basilar membrane transfer functions, shown in Fig. 6.23, has been obtained with the resistance value of the basilar membrane reduced by a factor of 3.3 and those of the stereocilia and the tectorial membrane, by a factor of 10. The magnitude functions are plotted on linear as well as log scales to better show their characteristic features. The reader should disregard the oscillation evident in the curves at low sound frequencies. It is artifactual and may be due to wave reflection at the model helicotrema as well as too coarse a representation of the cochlea in terms of finite sections. Of course, the magnitude functions are somewhat narrower and have greater amplitudes than for the passive resistance values, but they also exhibit secondary relative maxima right above the frequencies of their main peaks. The maxima occur about 40 to 50 dB below the peaks, in agreement with empirical data introduced in the preceding chapter. They have the same origin as discussed in the preceding section in connection with isolated cochlear sections but are much smaller. The reason for the difference stems from an energy loss associated with wave propagation along the cochlea, especially at the location of the resistance maximum produced by the parallel

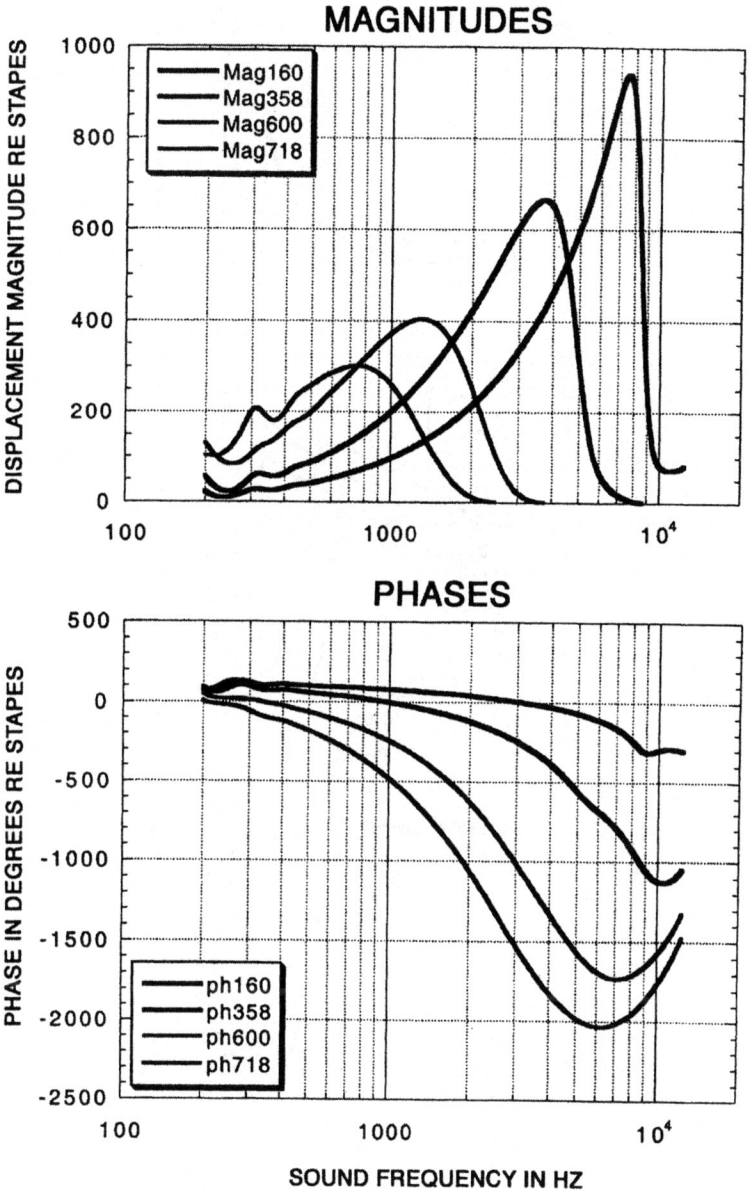

FIG. 6.22. Theoretical magnitudes and phases of human basilar membrane displacements at several cochlear locations with passive resistance values.

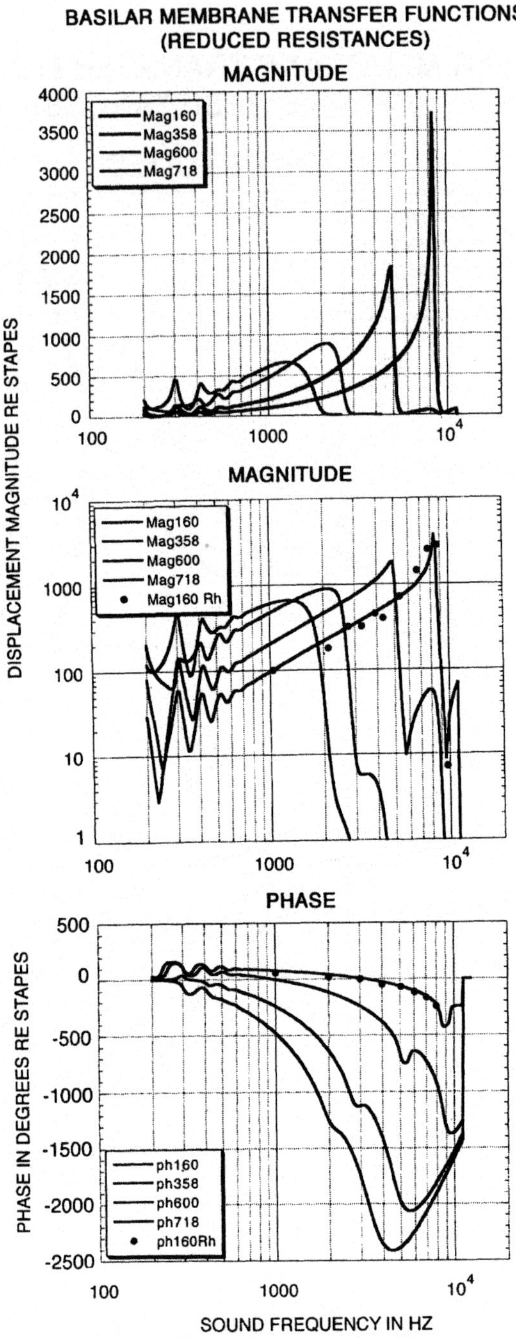

FIG. 6.23. Theoretical and empirical (circles; squirrel monkey) magnitudes and phases of basilar membrane displacements at several cochlear locations. In the middle panel, the magnitudes are plotted on a log scale. Reduced passive resistance values.

resonance in the stereocilia-tectorial membrane complex. As shown in the preceding section, their exact location depends the most directly on the stereocilia compliance. The secondary maxima were observed empirically for the first time by Rhode (1971) in squirrel monkeys. The data points associated with the highest frequency curve in the figure have been reproduced from Fig. 6 of his classical 1971 article. Because the maximum response of his results was almost at the same frequency as that of the theoretical curve, only a minor frequency adjustment was necessary. Rhode measured the tip portion of his curve at an SPL of 70 dB. The cochlear resistance values chosen for Fig. 6.23 should be suitable for this SPL. Although they have not been adjusted specifically to fit Rhode's data, the agreement with them appears to be good. Note in particular the height of the peak relative to the low frequency tail, the frequency of the high frequency notch and its vertical distance from the peak. However, the theoretical secondary peak is much more pronounced than the empirical one which is almost completely flattened. The likely explanation is that Rhode measured this part of the curve at a relatively high SPL of 90 dB where the damping must have been substantially greater than at 70 dB. It has been shown in the preceding section that the secondary peak is sensitive to the damping.

The phase curves corresponding to the magnitude curves in Fig. 6.23 are somewhat steeper than was true in the presence of the passive resistance values and exhibit a notch above the frequency of the magnitude peak. The notch is associated with the magnitude notch preceding the secondary maximum and results from the resistance peak produced by the parallel resonance in the stereocilia-tectorial membrane complex, as already discussed. It is necessary to note that the theoretical phase corresponding to the magnitude data reproduced from Rhode changes with sound frequency considerably more slowly than his. The total theoretical change to the notch, equivalent to his phase plateau, is on the order of 500°, whereas his changes somewhat in excess of 1000°. The difference is not due to a failure of the theory, however. A careful look at his phase data (e.g., Fig. 5) shows that the difference is simply due to a difference in the time delay, which was much greater in the squirrel monkey at the cochlear location of Rhode's measurements than is in the model for a corresponding frequency of the magnitude maximum. The delay can be derived from the slope of the rectilinear portion of the curve relating the phase to sound frequency. In this portion, the velocity of wave propagation is independent of sound frequency—a property of long waves compared to the canal depth, as has been discussed in chapters 4 and 5. In mathematical formulation, the phase for Rhode's measurements at sufficiently low sound frequencies can be expressed simply as $\Phi_R = 2\pi f \Delta_R t$, and for the theory, as $\Phi_Z = 2\pi f \Delta_Z t$, so that $\Phi_Z = (\Delta_Z t/\Delta_R t)\Phi_R$ for equal sound frequencies. The relationship remains valid for the portion of the curve where the wave velocity becomes dependent on sound frequency, provided the dependence is the same in both cases. The filled circles hugging the highest theoretical phase curve in the bottom panel of Fig. 6.23 have been obtained from Rhode's data (Fig. 5) by multiplying them by a factor of 0.22. Clearly, they follow closely its course, suggesting that they are controlled by the mechanism specified by the theory, which includes the stereocilia-tectorial membrane resonance. It is of some interest that the theory is limited to long waves. Apparently, the effect of relatively short waves is weak. This conclusion is confirmed below for the shear displacement between the tectorial membrane and the reticular lamina on a different animal species.

Transfer functions for the shear displacement, considered in the model as equivalent to the deflection of the OHC stereocilia, are shown in Figs. 6.24 and 6.25 for the same locations as for the basilar membrane in Figs. 6.22 and 6.23. The curves of Fig. 6.24 have been generated with passive resistance values, those of Fig. 6.25, with the basilar membrane resistance value reduced by a factor of 3.3 and the stereocilia and tectorial membrane resistance ones, by a factor of 10. The general patterns of the curves are similar to those of the basilar membrane. However, there are some significant differences. The magnitude curves for shear displacement have greater amplitudes, are narrower and more symmetrical. This is in agreement with the empirical OHC and Hensen's cells transfer functions of the preceding chapter. They change with the cochlear location at a greater rate. For example, the most basalward peak in the plot belonging to reduced resistances is about 10 times higher than the apical most peak. This ratio corresponds to a decibel difference of 20. Although this difference may appear large, a corresponding difference in neural tuning curves obtained on Mongolian gerbils was found to be even larger and reach over 30 dB (Schmiedt et al., 1980). It is likely that the greater difference resulted from even smaller resistance values prevailing near the neural firing thresholds. The plot showing the magnitude functions belonging to reduced resistances on a log scale shown in the middle panel of Fig. 6.25 reveals two additional differences with the corresponding basilar membrane functions. First, there appears a notch about two octaves below the frequency of the main peak. This notch is produced by the series resonance of the tectorial membrane, as already discussed and suggested previously by Allen (1980b). Second, the curves show plateau-like perturbations above the frequencies of their peaks. These perturbations do not exist in isolated cochlear sections and must originate in interactions among adjacent sections. As is shown further in this chapter, the perturbations can be demonstrated empirically and are associated with the stereocilia coupling between the reticular lamina and the tectorial membrane.

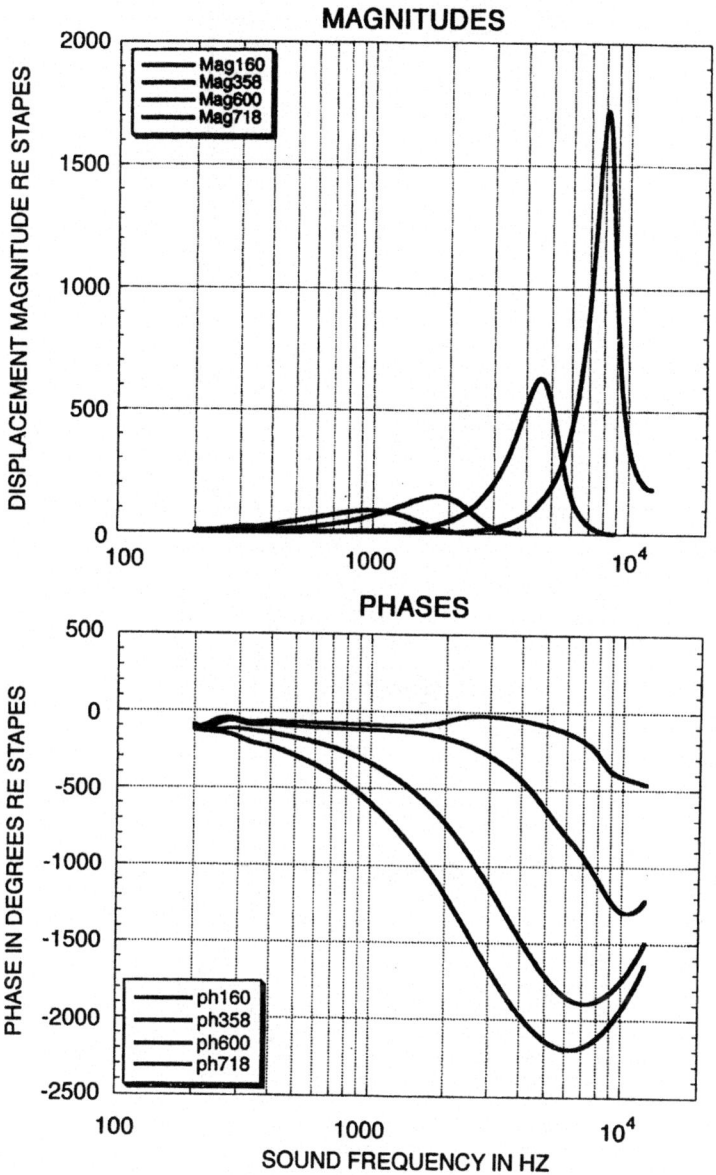

FIG. 6.24. Theoretical magnitudes and phases of shear displacements at several cochlear locations with human parameter values. Passive resistance values.

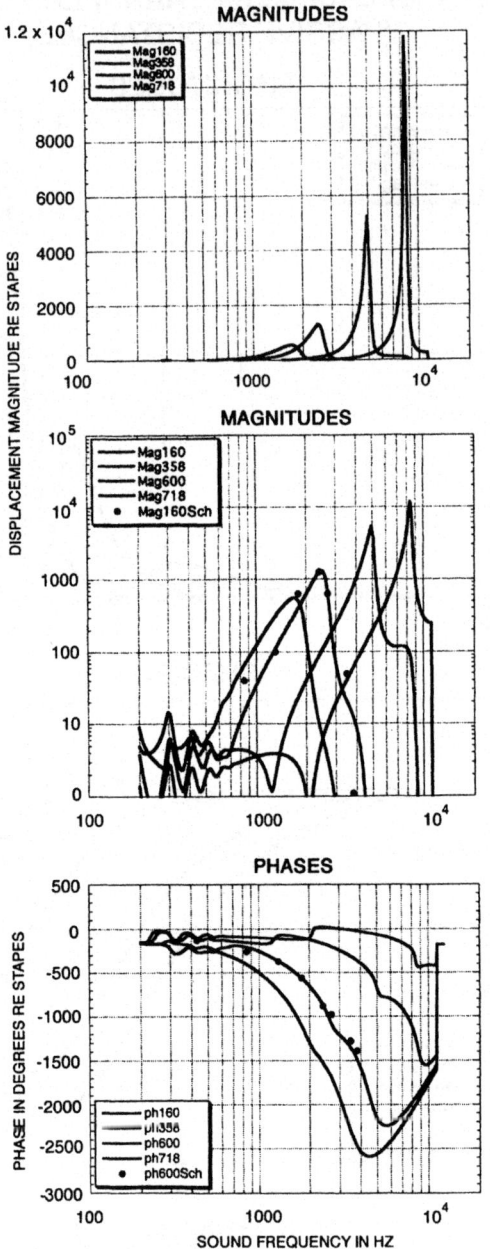

FIG. 6.25. Theoretical magnitudes and phases of shear displacements at several cochlear locations with human parameter values. Passive resistance values reduced. Corresponding CM responses are marked by circles (Mongolian gerbil).

To gain an impression of how well the magnitude transfer functions of Fig. 6.25 represent empirical data, a typical sample of such data has been indicated by the closed circles. The sample has been derived from CM tuning curves obtain on a Mongolian gerbil, using a criterion CM magnitude of 10 µV (Schmiedt & Zwislocki, 1977, Fig. 16, open triangles). It is assumed that the CM is directly proportional to shear motion, except at very high SPLs. The CF of the tuning curve was almost the same as the frequency of the maximum response of the second theoretical transfer function from the left, so that only a small frequency adjustment was required to obtain optimum coincidence. Although the CM data were somewhat noisy, a reasonable agreement between the theoretical and the empirical data should be apparent. If the empirical data give the impression of a slightly greater bandwidth than that of the theoretical ones, this should not be surprising because of the well known spread of the electrical field produced by the CM along the cochlea (e.g., Dallos, 1973)

The phase curves corresponding to the magnitude curves differ little from the basilar membrane phase curves when the resistance values are at the passive level, as evident in Figs. 6.22 and 6.24. However, when the resistance values are reduced, as in Figs. 6.23 and 6.25, the shear displacement curves exhibit perturbations not present in the basilar membrane curves. These perturbations amount to 180° phase jumps. As already discussed, they are produced by the resonances in the stereocilia-tectorial membrane complex and reverse the phase of hair cell stimulation. The phase reversal in its theoretical as well as empirical aspects is discussed extensively above in the section on the stereocilia-tectorial membrane complex and also further in this chapter.

Again, a sample of empirical phase data has been introduced in the bottom panel of Fig. 6.25 by means of closed circles to test the adequacy of the theory. The sample has been derived from the same CM recording as the magnitude points of the middle panel (Schmiedt & Zwislocki, 1977, Fig. 16, open triangles). Only a small correction for the apparent difference in the delay time was necessary—the absolute values of the empirical data have been increased by about 10%. The theoretical curve agrees with the empirical data over the whole frequency extent, including the high frequency nonlinear portion. The agreement confirms the conclusion reached above for the basilar membrane that the theory of this monograph, which includes the parallel resonance of the stereocilia-tectorial membrane complex, adequately accounts for the cochlear phase data.

The dependence of the basilar membrane and shear transfer functions on the numerical values of the basilar membrane, stereocilia and tectorial membrane resistances is discussed more extensively for the 600th section of the cochlear model at about the 63% of the cochlear length. The basilar membrane magnitude transfer functions are plotted in Fig. 6.26. The basilar membrane resistance in the upper panel amounts to 0.2 of its passive value. It increases by a factor of two in the middle panel and to the passive value in the bottom one. The resistances of the stereocilia and the tectorial membrane increase from 0.04 through 0.2, to 1.0 of their passive values. Note that all the resistances have similar effects. As their numerical values decrease the response amplitudes increase and their maxima move to higher sound frequencies. The frequency shift associated with increasing response sensitivity is consistent with empirical data discussed in the preceding chapter and demonstrated particularly convincingly by Ruggero's group (Ruggero et al., 1997).

The shear displacement transfer functions for the 63% location are plotted in Fig. 6.27. They show a similar but not identical dependence on the resistance values. The sensitivity to the resistances of the stereocilia and the tectorial membrane is increased compared to that for the basilar membrane resistance, and the curves are more symmetrical. A noteworthy feature is that they tend to converge at the feet of their high frequency skirts, a phenomenon that was observed empirically on many occasions and may be significant from the point of view of pitch coding, as discussed extensively in the last chapter (Zwislocki, 1991; Zwislocki & Nguyen, 1999).

The theoretical shear displacement transfer function based on passive resistance values and shown in the bottom panel of Fig 6.27 provides another chance for theory verification. For this purpose, empirical values derived from an intensity series of Hensen's cells transfer functions have been plotted in the panel by means of closed circles (Szymko, Zwislocki, & Hertig, 1997). They are samples of the curve determined at the 70 dB SPL at which the effect of the active feedback must have been week or entirely absent. Because the peak of the curve was at a somewhat lower sound frequency than the peak of the theoretical curve, the empirical points have been shifted accordingly by a constant log frequency distance. The empirical points suggest a curve shape quite similar to that of the corresponding theoretical curve, in particular, with respect to the slight asymmetry relative to the sound frequency of the peak. However, the empirical points suggest a somewhat broader bandwidth. The difference could be due to several factors concerning numerical values of the parameters involved, among them, too small a theoretical damping. In view of the agreement of the damping with several other characteristics, this seems unlikely, however. It is more likely that the difference stems from a difference between the sound frequencies of the peaks. As already mentioned, the peak of the empirical curve was located at a sound frequency almost one octave lower than is the peak of the theoretical curve. It is evident from Figs. 6.23 and 6.24 that the bandwidth increases as the frequency of the peak decreases.

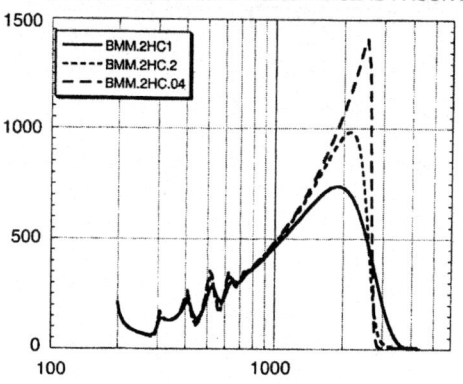
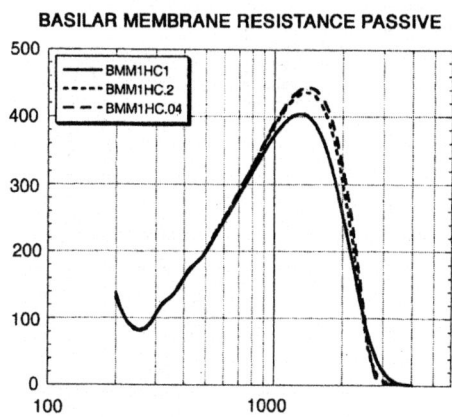

FIG. 6.26. Theoretical magnitudes of basilar membrane displacements at one cochlear location for several sets of resistance values.

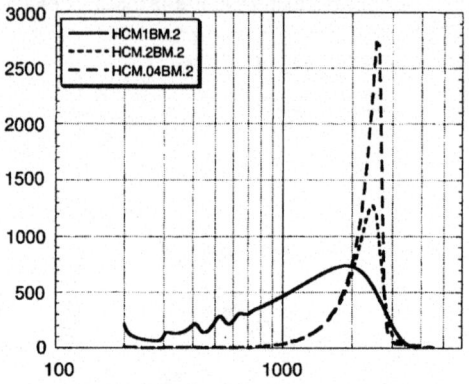
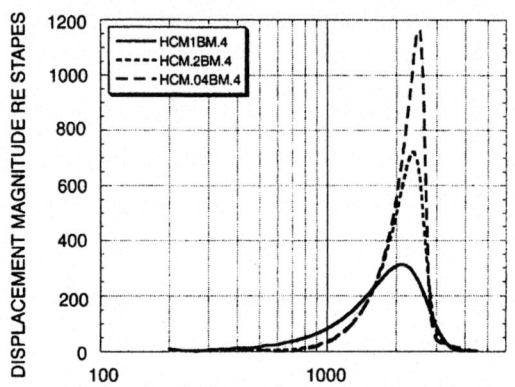
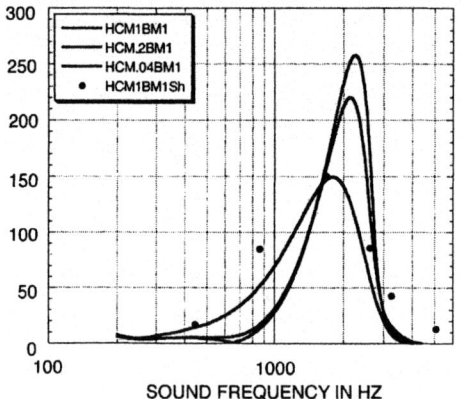

FIG. 6.27. Theoretical magnitudes of shear displacements at one cochlear location for several sets of resistance values. Circles indicate experimental results obtained at a lower frequency and shifted according to the log scale.

Changes in basilar membrane response phases as functions of SPL have been firmly established empirically and are discussed in the preceding chapter. Their outstanding features are an increasing phase lag associated with the cochlear wave propagation in the frequency region preceding the amplitude maximum and a decreasing phase lag in the frequency region following it, both occurring as SPL is increased. Geisler and Rhode (1982) suggested that, according to the theory of minimum phase filters, the phenomenon is associated with cochlear bandpass characteristics. The sharper is the response maximum, the greater a phase perturbation should occur. The following Fig. 6.28 demonstrates that the relationship holds for the cochlear model at hand. It shows for two cochlear locations, one at the 63% distance from the base (lower panel) and one closer to it, a slightly increased phase lag ahead of the frequency of the amplitude maximum and a more substantially decreased one beyond it when the resistances of the basilar membrane and of the stereocilia-tectorial membrane complex are increased. The phase changes are asymmetrical, presumably, because the magnitude transfer functions are asymmetrical, being much shallower at the lower frequencies than at the higher ones. The increasing resistances are assumed to be associated with increasing SPL in the live cochlea. In the figure, the intermittent curves correspond to the passive resistance values assumed to prevail at SPLs around 70 to 80 dB, the solid curves, to resistance values 5 times smaller, assumed to prevail at lower SPLs. It should be pointed that the substantial phase lead beyond the frequency of the magnitude maximum results to a large extent from the parallel resonance in the stereocilia-tectorial membrane complex, which produces a magnitude notch and a following secondary maximum in the basilar membrane transfer functions. When the notch is absent, the relative phase lead practically disappears, as was found in particular by Rhode and Reccio (2000) in chinchilla preparations.

The phase of shear displacement between the tectorial membrane and the reticular lamina depends on SPL in a very different way than does the phase of basilar membrane displacement, as is shown toward the end of the preceding chapter and in the section of this chapter dealing with the stereocilia-tectorial membrane complex. Here, the model dependence is described in connection with the cochlea as a whole and compared to empirical data. It is illustrated in Fig. 6.29. The same relationships are displayed in both panels of the figure but are shown over a somewhat more extended frequency range in the upper panel and compared to the empirical data of Fig. 5.61 in the lower panel. For the comparison, the data, obtained on a Mongolian gerbil, have been adjusted slightly to account for the difference in time delays between the gerbil and human cochleae and for additional time delays mentioned in the section on the stereocilia-tectorial

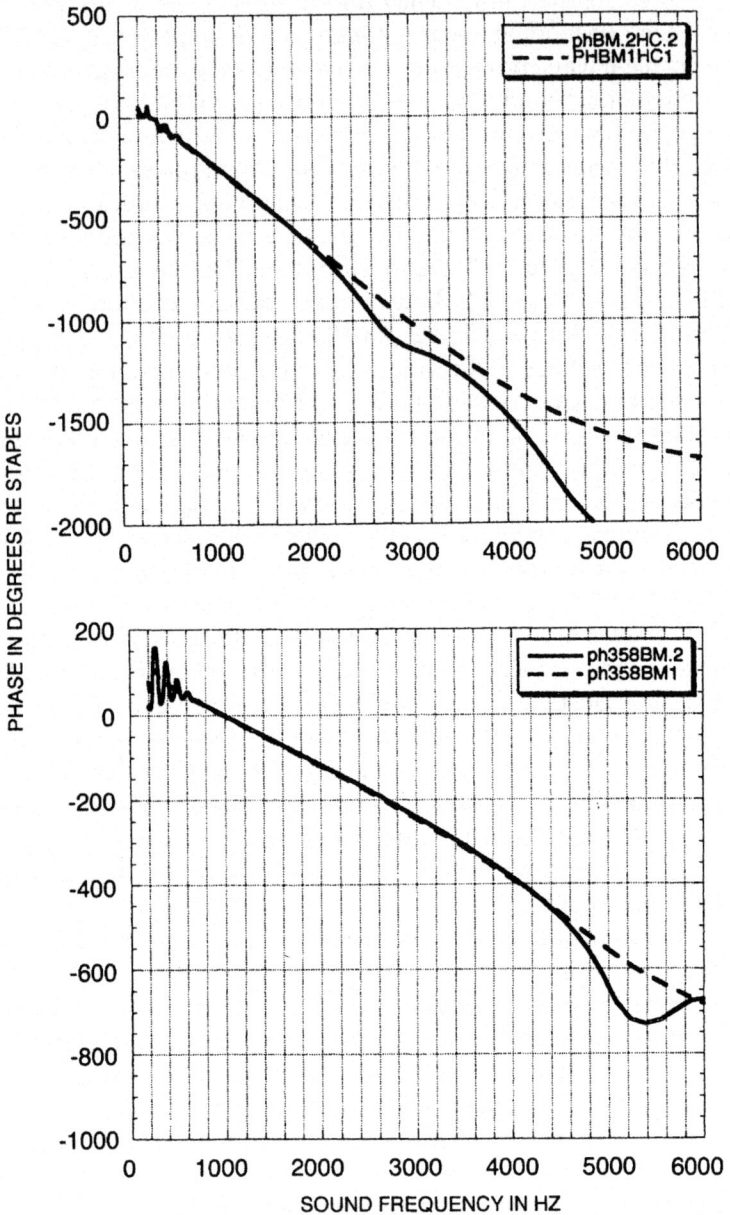

FIG. 6.28. Theoretical phases of basilar membrane displacements at two cochlear locations for passive (intermittent lines) and reduced resistance values.

membrane system. It should be noted that the empirical magnitude is very small beyond 4000 Hz, so that the theoretically derived phase can not be verified there with certainty. The three curves of each panel correspond to three sets of resistance values of the basilar membrane and of the stereocilia-tectorial membrane complex, each set believed to be associated with a different SPL. The dotted curve corresponds to the passive resistance values associated with an SPL of about 80 dB, the solid curve, to resistance values reduced for the stereocilia and the tectorial membrane to 0.04 of their passive values. According to the section on the stereocilia-tectorial membrane complex, they correspond to a SPL of about 40 dB. The dashed curve obeys the passive resistance values and, in addition, an increased effective compliance of the stereocilia by a factor of 3 and that of the basilar membrane by a factor of 1.3. These changes have been inferred from experimental data for high SPLs, on the order of 90 to 100 dB, as discussed in the section on the stereocilia-tectorial membrane complex. Due to the increased compliance values, the curve follows a somewhat steeper course than the 80 dB curve, signifying a reduced velocity of wave propagation and a faster phase change. Such a course is confirmed by the experimental data marked by solid squares in the lower panel, which may be regarded as typical. The curve associated with the passive resistance values but unchanged compliance values differs from the high SPL curve only by a reduced slope. Its course agrees roughly with the empirical data marked by the solid triangles and obtained at a SPL of 80 dB. The most complicated course is followed by the curve associated with the reduced resistance values. Over most of it, the curve is displaced by a phase lead of about 90° relative to the 80 dB curve and by about 180° relative to the 90 to 100 dB curve. The phase difference is confirmed by the experimental data marked by closed circles in the upper panel and obtained at a 40 dB SPL. Slightly above 500 Hz, the curve shows a phase jump of about 180°. Its nature has been explained in the section on the stereocilia-tectorial membrane complex and coincides with the series resonance of the tectorial membrane. As already explained, the phase jump signifies that, whereas the OHC are depolarized during basilar membrane displacement toward scala vestibuli at low sound frequencies, they are depolarized during its displacement toward scala tympani at the higher ones. Another feature of the theoretical low SPL curve worth noting is a shallow notch just below 3000Hz. It reflects the notch in basilar membrane phase occurring in the same frequency region and shown in Figs 6.23 and 6.28. The notch would produce a phase lead of the 80 dB curve relative to the 40 dB curve over its extent, similar to that found in the basilar membrane phase, were it not for the extended 90° difference between the two curves. This perturbation was found to be too small to be detected with certainty empirically.

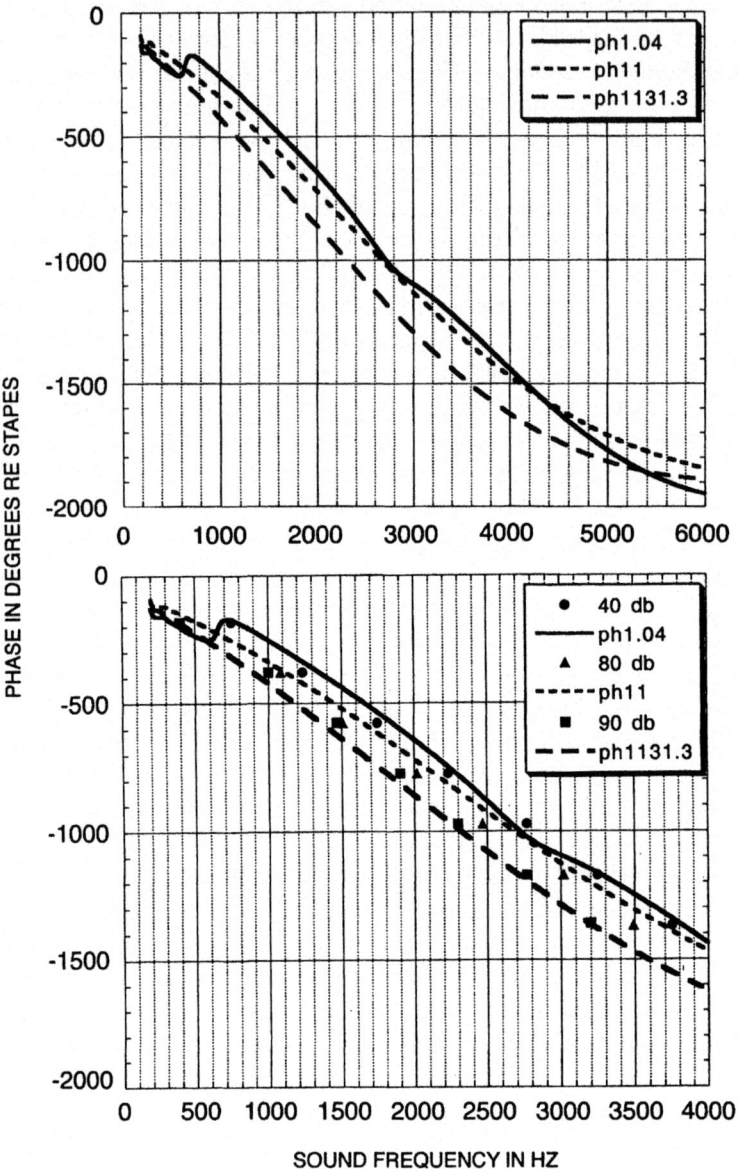

FIG. 6.29. Theoretical phases of shear displacement at one cochlear location for several values of resistance elements, believed to correspond to selected SPLs. In the lower panel, the theoretical curves are compared to measured values (Mongolian gerbil).

The magnitude and phase transfer functions can be shown in combination by taking the sin of the phase as a function of sound frequency and modulating it according to the magnitude function. The result is a wave pattern with variable amplitude. Such empirical patterns obtained for the alternating response potentials of a Hensen's cell at two SPL levels—40 and 90 dB have been shown in the preceding chapter. They are reproduced here in the upper panel of Fig. 6.30 for comparison with corresponding theoretical patterns shown in the lower panel. The latter have been obtained with the same parameter values as the solid and dashed phase curves of Fig. 6.29. These values have been found in the section on the stereocilia-tectorial membrane complex to best fit the empirical data obtained at the 40 and 90 dB SPLs, respectively. Except for some experimental artifacts, the two sets of patterns appear almost identical. For this to occur, both the amplitude and phase patterns have to be simultaneously similar. More specifically, this has to be true for the bandwidths and, more generally, the shapes of the magnitude curves and the SPL dependent peak shift, more over, for the positions and the wavelengths of the frequency domain waves. Note in particular the flat portions of the 40 dB curves at the low frequencies. According to the theory first introduced in the section on the stereocilia-tectorial membrane complex, the flat portion arises because the radial displacement amplitude of the tectorial membrane is almost as large as that of the reticular lamina, so that the shear displacement is minimized. In the same region, the series resonance of the tectorial membrane takes place and produces a phase reversal in the shear displacement. The empirical and theoretical patterns differ slightly in that the former is shifted somewhat toward lower frequencies and truncated below 0.25 kHz due to an equipment filter. Also, it contains a direct current component noticeable at the high frequency end. Finally, the phase shift between the low and high SPL curves is slightly greater in the theoretical than in the empirical curves. This is due to the specific choice of the numerical values of the theoretical parameters. The difference could be eliminated by choosing a slightly smaller multiplier for the basilar membrane compliance.

Up to this point, basilar membrane and shear displacement transfer functions have been examined separately in relation to the cochlear damping controlled by the basilar membrane, stereocilia and tectorial membrane resistances. But what is the relationship between the two kinds of transfer functions?—a question that has been discussed for many years. In the 1970s, there seemed to be little doubt that the frequency tuning was sharper at the level of the auditory nerve fibers than at that of the basilar membrane, and E. F. Evans (1977) even proposed a second filter interposed between the basilar membrane and the nerve fibers. Subsequent

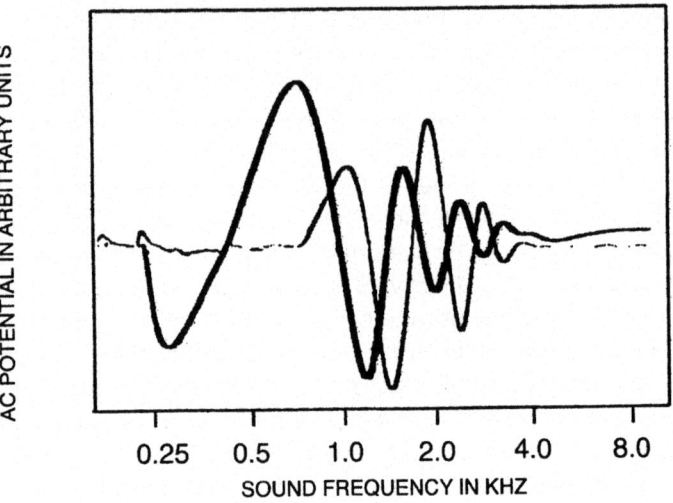

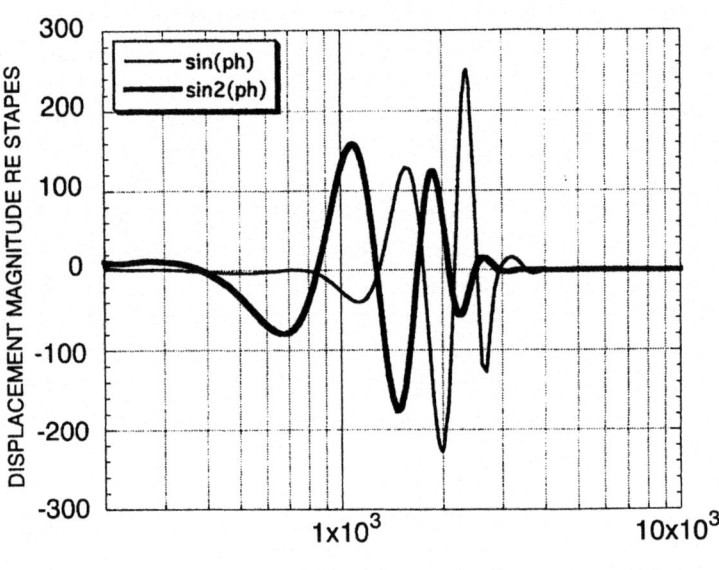

FIG. 6.30. Upper panel: sin functions of frequency sweeps at two SPLs (amplitudes adjusted) obtained on a Mongolian gerbil. Lower panel: corresponding theoretical functions.

measurements with improved methods tended to indicate sharper basilar membrane tuning, and the existence of the second filter begun to be questioned. A definitive decision was difficult to make because of the technical problems in obtaining corresponding data on the same animal specimen and referred to the same cochlear location. When this became possible in the end, a substantial difference between the basilar membrane and the neural tuning became clearly apparent (Narayan et al., 1998). The difference was produced by a relatively decreased neural sensitivity at low sound frequencies. A mechanism for such a decreased sensitivity, inherent in the model described in this monograph, is presented in the section on the stereocilia-tectorial membrane complex. Here, the difference itself between the basilar membrane and shear displacement tuning according to the model is shown. It is assumed that the neural response follows roughly the shear motion. As a typical example, two transfer functions, one for the basilar membrane and one for the shear displacement, both obtained at the same cochlear location, are displayed in Fig. 6.31. The curves are associated with basilar membrane, stereocilia and tectorial membrane resistances reduced by a factor of 5 from their passive values. The magnitudes of the shear motion have been multiplied by 0.77 to obtain a match with the basilar membrane curve at the maximum response. Note that, in agreement with the empirical results of Narayan et al., the shear displacement tuning is substantially sharper than the basilar membrane one, and that the shear displacement response is suppressed at the lowest frequencies by over 20 dB. Of course, the oscillation evident in both curves in this frequency region is artifactual, as has been mentioned already in connection with similar perturbations. Note that the peaks of the two curves do not coincide exactly but that the shear displacement peak is shifted slightly toward the higher frequencies. This is not artifactual but results from different mechanisms underlying the two peaks. The basilar membrane peak results from an interaction of the effective mass of the stereocilia-tectorial membrane complex with the effective compliance of the basilar membrane at a frequency somewhat below that of the parallel resonance of the complex. The shear displacement peak coincides more nearly with this resonance frequency. It should be pointed out that the numerical relationships between the two curves depend on the numerical values of the system's parameters, and that the values selected for Fig. 6.31 should correspond to SPLs between 40 and 60 dB.

A phenomenon of substantial theoretical significance is the secondary peak occurring somewhat above the frequency of the principal peak of the basilar membrane transfer functions. According to the literature, the exact location and size of this peak depend on the state of the preparation. In healthy preparations, the secondary peak appears to be well below the primary one and is often difficult to detect (e.g., Robles, Ruggero, & Rich, 1986). In others, it can be quite prominent (Wilson & Evans, 1983). It is shown here that its size and location depend on the effective compliance of the OHC stereocilia. An example is illustrated in Fig. 6.32 for small damping values of the system—0.2 of the passive value of the basilar membrane resistance and 0.04 of the passive value of the stereocilia and tectorial membrane resistances, which enhance the effects. The solid curve shows the basilar membrane transfer function for a stereocilia compliance considered as normal. For the dotted curve, the compliance has been increased by a factor of two and, for the dashed one by that of four. It is evident that, as the compliance increases, the secondary peak moves toward lower sound frequencies, and its magnitude increases. For the normal conditions, it is more than 40 dB below the principal peak. With the stereocilia compliance increased four times, it becomes practically equal to that of the main peak. As the secondary peak is shifted on the frequency axis and increased, the main peak is also shifted in the same direction, but its magnitude is decreased.

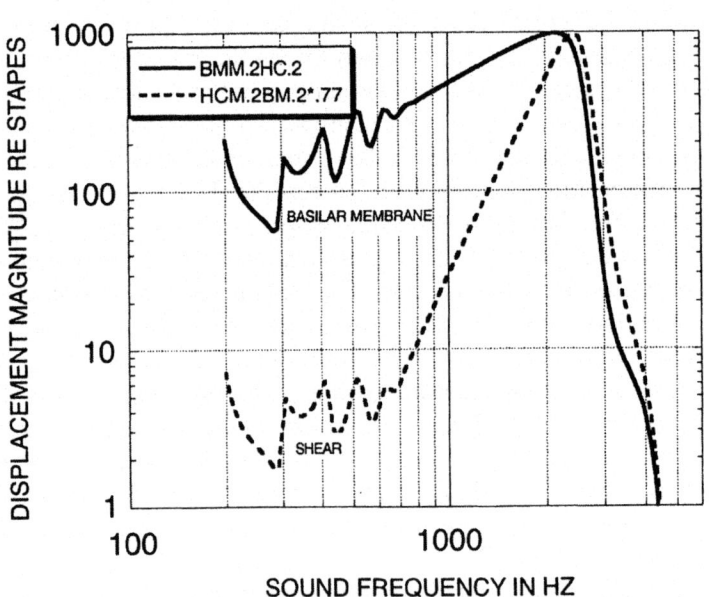

FIG. 6.31. Theoretical basilar membrane (solid curve) and shear displacement magnitudes at one cochlear location. Shear displacement magnitudes adjusted to equality with the basilar membrane magnitudes at maximum response. Passive resistance values reduced.

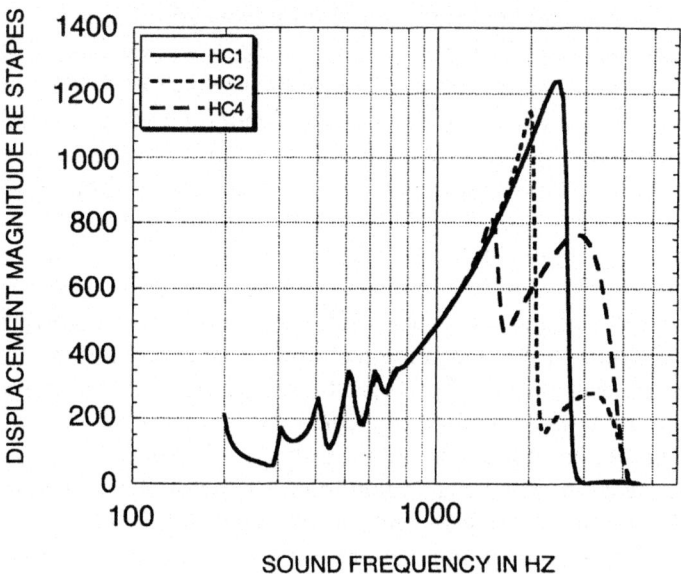

FIG. 6.32. Theoretical basilar membrane displacement magnitudes at one cochlear location for three values of the stereocilia compliance. The magnitude of the secondary maximum increases with increasing compliance. Passive resistance values reduced.

The characteristics of the secondary peak are difficult to study on the basilar membrane because of technical difficulties, but a related phenomenon can be observed in auditory nerve fibers and cochlear microphonics. For example, Liberman (1976), Schmiedt (1977) and Schmiedt et al. (1980) encountered double peaked tuning curves in their recordings of firing rates in single auditory nerve fibers. The secondary peak characteristics were also studied systematically by myself at the level of cochlear microphonics known to be produced by the OHCs (Zwislocki, 1986a, 1986b). The experiments were performed by mechanically loosening the coupling between the hair cell stereocilia and the tectorial membrane. A recording micropipette was introduced into scala media through a small opening in its lateral wall, in the same way as for recording hair cell potentials. However, the micropipette had a larger tip diameter, on the order of 5 µm, and was oriented at a somewhat different angle, aiming for the tectorial membrane. It was slowly advanced through scala media until it encounter the tectorial membrane. When this happened, the CM potential measured by the same micropipette and a second micropipette held stationary in the endolymphatic space dropped by a small but clearly detectable amount. The first micropipette was then advanced by an additional distance on the order of 50 µm and was moved back and forth parallel to the tectorial membrane surface and perpendicularly to it. At the same time sinusoidal sound with a logarithmically swept frequency was delivered to the cochlea to generate transfer functions, as in the experiments with hair cells and supporting cells. After sufficient tectorial membrane manipulation, the shape of the CM transfer function began to change. The change was progressive and increased with the duration of the manipulation. A typical example of the process is illustrated in Fig. 6.33. In the top panel, the thin line shows the CM transfer function just before the insertion of the micropipette into the tectorial membrane, and the thick line, just after its insertion. The small notch near the peak of each curve is artifactual and persists throughout the series of the transfer functions in the figure. The second panel from the top shows what happened to the transfer function after initial manipulation of the tectorial membrane. It should be evident that a small secondary peak was added at its high frequency skirt. With further manipulation, the peak grew and became as large as the main peak. This can be seen in the third panel which contains two curves. The thick one corresponds to the micropipette being lodged in the tectorial membrane, the thin one was obtained after its withdrawal. Finally, the bottom panel shows what happened after the micropipette had been reinserted into the tectorial membrane, moved back and forth some more and withdrawn. The secondary peak became still larger and overwhelmed the main peak.

The next figure, Fig. 6.34, demonstrates that the patterns of Fig. 6.33 can be duplicated in the model of this monograph by varying the effective compliance of the OHC stereocilia and leaving all the other parameters invariant. The top panel shows the shear displacement transfer function, assumed to be equivalent to the CM functions in the preceding figure, under conditions regarded to be normal. Note the low magnitude plateau at the foot of the high frequency skirt. The same plateau is more clearly apparent in the transfer functions of Fig. 6.25 plotted on a log magnitude scale. As already mentioned, this plateau is absent in isolated cochlear sections and must be due to interaction among the sections. Nevertheless, it is associated with the secondary peak of the basilar membrane transfer functions and depends on the effective compliance of the OHC stereocilia, as is clearly demonstrated in this and preceding figures. For the second panel from the top of Fig. 6.34, the stereocilia compliance was increased by a factor of 4, for the third, by a factor of 12 and for the bottom one, by a factor of

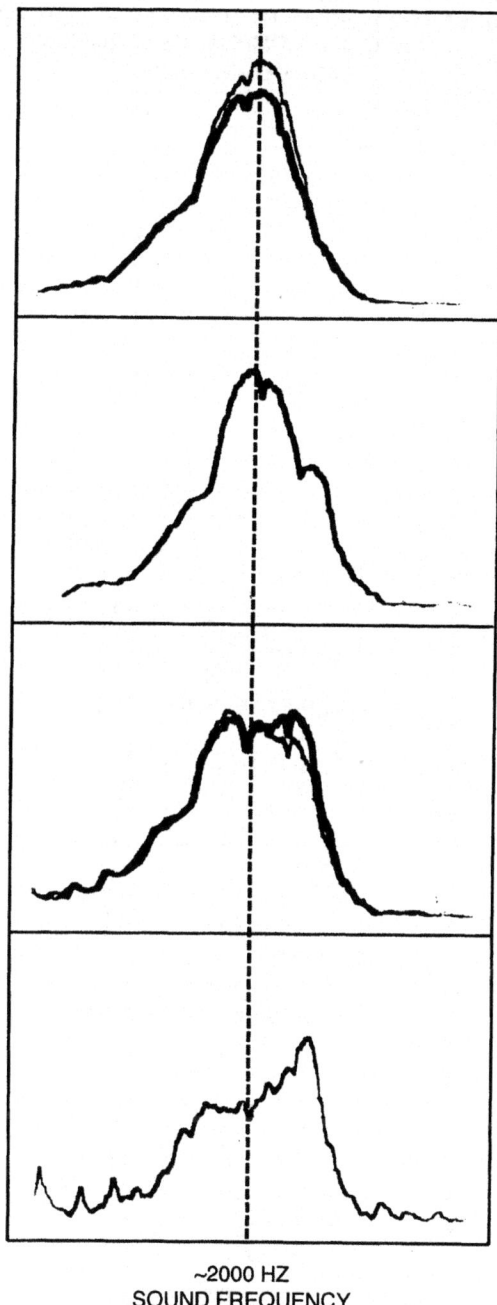

~2000 HZ
SOUND FREQUENCY

FIG. 6.33. Empirical CM responses measured on the spiral ligament of a Mongolian gerbil cochlea as functions of sound frequency for several stages of mechanical tectorial membrane manipulation.

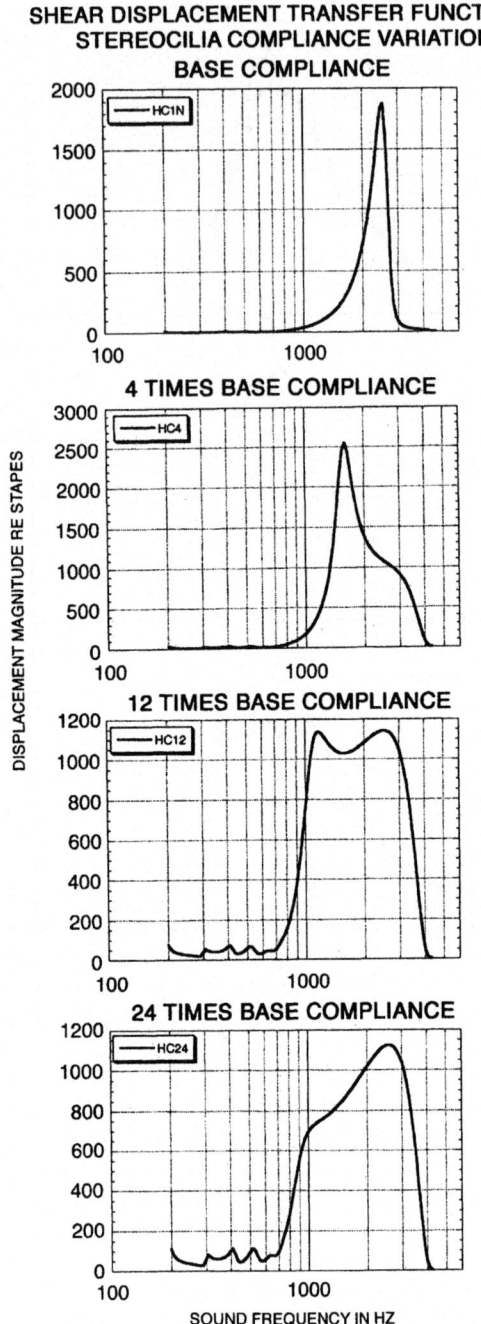

FIG. 6.34. Theoretical shear displacement magnitudes as functions of sound frequency for several values of stereocilia compliance, increasing from top to bottom.

24. Clearly, the resulting theoretical patterns mimic accurately the empirical patterns of the preceding figure. By doing so, they strengthen the validity of the model. One could go further and maintain that they indicate by how much the effective compliance in the empirical situation had to be increased to produce a given perturbation in the CM transfer function. Both the empirical results of Fig. 6.33 and the theoretical results of Fig. 6.34 are consistent with the results on the auditory nerve mentioned above.

On the occasion of the tectorial membrane manipulation, an additional rather surprising phenomenon was observed. After initial manipulation, producing a small secondary peak, the magnitude of the main peak was increased by about 20%, sometimes more, as can be seen in the second panel from the top of Fig. 6.33. One could have expected that decreasing the elastic coupling between the tectorial membrane and the reticular lamina would produce an opposite effect. However, increasing moderately the stereocilia compliance in the model produced a similar response enhancement, evident in the second panel from the top of Fig. 6.34. Of course, continued manipulation, leading to more severe perturbations of the transfer functions reversed the effect and produced magnitude decrements. The same reversal was obtained in the model on further increasing the stereocilia compliance. Thus, maximum OHC response appears to occur at a somewhat greater effective compliance of the OHC stereocilia than encountered under normal conditions. This suggests that the compliance is not tuned naturally for a maximum OHC output. The phenomenon of the response enhancement was studied by us systematically by exposing Mongolian gerbil cochleas to broad band sound at various SPLs and for various durations and measuring the response potentials of Hensen's cells. The response enhancement was clearly confirmed for moderate exposures at SPLs between 80 and 90 dB (Pietras & Zwislocki, 1993; Szymko et al., 1997)

Thus far, the analysis of shear displacement between the tectorial membrane and the reticular lamina has been applied tacitly to the location of the OHCs at which the displacement is maximum and where the tectorial membrane is driven radially by the OHC stereocilia. However, determination of the shear motion at the location of the IHCs is relevant for the stimulation of the auditory nerve fibers, most of which end there. The model of this monograph makes it possible to differentiate between the shear displacement at this location and the one at the OHC location (Zwislocki, 1994). It should be clear from the anatomical structures involved that the vibration amplitude of the basilar membrane must decrease toward the place of its attachment to the spiral osseous lamina and with it the radial vibration amplitude of the reticular lamina. For a similar reason, the radial vibration amplitude of the tectorial membrane must also decrease toward its anchor at the spiral limbus. If both amplitude decrements in the reticular lamina and the tectorial membrane are the same, the shear displacement between the two structures conserves its phase in agreement with the vector diagrams of Fig. 6.12. But, if the decrements are not the same, a phase rotation up to almost 180° may take place. This is particularly true at low sound frequencies where the amplitude of the radial tectorial membrane displacement at the location of the OHCs is almost as large as that of the reticular lamina. Under these conditions, a somewhat smaller amplitude decrement in the tectorial membrane may change the relationship between the two displacements to the point where the displacement of the tectorial membrane becomes the larger. As a consequence, the stereocilia of the IHCs would be deflected in the opposite direction from those of the OHCs. If, for example, the latter were deflected toward the modiolus, the inhibitory direction, the former would be deflected toward the spiral ligament, the excitatory direction. An example of the resulting IHC transfer functions is illustrated in Fig. 6.35 for resistance values assumed to be associated with a SPL of about 40 dB. In the upper panel are shown two magnitude transfer functions for the 63% location of the model cochlea. The solid curve has been obtained with equal amplitude decrements in the reticular lamina and the tectorial membrane and is in fact the same as for the OHCs. For the intermittent curve, the decrement in the tectorial membrane has been made 5% smaller than in the reticular lamina. This small difference changed the transfer function quite noticeably over its entire low frequency range up to the response maximum. In particular, in the low frequency tail, the magnitude was increased at some frequencies and decreased at others. The relative minimum at the location of the tectorial membrane series resonance around 700 Hz has disappeared. The most dramatic effects can be seen in the phase transfer functions in the lower panel of the figure. Here, three curves have been drawn—the lowest one, for equal amplitude decrements in the tectorial membrane and the reticular lamina, the middle one, for a decrement in the tectorial membrane decreased by 3%, and the upper one, for a decrement decreased by 5% and corresponding to the intermittent magnitude curve of the upper panel. The lowest curve is the same as for the OHCs and contains a phase jump of almost 180° at the location of the tectorial membrane series resonance. Note that the decrement reduction by only 3% has moved the phase jump substantially to a lower frequency, and that of 5% has eliminated it entirely. As a result, the response phase of the IHCs has been changed by almost 180°. The diagram shows in connection with the vector diagrams of Fig. 6.12 that, for a constant response phase of the OHCs, the response phase of the IHCs can assume any angle within the range of 180°. The critical dependence of the IHC model response phase on the fine tuning of the decrements of the radial displacement amplitudes in the reticular lamina and the tectorial mem-

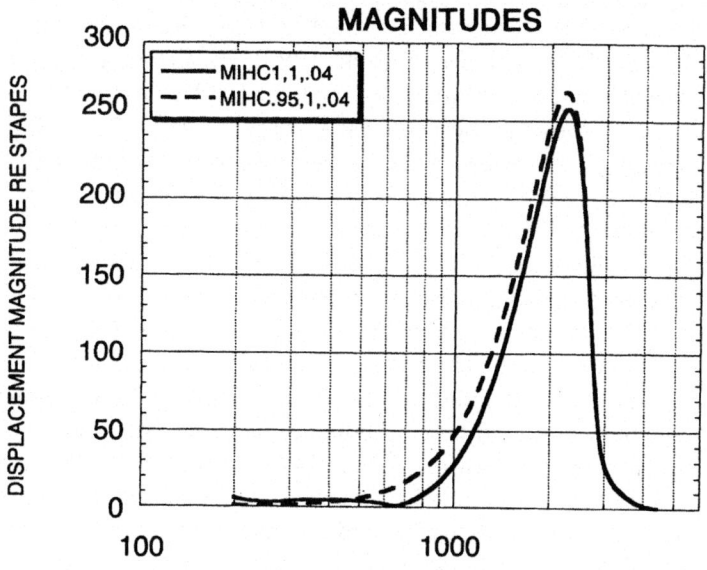
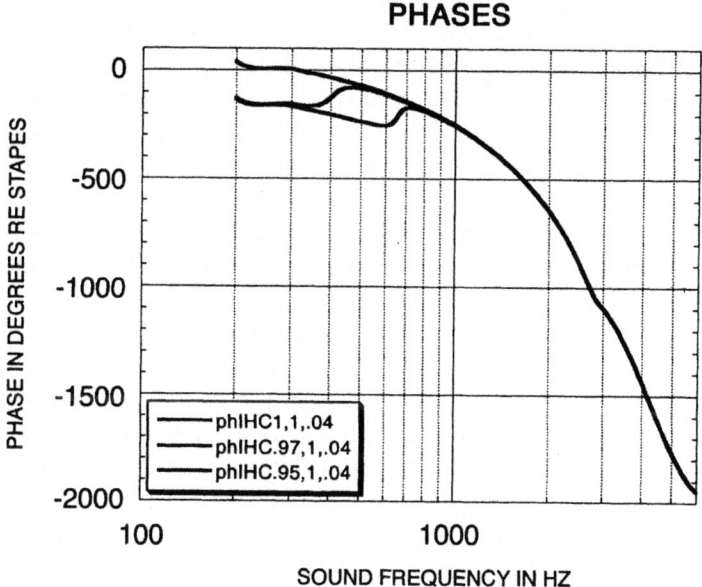

FIG. 6.35. Theoretical magnitudes and phases of shear displacements at the location of the IHCs for several radial displacement gradients in the tectorial membrane. The dashed line in the upper panel belongs to a reduced tectorial membrane gradient.

brane fully accounts for the range of response phases found empirically in auditory nerve fibers innervating the IHCs. (e.g., Sokolich et al., 1976; Schmiedt et al., 1980; Ruggero & Rich, 1983). It also can account for the finding that, at low sound frequencies, the IHCs tend to be depolarized during basilar membrane motion toward scala vestibuli while the OHC are depolarized during its displacement in the same direction (Dallos et al., 1972; Sellick & Russell, 1980; Nuttal, Brown, Masta, & Lawrence, 1981). The difference was explained in the literature by the anatomical observations that the stereocilia of the OHC are attached to the tectorial membrane and follow its relative displacements, whereas the stereocilia of the IHCs are free standing and are driven by fricative forces within the endolymph (Billone & Raynor, 1973). The hypothesis was never verified beyond reasonable doubt. It is made unnecessary by the displacements gradients in the tectorial membrane and the reticular lamina suggested here.

The theoretical mechanism of the shear displacement between the tectorial membrane and the reticular lamina together with the amplitude gradients, as described above, provide a resolution of the paradox observed empirically according to which a sensitivity enhancement can take place in the low frequency tails of neural tuning curves in damaged cochleas in which the population of the OHCs is depleted in the presence of a well preserved population of the IHCs (e.g., Liberman, 1976; Liberman & Kiang, 1978) According to the model of this monograph, depletion of OHCs increases the effective compliance of the stereocilia coupling between the reticular lamina and the tectorial membrane (Zwislocki, 1984). The effects on the corresponding magnitude and phase transfer functions are illustrated in Fig. 6.36. They have been plotted for the resistance values of the basilar membrane and the stereocilia-tectorial membrane complex assumed to prevail at SPLs around 40 dB and for a 5% smaller amplitude decrement in the tectorial membrane than in the reticular lamina. For the pathological cochlea, the stereocilia compliance has been increased arbitrarily by a factor of 10. No other changes have been introduced. The pathological IHC magnitude transfer function plotted by means of the intermittent line is compared to a normal one in the upper panel. Note that it traces higher magnitudes at the low sound frequencies than does the normal curve, whereas the reverse is true in the broad vicinity of the normal maximum response. This was found empirically on several cat preparations to be the prevailing pattern (e.g., Liberman, 1976). In the middle panel, is plotted the dB difference between the two curves. It increases as the sound frequency decreases and reaches 32 dB around 200 Hz. It becomes negative above 2000 Hz and follows a somewhat complicated pattern that appears characteristic, nevertheless, according to the empirical results of Liberman's. For comparison, data points derived from his results

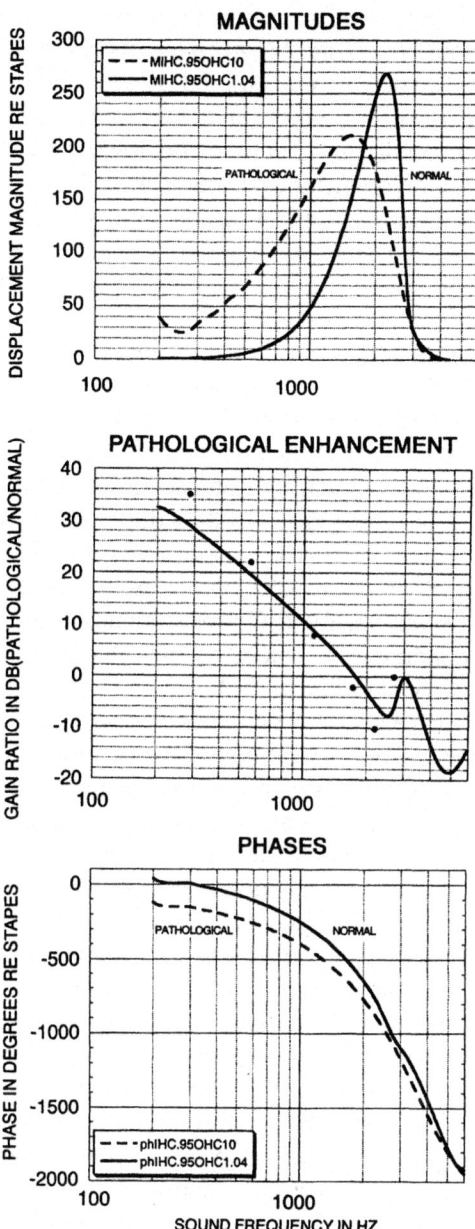

FIG. 6.36. Theoretical magnitudes and phases of shear displacements at the location of the IHCs for two values of the stereocilia compliance and two sets of passive resistance values (top and bottom panels). Middle panel: difference between the two magnitude curves in the upper panel. Difference values indicated by circles have been derived from empirical data.

at discrete sound frequencies are shown by means of the closed circles. Because the empirical results were obtained at a cochlear location corresponding to a CF of 4000 Hz, almost an octave higher than the theoretical results, the empirical points have been moved downward in frequency by a corresponding constant log distance. Clearly, there is good agreement between the theoretical and empirical results with respect to both the frequency pattern and the absolute magnitudes, although parameters of human cochleas have been used in the theory, and the empirical results were obtained on domestic cats.

The theoretical normal and pathological phases are compared in the bottom panel of Fig. 6.36. Over an extended low frequency range, there is a 180° difference between the resulting curves, the one for the normal cochlea indicated by the solid line, the one for the pathological one, by the intermittent line. Comparison with the phase curves of Fig. 6.35 indicates that the normal curve corresponds to IHC depolarization during basilar membrane displacement toward scala tympani, whereas the pathological curve is associated with their depolarization during basilar membrane displacement toward scala vestibuli. Corresponding empirical results were obtained on Mongolian gerbils whose cochleas were damaged by excessive sound exposure or kanamycin (Sokolich et al., 1976; Schmiedt et al., 1980). In morphologically confirmed cochlear portions with severely reduced or absent OHCs, but preserved IHCs, the nerve fibers were invariably excited during inferred basilar membrane displacements toward scala vestibuli, whereas, in portions with well preserved populations of OHCs, the excitation phase was variable, and the scala tympani excitation phase predominated. To the best of my knowledge, no other model of cochlear mechanics capable of explaining these relationships has been proposed.

It proved extremely difficult to measure cochlear wave propagation as a function of cochlear location for constant sound frequencies rather than as a function of sound frequency for constant cochlear locations. Thus far, only Békésy (1960) performed this feat on postmortem preparations and Russell and Nilsen (1997) on live guinea pigs but only in the very basal part of the cochlea. The model of this monograph makes it possible to fill the gap to some extent. As has been mentioned in preceding sections, it allows us to display cochlear response magnitudes either as functions of sound frequency or location. Thus far, the former possibility has been exploited but Fig. 6.37 shows model basilar membrane vibration magnitudes and phases displayed as functions of location for four sound frequencies in one octave ratios—1, 2, 4, and 8 kHz. They are plotted over abscissa scales expressed in numbers of network sections, 960 in all, so that 100 sections amount roughly to 10% of the cochlear length. As expected, the curves represent mirror images of related transfer functions in the frequency domain and do not reveal

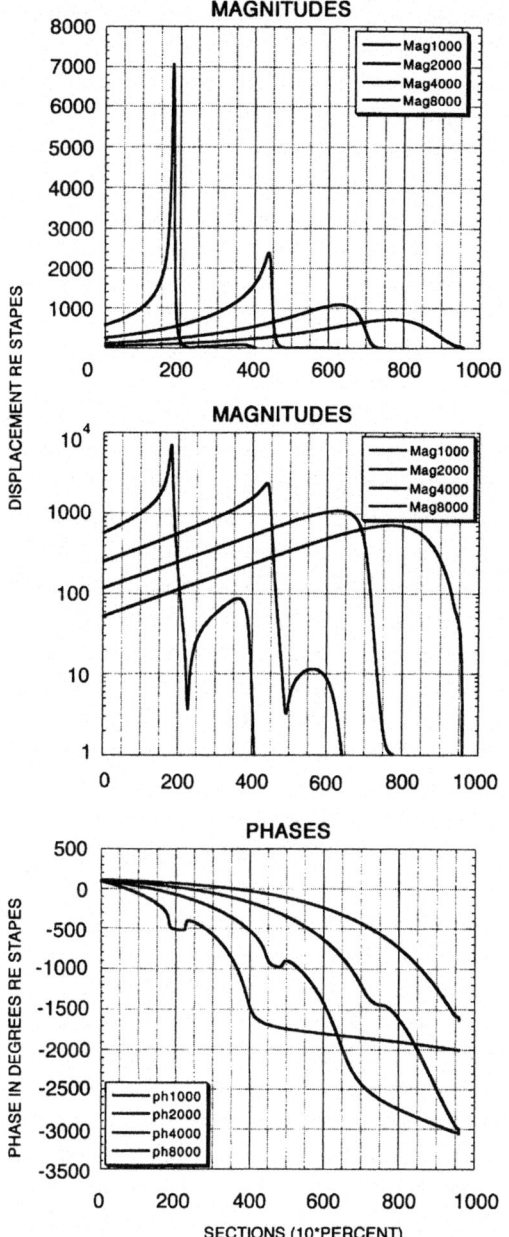

FIG. 6.37. Theoretical magnitudes and phases of basilar membrane displacements as functions of cochlear location for several sound frequencies. In the middle panel, the magnitudes are plotted on a log scale.

any fundamental new aspects. For the same nominal parameter values of the resistance elements, the high sound frequencies produce maximum responses more basally than the low sound frequencies and their spatial transfer functions have narrower bandwidths than the latter. The associated phase lags increase more rapidly for the high frequencies than for the low ones and exhibit notches at sound frequencies closely following the frequencies of the peaks. As in the instance of the frequency transfer functions, these notches, as well as the notches visible in the log representation of the magnitude functions, are produced by parallel resonances in the stereocilia-tectorial membrane complex. It should be pointed out that the course of the theoretical phase curves beyond the notches is meaningless, being associated with extremely small vibration magnitudes.

The measurements of basilar membrane vibration as functions of cochlear location performed by Russell and Nilsen provide an opportunity for comparing them to the theoretical results. Although their data were obtained near the very basal end of the cochlea, and the peaks of their magnitude functions were situated at a somewhat higher sound frequency than the highest frequency peak in Fig. 6.37, a close similarity between their data and the theoretical ones is unmistakable. When the cochlear distances are expressed in percent of the total cochlear length, their spatial transfer functions obtained at mid SPLs have about the same bandwidth as the theoretical ones, amounting to between 2% and 3% of the total length at 10 dB below the peak and their phase curves show similar phase lags. Because their data were determined on guinea pigs, and human numerical constants have been used in the theory, the agreement of the results further confirms the interspecies similarity. Of course, it also supports the theory.

In the next two figures, the effects of the basilar membrane, the stereocilia and the tectorial membrane resistances are investigated at the sound frequency of 2.5 kHZ. In the upper panel of Fig. 6.38, it is shown that decreasing the resistances successively by the factors of 0.4 and 0.2 increases the response peak almost in direct proportion and moves it apicalward by almost 20% of the cochlear length. Significantly, the apical skirts of the curves all converge to a common foot. This is reminiscent of the convergence of the frequency transfer functions both theoretical and empirical as has been discussed in chapter 5 and this section. In the lower panel, the effect of decreasing the resistances on the response phase is shown. Up to the frequency of the peak response, the effect is negligible. However, apicalward from it, decreasing the resistances increases the phase lag, in particular in the vicinity of the common foot of the apical skirts of the magnitude curves. This pattern is very much like the pattern found by Russell and Nilsen, if the assumption is maintained that the resistances decrease with SPL.

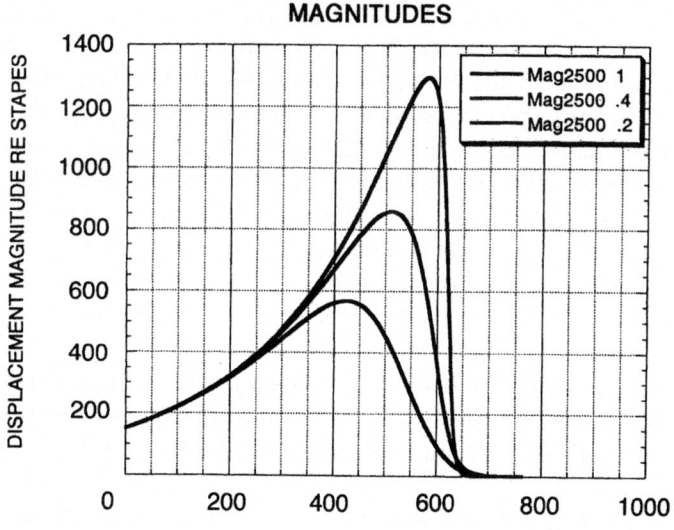
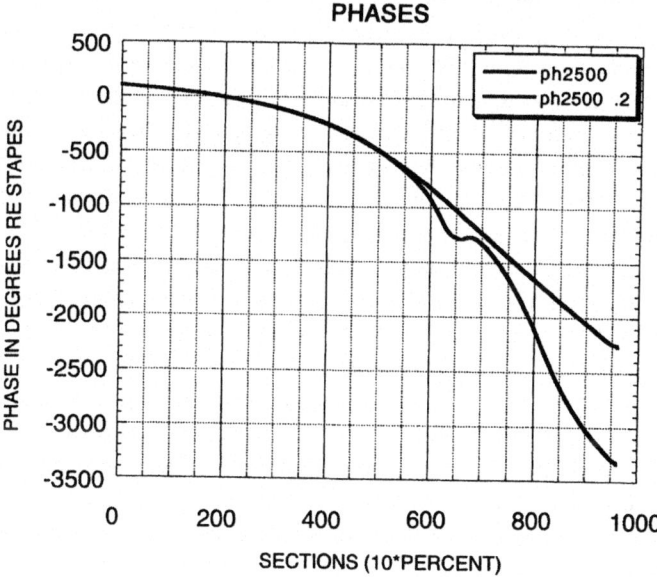

FIG. 6.38. Theoretical magnitudes and phases of shear displacements as functions of cochlear location for one sound frequency and several resistance values. The phases are only for two sets of resistance values, the upper curve belonging to the greater value.

In Fig. 6.39, the effects of varying the resistances of the stereocilia and the tectorial membrane on the shear displacement are investigated. The basilar membrane resistance is kept constant. Again, decreasing the resistances increases the response magnitude and moves the peak response toward the apex, as is evident in the upper panel. Again, the apical skirts of the functions converge at a common foot. However, the phase pattern in the lower panel is different from that seen on the basilar membrane in the preceding figure because of the phase jumps of about 180° in the phase curve associated with the small resistance of 0.04 of its passive value. These phase jumps are the same as have been seen in the frequency domain and are associated with the resonances in the stereocilia-tectorial membrane complex. They change the excitation phase of the OHC from scala vestibuli orientation near the cochlear base to scala tympani orientation further toward the apex. As discussed in connection with the frequency domain analysis, the 180° jumps produce a relative phase lag of about 90° at SPLs around 80 dB relative to the phase at lower SPLs. It is assumed that the increased SPL increases the numerical values of the stereocilia and tectorial membrane resistances. As has been already discussed, it was possible to demonstrate the 90° phase difference empirically. Past the frequency of the magnitude peak indicated in the upper panel of Fig. 6.39, the phase associated with the lower resistances dips below that associated with the higher ones and takes the same course as the phase of the basilar membrane shown in Fig 6.38. It may be worth noting that all the changes in the space domain are entirely consistent with the changes in the frequency domain and are produced solely by changes in numerical values of parameters that are explicit analogs of cochlear anatomical entities.

The analog electrical network mimicking the relevant structure of the cochlea and used as a basis for the computer program has also been realized in hardware for the purpose of verifying the computer results and seeing cochlear processing of input signals in real time. This has proved particularly useful in investigating cochlear speech processing. For practical reasons, the number of cascaded network sections has been reduced from 960 to only 96. This has the disadvantage that the cochlear analog waves produced by the model are not smooth but appear as sets of staircase steps. Otherwise, the results seem to be unaffected by comparison to the computer results.

The hardware model provides the same outputs as the computer model and allows us to look at the cochlear transfer functions in frequency as well as in space domains. For the latter, the section outputs are scanned electronically at a rate of $7\mu sec$ per section, fast enough to make the model waves appear stationary on the screen of a monitoring oscilloscope. Examples of the oscilloscope displays are shown in Figs. 6.40 and 6.41. For these

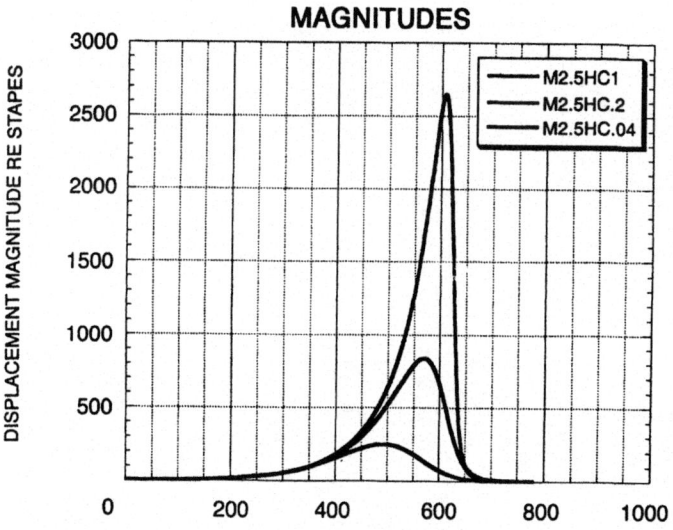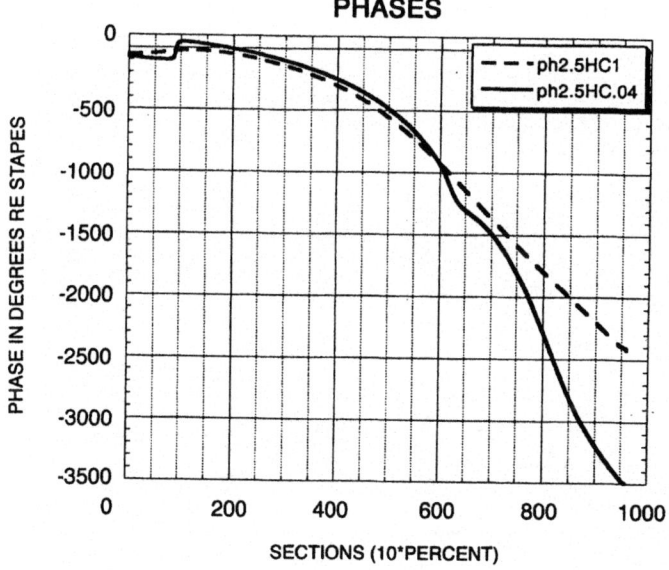

FIG. 6.39. Theoretical magnitudes and phases of shear displacements as functions of cochlear location for one sound frequency and several stereocilia resistance values. The phases are only for two resistance values, passive (intermittent line) and 0.04 passive.

HARDWARE MODEL
SINUSOIDAL SIGNAL

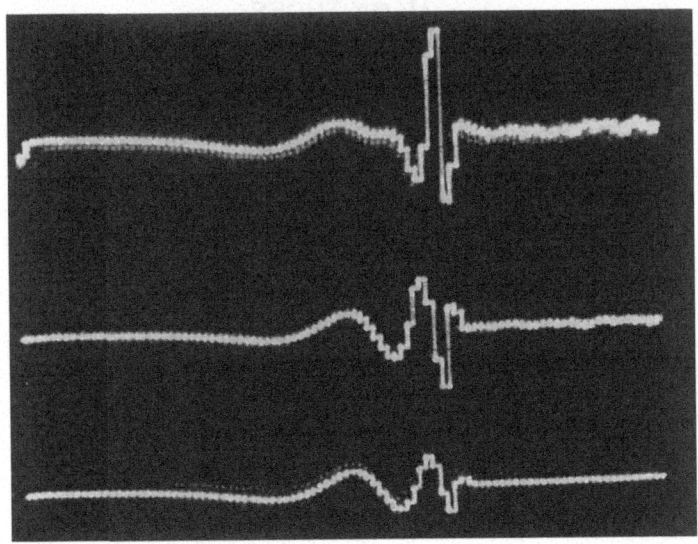

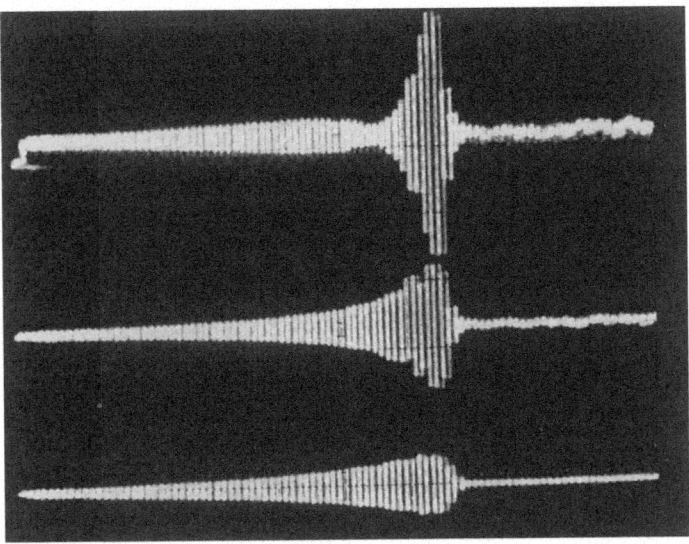

FIG. 6.40. Cochlear waves on the hardware model and their amplitude envelopes (lower panel) as functions of cochlear location. Lowest trace—basilar membrane (radial displacement of reticular lamina); middle trace—radial tectorial membrane displacement; upper trace—shear displacement. The cochlear base is to the left.

displays, the model was hooked up to an electrical network model of the middle ear described in chapter 3. The main effect of the middle ear model was to act as a broad bandpass filter. In the displays of Fig. 6.40 are shown responses to a sinusoidal signal in the mid-frequency range. The bottom trace of each panel shows the output at the basilar membrane level, the middle trace, the radial oscillation of the tectorial membrane, and the upper trace, the shear displacement between the tectorial membrane and the reticular lamina. The bottom panel shows the analog amplitude envelopes of the cochlear waves. Note that the one belonging to the basilar membrane is the flattest and that of the tectorial membrane is similar up to the vicinity of its maximum where the amplitude is enhanced. The envelope of the shear displacement reveals a much sharper peak than the preceding two envelopes. It follows a flat minimum which, according to the analysis of the preceding sections, is associated with the series resonance of the tectorial membrane. The secondary maximum on the apical side of the main maximum, extensively discussed in preceding sections, is too small to be discerned.

In the upper panel of Fig. 6.40 are displayed both the magnitudes and phases forming cochlear wave analogs. Note that the waves in the basilar membrane and tectorial membrane traces are in phase, but the shear displacement phase changes in the region of the tectorial membrane series resonance by about 180° and ends up by being in phase opposition to the lower two traces. How this happens has been explained in connection with the analysis of the stereocilia-tectorial membrane complex and the computer model.

In Fig. 6.41 are displayed the model responses to two vowel sounds—"i" in the upper panel and "a" in the lower one. The spatial separation between their formants both between the vowels and within their spectra is clearly apparent. This distinction makes it understandable why the two vowels sound so different. It is necessary to introduce a caveat, however. Evidence is introduced in the following chapter strongly suggesting that the cochlear vibration maxima are not suitable for a pitch code and that this code is more likely provided by the apical cutoffs of spectral envelopes. If this is so, the amplitude decrements evident in the envelopes of the vowel spectra displayed in Fig. 6.41 may be more significant than the amplitude maxima. This possibility may be worth investigating in phonemic studies.

At the end of this section, it may be worth pointing out that the described model, whose every element corresponds to an anatomical mechanical element within the cochlea, has made it possible to account for the most prominent characteristics of cochlear mechanics solely by varying the numerical values of its elements according to direct or at least indirect empirical evidence. This variation has included known cochlear pathology.

HARDWARE MODEL
SPEECH SIGNAL

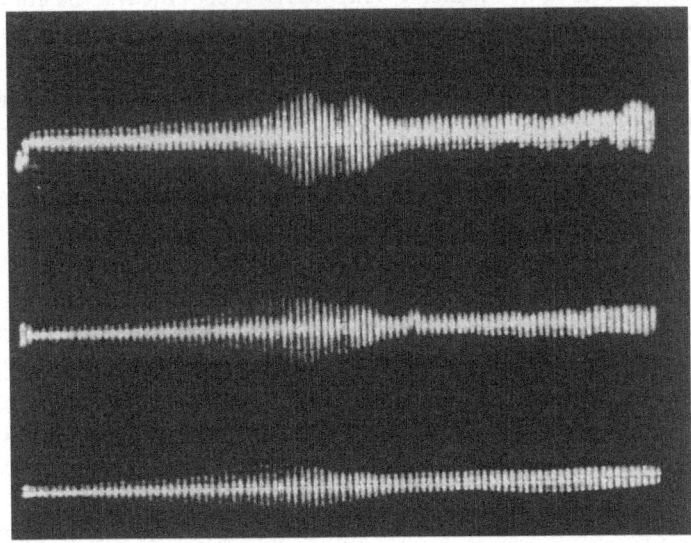

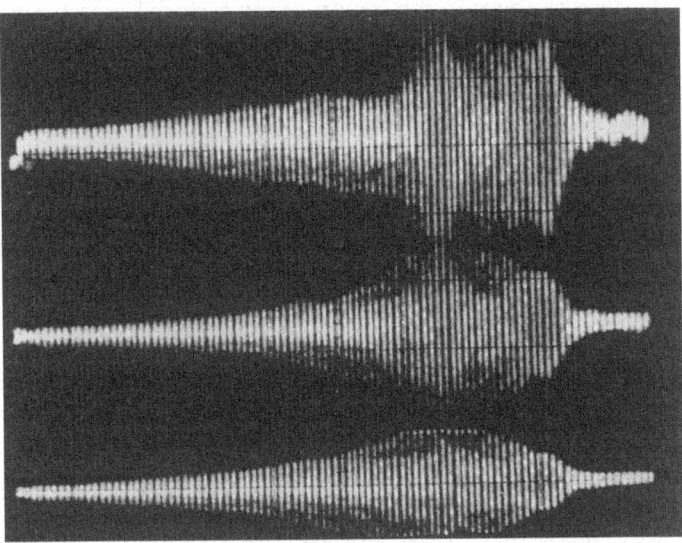

FIG. 6.41. Vowel spectra displayed on the hardware model in real time, "i" in the upper panel, "a" in the lower one. Lowest trace—basilar membrane (radial displacement of reticular lamina); middle trace—radial tectorial membrane displacement; upper trace—shear displacement.

ACTIVE FEEDBACK EFFECT

As already mentioned, there is little doubt that the mammalian cochlea is endowed with an active feedback that supplies metabolic energy to the process of wave propagation (see, e.g., Dallos, 1992; Geisler, 1998, for review). The feedback motor resides in OHCs that not only convert the deflections of their stereocilia into electrical potentials but also respond to the potentials they generate by changing their lengths. Deflection of the stereocilia toward the spiral ligament produces in them depolarizing receptor potentials that, in turn, shorten their lengths. Deflection of the stereocilia in the opposite direction hyperpolarizes the cells and increases their lengths. As was demonstrated on several occasions, these processes can occur quite rapidly, so that they are able to follow acoustic oscillations up to the highest audible sound frequencies (e.g., Dallos & Evans, 1995). When the cells change their lengths, they exert forces on the surrounding tissues. It is important to understand that a mechanical force must act between to points or surfaces. Because of its anatomical position, an OHC exerts its mechanical force between the basilar membrane and the reticular lamina. When the cell is shortened, it pulls the basilar membrane away from scala tympani and the reticular lamina away from the tectorial membrane. Because, according to the classical model of shear motion between the reticular lamina and the tectorial membrane, an OHC is depolarized during basilar membrane displacement toward scala vestibuli, the displacement is in phase with the pull of the hair cell, producing a positive feedback at the basilar membrane level. Superficially, this appears to be as it should be. However, a problem arises at the other end of the hair cell. Here, shortening of the cell pulls its apex away from the tectorial membrane, counteracting the push produced by the basilar membrane. As a result, the deflection of the stereocilia is decreased, producing a negative feedback. Various, in my view, contrived schemes were devised to overcome the problem. For example, arches of Corti with their extension, the reticular lamina, were assumed to absorb the apical force and lead it back to the basilar membrane near its fixation point to the spiral osseous lamina (e.g., Kolston, 1999). On the other hand, others recognized the feedback as being negative at low sound frequencies but changing to a positive one past the basilar membrane resonance frequency, due to the 180° phase shift (e.g., Mountain et al., 1983). In still other models, the mass of the organ of Corti was subdivided in two parts one being associated with the basilar membrane, the other with the reticular lamina (e.g., Markin & Hudspeth, 1995). The closest to my concept of cochlear micromechanics (Zwislocki, 1980) and the feedback described here came the feedback model of Neely and Kim (1986) in which they used the tectorial membrane as the second mass. Many other models could be cited (e.g., Geisler, 1998).

364 CHAPTER 6

It seems to me that all these models encounter one common fundamental difficulty—the refusal to consider the tectorial membrane in its natural state as highly flexible and endowed with a rather substantial mass nearing in magnitude the mass of the organ of Corti. The refusal seems to persist in spite of direct measurements in vivo of this flexibility (Zwislocki & Cefaratti, 1989), and the demonstration that the tectorial membrane is capable of substantial oscillation not only in the transversal but also the radial direction, when driven acoustically through the basilar membrane (Gummer et al., 1996). Very recent measurements confirmed the compliant nature of the tectorial membrane, although they referred to its internal deformation rather than to its deformation relative to its site of fixation at the spiral limbus (Abnet & Freeman, 2000). Perhaps, they will finally succeed in breaking down the surprisingly conservative nature of the scientific establishment in auditory physiology or biophysics.

The active feedback introduced in the model described in this monograph arises almost automatically from the structure of the model, which grossly but closely duplicates the anatomy of the cochlear partition, including the basilar membrane, the organ of Corti and the tectorial membrane. The feedback is analyzed here with the help of the electrical network version of the model, whose derivation has been described in this chapter in connection with Figs. 6.13 and 6.21. The network model with an explicit inclusion of the feedback is shown in Fig. 6.42. As indicated in a less formal way in Fig 6.13, the feedback is represented by the current source kVH, the symbol V having been borrowed from the acoustical system where it symbolizes the magnitude of volume velocity, placed between the ground and the junction between the basilar membrane and the stereocilia-tectorial membrane networks. This junction coincides in the cochlea with the reticular lamina, whereas the ground coincides with the outer bony wall of the cochlear canal. The feedback source itself coincides with the bodies of the OHCs in aggregate. It should be pointed out again that a parallel network in the electrical model corresponds to a mechanical series network. As a consequence, the feedback branch incorporating the kVH current source properly parallels the basilar membrane branch, which is mechanically in series with the OHCs. The justification for considering the kVH source as a current source has been already given. It results from the measured high stiffness of the organ of Corti elements, in particular, the arches of Corti, the Deiter's cells, and the OHCs themselves. It may be pointed out in addition that a smaller stiffness would reduce sound transmission from the basilar membrane to the OHC stereocilia on whose deflection depends the excitation of the hair cells, and the efficiency of the system would be reduced.

According to experimental evidence (e.g., Russell, Cody, & Richardson, 1986; Santos-Sacchi, 1992) and Geisler's (1998) review, the output of the

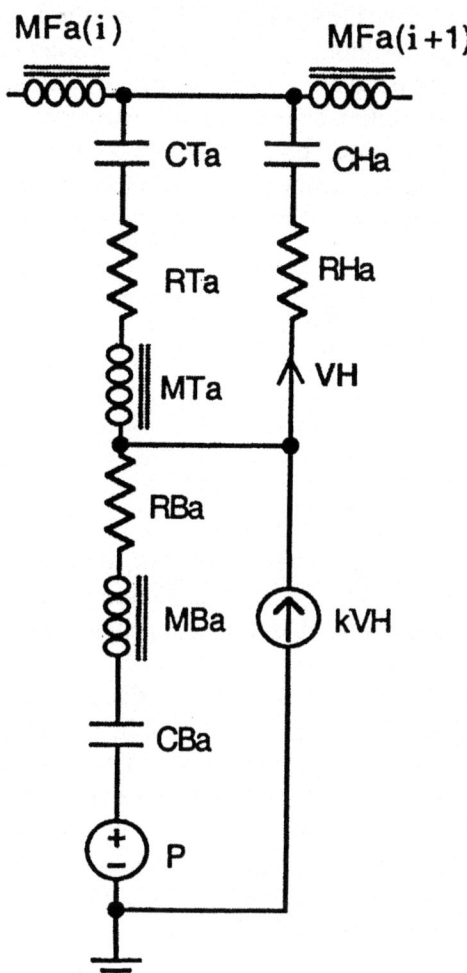

FIG. 6.42. Electrical analog of a cochlear incremental section with feedback. Explanation in text.

current source, kVH, can be considered for small signals as being directly proportional to the current VH in the stereocilia branch of the network model. This current represents the volume velocity of the OHC stereocilia deflection. Simple network analysis shows that, according to Fig. 6.42, the volume velocity of the stereocilia deflection, VH, can be related to the sound pressure at the basilar membrane by the following equation,

$$V_H = \frac{P}{Z_B} * \frac{1}{\frac{1}{Z_B} + \frac{1}{Z_H} + \frac{1}{Z_T}} * \frac{1}{1 + \frac{Z_H}{Z_T} - k} \qquad 6.2$$

where V_H means the volume velocity of the stereocilia, P—the sound pressure difference across the cochlear partition, Z_B —the acoustic impedance of the basilar membrane, Z_H —the acoustic impedance of the stereocilia aggregate and Z_T —the acoustic impedance of the tectorial membrane, all taken per unit length of the cochlea. In addition, k means a complex proportionality constant. With the help of equation 6.2, the relationship between the volume velocity of the basilar membrane, V_B, and the sound pressure, P, can be expressed as

$$V_B = V_H \left(1 + \frac{Z_H}{Z_T}\right) \qquad 6.3$$

It is significant that, for

$$k = 1 + \frac{Z_H}{Z_T} \qquad 6.4$$

both V_H and V_B become infinite. This means that, according to the model, very high amplifications and even spontaneous signal generation can be achieved in agreement with experimental evidence (e.g., Ruggero et al., 1997; Rhode & Recio, 2000).

The computer model already described contains the feedback loop and accepts input parameters defining the numerical values of the constant, k. When using the model, one should realize, however, that it converts volume velocities into displacements, so that the following model results are expressed as displacements rather than velocities. One should also realize that the constant, k, must be defined in terms of two parameters—a magnitude and a phase angle. In the analysis of the feedback considered below only the parameters of the constant k are varied. All other network parameters are kept constant, especially the resistances, which are maintained at their values corresponding to the passive cochlea. As already mentioned, the numerical value of the tectorial membrane compliance determined in the static measurements is used. It is about three times smaller than the compliance accepted in the analysis of the passive cochlea, increased artificially by this factor to compensate for the anticipated feedback effect.

As an example of the theoretical feedback effects, Fig. 6.43 shows basilar membrane magnitude and phase gain functions computed for several k magnitudes and a $-45°$ phase angle. For comparison, Fig. 6.44 reproduces corresponding data determined by Ruggero et al. (1997) on a chinchilla. Although direct numerical matching of the two sets of the data is not possible because of differences in cochlear lengths and the locations on the basilar membrane, similarities in essential features are striking, despite the differences in the coordinate scales. As the magnitude of the feedback in-

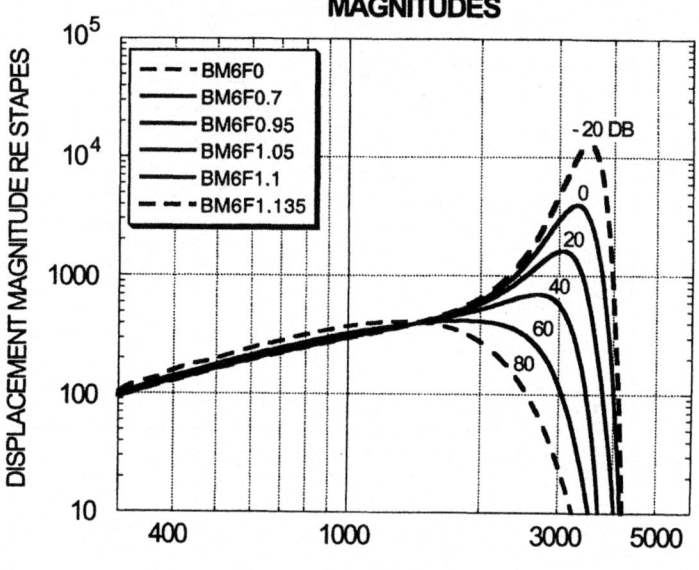
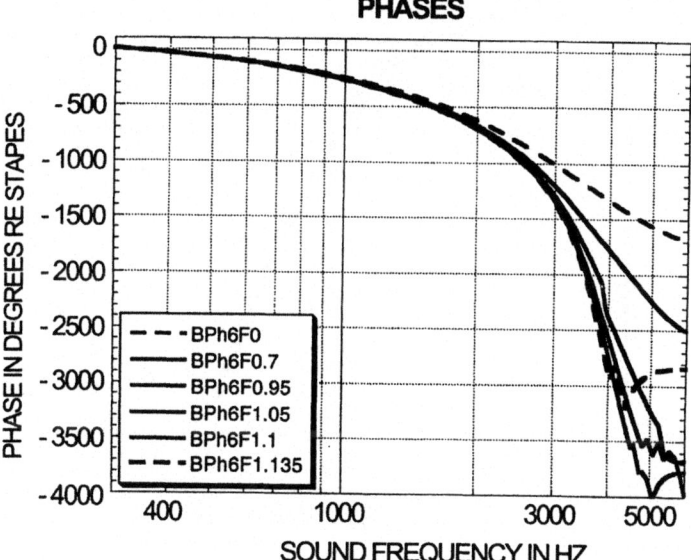

FIG. 6.43. Theoretical magnitude and phase gain functions of basilar membrane displacements at one cochlear location. Lower intermittent line in the upper panel and upper intermittent line in the lower panel—without feedback; solid lines—with variable feedback magnitude and constant –45° phase. The dB numbers by the curves suggest corresponding hypothetical SPLs.

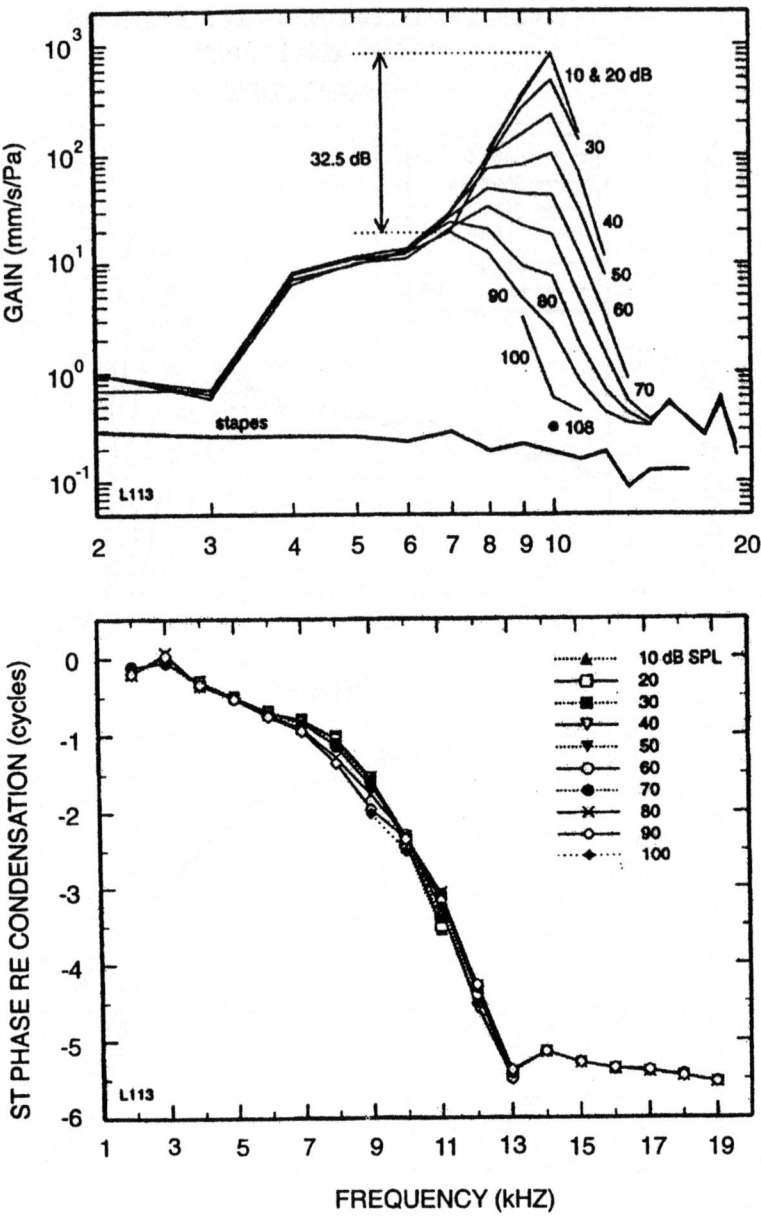

FIG. 6.44. Empirical magnitude and phase gain functions of basilar membrane displacement velocities of a chinchilla for a variable SPL (from Ruggero et al., 1997). From "Basilar Membrane Responses to Tones at the Base of the Chinchilla Cochlea," by M. A. Ruggero, N. C. Rich, A. Recio, S. Narayan, and L. Robles, 1997, *Journal of the Acoustical Society of America, 101,* pp. 2151–2163. Copyright © 1997 by Acoustical Society of America. Reprinted with permission.

creases, and the gain of the theoretical functions grows accordingly, the peak location moves upward in sound frequency. A similar relationship was found in Mongolian gerbil hair cell recordings in a comparable frequency range. The data obtained by Ruggero et al., in chinchilla at a three times higher CF in the basal part of the cochlea, show a somewhat smaller peak shift for a comparable gain increment. The difference is not of a fundamental nature, however, and it has been demonstrated that increasing the phase lag of the feedback decreases the peak shift. For example, for a phase lag of $-70°$ the peak shift is nearly abolished for all except the lowest gains. Generalizing, it is possible to state that the two variables—the gain and the amount of the peak shift make it possible to determine unambiguously the two parameters of the feedback constant k.

To facilitate comparison between the empirical data of Fig. 6.44 and the theoretical ones of Fig. 6.43, hypothetical SPLs have been indicated in the latter, next to the gain functions. They suggest a functional relationship between the feedback magnitudes and SPLs derived from the empirical results, the 80 dB SPL, at which the feedback is believed to vanish, serving as a common reference.

The similarity between the theoretical and empirical data is maintained in phase gain functions. Two common features are striking—the increased slopes of the functions above the frequency region of the magnitude peaks and a small dependence on the gain, especially at low sound frequencies. Note that the theoretical phase data are plotted on a log frequency scale, which emphasizes the increasing slope, whereas the empirical ones are plotted on a linear scale. A small difference between the theoretical and the empirical phase data of Ruggero et al. must be pointed out. Above the frequency region of the magnitude peaks, both the theoretical and empirical data are in agreement in revealing a gradually increasing phase lag with an increasing gain. Immediately below this frequency region, the empirical data show a perturbation in the opposite direction, however. This tendency is absent in the theoretical phase characteristics as well as in some other empirical results (e.g., Cooper, 2000). The difference may be due to a slightly increasing compliance of the basilar membrane with the SPL. This phenomenon has been discussed in preceding sections but not included in plotting the phase curves of Fig. 6.43. It would decrease the theoretical phase lag associated with an increasing feedback and produce a reversal between the phase lag and the phase lead below the CF region. It should be pointed out that the theoretical and empirical phase data end up in a plateau at high sound frequencies. As has been discussed in chapter 4, such a plateau must occur where the basilar membrane impedance becomes mass controlled.

From the point of view of the model validity, it is significant that the feedback constant, k, has only two degrees of freedom, as already mentioned.

They can be used to obtain gains and peak shifts in accordance with available empirical data. However, comparisons between the theoretical and empirical results can be made along several additional dimensions to test the validity of the model in its entirety. In doing so, all the numerical values of the model elements must be kept close to those determined for the passive cochlea.

The theoretical phase of the feedback constant, k, required for empirically correct gain functions can serve as a first test item. The phase has to be negative. This is in agreement with the physiological phase lag produced by the capacitance of the OHC membrane.

According to empirical evidence (e.g., Ruggero et al., 1997, Cooper, 2000), the feedback affects appreciably only the CF region of the basilar membrane transfer functions. At sound frequencies well below the CF, the gain remains practically constant, independent of SPL. This phenomenon is clearly evident in the theoretical family of the gain functions of Fig. 6.43.

The bandwidth of the gain functions decreases as the gain increases, so that the Q_{10dB} factor varies from about unity to about 5 or somewhat greater, depending on the CF region. This is exemplified by the empirical functions of Fig. 6.44. The Q_{10dB} factor of the theoretical curves of Fig. 6.43 varies from nearly unity in the absence of the feedback (intermittent curve) to about 5 for the maximum gain function shown. The feedback formula is able to account for very high gains, on the order of 50 dB, encountered in empirical measurements (Geisler, 1998, for review). The feedback formula also accounts for spontaneous cochlear emissions. Last but not least, the feedback loop in the model network simulates correctly the OHC feedback loop within the cochlear structure.

The theoretical feedback generates similar gain relationships for the shear displacement between the reticular lamina and the tectorial membrane, as shown in Fig. 6.45. They are entirely consistent with empirical transfer functions of OHC alternating receptor potentials and related transfer functions of the organ of Corti supporting cells discussed in the preceding chapter, as shown in Fig. 6.46. The latter have been obtained by multiplying the gain function by numbers corresponding to the levels marked in Fig. 6.45. In particular, a similar relationship is obtained between the size of the maximum response and its frequency shift. Furthermore, the theoretical gain depends on the feedback magnitude only in the frequency region of maximum response, the CF region, in agreement with patterns of the empirical dependence of the gain on SPL. At lower sound frequencies, both the theory and the relevant experiments indicate a constant gain. At these frequencies the absolute size of the gain is somewhat smaller for the shear displacement than for the basilar membrane. This is in agreement with indirect empirical evidence, as has been mentioned in a preceding section of this chapter and explained

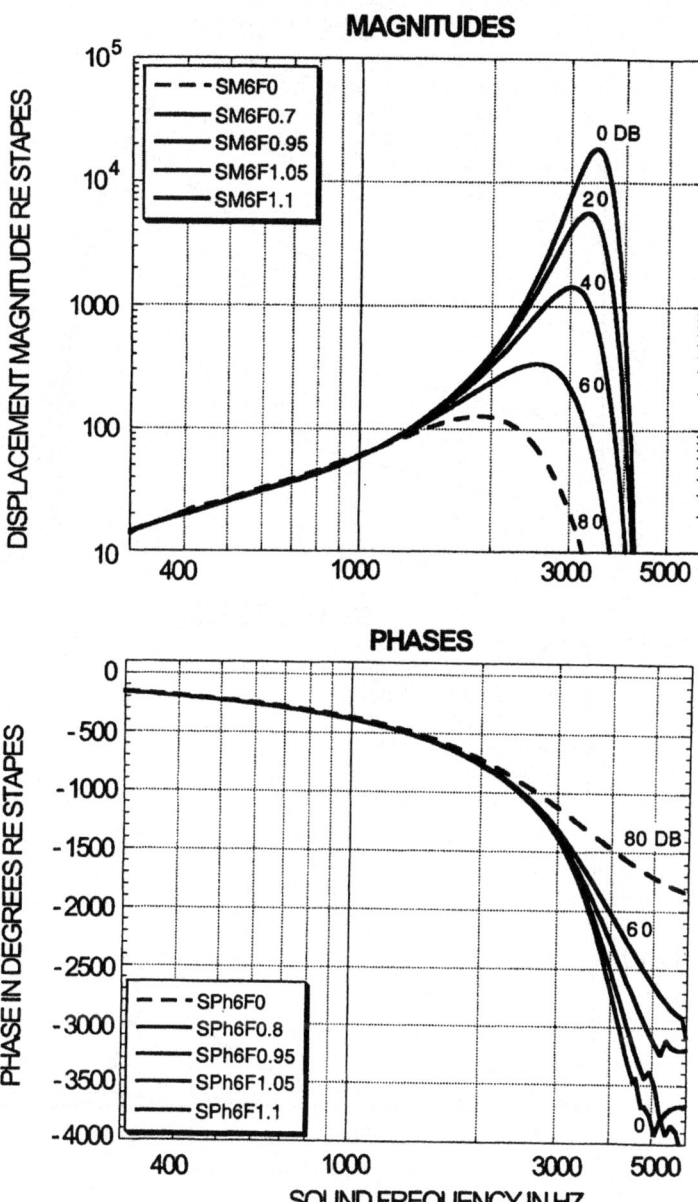

FIG. 6.45. Theoretical magnitude and phase gain functions of shear displacements at one cochlear location. Intermittent lines—without feedback; solid lines—with variable feedback magnitude and constant phase of –45°.

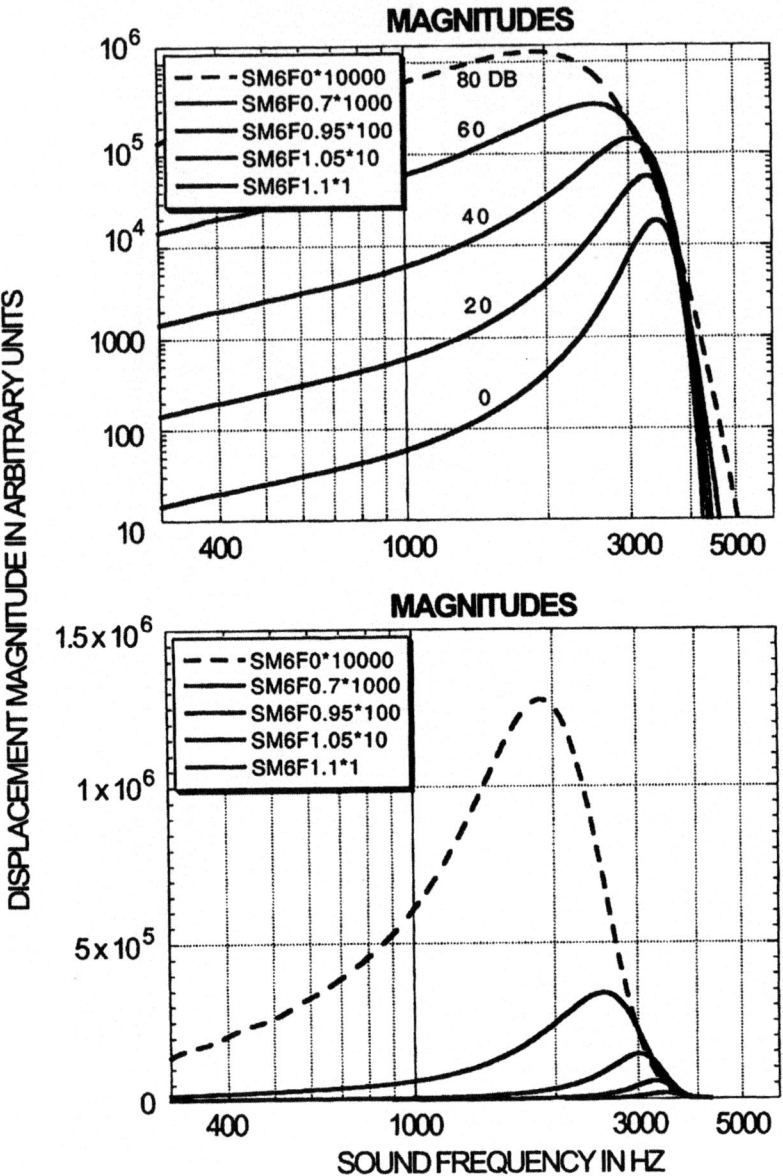

FIG. 6.46. Theoretical magnitude transfer functions of shear displacements at one cochlear location. Intermittent lines—without feedback; solid lines—with variable feedback magnitude and constant phase of −45°. Upper panel—log ordinates; lower panel—linear ordinates.

theoretically. On the other hand, the theoretical peak gain indicated in Fig. 6.45 is about 5 times larger for the shear displacement than for the basilar membrane. This happens because, according to the model, the positive feedback acts directly on the stereocilia-tectorial membrane complex and only indirectly through it on the basilar membrane, and because the elements of the complex have smaller numerical values than those of the basilar membrane. Finally, the convergence of the transfer functions at high sound frequencies, similar to the convergence of the transfer functions found empirically, must be mentioned.

The theoretical phase relationships for the shear displacement also parallel roughly those for the basilar membrane. Here too, the phase lag tends to increase with the gain above the CF but remains roughly constant below it. Small phase differences at low sound frequencies, such as discussed in connection with the passive cochlea, are not clearly resolved in Fig. 6.45. However, careful scrutiny reveals a crossing of the phase curves. Whereas, the higher gain curves lag the curve for the passive cochlea at high frequencies, the reverse is true at the very low ones. However, the difference is smaller than has been found without the feedback. The reason for this can be found in an enhanced magnitude of the tectorial membrane vibration. To compensate for it, a slightly increased compliance of the basilar membrane would have to be assumed, as has been already suggested.

The enhancement of the tectorial membrane vibration magnitude by the feedback is evident in the upper panel of Fig 6.47 where it is compared to that of the basilar membrane for a moderate feedback strength. Remember that the latter has been shown in a preceding section to be approximately equal in magnitude to the radial displacements of the reticular lamina. The tectorial membrane vibration magnitude exceeds that of the basilar membrane down to a sound frequency of about 500 Hz, where the corresponding curves cross. This is considerably below the passive series resonance of the tectorial membrane, which has been determined in chapter 5 to be around 1,400 Hz, and where the curves should cross in a passive cochlea. The shift in the crossing point makes the static measurements of the tectorial membrane compliance compatible with the dynamic cochlear characteristics discussed in the preceding chapter.

To complement the magnitude curves, the lower panel of Fig. 6.47 shows the corresponding phase characteristics. According to them, both elements vibrate practically in phase, except at high sound frequencies, where the tectorial membrane leads somewhat. As has been explained through the vector diagrams of Fig. 6.12, if both are approximately in phase, a radial displacement magnitude of the tectorial membrane exceeding that of the reticular lamina (basilar membrane in the model) signifies depolarization of the OHCs during basilar membrane displacement to-

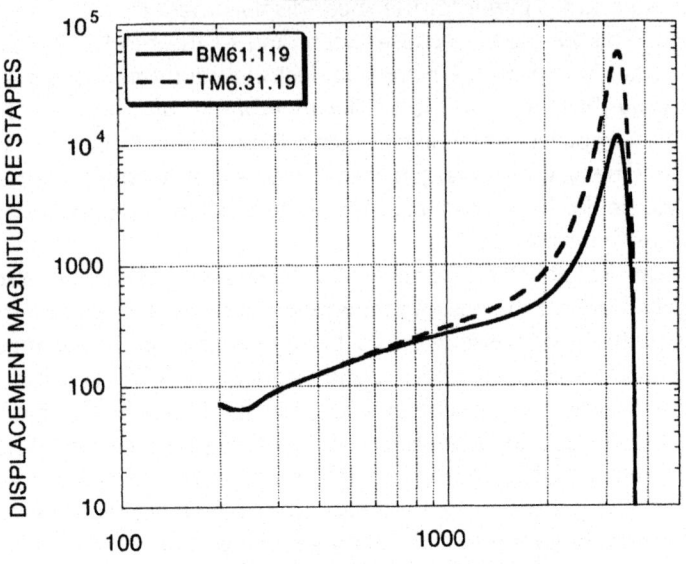
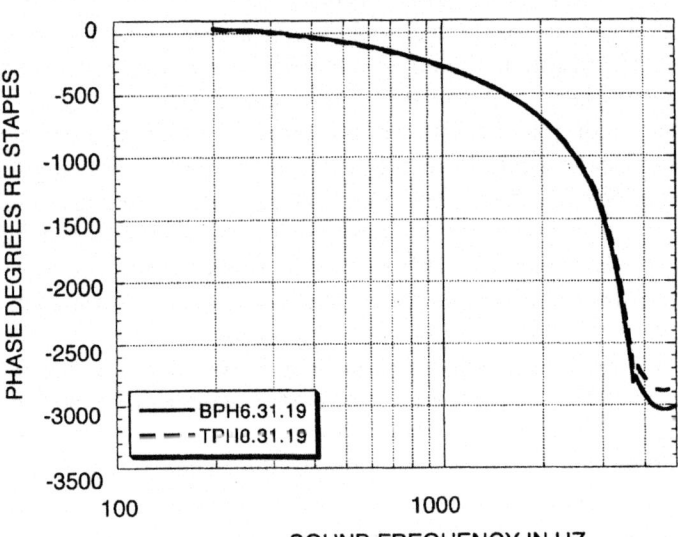

FIG. 6.47. Theoretical magnitude and phase gain functions for basilar membrane (radial displacement of reticular lamina—solid line) and radial tectorial membrane displacements with moderate feedback at −50° phase angle.

ward scala tympani. The comparisons of Fig. 6.47 mean, therefore, that the feedback introduced in this section has not altered the conclusion of preceding sections that the OHCs are depolarized during basilar membrane displacement toward scala tympani in the broad vicinity of the CF. Their depolarization in the opposite phase of the basilar membrane displacement occurs only at sound frequencies much lower than the CF.

Concluding this section, I feel justified in stating that the described feedback agrees with the empirical evidence from the point of view of its structure as well as its effect on the cochlear dynamic characteristics. Comparison of the last two sections indicates that the effect resembles in some but not all ways that of varying the resistance elements of the basilar membrane and the stereocilia-tectorial membrane complex.

COCHLEAR COMPRESSION

Beginning with the early measurements of basilar membrane vibration, it has been firmly established that the cochlea does not act as a linear system. For example, its output signal increases less than in direct proportion to the input signal (e.g., Rhode, 1971; Russell & Sellick, 1978; Sellick, Patuzzi, & Johnstone, 1982). It should be clear from the preceding section that the gain of the system decreases as the input signal increases, producing a compression of the output signal. It is not clear, however, how this comes about. In many technical systems, the compression is achieved by the so called automatic gain control, AGC. In physiological systems, the output reduction often occurs through adaptation—a gradually decreasing output when the input is maintained constant in time. However, in the cochlea, evidence for such an adaptive process seemed to be missing right from the pioneering experiments of Russell and Sellick (1977) on IHCs. Their initial observation was confirmed many times (e.g., Goodman et al. 1982) and extended to the basilar membrane (e.g., Ruggero & Rich, 1991a). The view became firmly established that the cochlear compression was not due to an adaptive process. Accordingly, efforts were made to explain the compression by an instantaneous nonlinearity that is stationary. The various models proposed in this connection run into a fundamental difficulty, however (see review by de Boer, 1993). They had to explain the lack of experimental evidence for strong wave distortions that are naturally associated with a stationary nonlinearity. Various filter schemes were proposed for the purpose.

In this section it is shown that empirically measured cochlear filter functions are insufficient to explain the absence of compatible nonlinearities in cochlear outputs, in addition, that the mammalian cochlea is endowed with adaptation that hides in the onset transients (Zwislocki, Szymko, & Hertig, 1996). The principal experiments were performed on 8 healthy Mongolian gerbils weighing 65 to 78 grams with the help of the methods described in the preceding chapter. Alternating potentials of Hensen's cells were recorded rather than those of OHCs for reasons already given. Hensen's cells are easier to find and to hold and require a shallower penetration of the organ of Corti, which was demonstrated in other experiments of ours to produce noticeable functional damage. On the other hand, the potentials are faithful copies of the OHC alternating receptor potentials. An example of an intensity series of Hensen's cell transfer functions is shown in Fig. 6.48, the SPL ranging from 40 to 90 dB. The functions were recorded intracellularly, and the alternating potentials picked up by the electrode were amplified and filtered in a Lock-in Amplifier with a bandwidth of 3 Hz. As a consequence, the functions are limited to first harmonics. However, it was verified on this and other preparations that, at least up to 80 dB, the effect of higher harmonics was negligible. The latter made themselves felt at high SPLs and low sound frequencies. This is documented in Fig. 6.49 showing Hensen's cell potentials obtained intracellularly at 90 dB and two sound frequencies, 900 and 500 Hz without any filtering. Whereas the wave form distortion of the 900 Hz potentials is relatively small, that of the 500 Hz potentials amounts to pick clipping. Two more caveats should be mentioned with respect to the curves of Fig. 6.48. The 90 dB curve is strongly distorted at low sound frequencies by a low pass filter of the Lock-in Amplifier, which spared the remaining curves. Furthermore, all curves contain a shallow notch around 1,500 Hz. This notch originated in the transfer characteristic of the middle ear.

Several features of the curves are noteworthy. First is the usual downward peak shift from a sound frequency of about 2,000 Hz at 40 dB to about 500 Hz at 90 dB; second, the peak ratio of about 3.25, equivalent to 10 dB, between the 70 and the 50 dB curves, pointing to a strong compression; third, the amplitude ratios of 1 (0 dB) and 1.5 (3.5 dB) between the 630 Hz location and its second and third harmonic locations of 1,260 and 1,890 Hz, respectively, as well the ratios of 3.8 (11.5 dB) and 12 (22 dB) between the 1,150 Hz location and its second and third harmonic locations of 2,300 and 3,450 Hz, respectively; finally, a small secondary maximum in the vicinity of 2,700 Hz.

For analytical purposes, the family of curves of Fig. 6.48 has been sectioned at two sound frequencies, 630 and 1,150 Hz, the first coinciding approximately with the response maximum at 90 dB, the second, with that at 60 dB. The sections have produced input–output functions plotted in Fig. 6.50. Note the strong compression in both curves, reaching down to at least 40 dB, especially in the 1,150 Hz one.

Multiplying a sinusoid by either of the compressed input-output functions produces a distorted wave form corresponding to the compression. The resulting distorted wave forms are plotted in Fig. 6.51 and compared

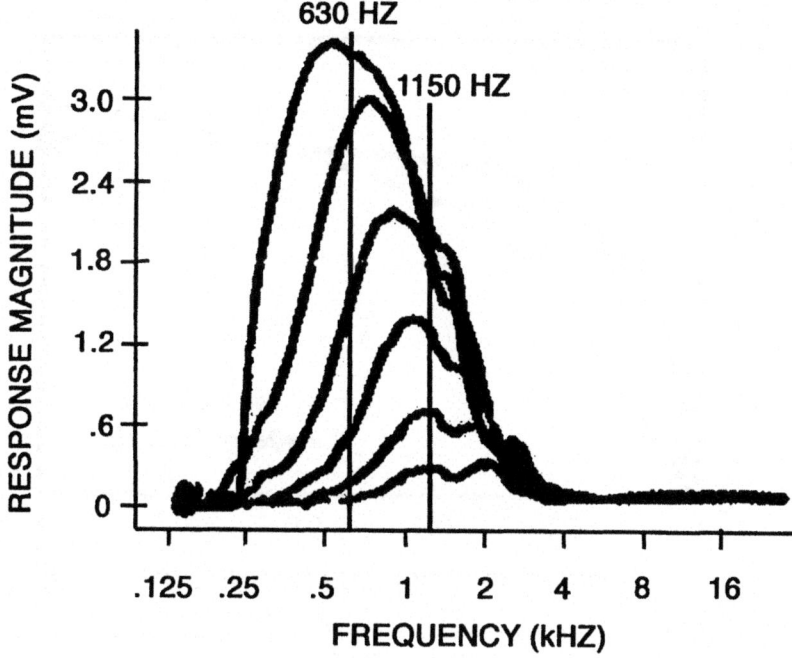

FIG. 6.48. Intensity series of intracellular Hensen's cell transfer functions for SPLs ranging from 40 to 90 dB in 10 dB steps. The notch in the curves is artifactual and stems from the middle ear.

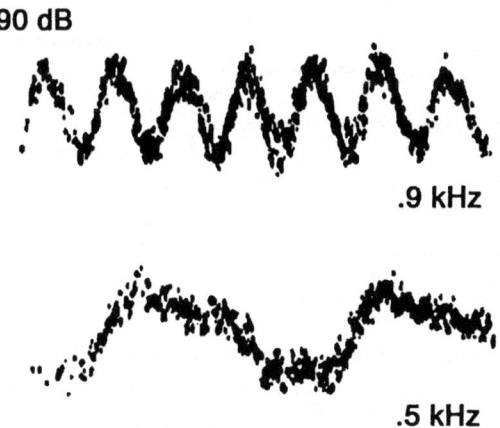

FIG. 6.49. Wave forms of the intracellular Hensen's cell potentials of Fig. 6.48 recorded at 90 dB SPL and two sound frequencies—0.5 and 0.9 kHz without any filtering.

377

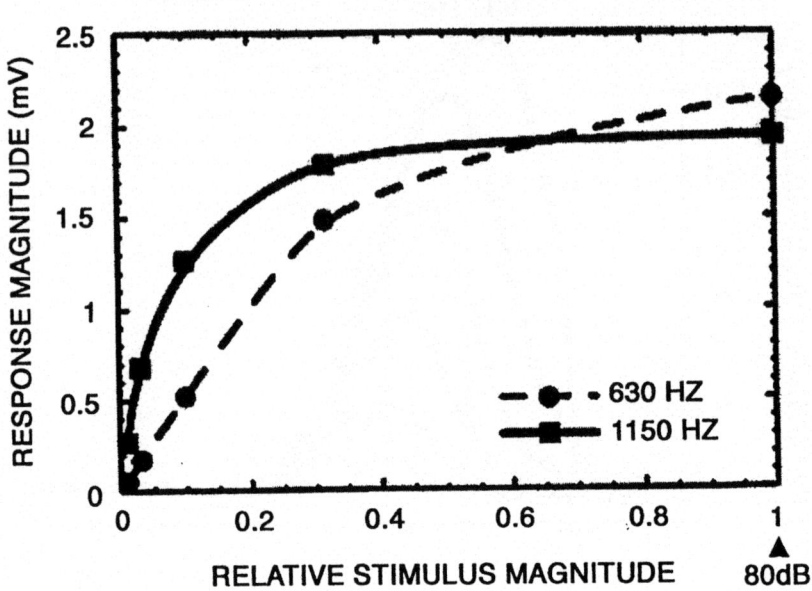

FIG. 6.50. Input–output functions constructed from the transfer functions of Fig. 6.48 at the sound frequencies of 630 and 1150 Hz.

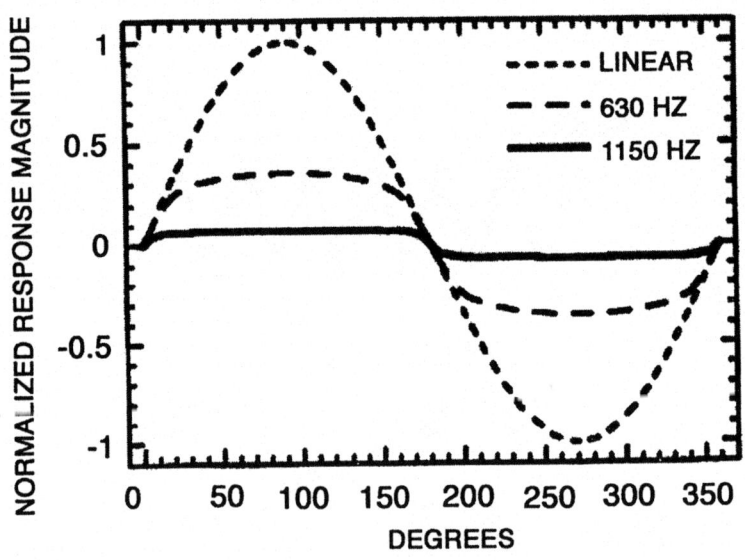

FIG. 6.51. Theoretical wave forms obtained by multiplying a sinusoid with the input–output functions of Fig. 6.50.

to an undistorted sinusoid. The distortion amounts in both curves to nearly peak clipping. According to the harmonic analysis sketched above, it would not be removed by the filter effect of the corresponding frequency characteristics shown in Fig. 6.48. On the other hand, measured waveforms at the same sound frequencies of 630 and 1,150 Hz were almost undistorted, as is evident in Fig. 6.52. These wave forms were obtained without filtering and were processed in time domain by RMS averaging over 50 to 250 repetitions. The input stimuli consisted of practically rectangular tone bursts 10 to 17 msec in duration repeated at 30 to 80 msec intervals. They were presented at a SPL of 70 dB at which Fig. 6.50 indicates strong compression. Accordingly, the practically undistorted empirical wave forms of Fig. 6.52 were obtained in the presence of this compression. It should be pointed out that the distortion noticeable at the offset of the lower frequency burst is not due to an input-output nonlinearity but to an interference of a signal at a nonharmonic frequency. The phenomenon is discussed in the following section.

Since no adaptation is evident in the burst responses of Fig. 6.52, how did the compression come about without producing wave form distortion? The answer lies in the asymmetry of the transients at the onsets and terminations of the burst response. The onset transients are substantially shorter than the termination ones. An even greater asymmetry is evident in Fig. 6.53 showing a Hensen's cell response to a tone burst in a different preparation. A pronounced asymmetry between the on-and-off transients was also found in basilar membrane vibration (Ruggero & Rich, 1991a). As an example, Fig. 6.54 shows some results of Ruggero and Rich (1991a) obtained at 50 dB SPL and several sound frequencies at one location of a chinchilla cochlea. The CF was around 8 kHz but the asymmetry is evident in the figure not only at this frequency but also at neighboring frequencies, ranging from 6 to 10 kHz.

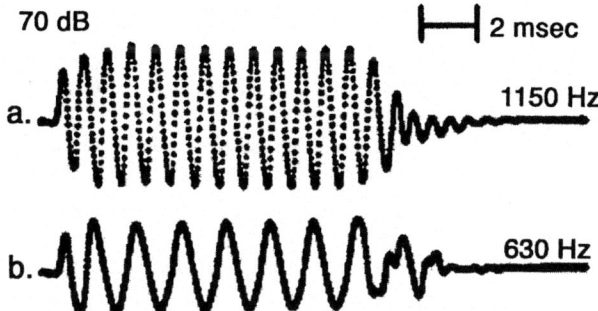

FIG. 6.52. Unfiltered intracellular responses of the Hensen's cell of Fig. 6.48 to short tone bursts at 70 dB SPL and two sound frequencies.

FIG. 6.53. Unfiltered intracellular response of another Hensen's cell to a short tone burst at a sound frequency near its maximum response.

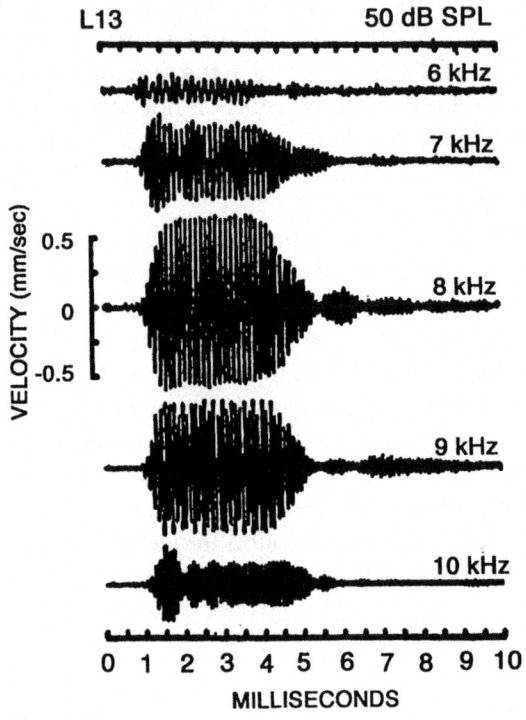

FIG. 6.54. Basilar membrane responses in the base of a chinchilla cochlea to short tone bursts at 50 dB SPL and several sound frequencies. (From Ruggero & Rich, 1991a). Reprinted from *Hearing Research 51*, M. A. Ruggero and N. C. Rich, Applications of a commercially manufactured Doppler-shift laser velocimeter to the measurement of basilar-membrane vibration, 215–230, copyright © 1991, with permission from Elsevier Science.

The asymmetry does not arise in cochlear responses in the absence of compression, as we have been able to show on our hardware model of the cochlea. The response with symmetrical transients, obtained at one location of the model is shown in Fig. 6.55 by the upper trace. The lower trace shows the rectangular input burst at the model stapes.

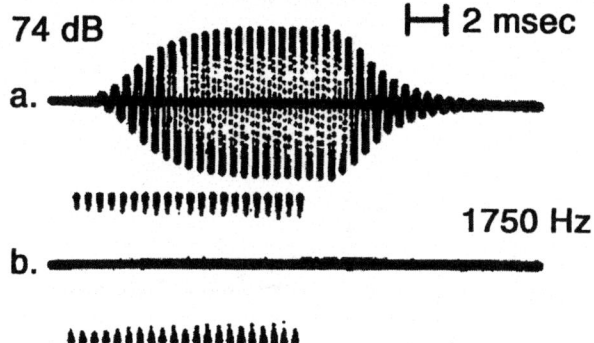

FIG. 6.55. Response of the hardware model of the cochlea to a short tone burst at 1750Hz (bottom trace).

The asymmetry between the onset and offset transients is a well-known phenomenon in electronics. It arises in electrical filters in the presence of AGC. A simple graphical explanation of the phenomenon is provided in Fig. 6.56 that contains two panels, the upper one, for an onset response, the lower one for a corresponding offset one. Hypothetical exponential transients in the absence of AGC are indicated by the open circles, the AGC time functions by the closed circles, and their products, by the solid lines. The former two functions have about the same time constant of 7 msec. The resultant onset transient, subject to AGC, has a time constant of only about 1 msec, the corresponding offset transient, that of 12 msec. The relatively long off transient is the prize the system has to pay for the short on transient. The prize may be worth paying because the on transient may contain important time information concerning such processes as stimulus precedence and sound source localization.

Indirect evidence for the effect of AGC on the cochlear transient asymmetry can be found in experiments of Ruggero and Rich (1991b), although they did not seem to pay attention to the phenomenon. They measured basilar membrane vibration in chinchilla in the presence of furosemide perfusion. Furosemide inhibits OHC motility, which is considered to be the mechanism responsible for the active feedback and, therefore, for the cochlear compression. Their recordings of basilar membrane responses to 9 kHz tone bursts presented at an SPL of 40 dB are shown in Fig. 6.57. The uppermost trace was obtained before the perfusion and shows a clear transient asymmetry. The amplitudes in the trace taken 14 minutes after the perfusion are too small to draw any reliable conclusions, but the traces taken 30 and 73 minutes after the perfusion appear to have quite symmet-

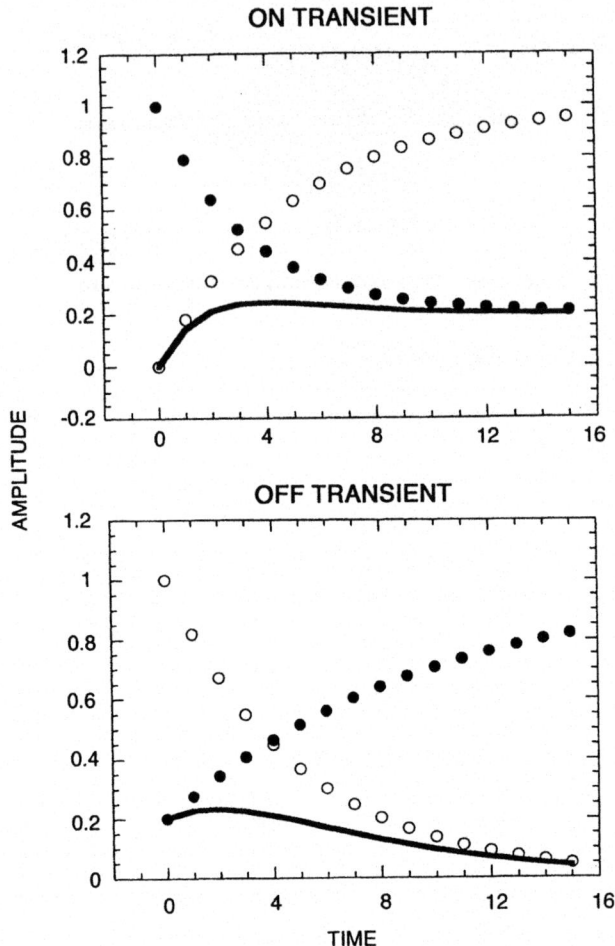

FIG. 6.56. Hypothetical functions representing three aspects of cochlear response to a short tone burst: transient linear on (upper panel) and off (lower panel) responses, shown by means of open circles; corresponding adaptation functions, shown by closed circles; their product functions, shown by means of solid lines.

rical transients. A slight asymmetry does not become noticeable until 105 minutes after the perfusion, as evidenced by the bottom trace of the figure.

The long cochlear off transients may be less important from the point of view of auditory communication but they contain valuable scientific information about the cochlea. For example, they allow us more easily to analyze their frequency contents. This is evident already in Fig. 6.52. Whereas, the sound frequency in the off transient of the higher frequency burst was almost the same as that of the burst, it was much higher in the transient of

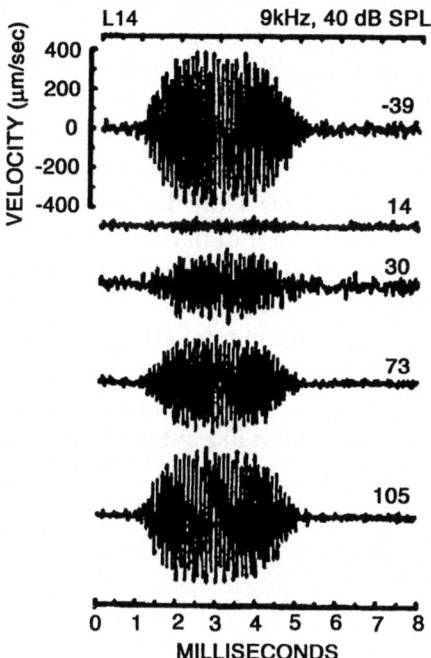

FIG. 6.57. Basilar membrane responses in the base of a chinchilla cochlea to short tone bursts at 40 dB SPL before (top trace) and at various times after furosemide perfusion (Ruggero & Rich, 1991b). From "Furosemide Alters Organ of Corti Mechanics: Evidence for Feedback of Outer Hair Cell Upon the Basilar Membrane," by M. A. Ruggero and N. C. Rich, 1991, *Journal of Neuroscience, 11*, pp. 1057–1067. Copyright © 1991 by the Society for Neuroscience. Reprinted with permission.

the lower frequency burst than in the burst itself. It is well known in physics that an oscillating system forced to oscillate at an imposed frequency returns to its natural frequency in the off transient. The frequency of the 1,150 Hz burst was near the response maximum at the SPL of 70 dB at which the input signal was delivered and, therefore, near the system's natural frequency. This was not true for the 630 Hz burst, and the higher natural frequency invaded the off transient and interfered with the frequency of the forced oscillation.

As is shown in Fig. 6.13, the cochlear partition may be regarded as two distributed resonance systems, one of basilar membrane and one of the tectorial membrane, coupled through the OHC stereocilia. Quite generally, two coupled resonance systems have two natural frequencies that depend not only on their resonance frequencies but also on the strength of the systems' coupling—the stronger the coupling the further the natural frequencies are driven apart. The natural frequencies, which are not usually harmonically related, produce a beat appearing as amplitude modulation. We were able to ascertain the presence of two beating frequency components in cochlear off transients, as exemplified in Fig. 6.58 (Zwislocki et al., 1996). This figure shows responses of one Hensen's cell to tone bursts at three different driving frequencies—1,700, 2000, and 3000 Hz. Note the long off transients whose frequency is about the same and independent of the driving frequency. Note also the amplitude modulation in the transients, caused by the presence of two component frequencies.

Two natural frequency components were also noticed by DeBoer and Nuttall (1996) in basilar membrane responses to clicks. Strong beating transients were measured in basilar membrane vibration by Ruggero and Rich (1991), as is evident in Fig. 6.54. These transients, as well as the ones shown in Fig. 6.58, strongly resemble transient cochlear otoacoustic emissions, and one has to wonder if the emissions do not originate in the off transients lasting for relatively long periods of time, due to the AGC effect.

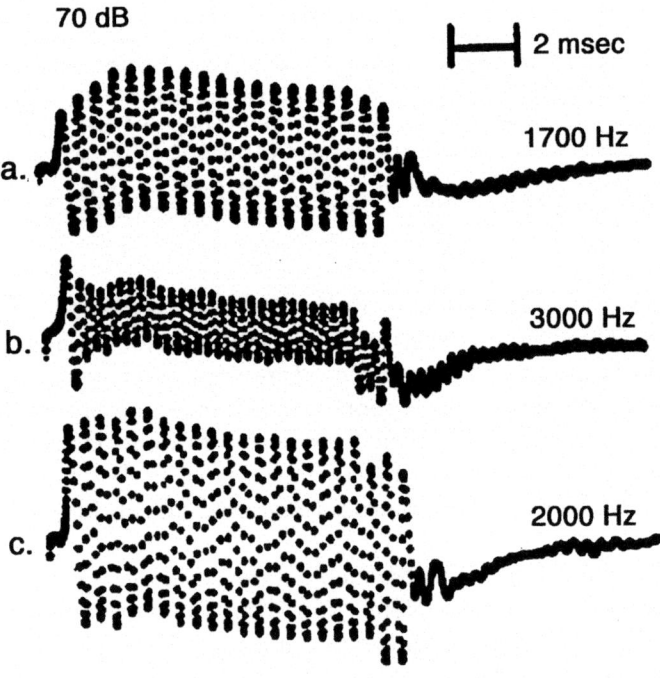

FIG. 6.58. Unfiltered intracellular responses of a Hensen's cell to short tone bursts at 70 dB SPL and three sound frequencies.

REFERENCES

Abnet, C. C., & Freeman, D. M. (2000). Deformations of the isolated mouse tectorial membrane produced by oscillatory forces. *Hearing Research 144*, 29–46.

Allen, J. B. (1980a). A cochlear micromechanic model of transduction. In G. V. D. Brink & F. A. Bilsen (Eds.), *Psychophysical, physiological and behavioral studies in hearing* (pp. 85–95). Delft, Netherlands: Delft University Press.

Allen, J. B. (1980b). Cochlear micromechanics—A physical model of transduction. *Journal of the Acoustical Society of America, 68*, 1660–1670.

Békésy, G. V. (1960). *Experiments in Hearing*. New York: McGraw Hill.

Billone, M., & Raynor, S. (1973). Transmission of radial shear forces to cochlear hair cells. *Journal of the Acoustical Society of America, 54*, 1143–1156.

Cooper, N. P. (2000). Radial variation in the vibrations of the cochlear partition: Recent developments in auditory mechanics. In H. Wade, T. Takasaker, Iheda, K. Ohyama, & T. Koike (Eds.), World Scientific, Singapore, pp. 109–115.

Dallos, P. (1973). *The Auditory Periphery*. New York: Academic Press.

Dallos, P. (1992). The active cochlea. *Journal of Neuroscience, 12*, 4575–4585.

Dallos, P., & Evans, B. N. (1995). High-frequency motility of outer hair cells and the cochlear amplifier. *Science, 267*, 2006–2009.

Dallos, P., Billone, M. C., Durrant, J. W., Wang, C.-Y., & Raynor, S. (1972). Cochlear inner and outer hair cells: Functional differences. *Science, 177*, 356–358.

Dallos, P., Ryan, A., Harris, D., McGee, T., & Özdamar, O. (1977). Cochlear frequency selectivity in the presence of hair cell damage. In E. F. Evans & J. P. Wilson (Eds.), *Psychophysics and physiology of hearing* (pp. 249–258). London: Academic Press.

De Boer, E. (1993). Some like it active. In H. Duifhuis, J. W. Horst, P. van Dijk, & S. M. van Netten (Eds.), *Biophysics of hair cell sensory systems* (pp. 3–22). Singapore: World Scientific.

De Boer, E., & Nuttall, A. L. (1996). On cochlear cross-correlation functions: Connecting nonlinearity and 'activity.' In E. R. Lewis, G. R. Long, R. F. Lyon, P. M. Navins, C. R. Steele, & E. Hecht-Poinar (Eds.), *Diversity in auditory mechanics* (pp. 291–297). Singapore: World Scientific.

Dieler, R., Shehata-Dieler, W. E., & Brownell, W. E. (1991). Concomitant salicylate-induced alterations of outer hair cell subsurface cisternae and electromotility. *Journal of Neurocytology, 20*, 637–653.

Evans, E. F. (1977). Frequency selectivity at high signal levels of single units in cochlear nerve and nucleus. In E. F. Evans & J. P. Wilson (Eds.), *Psychophysics and physiology of hearing* (pp. 185–192). London: Academic.

Evans, E. F., & Wilson, J. P. (1973). The frequency selectivity of the cochlea. In A. R. Møller (Ed.), *Basic mechanisms in hearing* (pp. 519–551). London: Academic Press.

Flock, Å. (1971). Sensory transduction in hair cells. In W. Lowenstein (Ed.), *Handbook of sensory physiology* (vol. 1, pp. 396–441). Berlin, New York: Springer.

Flock, Å. (1977). Physiological properties of sensory hairs in the ear. In E. F. Evans & J. P. Wilson (Eds.), *Psychophysics and physiology of hearing* (pp. 15–25). London: Academic Press.

Flock, Å., Flock, B., Fridberger, A., Scarfone, E., & Ulfendahl, M. (1999). Supporting cells contribute to control of hearing sensitivity. *The Journal of Neuroscience, 19*, 4498–4507.

Geisler, C. D. (1998). *From sound to synapse (physiology of the mammalian ear)*. New York: Oxford University Press.

Geisler, C. D., & Rhode, W. S. (1982). The phases of basilar membrane vibrations. *Journal of the Acoustical Society of America, 71*, 1201–1203.

Geisler, C. D., Rhode, W. S., & Kennedy, D. T. (1974). Responses to tonal stimuli of single auditory nerve fibers and their relationship to basilar membrane motion in the squirrel monkey. *Journal of Neurophysiology, 37,* 1156–1172.

Goodman, D. A., Smith, R. L., & Chamberlain, S. C. (1982). Intracellular and extracellular responses in the organ of Corti of the gerbil. *Hearing Research, 7,* 161–169.

Gummer, A. W., Hemmert, W., & Zenner, H. P. (1996). Resonant tectorial membrane motion in the inner ears: Its crucial role in frequency tuning. *Proceedings of the National Academy of Sciences, U.S.A., 93,* 8727–8732.

Hudspeth, A. J., & Corey, D. P. (1977). Sensitivity, polarity, and conductance change in the response of vertebrate hair cells to controlled mechanical stimuli. *Proceedings of the National Academy of Sciences, U.S.A., 74,* 2407–2411.

Kolston, P. J. (1999). Comparing in vitro, in situ, and in vivo experimental data in three-dimensional model of mammalian cochlear mechanics. *Proceedings of the National Academy of Sciences, U.S.A., 96,* 3676–3681.

Konishi, T., & Nielson, D. W. (1973). Temporal relationship between motion of the basilar membrane and initiation of nerve impulses in the auditory nerve fibers. *Journal of the Acoustical Society of America, 53,* 325.

Kronester-Frei, A. (1979). The effect of changes in endolymphatic ion concentrations on the tectorial membrane. *Hearing Research, 1,* 81–94.

LePage, E. L., & Johnstone, B. M. (1980a). Basilar membrane mechanics in the guinea pig cochlea. *Journal of the Acoustical Society of America, 67,* S45.

LePage, E. L., & Johnstone, B. M. (1980b). Basilar membrane mechanics in the guinea pig cochlea—Comparison of normals with kanamysin-treated animals. *Journal of the Acoustical Society of America, 67,* S46.

Liberman, M. C. (1976). *Abnormal discharge patterns of auditory-nerve fibers in acoustically traumatized cats.* Unpublished doctoral dissertation, Harvard University, Cambridge, MA.

Liberman, M. C., & Kiang, Y. S. (1978). Acoustic trauma in cats. *Acta Oto-Laryngologica Suppl., 358.*

Long, R. R., & Tubis, A. (1988). Modification of spontaneous and evoked otoacoustic emissions and associated psychoacoustic microstructure by aspirin consumption. *Journal of the Acoustical Society of America, 84,* 1343–1353.

Markin, V. S., & Hudspeth, A. J. (1995). Modeling the active process of the cochlea: Phase relations, amplification, and spontaneous oscillation. *Biophysical Journal, 69,* 138–147.

Mountain, D. C., Hubbard, A. E., & McMullen, T. A. (1983). Electromechanical processes in the cochlea. In E. De Boer & M. A. Viergever (Eds.), *Mechanics of Hearing* (pp. 119–126). Delft, Netherlands: Delft University Press.

Narayan, S. S., Temchin, A. N., Recio, A., & Ruggero, M. A. (1998). Frequency tuning of basilar membrane and auditory nerve fibers in the same cochleae. *Science, 282,* 1882–1884.

Neely, S. T., & Kim, D. O. (1986). A model for active elements in cochlear biomechanics. *Journal of the Acoustical Society of America, 79,* 1472–1480.

Nuttall, A. L., Brown, M. C., Masta, R. L., & Lawrence, M. (1981). Inner hair cell responses to the velocity of basilar membrane motion in the guinea pig. *Brain Research, 211,* 171–174.

Peake, W. T., & Ling, A., Jr. (1978). Absence of tonotopic organization in the motion of the basilar membrane of the alligator lizard. *Journal of the Acoustical Society of America, 63,* 566.

Pietras, B. W., & Zwislocki, J. J. (1993). Enhancement of cochlear single cell and CM responses after exposure to an intense tone or noise. *Journal of the Acoustical Society of America, 93,* 2368.

Rhode, W. S. (1971). Observations of the vibration of the basilar membrane in squirrel monkey using the Mössbauer technique. *Journal of the Acoustical Society of America, 49,* 1218–1231.

Rhode, W. S., & Recio, A. (2000). Study of mechanical motions in the basal region of the chinchilla cochlea. *Journal of the Acoustical Society of America, 107,* 3317–3332.

Robles, L., Ruggero, M. A., & Rich, N. C. (1986). Basilar membrane mechanics at the base of the chinchilla cochlea. I. Input-output functions, tuning curves, and response phases. *Journal of the Acoustical Society of America, 80,* 1364–1374.

Ruggero, M. A., & Rich, N. C. (1983). Chinchilla auditory nerve responses to low-frequency tones. *Journal of the Acoustical Society of America, 73,* 2096–2108.

Ruggero, M. A., & Rich, N. C. (1991a). Applications of a commercially manufactured Doppler-shift laser velocimeter to the measurement of basilar-membrane vibration. *Hearing Research, 51,* 215–230.

Ruggero, M. A., & Rich, N. C. (1991b). Furosemide alters organ of Corti mechanics: Evidence for feedback of outer hair cells upon the basilar membrane. *Journal of Neuroscience, 11,* 1057–1067.

Ruggero, M. A., Rich, N. C., Recio, A., Narayan, S., & Robles, L.. (1997). Basilar-membrane responses to tones at the base of the chinchilla cochlea. *Journal of the Acoustical Society of America, 101,* 2151–2163.

Russell, I. J., & Nilsen, K. E. (1997). The location of the cochlear amplifier: Spatial representation of a single tone on the guinea pig basilar membrane. *Proceedings of the National Academy of Sciences, 94,* 2660–2664.

Russell, I. J., & Sellick, P. M. (1977). The tuning properties of cochlear hair cells. In E. F. Evans & J. P. Wilson (Eds.), *Psychophysics and physiology of hearing* (pp. 71–84). London: Academic Press.

Russell, I. J., & Sellick, P. M. (1978). Intracellular studies of hair cells in the mammalian cochlea. *Journal of Physiology, 284,* 261–290.

Russell, I. J., & Sellick, P. M. (1983). Low-frequency characteristics of intracellularly recorded receptor potentials in guinea-pig cochlear hair cells. *Journal of Physiology, 338,* 179–206.

Russell, I. J., Cody, A. R., & Richardson, G. P. (1986). The responses of inner and outer hair cells in the basal turn of the guinea-pig cochlea and in the mouse cochlea grown in vitro. *Hearing Research, 22,* 199–216.

Santos-Sacchi, J. (1992). On the frequency limit and phase of outer hair cell motility: Effects of the membrane filter. *Journal of Neuroscience, 12,* 1906–1916.

Saunders, J. C., & Flock, Å. (1986). Recovery of threshold shift in hair-cell stereocilia following exposure to intense stimulation. *Hearing Research, 23,* 233–244.

Schmiedt, R. A. (1977). Single and two-tone effects in normal and abnormal cochleas: A study of cochlear microphonics and auditory nerve units. Unpublished doctoral dissertation, Syracuse University. Special Report, ISR-S-16, Institute for Sensory Research, Syracuse University, Syracuse, New York.

Schmiedt, R. A., & Zwislocki, J. J. (1977). Comparison of sound-transmission and cochlear-microphonic characteristics in Mongolian gerbil and guinea pig. *Journal of the Acoustical Society of America, 61,* 133–149.

Schmiedt, R. A., Zwislocki, J. J., & Hamernik, R. P. (1980). Effects of hair cell lesions on responses of cochlear nerve fibers. I. Lesions, tuning curves, two-tone inhibition, and responses to trapezoidal-wave patterns. *Journal of Neurophysiology, 43,* 1367.

Sellick, P. M., & Russell, I. J. (1980). The responses of inner hair cells to basilar membrane velocity during low frequency auditory stimulation in the guinea pig cochlea. *Hearing Research, 2,* 439–445.

Sellick, P. M., Patuzzi, R., & Johnstone, B. M. (1982). Measurement of basilar membrane motion in the guinea pig using the Mössbauer technique. *Journal of the Acoustical Society of America, 72,* 131–141.

Shambough, G. E. (1907). A restudy of the innate anatomy of the structures in the cochlea with conclusions bearing on the solution of the problem of tone perception. *American Journal of Anatomy, 7,* 245–258.

Shehata, W. E., Brownell, W. E., Cousillas, H., & Imredy, J. P. (1990). Salicylate alters membrane conductance of outer hair cells and diminishes rapid electromotile responses. *Association for Research in Otolaryngology Abstracts, 13,* 252–253.

Sokolich, W. G., Hamernik, R. P., Zwislocki, J. J., & Schmiedt, R. A. (1976). Inferred response, polarities of cochlear hair cells. *Journal of the Acoustical Society of America, 59,* 963.

Spoendlin, H. (1967). Innervation of the organ of Corti. *Journal of Laryngology and Otology,* 717–737.

Spoendlin, H. (1973). The innervation of the cochlear receptor. In A. R. Møller (Ed.), *Basic mechanisms in hearing* (pp. 185–230). New York: Academic.

Strelioff, D., & Flock, Å. (1984). Stiffness of sensory-cell hair bundles in the isolated guinea pig cochlea. *Hearing Research, 15,* 19–28.

Strelioff, D., Flock, Å., & Minser, K. E. (1985). Role of inner and outer hair cells in mechanical frequency selectivity of the cochlea. *Hearing Research, 18,* 169–175.

Stypulkowski, P. H. (1990). Mechanisms of salicylate ototoxicity. *Hearing Research, 46,* 113–164.

Szymko, Y. M., Nelson-Adesokan, P. M., & Saunders, J. C. (1995). Stiffness changes in chick hair bundles following in vitro overstimulation. *Journal of Comparative Physiology (A), 176,* 727–735.

Szymko, Y. M., Zwislocki, J. J., & Hertig, L. (1997). Enhanced cochlear responses after sound exposure. *Hearing Research, 110,* 164–178.

Tolomeo, J. A., & Holley, M. C. (1997). The function of the cytoskeleton in determining the mechanical properties of epithelial cells within the organ of Corti. In E. R. Lewis, G. R. Long, R. F. Lyon, P. M. Narins, C. R. Steele, & E. Hecht-Poinar, (Eds.), *Diversity in auditory mechanics* (pp. 556–562). Singapore: World Scientific.

Weiss, T. F., Mulroy, M. J., Turner, R. G., & Pike, C. L. (1976). Tuning of single fibers in the cochlear nerve of the alligator lizard: relation to receptor morphology. *Brain Research, 115,* 71–90.

Wersäll, J., Flock, Å., & Lindquist, P. G. (1965). Structural basis for directional sensitivity in cochlear and vestibular sensory receptors (pp.115–132). Cold Spring Harbor Symposium on Quantitative Biology, 30.

Wever, E. G. (1967). The tectorial membrane of the lizard ear: Species variations. *Journal of Morphology, 123,* 355–372.

Wilson, J. P., & Evans, E. F. (1983). Some observations on the 'passive' mechanics of the cat basilar membrane. In W. R. Webster & L. M. Aitkin (Eds.), *Mechanisms of hearing* (pp. 30–35). Clayton, Victoria, Australia: Monash University Press.

Zhang, M., & Zwislocki, J. J. (1995). OHC response recruitment and its correlation with loudness recruitment. *Hearing Research, 85,* 1–10.

Zwicker, E. (1971). Die Afmessungen des Innenohres das Hauschweines [The dimensions of the inner ear of domestic pig]. *Acustica, 25,* 232–239.

Zwislocki, J. J. (1974). A possible neuro-mechanical sound analysis in the cochlea. *Acustica, 31,* 354–359.

Zwislocki, J. J. (1980). Five decades of research on cochlear mechanics. *Journal of the Acoustical Society of America, 67,* 1679–1685.

Zwislocki, J. J. (1983). Cochlear micromechanics—a model and some of its consequences. In W. R. Webster & L. M. Aitkin (Eds.), *Mechanics of hearing* (pp. 21–26). Clayton, Victoria, Australia: Monash University Press.

Zwislocki, J. J. (1984). How OHC lesions can lead to neural cochlear hypersensitivity. *Acta Oto-Laryngologica, 97,* 529–534.

Zwislocki, J. J. (1986a). Changes in cochlear frequency selectivity produced by tectorial-membrane manipulation. In B. C. J. Moore & R. D. Patterson (Eds.), *Auditory frequency selectivity—A NATO advanced workshop* (pp. 3–11). New York: Plenum Press.

Zwislocki, J. J. (1986b). Analysis of cochlear mechanics. In Å. Flock & J. Wersäll (Eds.), *Cellular mechanisms in hearing* (pp. 155–169). Amsterdam: Elsevier Science.

Zwislocki, J. J. (1990). Active cochlear feedback: Required structure and response phase. In P. Dallos, C. D. Geisler, J. W. Matthews, M. A. Ruggero, & C. R. Steele (Eds.), The Mechanics and Biophysics of Hearing (pp. 114–120). Berlin, Germany: Springer.

Zwislocki, J. J. (1991). What is the cochlear place code for pitch? *Acta Oto-Laryngologica, 111,* 256–262.

Zwislocki, J. J. (1994). Differential intensity sensitivity in relation to subjective magnitudes: Experimental results and mathematical theory. In L. M. Ward (Ed.), *Proceedings of the Tenth Annual Meeting of the International Society for Psychophysics* (pp. 1–10). Vancouver, Canada: University of British Columbia Press.

Zwislocki, J. J. (2000). Response phases of cochlear hair cells. In H. Wada, T. Takasaka, K. Ikeda, K. Ohyama, & T. Koide (Eds.), Symposium on Recent Developments in Auditory Mechanics (pp. 158–164). Singapore: World Scientific.

Zwislocki, J. J., & Cefaratti, L. K. (1989). Tectorial membrane II: Stiffness measurements in vivo. *Hearing Research, 41,* 211–228.

Zwislocki, J. J., & Kletsky, E. J. (1979). Tectorial membrane: A possible effect on frequency analysis in the cochlea. *Science, 204,* 639–641.

Zwislocki, J. J., & Kletsky, E. J. (1981). Outer hair cells: Sharpness of tuning. *Acta Oto-Laryngologica, 91,* 481–485.

Zwislocki, J. J., & Kletsky, E. J. (1982). What basilar-membrane tuning says about cochlear micromechanics. *American Journal of Otolaryngology, 3,* 48–52.

Zwislocki, J. J., & Nguyen, M. (1999). Place Code for Pitch: A Necessary Revision. *Acta Oto-Laryngologica, 119,* 140–145.

Zwislocki, J. J., & Sokolich, W. G. (1973). Velocity and displacement responses in auditory-nerve fibers. *Science, 182,* 64–66.

Zwislocki, J. J., Szymko, Y. M., & Hertig, L. Y. (1996). The cochlea is an automatic gain control system after all. In E. R. Lewis, G. R. Long, R. F. Lyon, P. M. Narins, C. R. Steele, & E. Hecht-Poinar (Eds.), *Diversity in auditory mechanics* (pp. 354–360). Singapore: World Scientific.

chapter

Pitch and Loudness Codes

*O*ne of the main reasons for anatomical and physiological studies of sensory systems is the desire to understand the biological mechanisms underlying human sensations and perceptions. In hearing, pitch and loudness may be considered as the key subjective attributes of sound. It is not surprising, therefore, that a voluminous literature has been devoted to these sensations, and the curiosity about them may be traced to antiquity. This appears to be particularly true for the pitch, which is a unique subjective attribute of sound. It is not the intent of this chapter to trace the history of the concepts of pitch or loudness, of the various attempts at understanding the physical parameters controlling them or the physiological mechanisms underlying them. Attempts at determining their physiological codes in the modern sense seem to have begun in the 19th century—in particular through Helmholtz (1877), who proposed a physiological code for the pitch and studied extensively various subjective attributes of sound.

As all people involved professionally with the sense of hearing must know, Helmholtz proposed that the pitch of a pure tone was determined by the cochlear location at which the vibration amplitude of the basilar membrane was at its maximum. In the final version of his hypothesis, he ascribed the occurrence of this maximum to resonance of elastic fibers embedded radially in the basilar membrane. This mechanism was suggested by the already well-known vibration of strings in musical instruments. The vibration maximum had been entirely hypothetical until it was confirmed in pioneering experiments of Békésy's (1960), first indirectly in the first quarter of the 20th century, then directly, in cochlear postmortem preparations just before the mid century. Békésy had some doubts as to the nature of the maximum and seemed to think, it was produced by some sort of a mechanism not yet well understood in physics. I attempted to define this mechanism mathematically (Zwislocki-Moscicki, 1946, 1948), and the attempt is reproduced in a somewhat revised form in this monograph. Admittedly, I never thought of it as a fundamentally new mechanism but, rather, as a somewhat unique combination of well-known principles. In any event, Helmholtz may have been wrong with respect to the nature of the cochlear vibration maximum but not with respect to its existence. As has been mentioned in this monograph on several occasions, the existence of the maximum was confirmed experimentally many times not only in postmortem preparation but also in live animals of diverse species. I have attempted to define its nature both theoretically and experimentally. The latest version of these attempts is contained in this monograph. Perhaps their most important aspect from the point of view of pitch coding, which was entirely unexpected, is the finding that the cochlear vibration maximum cannot be an adequate pitch code, and that the classical place code of Helmholtz's, as it has become known in Békésy's times, has to be modified. The modification is discussed in this chapter.

Serious attempts at finding a physiological code for loudness had to wait until the neurophysiology of the auditory system was sufficiently developed at the beginning of the 20th century (Stevens & Davis, 1938, for review). The prevailing view has been that loudness is determined by the totality of the nerve impulses traveling up the auditory nerve. Until Spoendlin's (1967) discovery that almost all the afferent fibers of this nerve innervate the cochlear inner hair cells and only a few, the outer hair cells, it had been believed in analogy to the rods and cones in the mammalian eye that the outer hair cells, because of their anatomical position, conveyed information about weak sounds and the inner hair cells made their contributions when the sounds became sufficiently strong. This hypothesis was justified to a large extent by the enormous range of sound intensities processed by the auditory system and by the relatively small intensity ranges of individual nerve fibers. After Spoendlin's discovery, the task of accounting for the auditory intensity range became even more difficult. It has been solved thus far in part by the discovery of cochlear compression discussed in the preceding chapter and by the discovery that neurons innervating the IHCs have staggered sensitivities and varied dynamic ranges (Sachs & Abbas, 1974; Liberman, 1978). In the final section of this chapter, I discuss

PLACE CODE FOR PITCH

The cochlear place code for pitch, as proposed by Helmholtz (1877) and supported by Békésy's (1960) experiments, encounters two fundamental difficulties. One concerns our sensitivity to sound frequency changes, the other, the dependence of the vibration maximum in the cochlea on sound intensity. The first became masked to some extent by the prevailing manner of plotting the sensitivity of a nerve fiber as a function of sound frequency. Because of technical difficulties, it became easier to plot the sensitivity in terms of the SPL required to produce a criterion neural firing rate than to plot the firing rate for a constant SPL. The result of the former technique has been called a tuning curve. A typical example with a CF in the frequency region of Hensen's cell transfer functions of Fig. 7.1 (Zwislocki & Nguyen, 1999) is shown in Fig. 7.2 (Liberman & Kiang, 1978). In fact, the tuning curve is an inverse transfer function, in contrast to a forward transfer function used exclusively in this monograph. In a linear system, an inverse transfer function is a linear inversion of the forward transfer function and has the same, although inverted, shape. This is not true in a nonlinear system, however, as is made evident in Fig. 7.3 (Zwislocki, Solessio, & Cefaratti, 1991), in which forward transfer functions recorded on an OHC and shown in Fig. 7.4 (Zwislocki, 1991) are compared to the tuning curves (solid lines) constructed from them. The tuning curves were obtained simply by making horizontal cuts through the transfer functions of Fig. 7.4 at three levels coinciding with the peaks of the curves determined at 20, 30, and 40 dB SPL, respectively. To make the two families of curves comparable, both were plotted on log scales, and the transfer functions were inverted. The differences between the two curve families are clearly apparent. First, the tips of the tuning curves are sharper than those of the transfer functions, with the difference growing with the increasing tip SPL, as indicated by the Q_{10dB} factors entered in the figure. Second, the tuning curves have much steeper, almost vertical, high frequency skirts. The greater sharpness of the tuning curves by comparison to that of the transfer functions, which is already apparent by visual inspection of Figs. 7.1 and 7.2, even though the curves of Fig. 7.1 are plotted on a linear scale, whereas those of Fig. 7.2 are compressed by a log scale, may have given the impression that the sharpness of the peaks was sufficient to account for the extraordinarily keen frequency discrimination we are capable of. However, the auditory system receives its signals filtered according to the forward transfer functions whose peaks tend to be rounded, as in Figs. 7.1 and 7.4, and are entirely incompatible with sharp frequency discrimination.

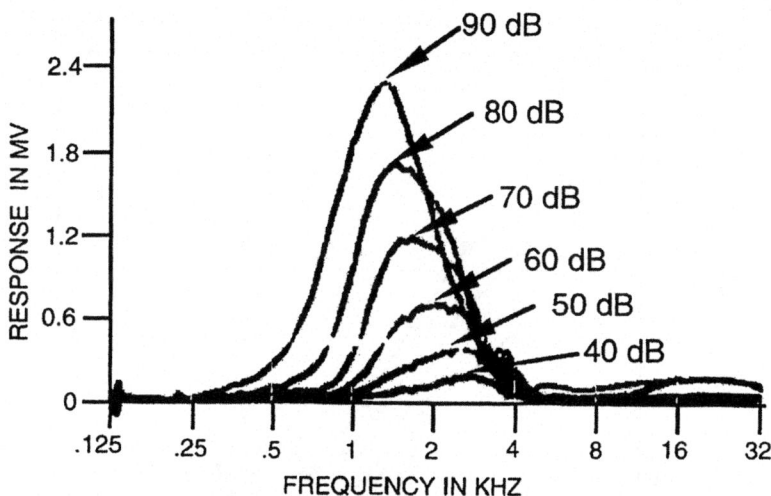

FIG. 7.1. Intensity series of Hensen's cell transfer functions at several SPLs (from Zwislocki & Nguyen, 1999). From "Place Code for Pitch: A Necessary Revision," by J. J. Zwislocki and M. Nguyen, 1999, *Acta Oto-Laryngologica, 119,* pp. 140–145. Copyright © 1999 by Taylor & Francis. Reprinted with permission.

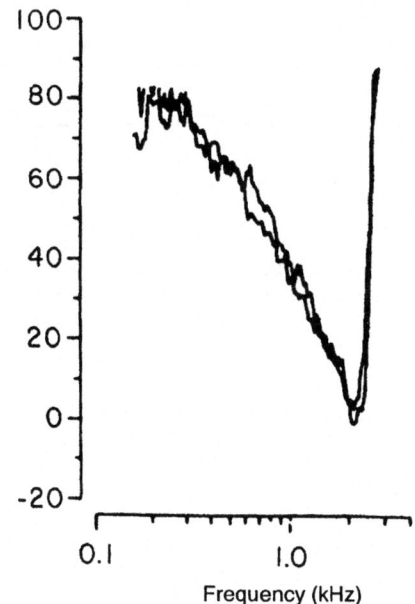

FIG. 7.2. Neural tuning curves of domestic cat in mid frequency range (from Liberman & Kiang, 1978). From "Acoustic Trauma in Cats," by M. C. Liberman and N. Y. S. Kiang, 1978, *Acta Oto-Laryngologica, Suppl. 358.* Copyright © 1978 by Taylor & Francis. Reprinted with permission.

393

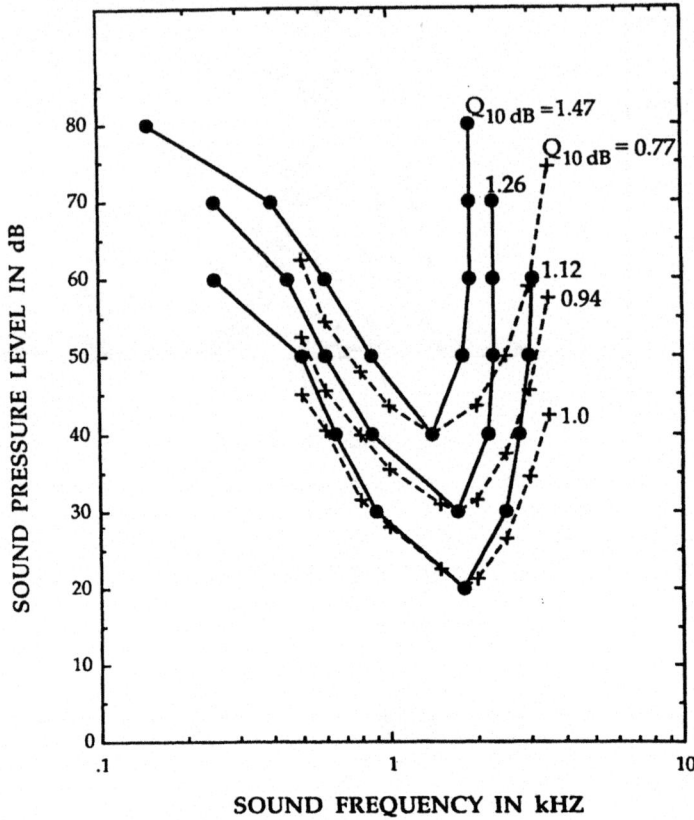

FIG. 7.3. Inverted low SPL forward transfer functions derived from the intensity series of OHC transfer functions of Fig. 7.4 (intermittent lines) and tuning curves constructed from the same (solid lines; from Zwislocki et al., 1991).

The second and probably more decisive difficulty with the classical place code for pitch results from the dependence of the frequency of the cochlear response maximum on sound intensity. Whether measured in terms of the vibration of the basilar membrane or hair cells' receptor potentials, as discussed in the preceding two chapters, or in terms of neural firing rates (e.g., Rose, Brugge, Anderson, & Hind, 1967, 1971; Møller, 1983), the maximum recorded at a given location in a healthy preparation always moves toward lower sound frequencies, as SPL is increased. The phenomenon is present in all cochlear turns (e.g., Chatterjee & Zwislocki, 1997; Rhode & Recio, 2000). As shown in Fig. 7.1 for a Hensen's cell, the shift of the maximum can be quite substantial and exceed an octave when the SPL is increased from 40 to 80 dB. Were the pitch coded by the location of the

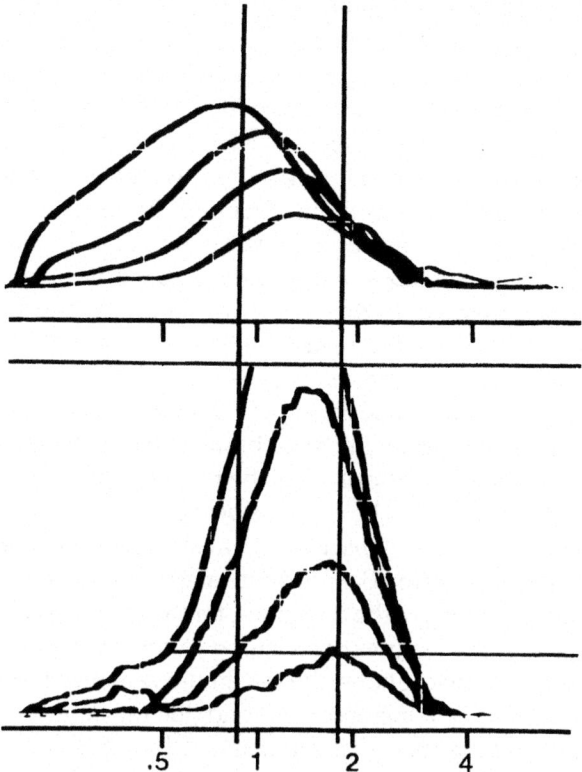

FIG. 7.4. Intensity series of OHC transfer functions at several SPLs (20, 30, 40, 50 dB in the lower panel; 50. 60, 70, 80 dB in the upper panel with reduced amplification; from Zwislocki, 1991). From "What Is the Cochlear Place Code for Pitch?" by J. J. Zwislocki, 1991, *Acta Oto-Laryngologica, Suppl.111*, pp. 256–262. Copyright © 1991 by Taylor & Francis. Reprinted with permission.

maximum, we would have to expect the sound frequency to vary substantially with SPL in order to keep the pitch invariant. This, of course, does not happen. As numerous experiments have shown, the same pitch can be maintained with a very small frequency variation over wide sound frequency ranges (e.g., Stevens, 1935; Gulick, 1971). This discrepancy by itself is sufficient to disqualify the location of the cochlear response maximum as a pitch code.

An even more direct demonstration that the location of the response maximum cannot be the place code for pitch is obtained by measuring cochlear responses at several cochlear locations. This was done by us for the Hensen's cell alternating potentials in the second turn of the Mongolian gerbil cochlea (Zwislocki & Nguyen, 1999). We used essentially the same techniques as described in chapter 5, except that up to three holes were drilled in the cochlear capsule over the lateral wall of scala media, as can be seen in Fig. 5.42. The distances between the holes were carefully measured with the help of the micromanipulator scale. Because we found the endolymphatic potential to be quite sensitive to the number of the holes, several variations of the procedure were used. We noticed that, when the more basally located hole was drilled first and the measurements made through it, then, the more apical hole was drilled and the measurement made through it, the difference in the response patterns between the two holes was greater than when the order was reversed. The relationships were checked by repeating the measurements through the first hole. Because of the order effect, in some experiments, both holes were drilled before making the measurements. Examples of our results are shown in Figs. 7.5 and 7.6. For the first, both the more basal hole and the measurements made through it preceded the drilling of the more apical hole and the measurements made through it. The distance between the holes amounted to 226 µm. At either hole, the Hensen's cell transfer functions were obtained at 6 SPLs ranging from 40 to 90 dB in 10 dB steps. For the more basal turn, they are plotted by means of the dark lines, for the more apical one, by the lighter gray lines. Three features of the curve patterns of Fig 7.5 are significant. First of all, at both locations, the maximum response moves to lower sound frequencies as SPL increases. Second, the maxima of one location overlap in sound frequency with the maxima of the other. For example, the 50 dB maximum belonging to the apical location coincides in sound frequency with the 60 dB maximum belonging to the basal location, the same is true for the 60 and 70 dB maxima belonging to the apical and basal locations, respectively, and for the 70 and 80 dB maxima too. Third, the high frequency skirts of the transfer functions belonging to the same location converge and have common feet, each foot being characteristic of each location. A similar pattern is evident in Fig. 7.6. However, the frequency difference between the bottom parts of the high frequency skirts is somewhat greater, corresponding to a greater separation between the experimental holes, which amounted to approximately 500 µm. Because of the approximately log relationship between cochlear distance and sound frequency, the approximately doubled distance between the holes in Fig. 7.6 relative to that in Fig. 7.5 should have produced a doubled log of the frequency ratios between the high frequency skirts at their feet. The ratios actually estimated from the figures came out to be only slightly smaller.

The overlap of the response maxima determined at two different cochlear locations should leave no doubt that the location of the response maximum in the cochlea cannot constitute a place code for pitch. If it did, the overlap would mean that the same pitch could be produced at two dif-

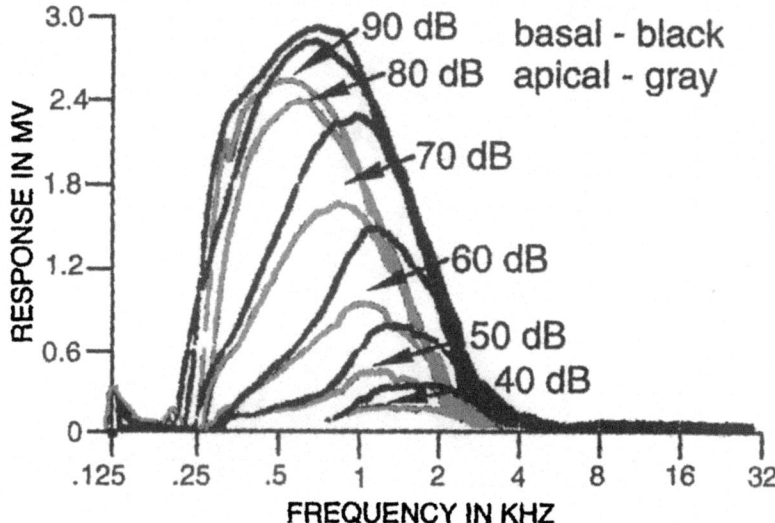

FIG. 7.5. Intensity series of Hensen's cell transfer functions recorded at two cochlear locations 226 μm apart (Zwislocki & Nguyen, 1999). From "Place Code for Pitch: A Necessary Revision," by J. J. Zwislocki and M. Nguyen, 1999, *Acta Oto-Laryngologica, Suppl. 119*, pp. 140–145. Copyright © 1999 by Taylor & Francis. Reprinted with permission.

ferent locations, depending on SPL. It would also mean that, for a constant sound frequency, the pitch changes with sound intensity. This is contrary to common experience.

If the location of the response maximum in the cochlea is not the place code for pitch, can any other feature of the cochlear excitation pattern assume this role. According to the patterns of Figs. 7.1, 7.4, 7.5, and 7.6, there is only one that could qualify. It is the high frequency cut off of the excitation, which has its spatial equivalent in the apical cut off. As is evident in the figures, the cut off is, like the pitch, approximately independent of SPL. Significantly, the independence is not complete. For sound frequencies above about 2,000 Hz and high SPLs, the pitch tends to increase slightly with SPL. This corresponds to a displacement of the high frequency cut off toward lower frequencies. Such a displacement is evident in the figures.

The near SPL independence of the high frequency, or apical cut off of cochlear excitation is a necessary but not sufficient condition for accepting the cut off as a cochlear place code for pitch. Additional evidence is needed. We were able to find one piece of such evidence quite incidentally. It concerns pitch changes following cochlear impairments. There is substantial literature indicating that hearing loss of cochlear origin produces an

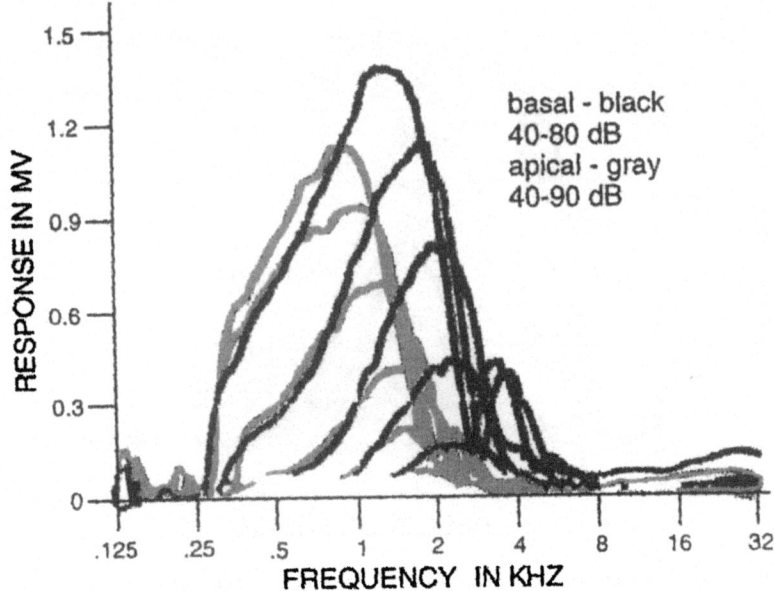

FIG. 7.6. Intensity series of Hensen's cell transfer functions recorded at two cochlear locations 500 μm apart (from Zwislocki & Nguyen, 1999). From "Place Code for Pitch: A Necessary Revision," by J. J. Zwislocki and M. Nguyen, 1999, *Acta Oto-Laryngologica, Suppl. 119*, pp. 140–145. Copyright © 1999 by Taylor & Francis. Reprinted with permission.

upward pitch shift even when the hearing loss is reasonably evenly distributed over the audible frequency range (Burns & Turner, 1986). The pitch shift is the most pronounced at low SPLs and tends to become smaller as SPL is increased. In the words of Burns and Turner: "Another manifestation of impaired pitch processing is in the form of exaggerated pitch-level effects. These exaggerated pitch-level effects appear, in many cases, to be a consequence of an increased pitch at low stimulus levels in the impaired ear which recruits toward 'normal' at higher levels" (p. 1540). Similar effects were found by us in frequency changes of the high frequency skirts of cochlear transfer functions after cochlear impairment produced by broad band noise at 100 dB SPL. An example concerning Hensen's cell alternating potentials is shown in Fig. 7.7 (Zhang & Zwislocki, 1992, 1996). In the upper panel are plotted transfer functions obtained before noise exposure at SPLs ranging from 20 to 60 dB in 10 dB steps, in the lower panel, those obtained after 140 minutes of exposure, the latter at SPLs of 40 to 70 dB. The recorded potentials of the lower panel were amplified 4 times relative to those of the upper panel. Clearly, the noise exposure shifted the bottom

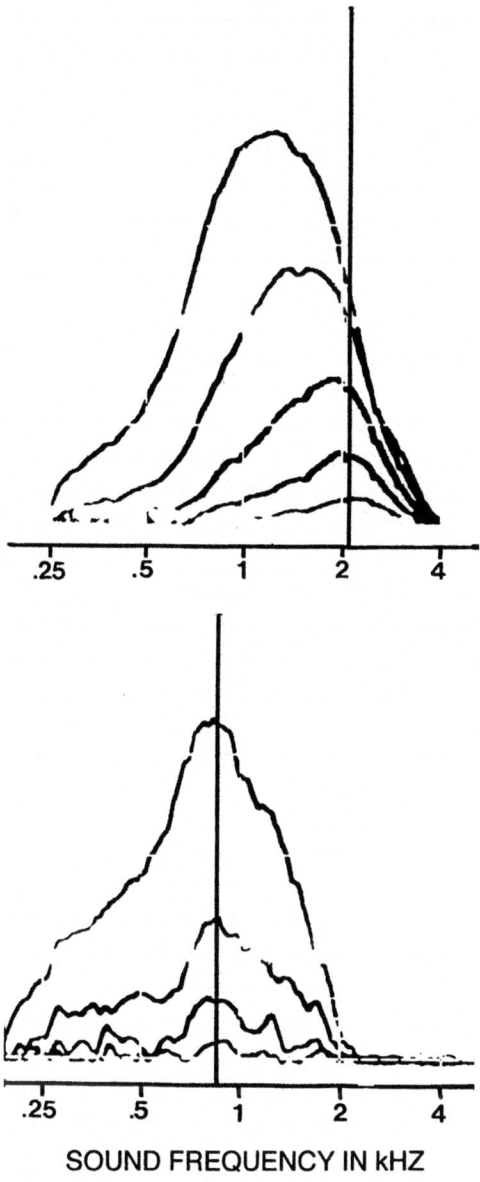

SOUND FREQUENCY IN kHz

FIG. 7.7. Intensity series of Hensen's cell transfer functions at SPLs ranging from 20 to 60 dB before noise exposure (upper panel). Intensity series of a neighboring Hensen's cell's transfer functions at SPLs ranging from 40 to 70 dB (lower panel; from Zhang, 1993). From "Effects of Noise on Cochlear Transfer Functions at the Cellular Level and Their Possible Underlying Mechanisms," by M. Zhang, 1993, *Special Report ISR-5-28*. Copyright © 1993 by Institute for Sensory Research. Reprinted with permission.

parts of the high frequency skirts from about 4,000 to about 2,000 Hz. This is equivalent to an upward pitch shift by an octave. Note that the postexposure skirts are spaced widely along the frequency scale in contrast to the preexposure ones. The wide spacing suggests a downward pitch change as SPL is increased. Thus, both features of pitch change found psychophysically in an impaired cochlea are reflected in the changes of the frequency positions of the high frequency skirts.

It should be pointed out that the entire transfer functions were shifted to lower sound frequencies after the noise exposure. This was particularly evident for the response peak. The peak shift was consistent with the shift of the high frequency skirt and with the upward pitch shift. However, the sound frequency of the peak did not increase with SPL, as did that of the skirt, but remained constant, in disagreement with the psychophysically determined pitch shift. This is another indication that the location of the excitation peak in the cochlea is not the place code for pitch.

With respect to the postexposure shift of the transfer functions to lower sound frequencies, it should be noted that this shift is consistent with a loss of the active feedback, as is evident in Fig. 6.43 (Zhang & Zwislocki, 1992, 1996). According to this figure, the frequency shift depends on the feedback—the greater the feedback the more are the transfer functions pushed toward higher sound frequencies. In the absence of the feedback, the frequency positions of the transfer functions should remain constant, as is evident in the lower panel of Fig. 7.7.

COCHLEAR LOUDNESS CODE

The two characteristic features of loudness that have to be accounted for are: the enormous range of sound intensity it encompasses and its nonlinear rate of growth. The former extends over at least 6 orders of magnitude of sound pressure, the latter follows a power function with an exponent on the order of 0.6 (e.g., Scharf, 1978, for review). The compression resulting from the 0.6 power reduces the physical range of the 6 orders of magnitude to roughly 4 orders—still very impressive. How can the auditory system handle such a range with receptor and neural units whose ranges are limited to roughly two orders of magnitude? Many answers have been proposed. I will not review them here but, rather limit this section to our own observations that may help in the future to find the ultimate solution.

All the pertinent cochlear information with respect to intensity coding is contained in transfer functions of the hair cells, in particular IHCs on which most of the auditory nerve fibers end. An example of an intensity series of such transfer functions is displayed in Fig. 7.8 (Chatterjee & Zwislocki, 1998). It has the usual characteristics already discussed on several occasions. With increasing SPL, the response magnitudes are increased, but unevenly, so that the response maximum is shifted gradually toward the lower frequencies, and the high frequency skirts converge on a common foot. By making vertical cuts through the curve family of Fig. 7.8, it is possible to obtain input–output functions at selected sound frequencies. Such curves are shown in Fig. 7.9 for the frequencies of 2,400, 2,000 and 1,400 Hz, which coincide respectively with the peaks of the transfer functions determined at the SPLs of 30, 40, and 50 dB (Chatterjee & Zwislocki, 1998). All three input–output functions tend toward saturation as the SPL is increased. Note, however, that the saturation response is relatively high at the frequency of the 50 dB peak and low at that of the 30 dB peak, which is close to the cochlear CF, where the response sensitivity is the greatest.

Studies of neurons of the auditory nerve often focus on the response characteristics obtained at the CFs of these neurons. According to experimental evidence, the neurons are classified in three groups according to their CF thresholds (Liberman, 1978). Of course, the threshold of every neuron must be the lowest at the frequency of the cochlear peak response

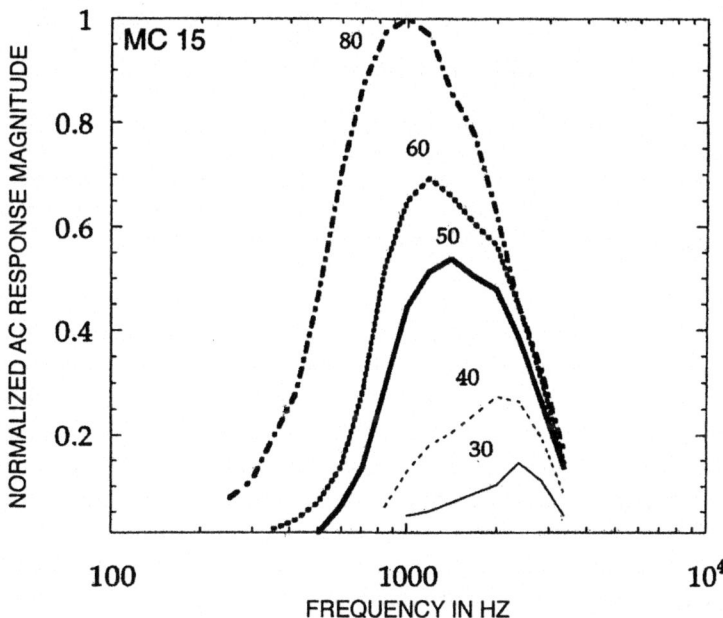

FIG. 7.8. Intensity series of IHC transfer functions at SPLs ranging from 30 to 80 dB. (from Chatterjee & Zwislocki, 1998). Reprinted from *Hearing Research, 124*, M. Chatterjee & J. J. Zwislocki, Cochlear Mechanics of Frequency and Intensity Coding II: Dynamic Range and the Code for Loudness, pp. 170–181, copyright © 1998, with permission from Elsevier Science.

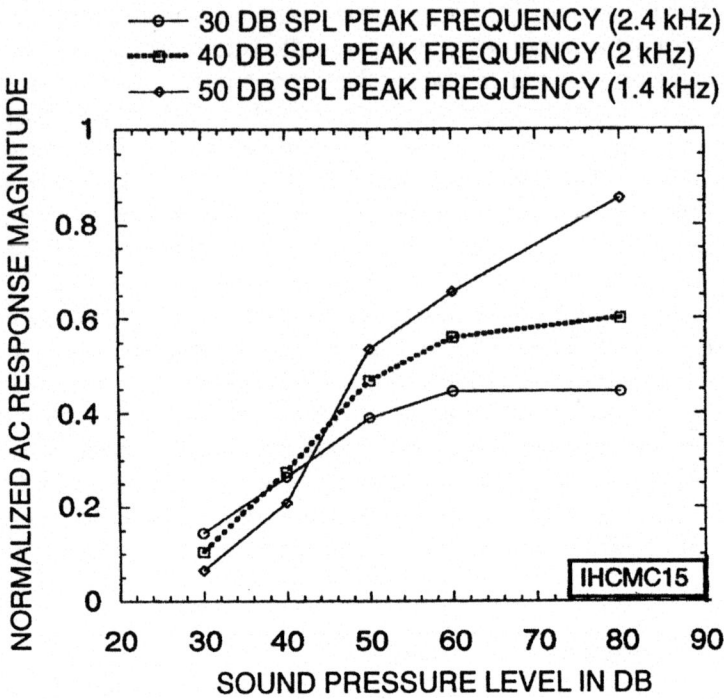

FIG. 7.9. Input–output functions constructed from the transfer functions of Fig. 7.8 at three sound frequencies (from Chatterjee & Zwislocki, 1998). Reprinted from *Hearing Research, 124,* M. Chatterjee and J. J. Zwislocki, Cochlear Mechanics of Frequency and Intensity Coding II: Dynamic Range and the Code for Loudness, pp. 170–181, copyright © 1998, with permission from Elsevier Science.

since its stimulus is the strongest there (Chatterjee & Zwislocki, 1998). But this frequency depends on SPL and becomes lower as the SPL is increased. As a consequence, neurons innervating the same IHC but having different thresholds must have different CFs. The ones with lower thresholds must have higher CFs than those with higher thresholds. On the other hand, neurons having the same CFs but different thresholds must innervate different IHCs. It should be pointed out, however, that these rules only apply to healthy cochleas (Zhang & Zwislocki, 1995, 1996). In impaired cochleas, where the location of the maximum response does not depend on SPL, neurons innervating the same IHCs can have the same CFs even if they have different thresholds.

Neurons with low-response thresholds are known to have small dynamic ranges. In the past, this was attributed to the properties of the neurons. However, the low-response saturation at the cochlear level near the low SPL maximum suggests that the small dynamic range seen in these neurons may be imposed by the dynamics of the cochlea (Chatterjee & Zwislocki, 1998). Accordingly, the cochlear input–output functions measured near the frequency of the greatest sensitivity for a given cochlear location look very much like the neuronal CF input–output functions, as exemplified in Fig. 7.10. As the neuronal thresholds become higher the neuronal CFs must coincide with cochlear response peaks belonging to higher SPLs and displaced, therefore, toward lower sound frequencies, where the cochlear response ranges are increased. This may be the reason why neurons with higher thresholds have greater dynamics ranges (Chatterjee & Zwislocki, 1998).

The small dynamic range around the frequency of the greatest sensitivity at a given cochlear location, which may be called the cochlear CF, makes it impossible for any one cochlear location to encode loudness over its entire dynamic range. According to our results, this can be accomplished in only one of two ways. Either different locations contribute their inputs

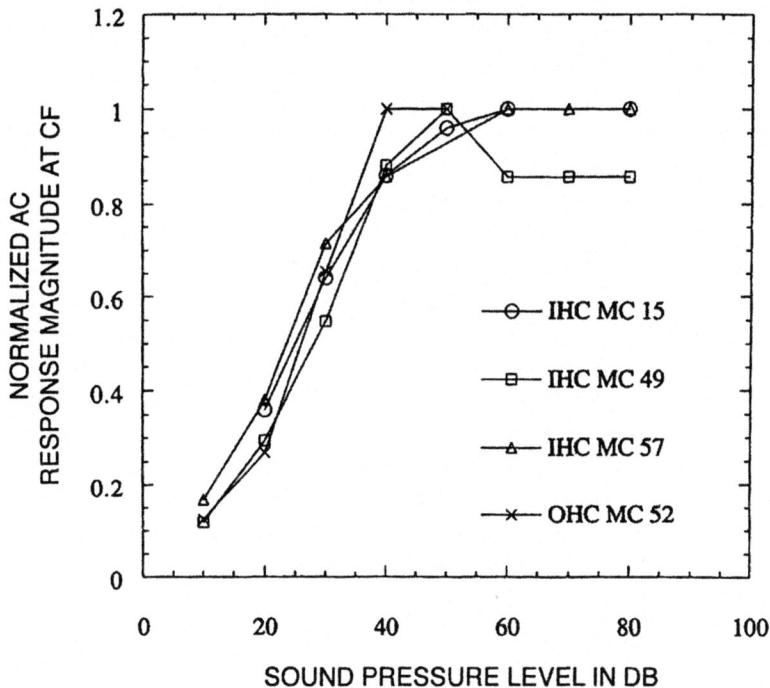

FIG. 7.10. Input–output functions of three IHCs and one OHC measured near the frequencies of their greatest sensitivities (Chatterjee & Zwislocki, 1998). Reprinted from *Hearing Research, 124,* M. Chatterjee and J. J. Zwislocki, Cochlear Mechanics of Frequency and Intensity Coding II Dynamic Range and the Code for Loudness, pp. 170–181, copyright © 1998, with permission from Elsevier Science.

consecutively when the maximum response moves toward lower sound frequencies or, reciprocally, toward the cochlear base, as SPL is increased, or all locations summate their inputs, and loudness is determined by what is sometimes called the response area. The input–output functions obtained according to the two schemes on 6 IHCs of 6 Mongolian gerbils are compared to each other and to a typical psychophysical loudness function in Fig. 7.11 (Chatterjee & Zwislocki, 1998). The physiological functions are quite similar up to almost 60 dB SPL and follow reasonably well the psychological one. However, they flatten out at higher SPLs, first the function belonging to the response peak, next, the function belonging to the response area. According to these results, the area function has the advantage of following the loudness function over a 20 dB greater intensity range than does the peak function, but what happens above the SPL of 80 dB?

We did not explore the cochlear responses above 80 dB SPL systematically enough to give a well documented answer. However, our research has pointed out a phenomenon that may be helpful in ultimately finding one. The flattening of the cochlear input–output functions, especially, of the area function, at high SPLs may not indicate saturation but, rather, a flat notch. In the section on the stereocilia-tectorial membrane complex of the preceding chapter, it has been pointed out that the damping of the complex increases at such levels, presumably, due to a decreased effect of the active feedback, and the radial vibration amplitude of the tectorial membrane is reduced. When the reduction is sufficient to make the amplitude first equal, then, smaller than the radial vibration amplitude of the reticular lamina, the shear displacement between them must go through a minimum, then reverse phase. According to our experiments discussed in the preceding two chapters, this has to happen around an SPL of about 80 dB where the area curve of Fig. 7.11 begins to flatten. If the flattening is due to a minimum in the shear displacement, it should not continue and become an asymptote, but should resume its ascent paralleling the growth of the basilar membrane vibration amplitude, if such growth continues. According to recent findings of Rhode and Recio (2000), it does. As has been discussed in the preceding chapter, the decreased radial tectorial membrane motion is due to a decreased stiffness of the OHC stereocilia, which takes place at very high SPLs, and reduces the radial drive of the tectorial membrane. As a result, the radial motion of the tectorial membrane ceases, and the shear motion is due exclusively to the radial motion of the reticular lamina, as prescribed by the classical model. Some measurements of neuronal responses in the auditory nerve support these conclusions (Liberman, 1976, Kiang, 1978). They show the firing rate to go through a minimum at SPLs around 90 dB, but to continue increasing at higher SPLs. In agreement with our cochlear findings, the minimum is associated with a 180° phase

PITCH AND LOUDNESS CODES   **405**

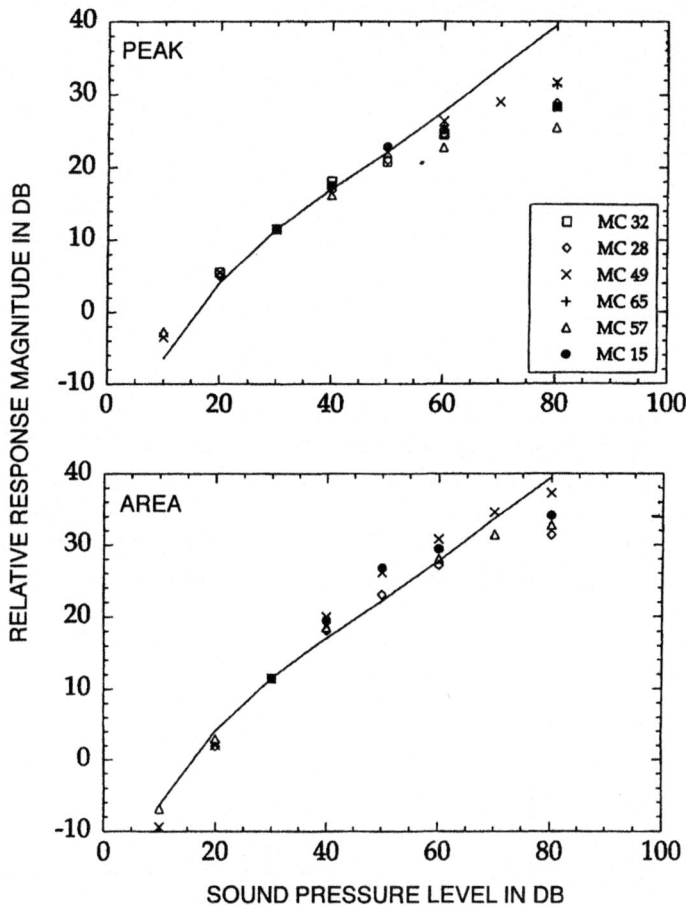

FIG. 7.11. Normalized IHC input–output functions of several specimens (symbols) following the peak response (varying sound frequency; upper panel), or the response area (lower panel). The solid lines show typical human loudness functions (Scharf, 1978, for review; from Chatterjee & Zwislocki, 1998). Reprinted from *Hearing Research, 124,* M. Chatterjee and J. J. Zwislocki, Cochlear Mechanics of Frequency and Intensity Coding II: Dynamic Range and the Code for Loudness, pp. 170–181, copyright © 1998, with permission from Elsevier Science.

shift. The phenomenon was discussed by me in the past in a different context (Zwislocki, 1986).

As has been shown in particular in chapter 3 on the middle ear, pathology can be helpful in understanding the function of a healthy organ. Experimental cochlear impairment produced by chemical or mechanical means in the cochlea has been used in several preceding chapters to reveal partic-

406 CHAPTER 7

ular mechanisms. Here, it is applied in connection with an otologically and audiologically well-known symptom called the *loudness recruitment*. Loudness recruitment occurs when the threshold of audibility is elevated and the loudness grows abnormally rapidly with SPL so that it becomes normal at sufficiently high SPLs.

We performed experiments on physiological mechanisms underlying the loudness recruitment (Zhang & Zwislocki, 1995). Mongolian gerbils were used for the purpose as animal models and were treated by the same methods as already described. To produce response impairment, the animals were exposed to broad band noise. For the results illustrated in Fig. 7.12, white noise was presented for 40 minutes at an SPL of 100 dB, and alternating potentials were recorded intracellularly in two neighboring Hensen's cells at several SPLs before and immediately after noise exposure. We should recall here that the transfer functions of Hensen's cells are quite similar to those of the OHCs as well as IHCs (e.g., Chatterjee & Zwislocki, 1997, 1998). Intensity series of the transfer functions obtained before the exposure are shown in the two panels on the left, those obtained after the exposure, on the right. In the upper panels, the SPLs extended

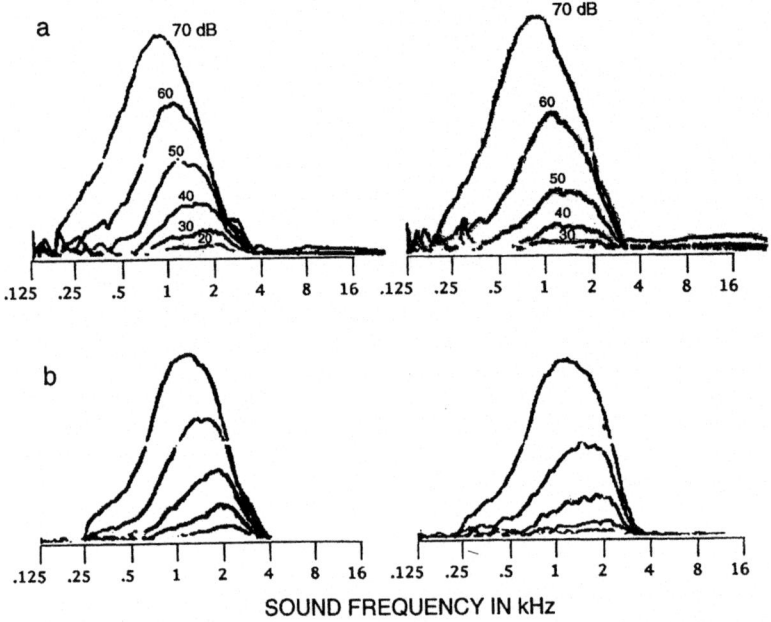

FIG. 7.12. Intensity series of Hensen's cell transfer functions recorded before (left column) and after noise exposure (right column; from Zhang & Zwislocki, 1995). Reprinted from *Hearing Research, 85,* M. Zhang and J. J. Zwislocki, OHC response recruitment and its correlation with loudness recruitment, 1–10, copyright © 1995, with permission from Elsevier Science.

from 20 to 70 dB, in the lower ones, from 20 to 60 dB in 10 dB steps. Note that, after the noise exposure, responses to the lowest level stimuli became undetectable and became reduced at all the other levels, except the highest ones. The reduction became gradually smaller, as the stimulus level increased, showing response recruitment similar to the loudness recruitment. The effect was evident at all sound frequencies at which a response could be measured, but was strongest near the maximum.

In addition to producing the response recruitment, the noise exposure abolished the SPL dependent peak shift at all levels at which the response was diminished (Zhang & Zwislocki, 1996). As should be apparent in Fig. 7.12, the response peaks are aligned at the same sound frequencies in both panels on the right, except those of the highest curves. Furthermore, the exposure increased the phase lag by about 90° at the lower stimulus levels, as is illustrated in Fig 7.13. Both phenomena have been shown in the preceding two chapters to occur when the active feedback is abolished or at least severely reduced. It is likely, therefore, that the cochlear response recruitment has the same origin.

It has been shown in several studies on basilar membrane vibration that, in a normal cochlea, the input–output functions exhibit strong compression

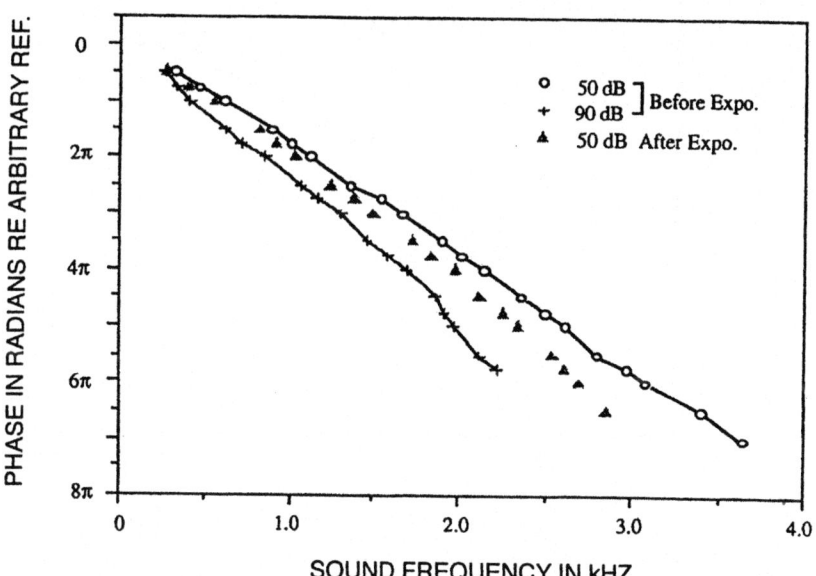

FIG. 7.13. Hensen's cell response phases as functions of sound frequency for three conditions: at 50 and 90 dB SPL before noise exposure and at 50 dB after the exposure (from Zhang & Zwislocki, 1995). Reprinted from *Hearing Research, 85*, M. Zhang and J. J. Zwislocki, OHC response recruitment and its correlation with loudness recruitment, 1–10, copyright © 1995, with permission from Elsevier Science.

in the broad vicinity of the maximum response but that this is not true in the absence of the active feedback. When the feedback is absent, the functions become linearized (e.g., Ruggero, Rich, & Recio, 1996). We found approximately the same relationships at the level of the OHC, as inferred from Hensen's cell responses. This is documented in Fig. 7.14 by the input–output functions derived from the transfer functions of Fig. 7.12 at their peaks (Zhang & Zwislocki, 1995). The solid lines correspond to the functions obtained before the noise exposure, the intermittent ones, to those obtained immediately after it. In the SPL range between about 30 and 60 dB, the postexposure curves are clearly steeper than the preexposure ones. More specifically, they are almost linear, whereas the preexposure ones follow roughly a power function with an exponent of approximately 0.6. The slope relationships are similar to those found in psychophysical experiments on loudness in the presence of normal hearing and mild hearing loss, respectively (Hellman & Zwislocki, 1964). It seems, therefore, that the response recruitment found in the cochlear responses at the hair cell level

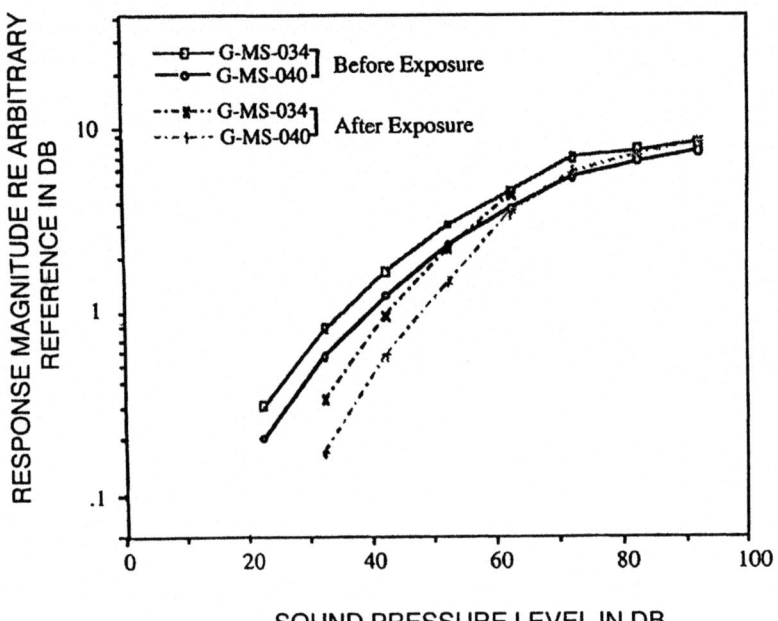

FIG. 7.14. Hensen's cell input–output functions constructed from the transfer functions of Fig. 7.12 and following the peak responses (frequency variable). The solid lines were obtained before noise exposure, the intermittent ones, after noise exposure in neighboring cells (from Zhang & Zwislocki, 1995). Reprinted from *Hearing Research, 85*, M. Zhang and J. J. Zwislocki, OHC response recruitment and its correlation with loudness recruitment, 1–10, copyright © 1995, with permission from Elsevier Science.

accounts for most if not all of the phenomenon of loudness recruitment (Zhang & Zwislocki, 1995).

It may be mentioned here that the empirical response patterns that have led to the conclusions of this chapter are entirely consistent with the theoretical results of the preceding chapter. This is true for the normal as well as the pathological cochlea with respect to both the pitch code and the loudness code. With respect to the pitch code, the theoretically accounted for peak shift and convergence of the high frequency skirts of the cochlear transfer functions should be particularly emphasized, as well as the lack of these features in the cochlea with damaged outer hair cells. With respect to the loudness code, the cochlear compression based on the active feedback, which, according to the theory, also accounts for the peak shift seems worth mentioning. This also applies to the phenomenon of loudness recruitment that, according to both our experiments and theory results from the missing active feedback and can be accounted for almost entirely by cochlear rather than neural processes. Perhaps the phase shift by about 180° at high SPLs, which must be associated with a perturbation in the cochlear input–output functions and mimics response saturation but allows further response growth at still higher SPLs, should also be mentioned. This phenomenon discovered in neural responses by others has been explained in chapter 6 by cochlear processes at the level of the shear motion between the tectorial membrane and the reticular lamina.

REFERENCES

Békésy, G. V. (1960). *Experiments in hearing*. New York: McGraw-Hill.

Burns, E. M., & Turner, C. (1986). Pure-tone pitch anomalies. II. Pitch-intensity effects and diplacusis in impaired ears. *Journal of the Acoustical Society of America, 79,* 1530–1540.

Chatterjee, M., & Zwislocki, J. J. (1997). Cochlear mechanics of frequency and intensity coding. I. The place code for pitch. *Hearing Research, 111,* 65–75.

Chatterjee, M., & Zwislocki, J. J. (1998). Cochlear mechanics of frequency and intensity coding. II. Dynamic range and the code for loudness. *Hearing Research, 124,* 170–181.

Gulick, W. L. (1971). *Hearing: Physiology and psychophysics*. New York: Oxford University Press.

Hellman, R. P., & Zwislocki, J. J. (1964). Loudness function of a 1000-cps tone in the presence of a masking noise. *Journal of the Acoustical Society of America, 36,* 1618–1627.

Helmholtz, H. L. F. (1954) *On the sensations of tone as a physiological basis for the theory of music* (E. Higgins, trans.). New York: Dover Publications. (Original work published in 1863.)

Kiang, N. Y. S. (1978). Peripheral neural processing of auditory information. In D. B. Tower (Ed.), *Handbook of physiology—The nervous system III* (pp. 639–674). New York: Raven Press.

Liberman, M. C. (1976). *Abnormal discharge patterns of the auditory-nerve fibers in acoustically-traumatized cats.* Unpublished doctoral dissertation, Harvard University, Cambridge, Massachusetts.

Liberman, M. C. (1978). Auditory nerve responses from cats raised in a low-noise chamber. *Journal of the Acoustical Society of America, 63,* 442–455.

Liberman, M. C., & Kiang, N. Y. S. (1978). Acoustic trauma in cats. *Acta Oto-Laryngologica, Suppl. 358.*

Møller, A. R. (1983). Frequency selectivity of phase-locking of complex sounds in the auditory nerve of the rat. *Hearing Research, 11,* 267–284.

Rhode, W. S., & Recio, A. (2000). Study of mechanical motions in the basal region of the chinchilla cochlea. *Journal of the Acoustical Society of America, 107,* 3317–3332.

Rose, J. E., Brugge, J. F., Anderson, D. J., & Hind, J. E. (1967). Phase-locked response to low-frequency tones in single auditory nerve fibers of the squirrel monkey. *Journal of Neurophysiology, 30,* 769–793.

Rose, J. E., Hind, J. E., Anderson, D. J., & Brugge, J. F. (1971). Some effects of stimulus intensity on response of auditory nerve fibers in the squirrel monkey. *Journal of Neurophysiology, 34,* 685–699.

Ruggero, M. A., Rich, N. C., & Recio, A. (1996). The effect of intense acoustic stimulation on basilar-membrane vibrations. *Auditory Neuroscience, 2,* 329–345.

Sachs, M. B., & Abbas, P. J. (1974). Rate versus level functions for auditory-nerve fibers in cats: Tone burst stimuli. *Journal of the Acoustical Society of America, 56,* 1835–1847.

Scharf, B. (1978). Loudness. In C. E. Carterette & M. P. Friedman (Eds.), Handbook of perception. IV. Hearing (pp. 187–242). California: Academic Press.

Spoendlin, H. (1967). Innervation of the organ of Corti. *Journal of Laryngology and Otology,* 717–737.

Stevens, S. S. (1935). The relation of pitch to intensity. *Journal of the Acoustical Society of America, 6,* 150–154.

Stevens, S. S., & Davis, H. (1938). *Hearing.* New York: Wiley.

Zhang, M. (1993). *Effects of noise on cochlear transfer functions at the cellular level and their possible underlying mechanisms.* Unpublished doctoral dissertation. Institute for Sensory Research, Syracuse University, Syracuse, New York.

Zhang, M., & Zwislocki, J. J. (1992). Hair cell magnitude and phase transfer functions: effects of overstimulation and their correlation with loudness functions. *ARO Abstract, 15,* 18.

Zhang, M., & Zwislocki, J. J. (1995). OHC response recruitment and its correlation with loudness recruitment. *Hearing Research, 85,* 1–10.

Zhang, M., & Zwislocki, J. J. (1996). Intensity-dependent peak shift in cochlear transfer functions at the cellular level, its elimination by sound exposure, and its possible underlying mechanisms. *Hearing Research, 96,* 46–58.

Zwislocki, J. J. (1986). Analysis of cochlear mechanics. In Å. Flock & J. Warsäll (Eds.), *Cellular mechanisms in hearing* (pp. 155–169). Amsterdam, Netherlands: Elsevier Science.

Zwislocki, J. J. (1991). What is the cochlear place code for pitch? *Acta Oto-Laryngologica (Stockholm), 111,* 256–262.

Zwislocki, J. J., & Nguyen, M. (1999). Place code for pitch: A necessary revision. *Acta Oto-Laryngologica (Stockholm), 119,* 140–145.

Zwislocki, J. J., Solessio, E., & Cefarratti, L. (1991). Hair-cell transfer functions versus tuning curves, and the place code for pitch. *ARO Abstract, 14,* 410.

Zwislocki-Moscicki, J. J. (1946). Über die mechanische Klanganalysedes Ohrs [On sound analysis in the ear]. *Experientia, 2,* 10–18.

Zwislocki-Moscicki, J. J. (1948). Theorie der Schneckenmechanik: Qualitative und Quantitative Analyse. [Theory of Cochlear Mechanics: Qualitative and Quantitative Analysis]. *Acta Oto-Laryngologica, Suppl. 72,* 1–76.

Author Index

A

Abbas, P. J., 391, *410*
Abnet, C. C., 364, *385*
Allen, J. B., 209, *274*, 291, 330, *385*
Anderson, D. J., 273, *274*, 394, *410*
Aritomo, H., 79, 80, *86*
Armstrong, N. J., 102, *173*
Ashmore, J. F., 176, *274*

B

Bader, C. R., 176, *274*
Ball, G., 80, *86*
Békésy, G. v., 4, 6, 7, 17, *25*, 27, 57, 58, 59, 65, 68, 73, 76, 79, 85, *86*, 89, 91, 93, 99, 100, 102, 109, 112, 113, 115, 120, 124, 128, 136, 137, 138, 139, 141, 145, 151, 152, 153, *172*, 175, 180, 188, 215, 217, 230, 234, 235, 247, 248, 250, 265, *274*, 319, 354, *385*, 391, 392, *409*
Benson, R. W., 8, *26*
Beranek, L. L., 17, 23, 25, 33, *86*
Bertrand, D., 176, *274*
Billone, M. C., 248, *275*, 352, *385*
Bogert, B. P., 114, *173*
Bohne, B. A., 220, *275*
Boyle, A. J. T., 175, *276*
Brink, P. R., 263, *276*
Brödel, M., 2, 7
Brown, M. C., 352, *386*
Brownell, W. E., 176, *274*, 305, *385*, *388*
Brugge, J. F., 273, *274*, 394, *410*
Burns, E. M., 398, *409*

C

Cambron, N. K., 39, *87*
Cefaratti, L. K., 177, 180, 188, 194, 195, 196, 197, 198, 200, 202, 203, 204, 206, 207, 209, 248, 250, 251, 252, 255, 256, 257, 260, 261, 262, 263, 264, 265, *278*, *279*, 294, 364, *389*, 392, 394, *410*
Chamberlain, S. C., 108, *174*, 177, 180, 184, 185, 188, 189, 190, 191, 192, 193, *279*, 375, *386*

411

Chatterjee, M., 269, 275, 394, 400, 401, 402, 403, 404, 405, 406, 409
Cody, A. R., 364, 387
Cook, J. P., 17, 22, 25
Cooper, N. P., 234, 238, 240, 241, 242, 243, 244, 245, 246, 247, 275, 369, 370, 385
Corey, D. P., 233, 276, 285, 386
Cousillas, H., 305, 388

D

Dallos, P., 73, 86, 91, 128, 138, 146, 172, 179, 247, 248, 254, 260, 261, 264, 267, 275, 276, 277, 293, 303, 310, 333, 352, 363, 385
Dancer, A., 230, 275
Davis, H., 92, 156, 165, 168, 172, 173, 410
De Boer, E., 375, 384, 385
Decraemer, W. F., 246, 278
Delany, M. E., 17, 22, 25
Dieler, R., 305, 385
Djupesland, G., 9, 12, 15, 25
Durrant, J. D., 248, 275, 352, 385
Duvall, A. J., III, 179, 181, 183, 211, 278

E

Eldredge, D. H., 220, 275
Emde, F., 131, 143, 173
Evans, B. N., 248, 275, 385
Evans, E. F., 282, 341, 344, 363, 385, 388

F

Feenstra, L., 80, 88
Feldman, A. S., 17, 26, 29, 38, 39, 44, 47, 75, 77, 86, 88
Fernandez, C., 217, 229, 275
Flanagan, J. L., 123, 128, 172
Flandermeyer, D. T., 28, 87
Fletcher, H., 109, 114, 172
Flock, Å., 176, 178, 179, 180, 194, 210, 234, 247, 254, 275, 276, 278, 282, 285, 286, 291, 301, 306, 310, 385, 387, 388

Flock, B., 303, 306, 385
Frank, O., 68, 86
Franke, K., 262, 276
Franke, R., 230, 275
Freeman, D. M., 364, 385
Frey, A. R., 60, 87, 115, 127, 173
Fridberger, A., 306, 385
Frommer, G. H., 188, 275
Furrer, W., 169, 173

G

Geisler, C. D., 93, 114, 172, 282, 337, 363, 364, 370, 385, 386
Gold, T., 176, 275
Goode, R. L., 79, 80, 86
Goodenough, D. H., 263, 277
Goodman, D. A., 375, 386
Grason, R. L., 8, 26
Greenwood, D. D., 177, 179, 223, 225, 275
Guinan, J., 138, 172
Gulick, W. L., 395, 409
Gully, R. L., 263, 275
Gummer, A. W., 102, 173, 209, 246, 247, 276, 280, 291, 364, 386
Gunderson, T., 80, 86
Gyo, K., 79, 80, 86

H

Hamernik, R. P., 281, 309, 310, 326, 330, 345, 352, 354, 388
Harris, D., 303, 385
He, D. Z. Z., 179, 276
Hellman, R. P., 242, 276, 408, 409
Helmholtz, H. L. F., 4, 5, 7, 59, 87, 90, 112, 173, 390, 392, 409
Hemmert, W., 209, 246, 247, 276, 280, 291, 364, 386
Hertig, L., 177, 233, 273, 278, 279, 334, 349, 375, 384, 388, 389
Hind, J. E., 273, 274, 394, 410
Holley, M. C., 108, 173, 292, 312, 388
Hubbard, A. E., 93, 114, 172, 363, 386
Hudspeth, A. J., 233, 276, 285, 363, 386

I

Imredy, J., 305, *388*
Ithell, A. H., 22, *25*
Iurato, S., 208, 262, *276*

J

Jahnke, E., 131, 143, *173*
Janisch, R., 95, *174*
Jaslove, S. W., 263, *276*
Johnson, E. G. T., 22, *25*
Johnstone, B. M., 102, *173*, 175, 217, 231, 234, *276*, *278*, 290, 375, *386*, *387*
Johnstone, J. R., 217, *278*

K

Kemp, D. T., 176, *276*
Kennedy, D. T., 282, *386*
Khanna, S. M., 51, 81, *87*, 176, 234, 246, *276*, *278*
Kiang, Y. S., 301, 303, 352, *386*, 392, 393, 404, *409*, *410*
Kim, D. O., 222, *277*, 363, *386*
Kinsler, L. E., 60, *87*, 115, 127, *173*
Kletsky, E. J., 282, 283, 325, *389*
Kohllöffel, L. U. E., 216, 218, *276*
Kolston, P. J., 363, *386*
Konishi, T., 280, *386*
Kössl, M., 176, *277*
Kringlebotn, M., 63, *87*
Kronester-Frei, A., 179, 207, 208, *276*, 285, *386*

L

Lamb, H., 109, 159, 160, 169, 172, *173*
Lawrence, M., 65, 68, *88*, 165, *174*, 352, *386*
Legouix, J. P., 156, 165, 168, *173*
Leonard, D. G. B., 176, *276*
LePage, E. L., 290, *386*
Liberman, M. C., 223, 247, *276*, 301, 303, 345, 352, *386*, 391, 392, 393, 401, 404, *409*, *410*

Lighthill, J., 159, *173*
Lim, D. J., 212, *277*
Lindquist, P. G., 303, *388*
Lindsay, R. B., 109, 159, *173*
Ling, A., Jr., 285, *386*
Lipes, R., 162, *174*
Long, R. R., 305, *386*
Luciano, L., 262, *276*
Lynch, T. J., 72, 73, 76, *87*, 91, 128, 134, 138, *173*, 230, *277*

M

Markin, V. S., 363, *386*
Masta, R. L., 352, *386*
McGee, T., 303, *385*
McMullen, T. A., 363, *386*
Merchant, S. N., 72, 73, 77, 78, 79, 80, *87*, 91, 128, 129, 130, 134, 138, 139, *173*, 230, *277*
Metz, O., 28. 29, 39, *87*
Miller, C. E., 102, 103, *173*
Miller, J. D., 220, *275*
Minser, K. E., 179, 210, *278*, 286, 310, *388*
Møller, A. R., 38, 48, 54, 57, 71, 74, *87*, 128, *173*, 394, *410*
Molvaer, O. I., 63, *87*
Mountain, D. C., 363, *386*
Mulroy, J. J., 263, *277*
Mulroy, M. J., 285, *388*
Mundie, J. R., 72, *87*, 128, *173*
Murugasu, E., 176, *277*

N

Nadol, J. B., Jr., 263, *277*
Nakamura, K., 79, 80, 86
Narayan, S., 176, 220, 221, 231, 239, 240, 242, 247, 269, 273, *277*, 299, 317, 334, 343, 366, 368, 370, *386*
Nedzelnitsky, V., 58, 72, 73, 76, *87*, 91, 128, 134, 138, *173*, 230, *277*
Neely, S. T., 363, *386*
Nelson-Adesokan, P. M., 301, *388*

Nguyen, M., 270, 271, *279*, 334, *389*, 392, 393, 395, 397, 398, *410*
Nielson, D. W., 280, *386*
Niemoeller, A. F., 8, *26*
Nilsen, K. E., 218, 234, 235, 237, 238, 240, 241, 269, *277*, 317, 354, *387*
Nishihara, S., 80, *86*
Nuttall, A. L., *385*, 352, 384, *386*

O

Oesterle, E. C., 261, *277*
Olsen, H. F., 15, *25*, 91, *173*
Onchi, Y., 35, 47, 54, 62, 63, 71, 77, 79, 80, *87*
Özdamar, O., 303, *385*

P

Pannese, E., 262, *276*
Patuzzi, R., 231, 234, *278*, 375, *387*
Peake, W. T., 72, 73, 76, *87*, 91, 128, 134, 138, *172*, *173*, 230, *277*, 285, *386*
Peterson, L. C., 114, *173*
Pfeiffer, R. R., 222, *277*
Pierce, J. R., 162, *174*
Pietras, B. W., 349, *386*
Pike, C. L., 285, *388*
Pumphery, R. J., 176, *275*

R

Rabinowitz, W. M., 34, 35, 39, 47, 48, 71, 75, *87*
Ranke, O. F., 168, *173*
Rasmussen, H. T., 3, 7
Rauch, S., 99, *173*, 212, *277*
Ravicz, M. E., 72, 73, 77, 78, 79, 80, 87, 91, 128, 129, 130, 134, 138, 139, 173, 230, *277*
Raynor, S., 248, *275*, 352, *385*
Reale, E., 262, *276*
Recio, A., 220, 221, 231, 239, 240, 242, 247, 269, 273, *277*, 299, 317, 334, 337, 343, 366, 368, 370, *386*, *387*, 394, 404, 408, *410*

Reese, T. S., 263, *275*
Reneau, J. P., 179, 181, 183, 211, *278*
Rhode, W. S., 58, *87*, 91, 123, 128, 130, 171, *173*, 175, 216, 218, 223, 224, 231, 234, 240, 241, 242, 243, 244, 245, 246, 247, *275*, *277*, 282, 289, 322, 329, 337, 366, 375, *385*, *386*, *387*, 394, 404, *410*
Ribaupiere, Y., 176, *274*
Rich, N. C., 176, 220, 221, 231, 234, 239, 240, 242, 247, 269, 273, *277*, 281, 317, 334, 344, 352, 366, 368, 370, 375, 379, 380, 381, 383, 384, *387*, 408, *410*
Richardson, G. P., 364, *387*
Robles, L., 176, 220, 221, 231, 234, 239, 240, 242, 247, 269, 273, *277*, 317, 334, 344, 366, 368, 370, *387*
Rose, J. E., 273, *274*, 394, *410*
Rosenblith, W. A., 17, *25*, 68, 73, *86*, 151, *172*
Rosowski, J. J., 28, 38, 72, 73, 77, 78, 79, 80, *87*, 91, 128, 129, 130, 134, 138, 139, *173*, 230, *277*
Ross, D. A., 9, 12, 14, 15, 16, *25*
Rudmose, W. F., 8, *26*
Rüedi, L., 169, *173*
Ruggero, M. A., 176, 220, 221, 231, 234, 239, 240, 242, 247, 269, 273, *277*, 281, 299, 317, 334, 343, 344, 352, 366, 368, 370, 375, 379, 380, 381, 383, 384, *386*, *387*, 408, *410*
Russell, I. J., 176, *218*, 234, 235, 237, 238, 240, 241, 269, *277*, 282, 283, 317, 352, 354, 364, 375, *387*
Ryan, A., 303, *385*

S

Sachs, M. B., 391, *410*
Santos-Sacchi, J., 176, 247, 254, 260, 262, *275*, *277*, 364, *387*
Saunders, J. C., 301, *387*, *388*
Scarfone, E., 306, *385*

Scharf, B., 400, *410*
Schmiedt, R. A., 76, *87*, 179, 212, 236, 249, *278*, 281, 309, 310, 326, 330, 333, 345, 352, 354, *387*, *388*
Schuknecht, H. F., 222, *278*
Scott, V., 17, 22, *25*
Sellick, P. M., 176, 231, 234, 277, *278*, 282, 283, *387*, 352, 375, *387*
Shambough, G. E., 285, *387*
Shanks, J. E., 39, *87*
Shaw, E. A. G., 8, 11, 12, 14, 15, 16, 17, *25*, 26, 48, 83, *87*
Shehata, W. E., 305, *388*
Shehata-Dieler, W. E., 305, *385*
Shivapuja, B. G., 234, *277*
Siebert, W. M., 159, *173*
Sivian, L. J., 84, *87*
Slepecky, N. B., 108, *174*, 177, 180, 184, 185, 188, 189, 190, 191, 192, 193, 248, 250, 251, 252, 255, 256, 257, 260, 261, 262, 263, 264, 265, *279*
Smith, R. L., 177, 248, 250, 251, 252, 255, 256, 257, 260, 261, 262, 263, 264, 265, 272, *279*, 375, *386*
Sokolich, W. G., 281, 309, 310, 352, 354, *388*, *389*
Solessio, E., 392, 394, *410*
Spoendlin, H., 175, *278*, 282, 283, *388*, 391, *410*
Steele, C. R., 159, *173*
Stevens, S. S., 391, 395, *410*
Strelioff, D., 178, 179, 180, 194, 210, 275, *278*, 286, 291, 310, *388*
Stypulkowski, P. H., 305, *388*
Szymko, Y. M., 177, 233, 273, *278*, *279*, 301, 334, 349, 375, 384, *388*, *389*

T

Tabor, L. A., 159, *173*
Tasaki, I., 156, 165, 168, *173*
Taylor, K., 217, *276*
Temchin, A. N., 234, *277*, 299, 343, *386*
Teoh, S. W., 28, *87*

Teranishi, R., 12, 14, 16, 17, *25*
Tolomeo, J. A., 292, 312, *388*
Tonndorf, J., 51, 81, *87*, 179, 181, 183, 211, *278*
Tubis, A., 305, *386*
Turner, C., 398, *409*
Turner, R. G., 285, *388*

U

Ulehlova, L., 95, *174*
Ulfendahl, M., 234, 246, *276*, *278*, 306, *385*

V

Vallersnes, F. M., 67, *87*
Viergever, M. A., 115, 159, 160, *174*
Vlaming, M. S. M. G., 80, *88*
Voldrich, L., 95, 115, *174*
von Gierke, H. E., 34, *88*

W

Wang, C.-Y., 248, *275*, 352, *385*
Warren, D., 34, *88*
Weiss, T. F., 263, *277*, 285, *388*
Wermbter, G., 262, *276*
Wersäll, J., 303, *388*
Wever, E. G., 65, 68, *88*, 89, 96, 102, 103, 105, 106, 165, *174*, 215, *278*, 285, *388*
White, S. D., 84, *87*
Whittle, L. S., 17, 22, *25*
Wiener, F. M., 9, 12, 14, 15, 16, *25*
Wilber, L. A., 39, *86*
Wilson, J. P., 176, 217, *278*, 282, 344, *385*, *388*
Wilson, Q. H., 39, *87*

Y

Yates, R. F., 22, *25*

Z

Zenner, H. P., 209, 246, 247, *276*, 280, 291, 364, *386*

AUTHOR INDEX

Zhang, M., 233, 234, 273, *278*, 293, 305, *388*, 398, 399, 400, 402, 406, 407, 408, *410*
Zurek, P. M., 176, *278*
Zweig, G., 162, *174*
Zwicker, E., 285, *388*
Zwislocki, J. J., 2, 3, 4, 7, 8, 9, 10, 12, 15, 17, 18, 20, 21, 22, *25*, 29, 30, 35, 36, 38, 39, 40, 41, 42, 43, 44, 45, 46, 47, 48, 49, 50, 51, 52, 53, 54, 55, 56, 57, 59, 62, 65, 67, 70, 72, 73, 75, 76, 77, 83, 85, *86, 87, 88*, 90, 91, 93, 96, 98, 100, 105, 107, 108, 109, 114, 119, 120, 125, 127, 128, 130, 131, 132, 156, 159, 166, *174*, 177, 179, 180, 184, 185, 188, 189, 190, 191, 192, 193, 194, 195, 196, 197, 198, 200, 202, 203, 204, 206, 207, 209, 212, 217, 229, 230, 233, 234, 236, 242, 248, 249, 250, 251, 252, 255, 256, 257, 260, 261, 262, 263, 264, 265, 266, 267, 269, 270, 271, 272, 273, *275, 276, 278, 279*, 281, 282, 283, 286, 293, 294, 303, 304, 305, 306, 309, 310, 312, 325, 326, 330, 333, 334, 345, 349, 352, 354, 363, 364, 375, 384, *386, 387, 388, 389*, 392, 393, 394, 395, 397, 398, 400, 401, 402, 403, 404, 405, 406, 407, 408, *409, 410*
Zwislocki-Moscicki, J. J., 27, 72, 73, *88*, 89, 90, 91, 94, 95, 96, 97, 99, 100, 105, 113, 114, 115, 119, 120, 123, 125, 128, 130, 131, 136, 141, 142, 151, 155, 156, 169, 172, *174*, 215, *279*, 391, *410*

Subject Index

C

Cochlea Simplified by Death (*also* cochlea postmortem), 89–172
  chronological (development), 89–92
  functional anatomy, 92–98, 101–102, 105–106
  mathematical analysis, 110–172
    basilar membrane mass, effect on wave velocity, 125–127
    Békésy's eddies, 168–172
    Békésy's paradoxical waves, 165–168
    canal depth, effect of, 158–165
    cochlear waves, partition's resistance and mass negligible, 130–133
    delay time (*also* travel time), 121–122, 123–124
    differential equation for long waves, 114–119
    input impedance, cochlear, 127–130
    input impedence at low sound frequencies, 133–141
    partition mass, effect on wave pattern, 146–151
    partition resistance, effect on wave pattern, 141–146
    phase of basilar membrane displacement, 122–123
    resonance of the basilar membrane, 112–114
    symbols used, 110–112
    transfer functions, cochlear, 151–158
    wavelength (first approximation), 124–125
    wave velocity, a first approximation, 119–121
  physical constants, 99–110

L

Live cochlea, analysis, 280–384
  active feedback effect, 363–375
    model, electrical network, 364–366
    feedback effects, theoretical, 366–375
  cochlear compression, 375–384
    burst response, *also* transients, 379–384
    filter functions, 375–379

417

418 SUBJECT INDEX

cochlear dynamics, passive, 322–362
  basilar membrane and shear displacement, relationships, 341–344
  basilar membrane, transfer functions, 326–330
  hardware model, 358–362
  IHCs, shear motion at the location of, 349–354
  resistances, dependence on, 333–342
  response enhancement (due to increased stereocilia compliance), 349
  secondary peak (characteristics), 343–349
  sensitivity enhancement, paradox of, 352–354
  shear displacement, transfer functions for, 330–344
  transmission line model (programmed on computer), 322–324
  wave propagation as a function of cochlear location, 354–358
dynamics of the cochlear partition (unit section), 310–322
  electrical analog, 313
  geometric relationships, 310–313
  Mongolian gerbil and human, 313–322
stereocilia–tectorial membrane complex, dynamics of, 291–310
  input impedance, 294–296
  mechanical network (*also* electrical analog), 292–294
  radial motion of the tectorial membrane, 296–298
  resonances, 294
  shear motion (between the reticular lamina and the tectorial membrane), 298–310
Live cochlea, physical constants and fundamental characteristics, 175–274
  basilar membrane, physical parameter values (*also* related cochlear variables), 212–229
  basilar membrane parameter values in live humans, 227–229

  input impedance in vivo—comparative considerations, 229–231
  stiffness and wave velocity in Mongolian gerbil cochlea, 212–215
  wave length, cochlear (mammals), 225–226
  wave travel time, cochlear (live humans), 228
  wave velocity, cochlear: a generalization, 215–225
  wave velocity, cochlear, in live humans, 227–229
cochlear parameter values for humans and Mongolian gerbils, 231–232
OHC stereocilia, physical parameter values, 210–212
  stiffness, effective, 210–212
  resistance, mechanical, 212
tectorial membrane, physical parameter values of the gerbil, 178–209
  elastic properties, 179–207
  mass, effective, 207–209
cochlear dynamics, some fundamental characteristics, 231–274
  basilar membrane vibration, 233–247
  ionic communication between OHCs and supporting cells, 260–265
  peak shift (of transfer functions), 269–271
  phase transfer functions (for shear motion), 271–274
  shear motion between tectorial membrane and reticular lamina (methods), 247–255
  transfer functions (magnitude, measured in OHC and Hensen's cells), 255–260, 266–269

M

Middle ear, 27–86
  acoustic impedance at the tympanic membrane, measurement of, 29–53
  acoustic bridge, with, 38–53

infinite resistance source, with, 29–38
middle ear disorders, on patients with, 49–54
(see also otosclerosis, and interrupted incudo-stapedial joint, 36–37)
middle ear function, analysis of the 54–83
cochlea, effect of, 72–74, 76–78
electrical network analog (electroacoustical analogies), 55–56
incus, middle ear without, 66–70
mechanical transformer, middle ear as a, 59–61
middle ear cavities, 61–64

models, classes of, 54–55
model (also network analog), general structure of, 56–59
normal middle ear, 72–83
otosclerotic middle ear, 70–72
structure of the model, general, (see also network analog), 56–59
transient stimuli, 82–83
tympanic membrane shunt, 64–66
threshold of audibility, versus middle ear transfer function, 83–86
transfer function (middle ear), 79–81

O

Outer ear, 8–25

modeling the outer ear, 15–25
acoustic variables, 16
adequacy, of ear simulator, 23–25
ear canal, electroacoustic analog of, 17–21
ear simulator, 21–23
electroacoustic analogies, 15–16
geometrical dimensions, 16–17
sound pressure transformation, measurement of, 9–15

P

Path of sound (through), 1–7
auricle and ear canal, 3
inner ear (also cochlea), 4–6
middle ear, 3–4
Pitch and Loudness Codes, 390–409
loudness code, cochlear, 400–409
hair cells, transfer functions of, 400–402
input–output functions, cochlear, 402–405
loudness recruitment, 405–408
pitch and loudness codes (historical remarks), 390–392
pitch, place code for, 392–400
high frequency (also apical) cut off of excitation (as a place code for pitch), 397–400
place code for pitch as proposed by Helmholtz (difficulties), 392–397